BY APPOINTMENT
TO H.M. QUEEN ELIZABETH THE QUEEN MOTHER
MOTOR CAR MANUFACTURERS

THE DAIMLER COMPANY LIMITED

Daimler

$2\frac{1}{2}$ LITRE V.8 AND 250 SALOON

SPARE PARTS CATALOGUE

SPARES DIVISION

ENGINE NUMBERS

2½ Litre Saloon	7A.1001 ONWARDS
250 Saloon	7K.1001 ONWARDS

CHASSIS NUMBERS

2½ Litre Saloon	RIGHT-HAND DRIVE	1A.1001 ONWARDS
	LEFT-HAND DRIVE	1A.20001 ONWARDS
250 Saloon	RIGHT-HAND DRIVE	1K.1001 ONWARDS
	LEFT-HAND DRIVE	1K.30001 ONWARDS

THE DAIMLER COMPANY LIMITED, COVENTRY, ENGLAND

Telephone: Allesley 2121 (P.B.X.)

Code: BENTLEY'S SECOND

Telegraphic Address: Daimler 'Phone Coventry Telex. 31622

Brooklands Books Ltd., PO Box 904, Amersham
Bucks., HP6 9JA, UK
brooklands-books.com

Part Number: D.4

ISBN 9781855200098 Ref: D68PH 1W6/2371

INDEX TO SECTIONS

CYLINDER BLOCK

Part No.	Description	Plate No.	No. per Unit	Remarks
C.21034	**CYLINDER BLOCK ASSEMBLY**	1–1	1	Fitted from Engine No. 7A.1001 and subs 7K.1001 to 7K.1342
C.27419	**CYLINDER BLOCK ASSEMBLY**		1	Fitted to Engine No. 7K.1343 and subs
C.22660	**Plug,** in each end of Oil Gallery		2	
C.2296/1	**Washer,** (Copper) for Plugs		2	
C.26004	**Welch Washer,** at each side of Cylinder Block	1–2	6	
C.317	**Welch Washer,** at front of Cylinder Block	1–3	2	
C.21631	**Welch Washer** (small) at rear of Cylinder Block	1–4	2	
C.26103	**Welch Washer** (large) at rear of Cylinder Block	1–5	1	
420304	**Dowel** (Short) at rear of Cylinder Block	1–6	1	
428423	**Dowel** (Long) at rear of Cylinder Block	1–7	1	
430381	**Ring Dowel,** locating Cylinder Heads on Cylinder Block	1–8	4	
430582	**Bearing,** for Camshaft (Front)	1–9	1	Fitted from Engine No. 7A.1001 and subs 7K.1001 to 7K.1342
430583	**Bearing,** for Camshaft (Intermediate)	1–10	4	
430584	**Bearing,** for Camshaft (Rear)	1–11	1	
C.27414	**Bearing,** for Camshaft (Front)		1	Fitted to Engine No. 7K.1343 and subs
C.27415	**Bearing,** for Camshaft (Intermediate)		4	
C.27416	**Bearing,** for Camshaft (Rear)		1	
422583	**Bush,** for Oil Pump Spindle		1	Superseded by C.23331
C.23331	**Bush,** for Oil Pump Spindle		1	Supersedes 422583
C.22062	**CYLINDER BLOCK ASSEMBLY**		1	**FOR USE ONLY ON CARS EXPORTED TO AUSTRIA**
C.22535	**Liner,** for Cylinder Block		8	Fitted up to and including Engine No. S7A.5293 only
	All other items are as for Cylinder Block Assembly C.21034			**NOTE: These Engines are identified by the addition of the prefix letter 'S' to the Engine Number**
C.21360	**Adaptor,** in Cylinder Block, for Oil Pressure Element	1–12	1	
C.4146	**Washer,** under Adaptor	1–13	1	
C.21361	**Banjo,** on Adaptor	1–14	1	
FW.105/E	**Washer,** Copper, between Banjo and Adaptor	1–15	1	
C.21505	**Dome Nut,** securing Banjo to Adaptor	1–16	1	
C.2296/4	**Washer,** Copper, under Dome Nut	1–17	1	
430680	**Dowel,** in Bearing Caps		10	
430420	**Bolt,** securing Bearing Caps		10	
430366	**Lockplate,** securing Bolts		10	2 only required (for Rear Bearing Cap Bolts) at Engine No. 7A.8200 and subs
C.25727	**Lockplate,** securing Bolts at Front, Centre, and Intermediate Bearing Caps		4	Fitted to Engine No. 7A.8200 and subs
UCB137/16R	**Bolt,** securing Rear Bearing Cap		2	
FG.106/X	**Washer,** Spring, on Bolts		2	
UCS.131/4R	**Plug,** in Cylinder Block Water Drain Holes	1–18	2	
400352	**Washer,** on Plugs	1–19	2	
135873	**CAMSHAFT**		1	Fitted from Engine No. 7A.1001 to 7A.1147
C.21282	**CAMSHAFT**	1–20	1	Fitted from Engine No. 7A.1148 and subs 7K.1001 to 7K.1342
C.27413	**CAMSHAFT**		1	Fitted to Engine No. 7K.1343 and subs
430365	**Thrust Plate,** for Camshaft Bearing	1–21	1	
UCS.131/5R	**Screw,** Set, securing Thrust Plate to Cylinder Block		2	
C.22295	**Lockplate,** securing Setscrews		2	
430364	**Thrust Washer,** for Camshaft	1–22	1	
307925	**CHAIN WHEEL ON CAMSHAFT**	1–23	1	
422510/15	**Key,** locating Chain Wheel	1–24	1	
430399	**Nut,** securing Chain Wheel	1–25	1	
430400	**Lockplate,** under Nut	1–26	1	
307363	**TIMING CHAIN**	1–27	1	

Part No.	Description	Plate No.	No. per Unit	Remarks
C.10332	**TIMING CHAIN TENSIONER (622302)**	1–28	1	
UCB.125/11R	**Bolt,** securing Tensioner to Cylinder Block		2	
430421	**Lockplate,** securing Bolts		1	
307920	**TIMING CHAIN VIBRATION DAMPER**	1–29	1	
UCS.125/6R	**Screw,** Set, securing Damper to Cylinder Block	1–30	2	
400000	**Lockplate,** securing Setscrews	1–31	2	
C.21042	**TIMING CHAIN COVER ASSEMBLY**	1–32	1	
400301	**Bush,** in Cover		1	
C.22611	**Gasket,** between Cover and Cylinder Block	1–33	1	
UCS.125/5R	**Screw,** Set, securing Cover to Cylinder Block		10	
UCS.125/7R	**Screw,** Set, securing Cover, Oil Cooler Hose Support Bracket, and Ignition Timing Indicator to Cylinder Block		2	
FW.104/E	**Washer** (Copper) on Setscrews		16	
C.21891	**IGNITION TIMING INDICATOR**	1–34	1	
C.20982	**TAPPET BLOCK AND BUSH ASSEMBLY**		1	Fitted from Engine No. 7A.1001 to 7A.3546
432997	**Bush,** in Tappet Guides (L.6462/1)		16	
C.22079	**TAPPET BLOCK AND BUSH ASSEMBLY**	1–35	1	Fitted to Engine No. 7A.3547 and subs
432997	**Bush,** in Tappet Guides (L.6462/1)		16	
430388	**Tappet Assembly,** (comprising Tappet Body and Tip)	1–36	16	
430563	**Ball End,** for Tappet	1–37	16	
C.22035	**Cover Plate Assembly,** for Tappet Block		1	Fitted from Engine No. 7A.1001 to 7A.8899
C.22030	**Fibreglass Pad,** under Cover Plate		1	Fitted from Engine No. 7A.1001 to 7A.8899
UCS.131/5R	**Screw,** Set, securing Cover Plate to Tappet Block		1	Fitted from Engine No. 7A.1001 to 7A.3546
FG.105/X	**Washer,** Spring, on Setscrew		1	Fitted from Engine No. 7A.1001 to 7A.3546
UCB.125/8R	**Bolt,** securing Cover Plate to Tappet Block		2	Fitted from Engine No. 7A.3547 to 7A.8899
418310	**Washer** (Copper) on Bolts		2	Fitted from Engine No. 7A.3547 to 7A.8899
135891	**Gasket,** between Tappet Block and Cylinder Block	1–38	1	
UCS.131/7R	**Screw,** Set, securing Tappet Block to Cylinder Block		9	
UCS.131/10R	**Screw,** Set, securing Tappet Block to Cylinder Block		2	
FG.105/X	**Washer,** Spring, on Setscrews		11	
430680	**Dowel,** locating Tappet Block on Cylinder Block	1–39	2	
34005/23	**Stud,** in Tappet Block, for Dynamo Mounting Bracket	1–40	4	
C.8511	**Stud,** in Tappet Block, for Alternator Mounting Bracket		4	Fitted to Engine No. 7K.1001 and subs
C.21325	**DYNAMO MOUNTING BRACKET ASSEMBLY**	1–41	1	
431972/01	**Shim** (.010" thick) for adjustment of Dynamo Bracket	1–42	As req'd	
431972/02	**Shim** (.015" thick) for adjustment of Dynamo Bracket	1–42	As req'd	
UFN.131/L	**Nut,** securing Bracket to Studs in Tappet Block		4	
FG.105/X	**Washer,** Spring, under Nuts		4	

CYLINDER BLOCK

Part No.	Description	Plate No.	No. per Unit	Remarks
C.21033	**CRANKSHAFT ASSEMBLY**	1–43	1	
C.2226	**Bush,** in rear end of Crankshaft, for Gearbox Primary Shaft		1	**REQUIRED FOR CARS FITTED WITH MANUAL TRANSMISSION**
1216/14	**Bush,** in Crankshaft Flange		1	Fitted from Engine No. 7A.1001 to 7A.9317
430238	**Dowel,** in Crankshaft Flange		2	
C.22996	**Plug,** in Crankshaft		4	
307359	**Main Bearing,** Front Journal (VP.4168)	1–44	2	**SUPPLIED ONLY AS A KIT, TO PART NUMBER 11228**
C.26663	**Main Bearing** (Top), Intermediate Journals (L.7782/1)	1–45	3	
C.26664	**Main Bearing** (Bottom), Intermediate Journals (L.7782/2)	1–46	3	
C.23289	**Main Bearing,** Rear Journal (L.7182/2)	1–47	2	
	CRANKSHAFT MAIN BEARINGS are obtainable in the following undersizes: –.010", –.020", –.030" **When ordering, please quote Part Number and undersize required (e.g. 11228/20 for .020" undersize**			
307680/1	**Thrust Washer** (Standard) for Crankshaft, (B.5145)	1–48	2	**ALTERNATIVES**
307680/2	**Thrust Washer** (Oversize .0025") for Crankshaft (B.5145)	1–48	2	
307680/3	**Thrust Washer** (Oversize .005") for Crankshaft (B.5145)	1–48	2	
C.22996	**Oil Seal,** Rear, on Crankshaft	1–49	1	
307339	**CHAINWHEEL ON CRANKSHAFT**	1–50	1	
422510/39	**Key,** locating Chainwheel on Crankshaft	1–51	1	
C.20973	**PULLEY, ON CRANKSHAFT**	1–52	1	**NOT REQUIRED WHEN POWER STEERING IS FITTED**
C.20974	**PULLEY, ON CRANKSHAFT**		1	**FOR USE ONLY WHEN POWER STEERING IS FITTED**
422510/39	**Key,** locating Pulley on Crankshaft	1–53	1	
307644	**CRANKSHAFT DAMPER**	1–54	1	
C.20911	**Hub,** for Crankshaft Damper	1–55	1	
C.21272	**Bolt,** securing Damper to Pulley	1–56	6	
C.21015	**Lockplate,** on Bolts	1–57	3	
C.21026	**Nut,** securing Damper and Pulley to Crankshaft	1–58	1	
425917	**Lockplate,** under Nut	1–59	1	
425901	**Oil Shield,** on Crankshaft	1–60	1	
307332	**CONNECTING ROD ASSEMBLY**	1–61	8	**NOT REQUIRED FOR CARS EXPORTED TO AUSTRIA PRIOR TO** Engine No. 7A.5294
307349	**Bearing,** Small End (B.5146)		8	
C.22236	**Bolt,** securing Bearing Caps to Connecting Rods		16	
430424	**Nut,** Nyloc, on Bolts		16	
C.22365	**CONNECTING ROD ASSEMBLY**		8	**FOR USE ONLY ON CARS EXPORTED TO AUSTRIA UP TO AND INCLUDING** Engine No. S7A.5293
307349	**Bearing,** Small End (B.5146)		8	
C.22236	**Bolt,** securing Bearing Caps to Connecting Rods		16	
C.28535	**Nut,** Nyloc, on Bolts		16	
307715	**Bearing** (Standard) Big End (VP.4166)	1–62	16 halves	
	BIG END BEARINGS are obtainable in the following undersizes : –.010", –.020", –.030" **When ordering, please quote Part Number and undersize required (e.g. 307715/20 for a .020" undersize bearing)**			

Part No.	Description	Plate No.	No. per Unit	Remarks
C.24286	**PISTON, COMPLETE**	1–63	8	Fitted from Engine No. 7A.1001 to 7A.10624 7A.10731 to 7A.12586 **BUT NOT REQUIRED FOR CARS EXPORTED TO AUSTRIA UP TO AND INCLUDING** Engine No. S7A.5293
C.23837	**Pressure Ring** (Upper) Chromium Plated	1–64	8	
C.23838	**Pressure Ring** (Lower)	1–65	8	
C.23839	**Scraper Ring**	1–66	8	
431449	**Gudgeon Pin**, retaining Piston on Connecting Rods	1–67	8	
C.23846	**Circlip**, securing Gudgeon Pins in Pistons	1–68	16	
C.26390	**PISTON, COMPLETE**		8	Fitted to Engine No. 7A.10625 to 7A.10730 7A.12587 and subs
	Individual items are as for Piston C.24286			
	PISTONS AND RINGS are obtainable in the following oversizes: +.010", +.020", +.030", +.040"			
	When ordering, please quote Part Number and oversize required (e.g. C.24286/20 for .020" oversize)			
C.24861	**PISTON, COMPLETE**		8	**FOR USE ONLY ON CARS EXPORTED TO AUSTRIA UP TO AND INCLUDING** Engine No. S7A.5293
C.22047	**Pressure Ring** (Upper), Chromium Plated (RM.4729)		8	
C.22048	**Pressure Ring** (Lower), Taper (RM.4728)		8	
C.22455	**Scraper Ring**		8	
431449	**Gudgeon Pin**, retaining Pistons on Connecting Rods		8	
430385	**Circlip**, securing Gudgeon Pins in Pistons		16	
	PISTONS AND RINGS ARE NOT OBTAINABLE IN OVERSIZES			
	CYLINDER BLOCK LINERS (Standard Bore) are available for Reconditioning purposes			
C.21036	**OIL SUMP ASSEMBLY**	1–69	1	Fitted from Engine No. 7A.1001 and subs 7K.1001 to 7K.1255
C.27488	**OIL SUMP ASSEMBLY**		1	Fitted to Engine No. 7K.1256 and subs Also fitted to certain engines prior to the number quoted above
C.20904	**Tray**, for Oil Sump	1–70	1	
UCS.125/4R	**Screw**, Set, securing Tray to Sump		6	
FG.104/X	**Washer**, Spring, on Setscrews		6	
C.21554	**Tube Assembly**, for Oil Sump Dipstick	1–71	1	
416380	**Washer**, on Dipstick Tube	1–72	1	
UCS.131/7R	**Screw**, Set, securing Oil Sump to Cylinder Block		16	
FG.105/X	**Washer**, Spring, on Setscrews		16	
C.23435	**Drain Plug**, for Oil Sump	1–73	1	
C.2296/1	**Washer**, on Drain Plug	1–74	1	
135870	**OIL SUMP BRIDGE PIECE**	1–75	1	
425938	**Bolt**, securing Bridge Piece to Cylinder Block		2	
FG.105/X	**Washer**, Spring, on Bolts		2	
308402	**Oil Seal**, Front, between Bridge Piece and Oil Sump	1–76	1	
308470	**Oil Seal**, Rear, between Rear Bearing Cap and Oil Sump	1–77	1	
C.21551	**OIL SUMP DIPSTICK**	1–78	1	

CYLINDER BLOCK

Part No.	Description	Plate No.	No. per Unit	Remarks
C.27426	**FLYWHEEL**		1	Fitted to Engine No. 7A.13231 and subs
C.8631	**Dowel**, on outer face of Flywheel, locating Clutch Cover		3	(Also to 7A.12185 and 7A.12186) 7K.1001 and subs
430465	**Bolt**, securing Flywheel to Crankshaft		6	**See also Clutch Unit**
430238	**Bolt**, securing Flywheel to Crankshaft		2	
430237	**Lockplate**, securing Bolts		3	
C.22144	**TORQUE CONVERTER DRIVE PLATE ASSEMBLY**	1–79	1	**FITTED TO CARS WITH AUTOMATIC TRANSMISSION ONLY**
C.22147	**Spacing Washer**	1–80	2	
430239	**Bolt**, securing Drive Plate to Crankshaft Flange	1–81	6	
430237	**Lockplate**, on Bolts	1–82	3	
C.21163	**ROCKER OIL FEED PIPE ASSEMBLY**	1–83	1	
430417	**Banjo Bolt**, (Long)	1–84	1	
430415	**Banjo Bolt**, (Short)	1–85	4	
416490	**Washer**, (Soft Copper) on each face of Feed-Pipe Banjos	1–86	10	
C.21455	**OIL FILLER TUBE ASSEMBLY**	1–87	1	Fitted to Engine No. 7A.1001 and subs
C.27402	**OIL FILLER TUBE ASSEMBLY**		1	Fitted from Engine No. 7K.1001 and subs
C.21667	**Locknut**, securing Filler Tube	1–88	1	
431288	**CAP, ON OIL FILLER TUBE**	1–89	1	
C.22665	**GEARBOX SUPPORT PLATE**	1–90	1	
UCS.131/7R	**Screw**, Set, securing Support Plate to Cylinder Block		3	
UCB.131/8R	**Bolt**, securing Support Plate to Cylinder Block		1	
FG.105/X	**Washer**, Spring, on Bolt and Setscrews		4	
UCB.131/12R	**Bolt**, securing Support Plate and Converter Housing to Cylinder Block		6	
UCB.131/13R	**Bolt**, securing Support Plate and Converter Housing to Cylinder Block		1	
FW.105/T	**Washer**, Plain, on Bolts		1	
FG.105/X	**Washer**, Spring, on Bolts		7	
UFB.137/17R	**Bolt**, securing Clutch Housing to Sump		1	
UFN.137/L	**Nut**, on Bolt		1	
FW.106/T	**Washer**, Plain, on Bolt		1	
FG.106/X	**Washer**, Spring, on Bolt		1	
C.21663	**Plug**, for Gearbox Support Plate	1–91	2	1 only required for later Engines
C.20842	**WATER PUMP INLET PIPE**	1–92	1	
430429	**Gasket**, between Inlet Pipe and Cylinder Block	1–93	1	
34005/12	**Stud**, in Cylinder Block, for Water Inlet Pipe	1–94	4	
UFN.131/L	**Nut**, on Studs, securing Water Inlet Pipe to Cylinder Block		4	
FG.105/X	**Washer**, Spring, under Nuts		4	
UCS.125/8R	**Screw**, Set, securing Water Inlet Pipe to Water Outlet Pipe		2	
FG.105/X	**Washer**, Spring on Setscrews		2	

Part No.	Description	Plate No.	No. per Unit	Remarks
C.21228	**R.H. ENGINE MOUNTING BRACKET ASSEMBLY**	1–95	1	Fitted from Engine No. 7A.1001 to 7A.13052
C.27445	**R.H. ENGINE MOUNTING BRACKET ASSEMBLY**		1	Fitted to Engine No. 7A.13053 and subs
C.21018	**L.H.ENGINE MOUNTING BRACKET ASSEMBLY**	1–96	1	NOT REQUIRED WHEN POWER STEERING IS FITTED UP TO AND INCLUDING ENGINE NO. 7A.13052
C.27702	**L.H. ENGINE MOUNTING BRACKET ASSEMBLY**		1	NOT REQUIRED WHEN POWER STEERING IS FITTED AT ENGINE NO. 7A.13053 AND SUBS
C.21063	**L.H. ENGINE MOUNTING BRACKET ASSEMBLY**		1	FOR USE ONLY WHEN POWER STEERING IS FITTED UP TO AND INCLUDING ENGINE NO. 7A.13052
C.28322	**L.H. ENGINE MOUNTING BRACKET ASSEMBLY**		1	FOR USE ONLY WHEN POWER STEERING IS FITTED
UCS.137/6R	**Screw**, Set, securing Engine Mounting Brackets to Cylinder Block		8	
FG.106/X	**Washer**, Spring, on Setscrews		8	
C.4794	**FRONT ENGINE MOUNTING**	1–97	2	
UFS.150/7R	**Screw**, Set, securing Front Engine Mountings to Engine Brackets		2	
FW.108/T	**Washer**, Plain, on Setscrews		2	
FG.108/X	**Washer**, Shakeproof, on Setscrews		2	
UFB.131/8R	**Bolt**, securing Front Engine Mountings to Engine Support Bracket Assembly		4	
FG.105/X	**Washer**, Spring, on Bolts		4	
UFN.131/L	**Nut**, on Bolts		4	
C.11529	**R.H. FRONT ENGINE SUPPORT BRACKET**	1–98	1	
C.11528	**L.H. FRONT ENGINE SUPPORT BRACKET**	1–99	1	
UFS.131/6R	**Screw**, Set, securing Brackets to Body Underframe		4	
UFB.131/30R	**Bolt**, securing Brackets to Body Underframe		4	
FG.105/X	**Washer**, Spring, on Setscrews and Bolts		8	
UFN.131/L	**Nut**, on Setscrews and Bolts		8	
C.27458	**REAR ENGINE MOUNTING BRACKET ASSEMBLY**		1	FOR CARS FITTED WITH STANDARD OR OVERDRIVE TRANSMISSION
C.27457	**Mounting Bracket** only		1	
C.12335	**Spring Seat**, (Rubber) cemented to inside of Mounting Bracket		1	Fitted to Engine No. 7A.13231 and subs 7K.1001 and subs
C.12265	**Centre Bush** (Rubber)		1	
UFB.131/10R	**Bolt**, securing Mounting Bracket to Body		4	
C.12027	**Spacer**, on Bolts		4	
FG.105/X	**Washer**, Spring, on Bolts		4	
FW.105/T	**Washer**, Plain, under Spring Washers		4	
C.19988	**COIL SPRING, FOR REAR ENGINE MOUNTING**		1	
C.12335	**Spring Seat** (Rubber) at top of Coil Spring		1	

CYLINDER BLOCK

Part No.	Description	Plate No.	No. per Unit	Remarks
C.20885	**BRACKET ASSEMBLY, MOUNTING JOCKEY PULLEY CARRIER TO CYLINDER BLOCK**	1–100	1	
C.791	**Packing Washer**, for Bracket		As req'd	
UCS.137/8R	**Screw**, Set, securing Bracket		2	
FG.106/X	**Washer**, Spring, on Setscrews		2	
C.18835	**CARRIER FOR JOCKEY PULLEY**		1	Fitted from Engine No. 7A.1001 to 7A.1110
C.8737/2	**Nut**, Self-Locking, retaining Carrier in Bracket		1	
C.18454	**Thrust Washer**, under Nut		1	
C.21330	**CARRIER FOR JOCKEY PULLEY**	1–101	1	Fitted to Engine No. 7A.1111 and subs
C.8737/2	**Nut**, Self-Locking, retaining Carrier in Bracket		1	
C.21326	**Thrust Washer and Stop**, under Nut	1–102	1	
C.18447	**TORSION SPRING, FOR FAN BELT TENSION**	1–103	1	
C.18733	**PULLEY ASSEMBLY, APPLYING TENSION TO FAN BELT**	1–104	1	
C.3968/2	**Nut**, Self-Locking, securing Pulley to Carrier		1	
C.20901	**DISTRIBUTOR BRACKET**		1	Fitted from Engine No. 7A.1001 to 7A.2340
C.22139	**DISTRIBUTOR BRACKET**	1–105	1	Fitted to Engine No. 7A.2341 and subs
UCS.125/7R	**Screw**, Set, securing Distributor Bracket to Tappet Block		1	
FG.104/X	**Washer**, Spring, on Setscrew		1	
UCS.131/10R	**Screw**, Set, securing Distributor Bracket to Tappet Block		1	
FG.105/X	**Washer**, Spring, on Setscrew		1	
UCS.125/5/R	**Screw**, Set, retaining Distributor in Bracket		2	
C.724	**Washer**, Shakeproof, on Setscrews		2	
FW.104/T	**Washer**, Plain, under Shakeproof Washers		2	
C.20975	**REV. COUNTER PINION**	1–106	1	
432524	**Thrust Washer**, on Pinion	1–107	1	
C.20978	**REV. COUNTER PINION HOUSING ASSEMBLY**		1	Fitted from Engine No. 7A.1001 to 7A.2340
430690	**Screw**, locating Housing Assembly in Distributor Mounting Bracket		1	
FG.105/X	**Washer**, Spring, on Screw		1	
C.22143	**REV. COUNTER PINION HOUSING ASSEMBLY**	1–108	1	Fitted to Engine No. 7A.2341 and subs
C.8011	**Screw**, locating Housing Assembly in Distributor Mounting Bracket	1–109	1	
UFN.225/L	**Locknut**, securing Locating Screw	1–110	1	
C.20976	**ADAPTOR, ON REV. COUNTER PINION**	1–111	1	
C.25426	**Driving Dog**, coupling Adaptor to Rev. Counter Generator	1–112	1	
C.16773	**Circlip**, retaining Driving Dog in Adaptor	1–113	1	
C.14996	**REV. COUNTER GENERATOR (TV.1100/00)**	1–114	1	
C.14990	**'O' Ring**, on Generator	1–115	1	
C.14992	**Screw**, Set, securing Rev. Counter Generator to Housing Assembly	1–116	3	
C.15918	**Plate Washer**, securing Setscrews	1–117	1	
C.15919	**Lockwasher**, on Setscrews	1–118	3	
9990	**GASKET PACKAGE FOR COMPLETE ENGINE OVERHAUL (SD1124)**		1	

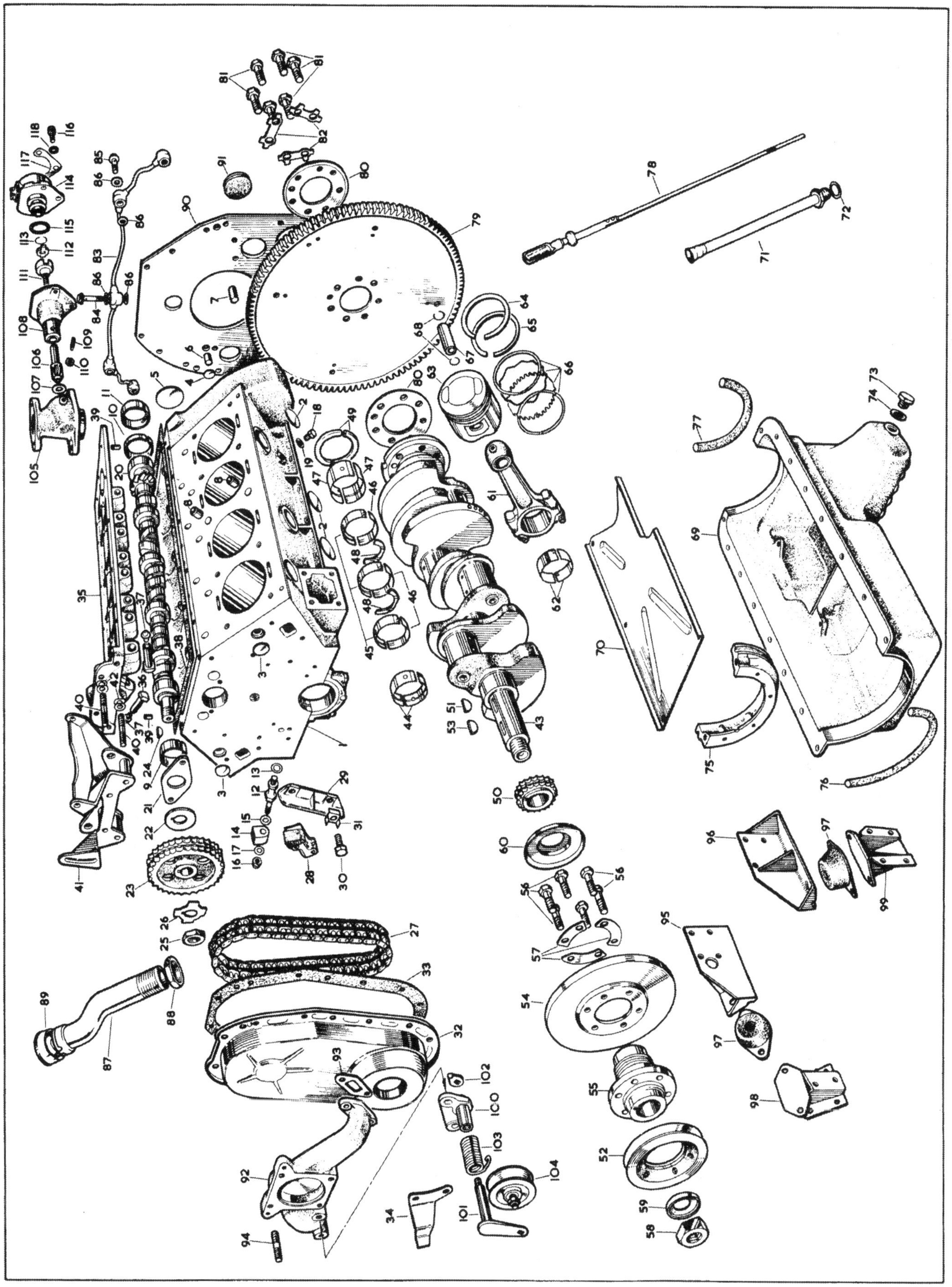

PLATE 1

OIL PUMP

Part No.	Description	Plate No.	No. per Unit	Remarks
C.21975	**OIL PUMP ASSEMBLY**		1	
430360	**Body Assembly**	2–1	1	
C.20971	**Spindle,** for Oil Pump	2–2	1	
430363	**Spiral Gear,** on Spindle	2–3	1	
422510/11	**Key,** locating Gear on Spindle	2–4	1	
401831	**Thrust Washer,** under Gear	2–5	1	
C.20972	**Driving Gear,** for Pump	2–6	1	
43037	**Driven Gear and Bush,** for Pump	2–7	1	
C.21120	**Cover and Strainer Assembly**	2–8	1	
UCS.125/6R	**Screw,** Set, securing Cover	2–9	4	
423044	**Lockplate,** securing Setscrews	2–10	2	
C.22756	**Relief Valve,** in Pump Body	2–11	1	
431498	**Spring,** for Relief Valve	2–12	1	
431948	**Cap,** retaining Spring	2–13	1	
431949	**Lockplate,** securing Cap	2–14	1	
430374	**Gasket,** between Oil Pump and Rear Bearing Cap	2–15	1	
UCS.131/6R	**Screw,** Set, securing Oil Pump to Rear Bearing Cap	2–16	3	
430356	**Lockplate,** securing Setscrews	2–17	3	

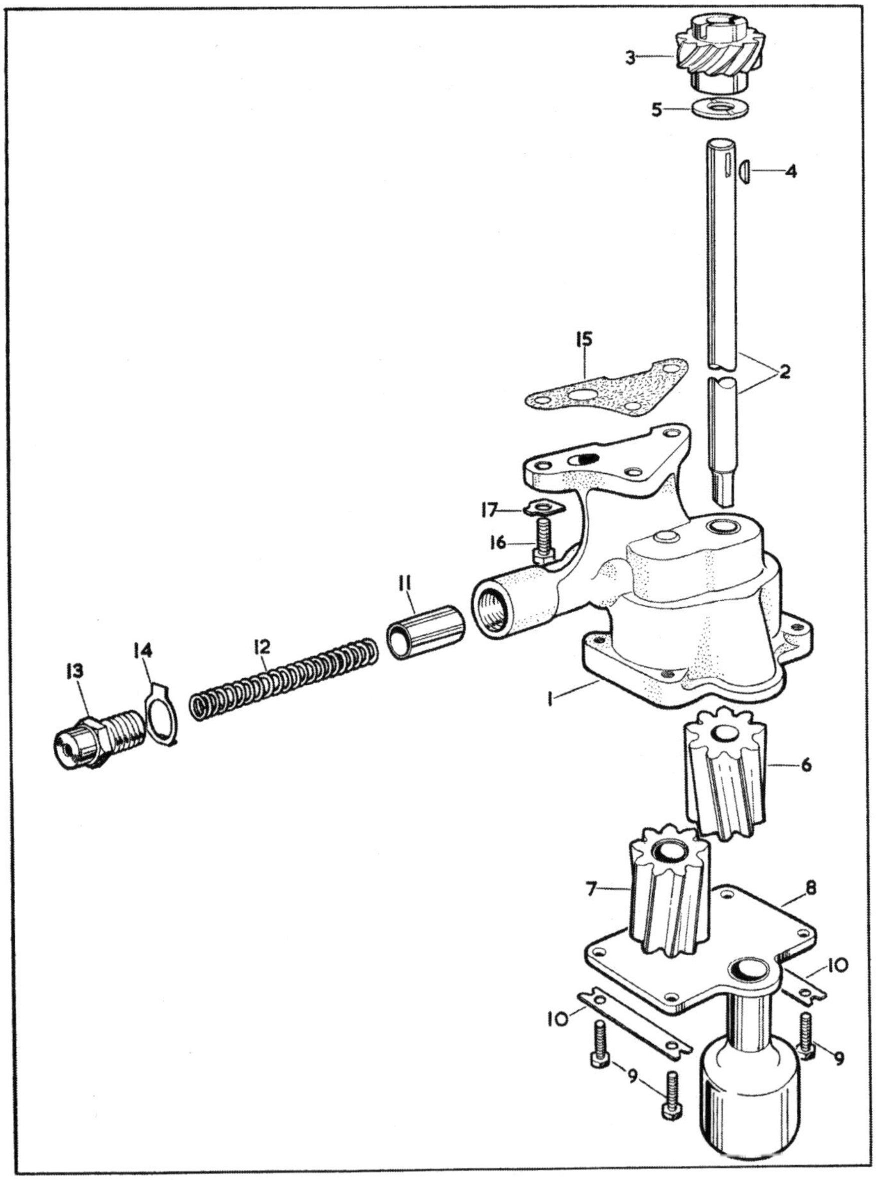

PLATE 2

Part No.	Description	Plate No.	No. per Unit	Remarks
C.20917	**OIL CLEANER COMPLETE (FA.2116)**	3–1	1	
9672	**Canister Assembly** (125528)	3–2	1	
6159	**Spring,** at bottom of Canister (160441)	3–3	1	
6160	**Washer,** Plain, at top of Spring (174660)	3–4	1	
6161	**Washer,** Felt, under Pressure Plate (141251)	3–5	1	
9675	**Pressure Plate** (134482)	3–6	1	
10454	**Bolt,** through Canister (46612)	3–7	1	
6158	**Washer** (Rubber) under Head of Bolt (137574)	3–8	1	
9674	**Spring Clip,** on Bolt (175012)	3–9	1	
9673	**Element** (FP.3317)	3–10	1	
9348	**Anchor Insert** (105360)	3–11	1	
9671	**Clamping Plate,** on Filter Head (121822)	3–12	1	
6883	**Sealing Ring,** between Filter Head and Canister (137493)	3–13	1	
9670	**Filter Head** (41091)	3–14	1	
430427	**Gasket,** between Cylinder Block and Oil Cleaner	3–15	1	
UFB.131/8R	**Bolt,** securing Oil Cleaner to Cylinder Block		4	
FG.105/X	**Washer,** Spring, on Bolts		4	

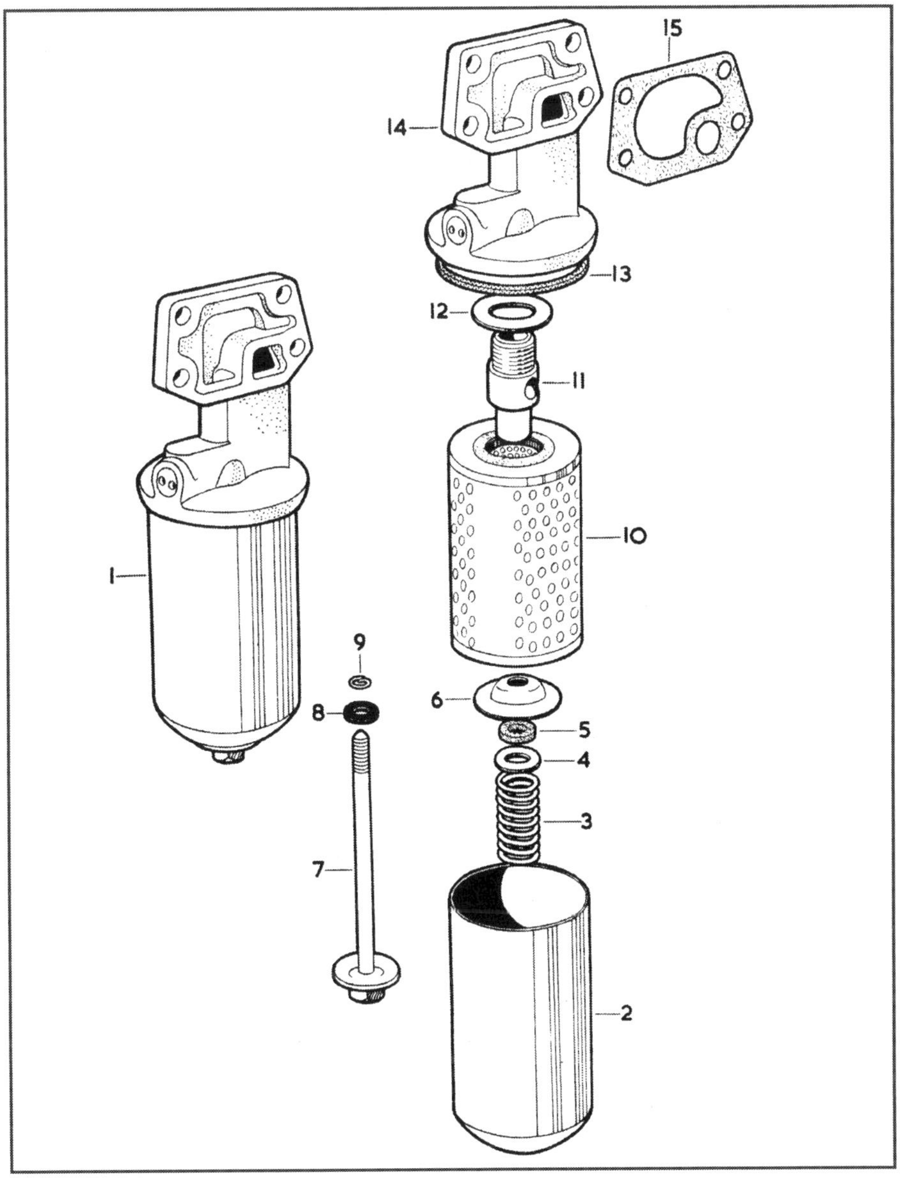

PLATE 3

WATER PUMP

Part No.	Description	Plate No.	No. per Unit	Remarks
C.21940	**WATER PUMP ASSEMBLY**		1	Fitted from Engine No. 7A.1001 to 7A.1098
C.20913	**Pump Body**		1	
C.8167	**Bearing,** in Body		1	
C.8011	**Screw,** locating Bearing in Body		1	
UFN.225/L	**Nut,** locking Screw		1	
C.724	**Washer,** Shakeproof, under Nut		1	
C.6773	**Seal,** for Bearing		1	
C.8256	**Thrower Ring,** on Spindle		1	
C.7675	**Impellor,** on Spindle		1	
C.20891	**Fan Pulley,** on Spindle		1	
C.21941	**WATER PUMP ASSEMBLY**		1	Fitted to Engine No. 7A.1099 and subs
C.20913	**Pump Body**	4–1	1	
C.8167	**Bearing,** in Body	4–2	1	
C.8011	**Screw,** locating Bearing in Body	4–3	1	
UFN.225/L	**Nut,** locking Screw	4–4	1	
C.724	**Washer,** Shakeproof, under Nut	4–5	1	
C.6773	**Seal,** for Bearing	4–6	1	
C.8256	**Thrower Ring,** on Spindle	4–7	1	
C.7675	**Impellor,** on Spindle	4–8	1	
C.21738	**Fan Pulley,** on Spindle	4–9	1	
C.16751	**Stud,** for Fan Pulley	4–10	4	
C.21117	**Gasket,** between Pump Body and Water Inlet Pipe	4–11	1	
UCS.131/8R	**Screw,** Set, securing Pump Body to Water Inlet Pipe		4	
FG.105/X	**Washer,** Spring, on Setscrews		4	
C.28329	**WATER PUMP ASSEMBLY**		1	Fitted from Engine No. 7K.1001 and subs
C.20913	**Pump Body**		1	
C.8167	**Bearing,** in Body		1	
C.8011	**Screw,** locating Bearing in Body		1	
UFN.225/L	**Nut,** locking Screw		1	
C.724	**Washer,** under Nut		1	
C.6773	**Seal,** for Bearing		1	
C.8256	**Thrower Ring,** on Spindle		1	
C.7675	**Impellor,** on Spindle		1	
C.27434	**Pulley,** for Water Pump		1	
C.16751	**Stud,** for Fan Pulley		4	
C.21117	**Gasket,** between Pump Body and Water Inlet Pipe		1	
UCS.131/8R	**Screw,** Set, securing Pump Body to Water Inlet Pipe		4	
FG.105/X	**Washer,** Spring, on Setscrews		4	
C.21721	**FAN ASSEMBLY**		1	Fitted from Engine No. 7A.1001 to 7A.1098
C.21661	**Balance Plate,** for Fan		As req'd	
UCS.125/7R	**Screw,** Set, securing Fan Assembly to Pulley		4	
C.740	**Washer,** Shakeproof, on Setscrews		4	
C.20887	**BELT, DRIVING FAN**	4–12	1	
C.21025	**BELT, DRIVING DYNAMO**	4–13	1	
C.27354	**BELT, DRIVING ALTERNATOR**		1	Fitted to Engine No. 7K.1001 and subs
C.21819	**AUTOMATIC FAN UNIT (TORQATROL)**	4–14	1	Fitted from Engine No. 7A.1099 to 7A.7231 except 7A.6840 to 7A.6940
C.24257	**AUTOMATIC FAN UNIT (HOLSET)**		1	Fitted to Engine No. 7A.7232 and subs also to 7A.6840 to 7A.6940
UFN.131/L	**Nut,** securing Automatic Fan Unit to Fan Pulley		4	Fitted to Engine No. 7A.1099 and subs
C.740	**Washer,** Shakeproof, under Nuts		4	
C.21735	**Fan Assembly**	4–15	1	
UFN.131/L	**Nut,** securing Fan Assembly to Fan Unit		4	
FG.105/X	**Washer,** Shakeproof, under Nuts		4	

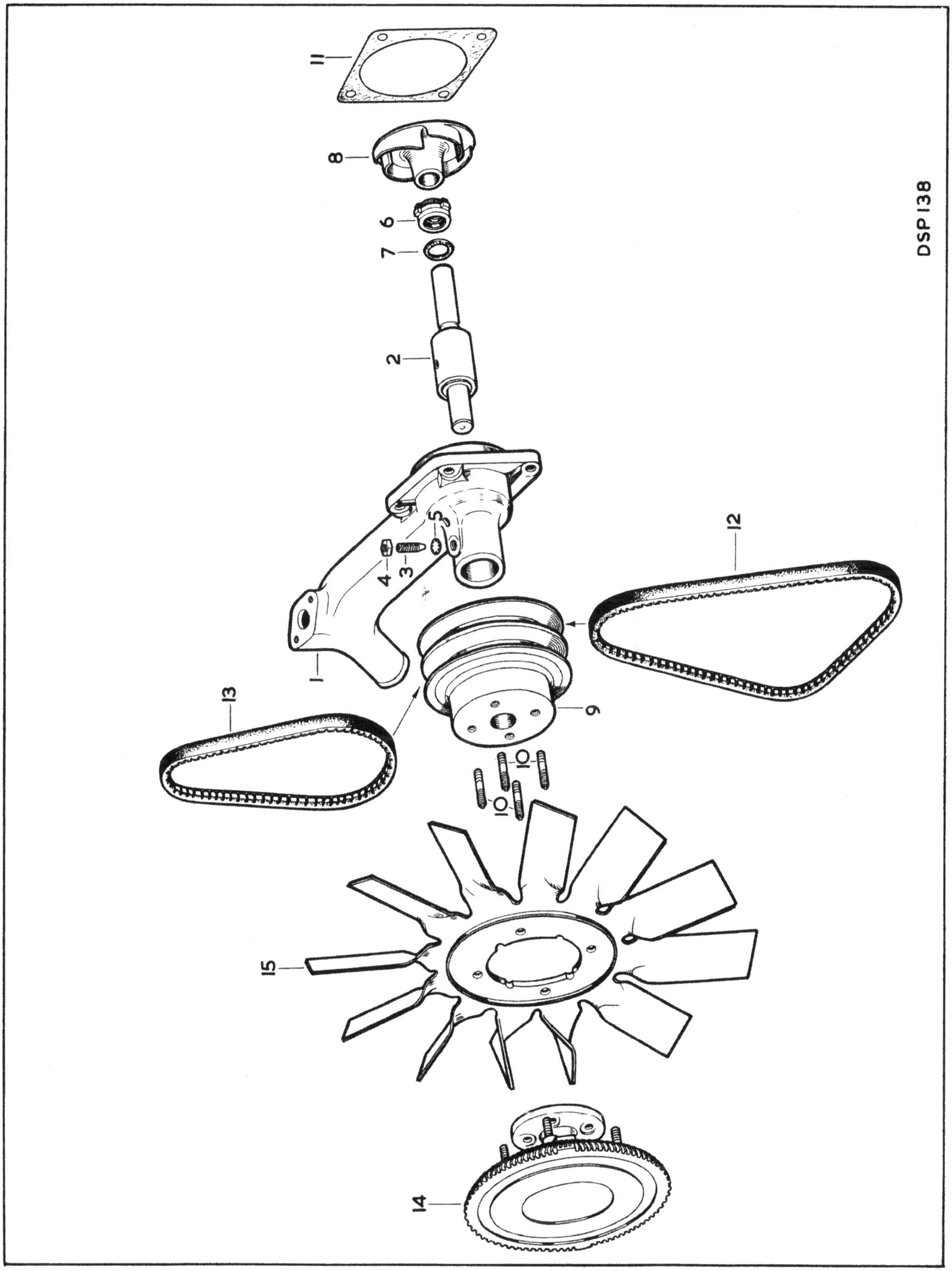

PLATE 4

CYLINDER HEADS

Part No.	Description	Plate No.	No. per Unit	Remarks
135893	**CYLINDER HEAD ASSEMBLY**	5–1	2	Fitted from Engine No. 7A.1001 to 7A.11610
C.315	**Welch Washer,** in each end of Cylinder Heads	5–2	4	**NOTE: On some early Cylinder Heads, Taper Plugs are fitted. When replacement is necessary fit Taper Plug Part No. 432099**
431000	**Guide,** for Inlet Valves	5–3	8	
431000/2	**Guide,** for Inlet Valves, .005" oversize		8	
431000/3	**Guide,** for Inlet Valves, .010" oversize		8	
431251	**Guide,** for Exhaust Valves	5–4	8	
431251/2	**Guide,** for Exhaust Valves, .005" oversize		8	
431251/3	**Guide,** for Exhaust Valves, .010" oversize		8	
34005/13	**Stud,** for Exhaust Manifolds	5–5	16	
430741	**Tube,** for Sparking Plugs	5–6	8	
430742	**Washer,** under Tubes	5–7	8	
C.26312	**CYLINDER HEAD ASSEMBLY**		2	Fitted to Engine No. 7A.11611 and subs 7K.1001 and subs
C.26302	**Insert,** for Valves		16	
	All other items are as for Cylinder Head 135893			
C.22245	**Gasket,** between Cylinder Heads and Cylinder Block	5–8	2	
UCS.131/4R	**Plug,** sealing Rocker Oil Feed holes in Cylinder Heads	5–9	4	
400352	**Washer,** on Plugs	5–10	4	
430444	**Blanking Plate,** sealing Water apertures in Cylinder Heads	5–11	2	
430429	**Gasket,** under Blanking Plates	5–12	2	
UCS.131/6R	**Screw,** Set, securing Blanking Plates		4	
FG.105/X	**Washer,** Spring, on Setscrews		4	
C.21037	**Bolt,** securing Rocker Brackets and Cylinder Heads to Cylinder Block	5–13	20	
C.23586	**Washer,** Plain on Bolts		20	
9989	**GASKETS PACKAGE FOR DECARBONISING ENGINE (SD.1123)**		1	
307358	**INLET VALVE**	5–14	8	
307620	**EXHAUST VALVE**	5–15	8	
C.21687	**VALVE SPRING (INNER)**	5–16	16	
C.21688	**VALVE SPRING (OUTER)**	5–17	16	
430980	**Seating Washer,** for Valve Springs	5–18	16	Fitted from Engine No. 7A.1001 to 7A.7332
C.24050	**Seating Washer,** for Valve Springs		16	Fitted to Engine No. 7A.7333 and subs 7K.1001 and subs
C.24051	**Distance Washer,** under Seating Washers	5–19	16	
430452	**Collar,** at top of Valve Springs	5–20	16	
432971	**Split Cone** (Plain Half) for Valves	5–21	16	
432971/1	**Split Cone** (Padded Half) for Valves	5–22	16	
307368	**ROCKER SHAFT ASSEMBLY (INLET)**	5–23	2	
307378	**ROCKER SHAFT ASSEMBLY (EXHAUST)**	5–24	2	

Part No.	Description	Plate No.	No. per Unit	Remarks
C.24010	**ROCKER SHAFT BRACKET ASSEMBLY (INTERMEDIATE)**	5–25	12	
430381	**Dowel**, in Rocker Shaft Bracket		12	
C.24011	**ROCKER SHAFT BRACKET (AT EACH END OF SHAFTS)**	5–26	8	
430551	**Screw**, locating Rocker Shaft in End Brackets	5–27	4	
430552	**Lockplate**, on Screws		4	
307352	**VALVE ROCKER (INLET)**	5–28	8	
307353	**VALVE ROCKER (EXHAUST)**	5–29	8	
430411	**Spacing Spring**, between Rockers and Brackets	5–30	16	
430412	**Washer**, between Rockers and Springs and Rockers and Brackets	5–31	48	
431011	**Ball End**, on Rockers	5–32	16	
430419	**Nut**, securing Ball Ends to Rockers	5–33	16	
430754	**PUSH ROD ASSEMBLY (INLET)**	5–34	8	
430756	**PUSH ROD ASSEMBLY (EXHAUST)**	5–35	8	
307372	**INDUCTION MANIFOLD**		1	Fitted from Engine No. 7A.1001 to 7A.2263
C.22412	**INDUCTION MANIFOLD**	5–36	1	Fitted to Engine No. 7A.2264 and subs
34005/13	**Stud**, mounting Carburetters	5–37	8	
UFN.131/L	**Nut**, securing Carburetters to Induction Manifold		8	
FG.105/X	**Washer**, Spring, under Nuts		8	
307876	**Gasket**, between Induction Manifold and Cylinder Heads	5–38	2	
UCS.131/9R	**Screw**, Set, securing Induction Manifold to Cylinder Heads		16	
FG.105/X	**Washer**, Spring, on Setscrews		16	
C.21452	**ADAPTOR, IN INDUCTION MANIFOLD, FOR SERVO VACUUM PIPE**	5–39	1	
431365	**Copper Washer**, on Adaptor	5–40	1	
C.20909	**EXHAUST MANIFOLD, R.H.**	5–41	1	
C.20910	**EXHAUST MANIFOLD, L.H.**		1	
307354	**Gasket**, between Exhaust Manifolds and Cylinder Heads	5–42	2	
422804	**Nut**, securing Exhaust Manifolds to Cylinder Heads		16	
FG.105/X	**Washer**, Plain, under Nuts		16	
C.20848	**WATER OUTLET PIPE CONNECTING CYLINDER HEADS**	5–43	1	
430429	**Gasket**, between Water Outlet Pipe and Cylinder Heads	5–44	2	
UCS.131/8R	**Screw**, Set, securing Water Outlet Pipe to Cylinder Heads		4	
FG.105/X	**Washer**, Spring, on Setscrews		4	

CYLINDER HEADS

Part No.	Description	Plate No.	No. per Unit	Remarks
C.27650	**THERMOSTAT UNIT IN WATER OUTLET PIPE (3B.1030.74) (OPENING TEMPERATURE 160°–170°F) (71°–76.5°C)**	5–45	1	**Supersedes C.20766**
C.27650/1	**THERMOSTAT UNIT IN WATER OUTLET PIPE (3B.1030.82) (OPENING TEMPERATURE 174°–183°F) (79°–84°C)**	5–45	1	**HIGHER SETTING FOR COLD COUNTRIES Supersedes C.20766/1**
C.15472	**WATER TEMPERATURE ELEMENT IN WATER OUTLET PIPE (TT4801/00)**	5–46	1	
C.21049	**WATER OUTLET PIPE EXTENSION FOR TOP RADIATOR HOSE**	5–47	1	
C.4985	**Gasket,** between Outlet Pipe and Extension	5–48	1	
UCS.131/7R	**Screw,** Set, securing Extension to Outlet Pipe		2	
FG.105/X	**Washer,** Spring, on Setscrews		2	
432237	**WATER FLANGE ASSEMBLY, ON OUTLET PIPE**	5–49	1	
432236	**Gasket,** between Flange and Outlet Pipe	5–50	1	
UCS.125/6R	**Screw,** Set, securing Flange to Outlet Pipe		2	
FG.104/X	**Washer,** Spring, on Setscrews		2	
432240	**HOSE, CONNECTING FLANGE ASSEMBLY TO WATER PIPE**	5–51	1	
C.2905/3	**Clip,** securing Hose	5–52	2	
C.21012	**WATER PIPE BETWEEN HOSE AND TEE PIECE**	5–53	1	
C.21170	**TEE PIECE BETWEEN WATER PIPE AND INDUCTION MANIFOLD INLET**	5–54	1	
C.2905/3	**Clip,** securing Tee Piece to Water Pipe and Induction Manifold Inlet	5–55	2	
432237	**WATER PIPE FLANGE ASSEMBLY ON WATER PUMP HOUSING**	5–56	1	
432236	**Gasket,** between Flange and Water Pump Body	5–57	1	
UCS.125/6R	**Screw,** Set, securing Flange to Water Pump Body		2	
FG.104/X	**Washer,** Spring, on Setscrews		2	
432240	**HOSE CONNECTING FLANGE ASSEMBLY TO WATER PIPE**	5–58	1	
C.2905/3	**Clip,** securing Hose to Flange and Water Pipe	5–59	2	
C.20995	**WATER PIPE BETWEEN HOSE AND TEE PIECE**	5–60	1	
C.21170	**TEE PIECE**	5 61	1	
C.2905/3	**Clip,** securing Tee Piece to Water Pipes	5–62	2	

Part No.	Description	Plate No.	No. per Unit	Remarks
431493	**WATER PIPE FROM TEE PIECE TO ELBOW HOSE**		1	Fitted from Engine No. 7A.1001 to 7A.2263
431495	**ELBOW HOSE CONNECTING WATER PIPE TO INDUCTION MANIFOLD OUTLET**		1	
C.2905/3	**Clip,** securing Elbow Hose to Water Pipe and Induction Manifold Outlet		2	
C.22390	**WATER PIPE ASSEMBLY FROM TEE PIECE TO INDUCTION MANIFOLD**	5–63	1	Fitted to Engine No. 7A.2264 and subs
432236	**Gasket,** between Pipe Flange and Induction Manifold	5–64	1	
UCS.125/5R	**Screw,** Set, securing Pipe Flange to Manifold		2	
FG.104/X	**Washer,** Spring, on Setscrews		2	
136769	**TOP COVER ON R.H. CYLINDER HEAD**	5–65	1	
136404	**TOP COVER ON L.H. CYLINDER HEAD**		1	
307827	**Gasket,** between Top Covers and Cylinder Heads	5–66	2	
432136	**Nut,** Special, securing Top Covers to Sparking Plug Tubes	5–67	8	
425280/25	**'O' Ring,** under Nuts	5–68	8	
C.27515	**TOP COVER ON CYLINDER HEAD**		2	Fitted to Engine No. 7K.1001 and subs
433099	**BREATHER PIPE ASSEMBLY ON CYLINDER HEAD COVERS**	5–69	2	
C.5634	**SPARKING PLUG (CHAMPION N.8)**		8	
432665	**EXTENSION ASSEMBLY FOR SPARKING PLUGS**		8	
432663	**Contact Clip,** for Extension		8	
432858	**Spring,** for Contact Clips		8	
	HIGH TENSION LEADS			
432659/12	**Lead to No. 1 R.H. Sparking Plug**		1	
432659/12	**Lead to No. 2 R.H. Sparking Plug**		1	
432659/07	**Lead to No. 3 R.H. Sparking Plug**		1	
432659/04	**Lead to No. 4 R.H. Sparking Plug**		1	
432659/12	**Lead to No. 1 L.H. Sparking Plug**		1	
432659/07	**Lead to No. 2 L.H. Sparking Plug**		1	
432659/07	**Lead to No. 3 L.H. Sparking Plug**		1	
432659/04	**Lead to No. 4 L.H. Sparking Plug**		1	
C.21185/04	**HIGH TENSION CABLE ASSEMBLY FROM COIL TO DISTRIBUTOR**		1	
C.30191	**HIGH TENSION CABLE ASSEMBLY FROM COIL TO DISTRIBUTOR**		1	For use when Coils with Push-In Connectors are fitted. At Chassis Nos. 1K.2712 and subs 1K.30067 and subs
431404	**RUBBER RING HOLDING HIGH TENSION CABLES (SIC.5986)**		2	

CYLINDER HEADS

Part No.	Description	Plate No.	No. per Unit	Remarks
431308	**CARRIER FOR SPARKING PLUG LEADS**		8	(4 per unit required Engine No. 7K.1001 and subs)
431309	**CARRIER FOR SPARKING PLUG LEADS**		2	
UCS.525/10H	**Screw,** Set, securing Carriers		2	
UCS.525/7H	**Screw,** Set, securing Carriers		2	

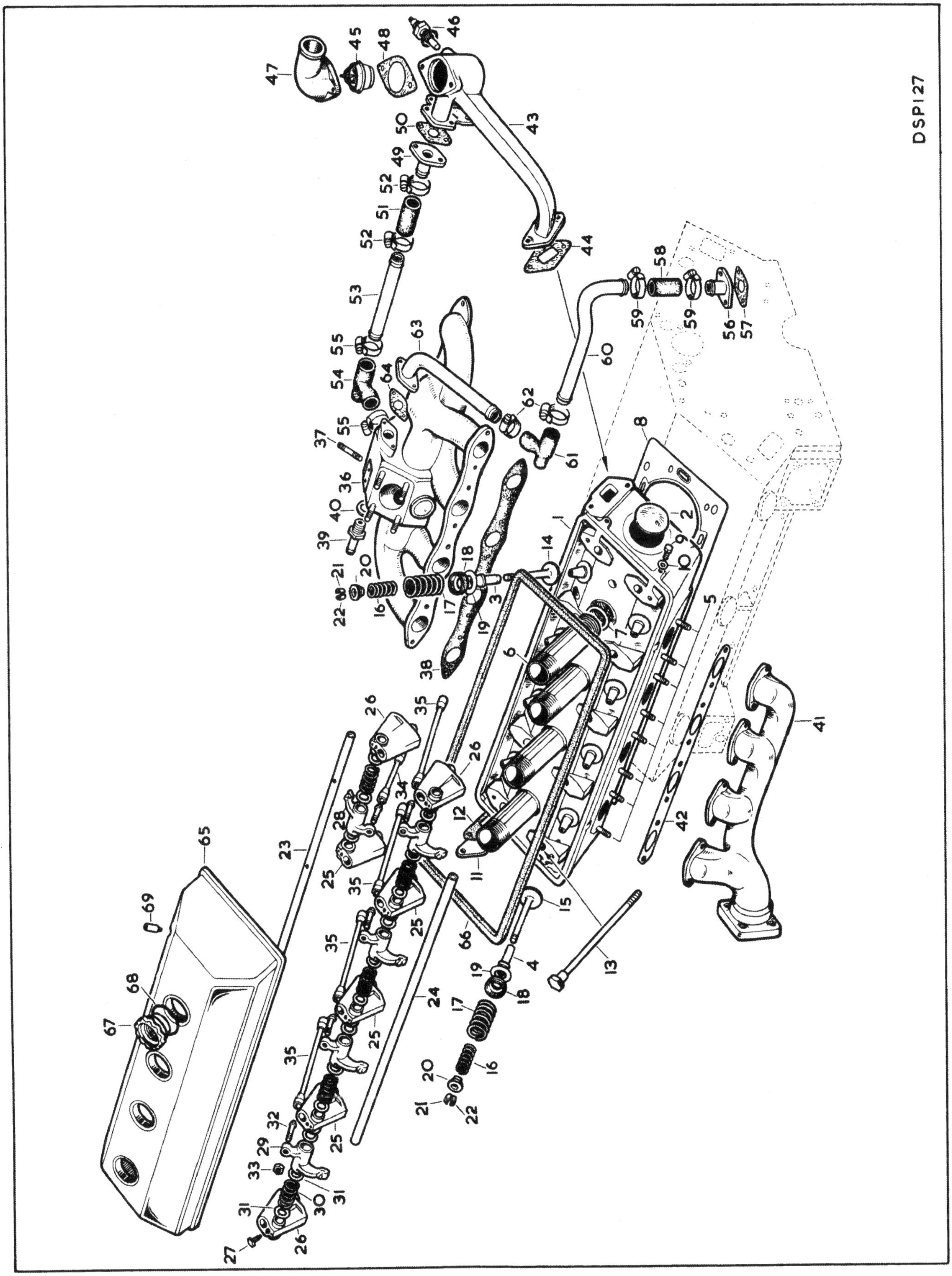

PLATE 5

CARBURETTERS

Part No.	Description	Plate No.	No. per Unit	Remarks
C.21155/2	**L.H. CARBURETTER ASSEMBLY (HD.6)**		1	Fitted from Engine No. 7A.1001 to 7A.3073
9690	**Throttle Lever** (AUC.8239)		1	
5803	**Bolt,** securing Throttle Lever to Throttle Spindle (AUC.2694)		1	
5614	**Nut,** on Bolt (AUC.2156)		1	
7966	**Washer,** under Nut (AUC.8396)		1	
	All other items are as for L.H. Carburetter C.25207/2			
C.21155/1	**R.H. CARBURETTER ASSEMBLY (HD.6)**		1	Fitted from Engine No. 7A.1001 to 7A.3073
9689	**Body only** (AUC.8243)		1	
9694	**Jet Housing Assembly** (AUD.9320)		1	
9696	**Float Chamber** (AUC.2062)		1	
10059	**Lid,** for Float Chamber (AUD.9235)		1	
10028	**Needle and Seat,** controlling flow of petrol		1	
9690	**Throttle Lever** (AUC.8239)		1	
5803	**Bolt,** securing Throttle Lever to Throttle Spindle (AUC.2694)		1	
5614	**Nut,** on Bolt (AUC.2156)		1	
7966	**Washer,** under Nut (AUC.8396)		1	
	All other items are as for R.H. Carburetter C.25207/1			
C.23579/2	**L.H. CARBURETTER ASSEMBLY (HD.6)**		1	Fitted from Engine No. 7A.3074 to 7A.6890
10059	**Lid,** for Float Chamber (AUD.9581)		1	
	All other items are as for L.H. Carburetter C.25207/2			
C.23579/1	**R.H. CARBURETTER ASSEMBLY (HD.6)**		1	Fitted from Engine No. 7A.3074 to 7A.6890
10059	**Lid,** for Float Chamber (AUD.9235)		1	
10028	**Needle and Seat,** controlling flow of petrol to Float Chamber		1	
C.25207/2	**L.H. CARBURETTER ASSEMBLY (HD.6)**	6–1	1	Fitted to Engine No. 7A.6891 and subs 7K.1001 and subs
9688	**Body only** (AUC.8250)	6–2	1	
9691	**Adaptor,** for Vacuum Suction Pipe (AUC.1323)	6–3	1	
8091	**Gasket,** under Adaptor (AUC.2014)	6–4	1	
7961	**Screw,** securing Adaptor (AUC.5156)		2	
C.723	**Washer,** Shakeproof, on Screws (AUA.4643)		2	
9692	**Suction Chamber Assembly** (AUD.9184)	6–5	1	
7093	**Screw,** holding Jet Needle in position (AUC.2057)		1	
6961	**Damper Assembly,** in Suction Chamber (AUC.8102)	6–6	1	
7095	**Washer,** for Damper (AUC.4900)	6–7	1	
7094	**Spring,** assisting return of Piston (AUC.4387)	6–8	1	
7038	**Skid Washer,** at bottom of Spring (AUC.3071)	6–9	1	
5618	**Screw,** securing Suction Chambers to Carburetter Body (AUC.2175)		3	
5980	**Jet Needle** (T.Z)	6–10	1	
6957	**Jet Assembly** (AUC.8155)	6–11	1	
6958	**Jet Bearing** (AUC.2001)	6–12	1	
6959	**Nut,** locking Jet Bearing (AUC.2002)	6–13	1	
8765	**Spring,** under Jet Diaphragm (AUD.2093)	6–14	1	
9693	**Jet Housing Assembly,** (AUD.9321)	6–15	1	
9697	**Shoe and Rod Assembly,** operating Throttle during cold starting (AUC.8964)	6–16	1	
8760	**Return Spring,** for Shoe and Rod Assembly (AUC.2020)	6–17	1	
8761	**Plate,** at top of Shoe and Rod Assembly (AUC.2019)	6–18	1	
7879	**Screw,** securing Plate (AUC.4790)		1	
C.723	**Washer,** Shakeproof, on Screw (AUA.4643)		1	
5780	**Screw,** in Plate, operating Choke Throttle Control (AUC.2521)	6–19	1	
5862	**Spring,** on Screw (AUC.2451)	6–20	1	
9695	**Float Chamber** (AUC.2067)	6–21	1	
7115	**Bolt,** securing Float Chamber to Carburetter Body (AUC.2110)		4	

Part No.	Description	Plate No.	No. per Unit	Remarks
C.723	**Washer,** Shakeproof, on Bolts (AUA.4643)		4	
10060	**Lid,** for Float Chamber (AUD.9238)	6–22	1	
1090	**Float** (AUC.1123)	6–23	1	
10028	**Needle and Seat,** controlling flow of Petrol to Float Chamber (AUD.9096)	6–24	1	
10027	**Lever,** operating Float Needle (AUD.2285)	6–25	1	
1153	**Pin,** securing Lever to Float Chamber Lid (AUC.1152)	6–26	1	
7119	**Gasket,** under Float Chamber Lid (AUC.1147)	6–27	1	
1163	**Cap Nut,** securing Lid (AUC.1163)	6–28	1	
7120	**Aluminium Washer,** under Cap Nut (AUC.1557)	6–29	1	
6967	**Valve,** for Slow Running Adjustment (AUC.2028)	6–30	1	
6966	**Spring,** for Slow-run Valve (AUC.2027)	6–31	1	
6968	**Gland Washer,** for Slow-run Valve (AUC.2029)	6–32	1	
6969	**Dished Washer,** (Brass) at top of Gland Washer (AUC.2030)	6–33	1	
C.25207/1	**R.H. CARBURETTER ASSEMBLY (HD.6)**	6–34	1	Fitted to Engine No. 7A.6891 and subs 7K.1001 and subs
9689	**Body only** (AUC.8243)		1	
9694	**Jet Housing Assembly** (AUD.9320)		1	
9696	**Float Chamber** (AUC.2062)		1	
10940	**Lid,** for Float Chamber (AUD.9579)		1	
10532	**Needle and Seat,** controlling flow of Petrol to Float Chamber (AUD.9585)		1	
	All other items are as for L.H. Carburetter C.25207/2			
432607	**THROTTLE LEVER ON CARBURETTERS**	6–35	2	Fitted to Engine No. 7A.3074 and subs
UFS.119/6R	**Screw,** Set, securing Throttle Levers to Carburetter Spindles	6–36	2	Fitted to Engine No. 7A.3074 and subs
UFN.119/L	**Nut,** on Setscrews	6–37	2	Fitted to Engine No. 7A.3074 and subs
AW.102/T	**Washer,** Plain, under Nuts	6–38	2	Fitted to Engine No. 7A.3074 and subs
432586	**RETURN SPRING FOR THROTTLE LEVERS**	6–39	2	Fitted to Engine No. 7A.3074 and subs
C.22435	**Clip,** on Induction Manifold Studs, anchoring Return Springs		2	Fitted to Engine No. 7A.3074 and subs
430979	**INSULATOR BETWEEN CARBURETTERS AND INDUCTION MANIFOLD**	6–40	2	
C.21602	**KICKDOWN LEVER ON L.H. CARBURETTER THROTTLE SPINDLE**	6–41	1	Fitted from Engine No. 7A.1001 to 7A.3046 **Note: Later items are shown under "Accelerator Controls"**
C.21603	**Screw,** Set, securing Kick-down Lever to Spindle		1	Fitted from Engine No. 7A.1001 to 7A.3046
C.4202/9	**Nut,** on Setscrew		1	Fitted from Engine No. 7A.1001 to 7A.3046
BD.541/28	**Washer,** Plain, under Nut		1	Fitted from Engine No. 7A.1001 to 7A.3046
C.21514	**ABUTMENT BRACKET FOR KICKDOWN CABLE AT L.H. CARBURETTER**	6–42	1	Fitted from Engine No. 7A.1001 to 7A.3046
C.8737/9	**Nut,** securing Cable Post to Abutment Bracket		1	Fitted from Engine No. 7A.1001 to 7A.3046
C.22129	**Washer,** on Cable Post		1	Fitted from Engine No. 7A.1001 to 7A.3046
C.15830	**Washer,** under Nut		1	Fitted from Engine No. 7A.1001 to 7A.3046
J.103/8S	**Clevis Pin,** securing Cable Fork-end to Lever on L.H. Carburetter Spindle	6–43	1	Fitted from Engine No. 7A.1001 to 7A.3046
BD.541/28	**Washer,** on Clevis Pin (under Head)		1	Fitted from Engine No. 7A.1001 to 7A.3046
L.102/4U	**Split Pin,** securing Clevis Pin		1	Fitted from Engine No. 7A.1001 to 7A.3046
C.15830	**Washer,** under Split Pin		1	Fitted from Engine No. 7A.1001 to 7A.3046
C.14212	**RUBBER CONNECTION BETWEEN L.H. CARBURETTER AND DISTRIBUTOR SUCTION PIPE**	6–44	1	
307991	**SUCTION PIPE ASSEMBLY FOR DISTRIBUTOR VACUUM UNIT**	6–45	1	

CARBURETTERS

Part No.	Description	Plate No.	No. per Unit	Remarks
C.21241	**PETROL PIPE ASSEMBLY BETWEEN CARBURETTERS & PETROL FILTER**	6–46	1	
C.20294	**Adaptor Nut,** securing Petrol Pipe Assembly to Carburetters	6–47	2	
C.20295	**Olive,** for Adaptor Nuts	6–48	2	
C.13681	**PETROL FILTER ASSEMBLY (AC.7950001)**		1	
7297	**Filter Casting**		1	
7298	**Sealing Washer,** between Filter Casting and Glass Bowl		1	
7299	**Filter Gauze**		1	
7300	**Glass Bowl**		1	
7301	**Retaining Strap Assembly**		1	
C.13705	**Banjo Bolt,** securing Petrol Pipes to Filter		1	
C.784	**Fibre Washer,** on Banjo Bolts		2	
C.14726	**BRACKET, MOUNTING PETROL FILTER TO R.H. FRONT WING VALANCE**	6–49	1	
UFS.131/6R	**Screw,** Set, securing Filter to Bracket and Bracket to Wing Valance		4	
UFN.131/L	**Nut,** on Setscrews		4	
FG.105/X	**Washer,** Spring, under Nuts		4	
C.27588	**PETROL FILTER ASSEMBLY (AC.7971243)**	6–50	1	SUPERSEDES C.13681
11450	**Filter Casting** (7971137)	6–51	1	
11951	**Sealing Washer,** between Filter Casting and Element (854471)	6–52	1	
C.28080	**Filter Element,** (Fibre) complete with Sealing Washers (854347)	6–53	1	
11950	**Sealing Washer,** between Filter Casting and Glass Bowl (854361)	6–54	1	
7300	**Glass Bowl** (1523620)	6–55	1	
7301	**Retaining Strap Assembly** (5592050)	6–56	1	
C.13705	**Banjo Bolt,** securing Petrol Pipes to Filter	6–57	1	
C.784	**Fibre Washer,** on Banjo Bolts	6–58	2	
C.22385	**OVERFLOW PIPE FOR L.H. CARBURETTER**		1	Fitted from Engine No. 7A.1001 to 7A.6890
C.22384	**OVERFLOW PIPE FOR R.H. CARBURETTER**		1	
C.25237	**OVERFLOW PIPE FOR CARBURETTERS**		2	Fitted from Engine No. 7A.6891 to 7A.10151
C.22386	**CORAGRIP PIPE, CONNECTING OVERFLOW PIPES TO CARBURETTERS**		2	Fitted from Engine No. 7A.1001 to 7A.10151
431354	**Clip,** securing Overflow Pipes		4	
C.26154	**OVERFLOW PIPE (P.V.C.) FOR CARBURETTERS**		2	Fitted to Engine No. 7A.10152 and subs
C.1040/12	**Clip,** securing Overflow Pipes		2	

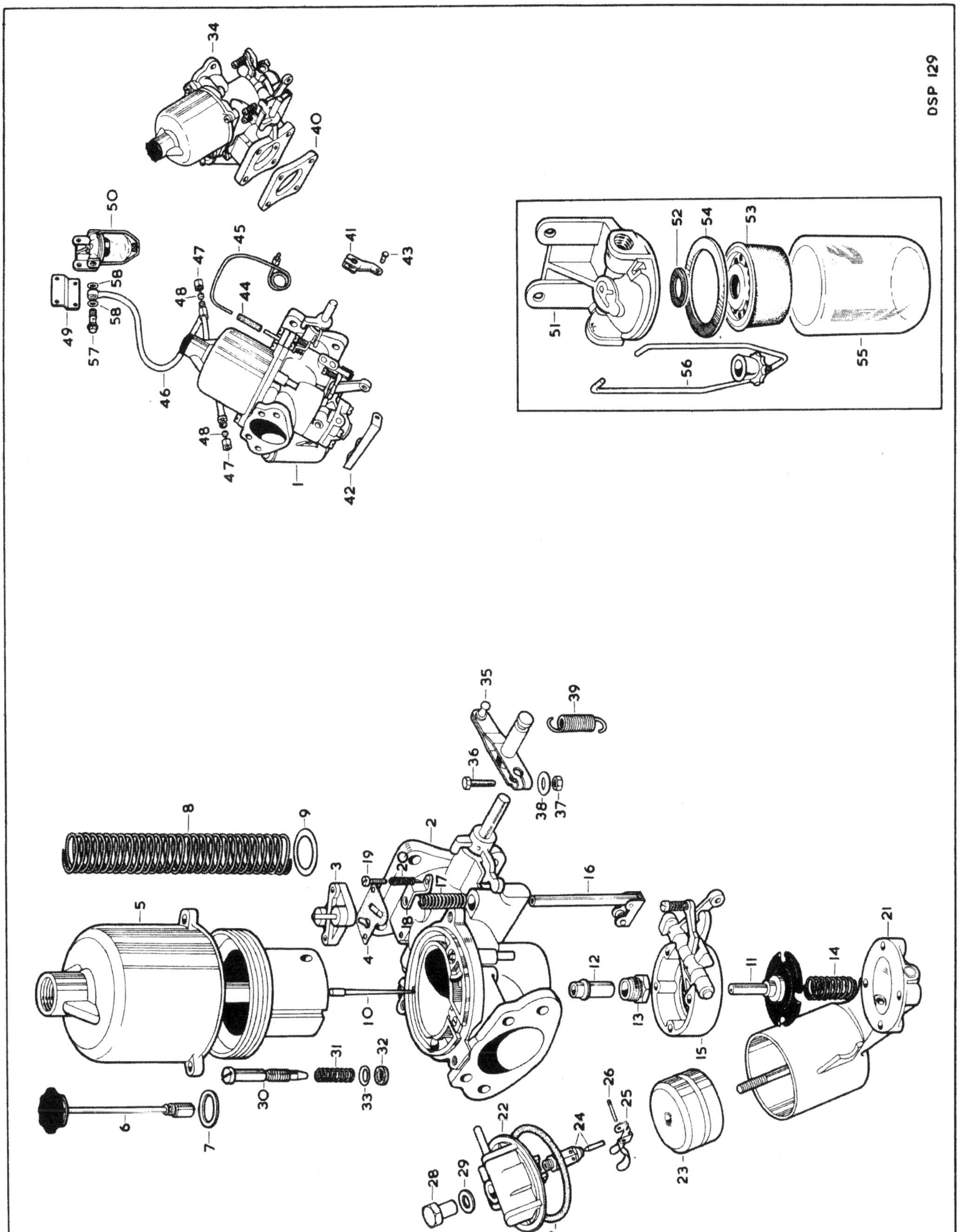

PLATE 6

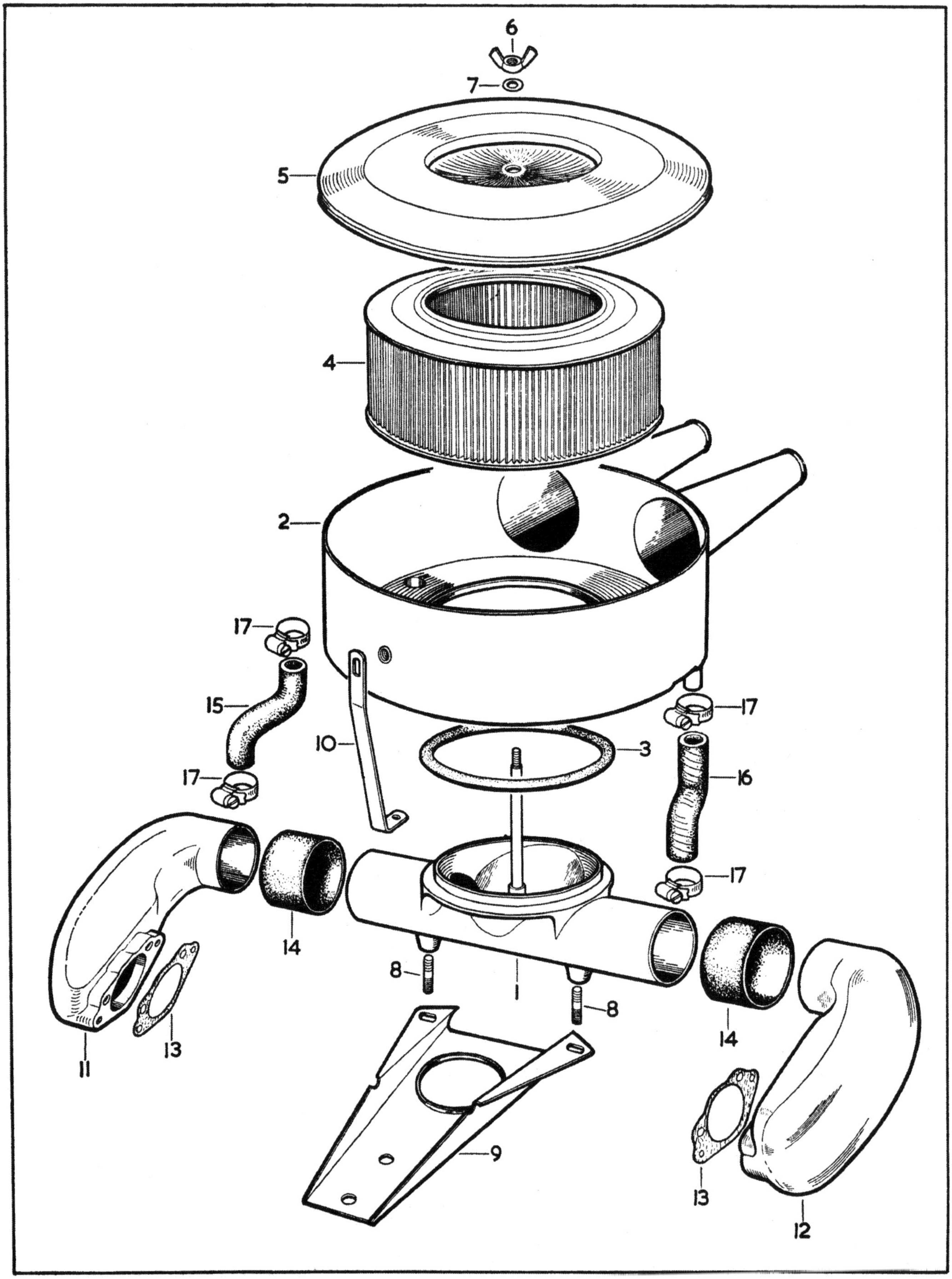

PLATE 7

Part No.	Description	Plate No.	No. per Unit	Remarks
C.21644	**AIR CLEANER ASSEMBLY (PD.2749)**		1	Fitted from Chassis No. 1A.1001 to 1A.1204 1A.20001 to 1A.20031
C.21645	**Casting Assembly** (6914)		1	
C.21646	**Body Assembly** (6911)		1	
C.21653	**Gasket,** between Body and Casting (6907)		1	
C.21648	**Element** (6913T)		1	
C.21647	**Top Cover Assembly** (6912)		1	
C.21652	**Wing Nut,** securing Top Cover		1	
C.21651	**Washer,** Fibre, under Wing Nut (6819)		1	
C.21909	**AIR CLEANER ASSEMBLY**		1	Fitted to Chassis No. 1A.1205 and subs 1A.20032 and subs
C.21645	**Casting Assembly** (6914)	7–1	1	
C.21910	**Body Assembly**	7–2	1	
C.21653	**Gasket,** between Body and Casting (6907)	7–3	1	
C.21648	**Element** (6913T)	7–4	1	
C.21647	**Top Cover Assembly** (6912)	7–5	1	
C.21652	**Wing Nut,** securing Top Cover	7–6	1	
C.21651	**Washer,** Fibre, under Wing Nut (6819)	7–7	1	
C.16751	**Stud,** fixing Air Cleaner to Bracket	7–8	2	Superseded by UCS.131/6R
UFN.131/L	**Nut,** on Studs		2	
UCS.131/6R	**Screw,** Set, securing Air Cleaner to Bracket		2	Supersedes C.16751 and UFN.131/L
FG.105/X	**Washer,** Spring, under Nuts or on Setscrews		2	
C.15829	**Washer,** Plain, under Spring Washers		2	
C.21640	**BRACKET, MOUNTING AIR CLEANER ASSEMBLY**	7–9	1	
C.21641	**BRACKET SUPPORTING AIR CLEANER ASSEMBLY**	7–10	1	
UFS.131/4R	**Screw,** Set, securing Air Cleaner to Support Bracket		1	
C.15829	**Washer,** Plain, on Setscrew		1	
FG.105/X	**Washer,** Spring, on Setscrew		1	
C.21636	**AIR INTAKE PIPE FOR L.H. CARBURETTER**		1	Fitted from Chassis No. 1A.1001 to 1A.1204 1A.20001 to 1A.20031
C.21637	**AIR INTAKE PIPE FOR R.H. CARBURETTER**		1	
C.21907	**AIR INTAKE PIPE FOR L.H. CARBURETTER**	7–11	1	Fitted to Chassis No. 1A.1205 and subs 1A.20032 and subs
C.21908	**AIR INTAKE PIPE FOR R.H. CARBURETTER**	7–12	1	
C.4475	**Gasket,** between Air Intake Pipes and Carburetters	7–13	2	
UCS.131/7R	**Screw,** Set, securing Air Intake Pipes to Carburetters		4	
FG.105/X	**Washer,** Spring, on Setscrews		4	
C.21642	**RUBBER SEAL BETWEEN AIR INTAKE PIPES AND AIR CLEANER**	7–14	2	
C.21749	**BREATHER HOSE, BETWEEN L.H. AIR INTAKE PIPE AND L.H. CYLINDER HEAD COVER**		1	Fitted from Chassis No. 1A.1001 to 1A.1204 1A.20001 to 1A.20031
C.21643	**BREATHER HOSE, BETWEEN R.H. AIR INTAKE PIPE AND R.H. CYLINDER HEAD COVER**		1	

AIR CLEANER

Part No.	Description	Plate No.	No. per Unit	Remarks
C.21913	**BREATHER HOSE, BETWEEN AIR CLEANER BODY AND L.H. CYLINDER HEAD COVER**	7–15	1	Fitted to Chassis No. 1A.1205 and subs 1A.20032 and subs
C.21914	**BREATHER HOSE, BETWEEN AIR CLEANER BODY AND R.H. CYLINDER HEAD COVER**	7–16	1	
C.2905/2	**Clip,** securing Hoses	7–17	4	
C.27393	**AIR CLEANER ASSEMBLY (7972431)**		2	Fitted to Chassis No. 1K.1001 and subs 1K.30001 and subs
12332	**Body Assembly** (7972434)		2	
C.4475	**Gasket,** between Air Cleaner Body and Carburetter Flange		2	
11875	**Element** (A.2447)		2	
12333	**Top Cover Assembly** (7964223)		2	
6818	**Wing Nut,** securing Top Cover		2	
10072	**Washer,** Fibre, under Wing Nut		2	
BD.30541/1	**Plastic Plug,** in Air Intake Tube		4	
UFS.131/6R	**Screw,** Set, securing Air Cleaner Assembly to Carburetter Flange		4	
FG.105/X	**Washer,** Spring, on Setscrews		4	
C.27460	**BREATHER PIPE, BETWEEN AIR CLEANER BODY AND CYLINDER HEAD COVER**		2	
C.18591	**HOSE, ELBOW, CONNECTING BREATHER PIPE TO AIR CLEANER BODY**		2	
C.26538	**HOSE, CONNECTING BREATHER PIPE TO CYLINDER HEAD COVER**		2	
C.15886/12	**Clip,** securing Hoses		8	

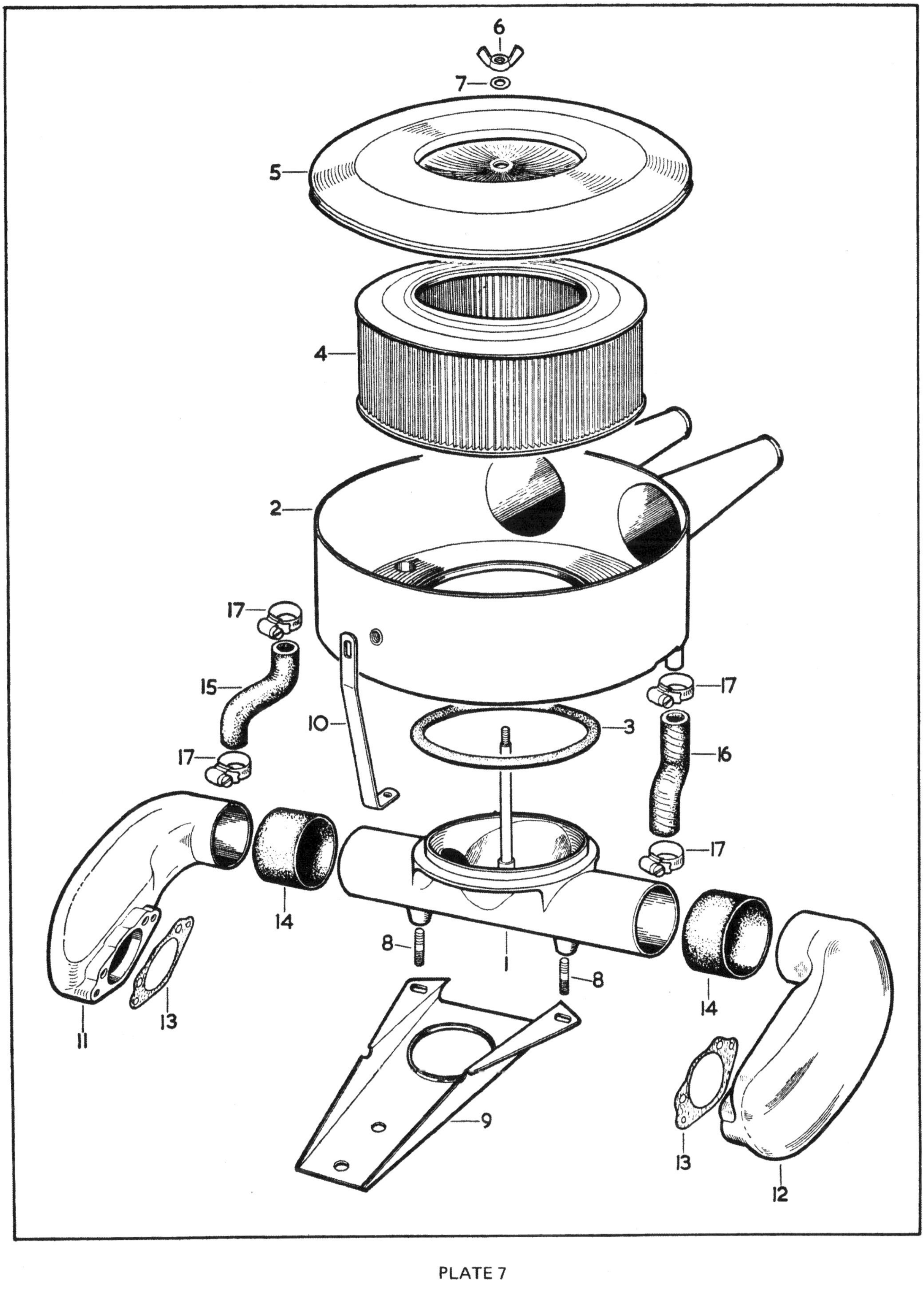

PLATE 7

ACCELERATOR CONTROLS
RIGHT-HAND DRIVE CARS ONLY

Part No.	Description	Plate No.	No. per Unit	Remarks
C.13622	**ACCELERATOR PEDAL ASSEMBLY**	8–1	1	
UFS.125/11R	**Screw,** Set, restricting Pedal Travel	8–2	1	
UFN.125/L	**Nut,** on Setscrew	8–3	1	
C.3479	**Felt Washer,** at each side of Pedal Boss	8–4	2	
C.13387	**BRACKET ASSEMBLY, MOUNTING ACCELERATOR PEDAL TO FLOOR**	8–5	1	
C.11908	**Distance Piece,** under Mounting Bracket	8–6	4	
BD.8291/4	**Screw,** Set, securing Mounting Bracket to Floor		4	
FG.104/X	**Washer,** Spring, on Setscrews		4	
C.6969	**Fulcrum Pin,** securing Accelerator Pedal to Mounting Bracket	8–7	1	
C.8737/1	**Nut,** on Fulcrum Pin		1	
FW.104/T	**Washer,** Plain under Nut		1	
C.11671	**RUBBER PAD ON ACCELERATOR PEDAL**	8–8	1	
C.13532	**ACCELERATOR PEDAL SHAFT ASSEMBLY**	8–9	1	
C.12486	**Roller,** on Accelerator Pedal Shaft	8–10	1	
L.102.5/8U	**Split Pin,** securing Roller to Shaft	8–11	1	
C.386	**Special Washer,** behind Split Pin	8–12	1	
C.10940/1	**RUBBER BUSH FOR ACCELERATOR PEDAL SHAFT**	8–13	1	
C.11678	**Bearing,** inside Rubber Bush	8–14	2	
C.386	**Special Washer,** behind Bush	8–15	1	
L.102.5/8U	**Split Pin,** retaining Washer	8–16	1	
C.13316	**Distance Washer,** at front end of Rubber Bush	8–17	1	
C.13545	**Distance Collar,** behind Washer	8–18	1	
C.13537	**Split Collar,** at end of Pedal Shaft	8–19	1	
C.750	**Key,** retaining Split Collar on Pedal Shaft	8–20	1	
BD.1708/2	**Washer,** Double Coil Spring, behind Split Collar		1	
C.386	**Special Washer,** at end of Pedal Shaft		1	
L.102.5/8U	**Split Pin,** through Pedal Shaft		1	
C.11672	**BRACKET, MOUNTING ACCELERATOR PEDAL SHAFT**	8–24	1	
C.13395	**Spacer,** for Mounting Bracket	8–25	2	
UFS.125/9R	**Screw,** Set, securing Mounting Bracket to Dash		2	
FG.104/X	**Washer,** Spring, on Setscrews		2	
C.8016	**Rubber Grommet,** in Mounting Bracket	8–26	1	
C.11678	**Bearing,** in Rubber Grommet	8–27	1	
C.13536	**LEVER ASSEMBLY AT END OF PEDAL SHAFT AND AT END OF SLAVE SHAFT**	8–28	2	
UFS.131/7R	**Screw,** Set, clamping Levers to Shafts		2	
FG.105/X	**Washer,** Spring, on Setscrews		2	
C.13538	**CONNECTING LINK ASSEMBLY BETWEEN LEVERS ON PEDAL AND SLAVE SHAFTS**	8–29	1	
BD.1708/5	**Washer,** Double Coil Spring, on Connecting Link	8–30	2	
FW.104/T	**Washer,** Plain, on Connecting Link	8–31	2	
L.102.5/7U	**Split Pin,** retaining Connecting Link	8–32	2	

ACCELERATOR CONTROLS

RIGHT-HAND DRIVE CARS ONLY

Part No.	Description	Plate No.	No. per Unit	Remarks
C.21144	**SLAVE CROSS SHAFT ASSEMBLY**	8–33	1	
UFS.125/5R	**Screw,** Set, securing Shaft Bracket to Dash		2	
C.8737/1	**Nut,** Self Locking, on Setscrews		2	
FW.104/T	**Washer,** Plain, on Setscrews		4	
C.13308	**BEARING AND HOUSING ASSEMBLY FOR SLAVE CROSS SHAFT**	8–34	2	
BD.1814/2	**Pop Rivet,** securing Bearing Assemblies to Body		6	
C.21152	**Distance Tube,** between Lever Bush and Bearing	8–35	1	
C.13316	**Distance Piece,** behind Bearing	8–36	2	
BD.1708/2	**Washer,** Double Coil Spring, behind Distance Piece	8–37	1	
FW.106/T	**Washer,** Plain, behind Double Coil Spring Washer	8–38	1	
L.102.5/6U	**Split Pin,** through Slave Shaft	8–39	1	
C.21473	**THROTTLE ROD LEVER ASSEMBLY, ON SLAVE SHAFT ASSEMBLY**	8–40	1	
UFS.131/7R	**Screw,** Set, securing Lever to Slave Shaft Assembly		1	
FG.105/X	**Washer,** Spring, on Setscrew		1	
C.20868	**CONTROL ROD ASSEMBLY, LINKING LEVER TO THROTTLE AND MIXTURE CONTROL BRACKET**	8–41	1	
C.16809	**Ball Joint** (Amal), at Lower End of Control Rod		1	
420963	**Ball Joint,** at Upper End of Control Rod		1	
UFN.225/L	**Locknut,** securing Ball Joints		2	
C.21507	**THROTTLE AND MIXTURE CONTROL BRACKET ASSEMBLY**	8–42	1	Fitted from Engine No. 7A.1001 to 7A.3073
C.23453	**THROTTLE AND MIXTURE CONTROL BRACKET ASSEMBLY**		1	Fitted to Engine No. 7A.3074 and subs
C.22433	**Abutment Bracket,** on Throttle and Mixture Control Bracket, for Kickdown Cable		1	
UCS.131/5R	**Screw,** Set, securing Abutment Bracket to Control Bracket		2	
FG.105/X	**Washer,** Spring, on Setscrews		2	
C.8737/9	**Nut,** Self Locking, securing Cable Post to Abutment Bracket		1	}
C.22129	**Washer,** on Cable Post		1	}
C. i5830	**Washer,** under Nut		1	}
J.103/8S	**Clevis Pin,** securing Cable Fork-end to Kickdown and Throttle Lever		1	} Fitted to Engine No. 7A.3074 and subs
BD.541/28	**Washer,** on Clevis Pin		1	}
L.102/4U	**Split Pin,** securing Clevis Pin		1	}
C.15830	**Washer,** under Split Pin		1	}
C.21619	**Anchor Bracket,** for Throttle Spring	8–43	1	
405315	**Clip,** securing Mixture Control Wire	8–44	1	
UFS.119/5R	**Screw,** Set, securing Clip to Bracket		1	
UFN.119/L	**Nut,** on Setscrew		1	
AG.102/X	**Washer,** Spring, under Nut		1	
UCS.131/9R	**Screw,** Set, securing Bracket and Throttle Spring Anchor Bracket to Induction Manifold		2	
UCS.131/8R	**Screw,** Set, securing Bracket and Throttle Spring Anchor Bracket to Induction Manifold		2	Fitted to Engine No. 7K.1001 and subs
FG.105/X	**Washer,** Spring, on Bolts		2	
424813/2	**BALL PIN, ON THROTTLE LEVERS**	8–45	2	
UFN.119/L	**Nut,** securing Ball Pins to Levers		2	
AG.102/X	**Washer,** Spring, under Nuts		2	

ACCELERATOR CONTROLS
RIGHT-HAND DRIVE CARS ONLY

Part No.	Description	Plate No.	No. per Unit	Remarks
C.21620	**THROTTLE RETURN SPRING**	8–46	1	SUPERSEDED BY C.24536
C.24536	**THROTTLE RETURN SPRING**		1	SUPERSEDES C.21620
431097	**THROTTLE CONTROL ROD FROM THROTTLE AND MIXTURE CONTROL BRACKET TO CARBURETTERS**	8–47	2	
424822/3	**Ball Socket,** on Control Rods	8–48	4	
UFN.119/L	**Nut,** securing Ball Sockets		4	
431074	**MIXTURE CONTROL ROD FROM THROTTLE AND MIXTURE CONTROL BRACKET TO CARBURETTERS**	8–49	2	SUPERSEDED BY 431862
431862	**MIXTURE CONTROL ROD FROM THROTTLE AND MIXTURE CONTROL BRACKET TO CARBURETTERS**		2	SUPERSEDES 431074
42822/3	**Ball Socket,** on Control Rod at Carburetter End	8–50	2	
UFN.119/L	**Nut,** securing Ball Sockets		2	
L.102/5U	**Split Pin,** retaining Control Rods to Levers on Throttle and Mixture Control Bracket	8–51	2	
427461	**Washer,** behind Split Pins	8–52	2	
432539/1	**Ball End,** on Carburetter Mixture Levers	8–53	2	
UFN.119/L	**Nut,** securing Ball Ends to Mixture Levers		2	
AG.102/X	**Washer,** Spring, under Nuts		2	

PLATE 8

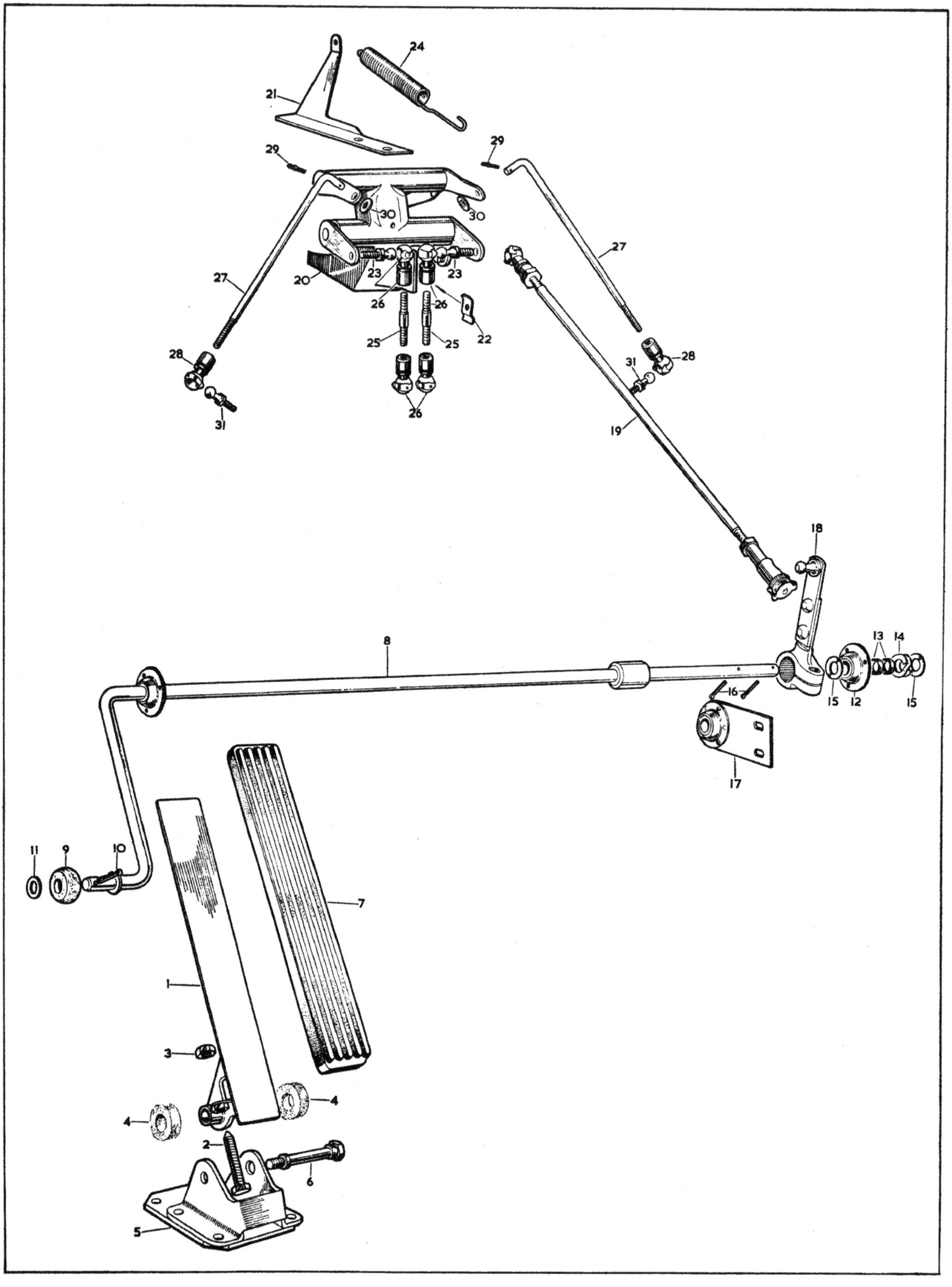

24
21
29
29
30
30
27
23
23
20
26
26
27
22
25
25
28
28
31
31
26
19
18
8
13
14
16
15
12
15
17
11
9
10
7
1
3
4
4
2
6
5

PLATE 9

Part No.	Description	Plate No.	No. per Unit	Remarks
C.13934	**ACCELERATOR PEDAL ASSEMBLY**	9–1	1	
UFS.125/11R	**Screw**, Set, restricting Pedal travel	9–2	1	
UFN.125/L	**Nut**, on Setscrew	9–3	1	
C.3479	**Felt Washer**, at each side of Pedal Boss	9–4	2	
C.13387	**BRACKET ASSEMBLY, MOUNTING ACCELERATOR PEDAL TO FLOOR**	9–5	1	
BD.8291/2	**Screw**, Set, securing Mounting Bracket to Floor		4	
FG.104/X	**Washer**, Spring, on Setscrews		4	
C.6969	**Fulcrum Pin**, securing Accelerator Pedal to Mounting Bracket	9–6	1	
C.8737/1	**Nut**, on Fulcrum Pin		1	
FW.104/T	**Washer**, Plain, under Nut		1	
C.11671	**RUBBER PAD ON ACCELERATOR PEDAL**	9–7	1	
C.21143	**ACCELERATOR PEDAL SHAFT ASSEMBLY**	9–8	1	
C.12486	**Roller**, on Accelerator Pedal Shaft	9–9	1	
L.102.5/8U	**Split Pin**, securing Roller to Shaft	9–10	1	
C.386	**Special Washer**, behind Split Pin	9–11	1	
BD.1814/2	**Pop Rivet**, securing Pedal Housing to L.H. side of Dash		3	
C.13308	**BEARING AND HOUSING ASSEMBLY, FOR PEDAL SHAFT, ON R.H. SIDE OF DASH**	9–12	1	
BD.1814/2	**Pop Rivet**, securing Bearing Housing to R.H. side of Dash		3	
C.13316	**Distance Piece**, behind Bearing on Pedal Shaft, at R.H. side of Dash	9–13	2	
BD.1708/2	**Washer**, Double Coil Spring, behind Distance Piece	9–14	1	
FW.106/T	**Washer**, Plain, behind Spring Washer	9–15	2	
L.102.5/6U	**Split Pin**, through Pedal Shaft	9–16	2	
C.21460	**STEADY BRACKET ASSEMBLY, ON ACCELERATOR PEDAL SHAFT**	9–17	1	
UFS.125/4R	**Screw**, Set, securing Steady Bracket to Bracket on Dash		2	
C.8737/1	**Nut**, Self-Locking, on Setscrews		2	
FW.104/T	**Washer**, Plain, on Setscrews		4	
C.21473	**THROTTLE ROD LEVER ASSEMBLY ON PEDAL SHAFT**	9–18	1	
UFS.131/7R	**Screw**, Set, securing Lever to Pedal Shaft		1	
FG.105/X	**Washer**, Spring, on Setscrew		1	
C.20868	**CONTROL ROD ASSEMBLY, LINKING LEVER TO THROTTLE AND MIXTURE CONTROL BRACKET ASSEMBLY**	9–19	1	
C.16809	**Ball Joint** (Amal) at Lower End of Control Rod		1	
420963	**Ball Joint**, at Upper End of Control Rod		1	
UFN.225/L	**Locknut**, securing Ball Joints		2	
C.21507	**THROTTLE AND MIXTURE CONTROL BRACKET ASSEMBLY**	9–20	1	Fitted from Engine No. 7A.1001 to 7A.3073
C.23453	**THROTTLE AND MIXTURE CONTROL BRACKET ASSEMBLY**		1	Fitted to Engine No. 7A.3074 and subs
C.22433	**Abutment Bracket**, on Throttle and Mixture Control Bracket, for Kickdown Cable		1	
UCS.131/5R	**Screw**, Set, securing Abutment Bracket to Control Bracket		2	
FG.105/X	**Washer**, Spring, on Setscrews		2	

ACCELERATOR CONTROLS

LEFT-HAND DRIVE CARS ONLY

Part No.	Description	Plate No.	No. per Unit	Remarks
C.8737/9	**Nut,** Self Locking, securing Cable Post to Abutment Bracket		1	Fitted to Engine No. 7A.3074 and subs
C.22129	**Washer,** on Cable Post		1	Fitted to Engine No. 7A.3074 and subs
C.15830	**Washer,** under Nut		1	Fitted to Engine No. 7A.3074 and subs
J.103/8S	**Clevis Pin,** securing Cable Fork-end to Kickdown and Throttle Lever		1	Fitted to Engine No. 7A.3074 and subs
BD.541/28	**Washer,** on Clevis Pin		1	Fitted to Engine No. 7A.3074 and subs
L.102/4U	**Split Pin,** securing Clevis Pin		1	Fitted to Engine No. 7A.3074 and subs
C.15830	**Washer,** under Split Pin		1	Fitted to Engine No. 7A.3074 and subs
C.21619	**Anchor Bracket,** for Throttle Spring	9–21	1	
405315	**Clip,** securing Mixture Control Wire	9–22	1	
UFS.119/5R	**Screw,** Set, securing Clip to Bracket		1	
UFN.119/L	**Nut,** on Setscrew		1	
AG.102/X	**Washer,** Spring, under Nut		1	
UFS.131/8R	**Screw,** securing Bracket and Throttle Spring Anchor Bracket to Induction Manifold		2	
FG.105/X	**Washer,** Spring, on Bolts		2	
424813/2	**BALL PIN, ON THROTTLE LEVERS**	9–23	3	
UFN.119/L	**Nut,** securing Ball Pins to Levers		3	
AG.102/X	**Washer,** Spring, under Nuts		3	
C.21620	**THROTTLE RETURN SPRING**	9–24	1	**SUPERSEDED BY C.24536**
C.24536	**THROTTLE RETURN SPRING**		1	**SUPERSEDES C.21620**
431097	**THROTTLE CONTROL ROD FROM THROTTLE AND MIXTURE CONTROL BRACKET TO CARBURETTERS**	9–25	2	
424822/3	**Ball Socket,** on Control Rods	9–26	4	
UFN.119/L	**Nut,** securing Ball Sockets		4	
431074	**MIXTURE CONTROL ROD, FROM THROTTLE AND MIXTURE CONTROL BRACKET TO CARBURETTERS**	9–27	2	**SUPERSEDED BY 431862**
431862	**MIXTURE CONTROL ROD, FROM THROTTLE AND MIXTURE CONTROL BRACKET TO CARBURETTERS**		2	**SUPERSEDES 431074**
424822/3	**Ball Socket,** on Control Rods at Carburetter End	9–28	2	
UFN.119/L	**Nut,** securing Ball Sockets		2	
L.102/5U	**Split Pin,** retaining Control Rods to Levers on Throttle and Mixture Control Bracket	9–29	2	
427461	**Washer,** Plain, behind Split Pin	9–30	2	
432539/1	**Ball End,** on Carburetter Mixture Levers	9–31	2	
UFN.119/L	**Nut,** securing Ball Ends to Mixture Levers		2	
AG.102/X	**Washer,** Spring, under Nuts		2	

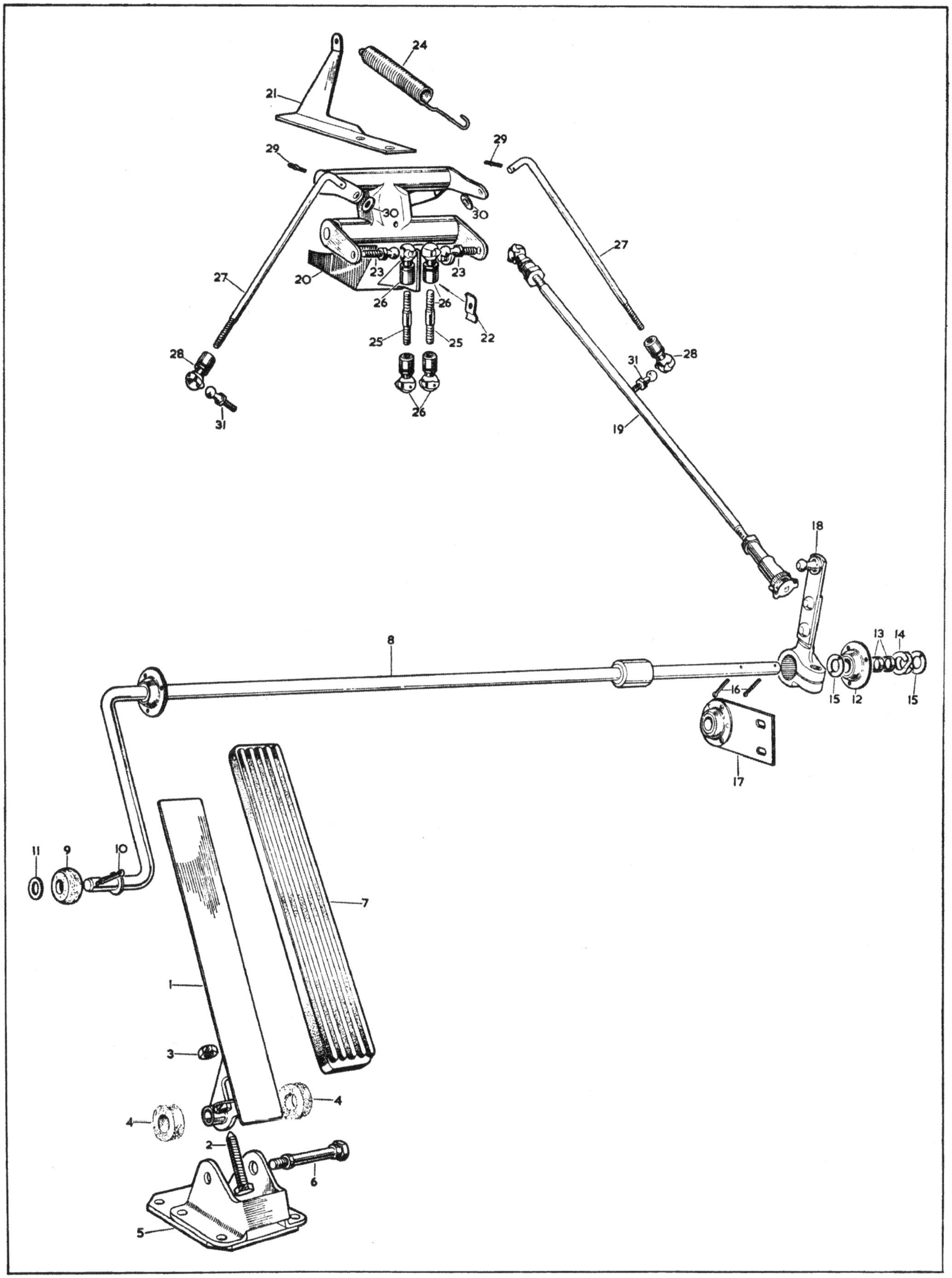

24
21
29
29
30
30
27
27
23
23
20
26
26
22
25
25
28
28
31
31
26
19
18
8
13
14
16
15
12
15
17
11
9
10
7
1
3
4
4
2
6
5

PLATE 9

CLUTCH UNIT

Part No.	Description	Plate No.	No. per Unit	Remarks
C.27427	**CLUTCH COVER ASSEMBLY (76251/18)**	10–1	1	Fitted to Engine No. 7A.13231 and subs. (Also to 7A.12185 and 7A.12186) 7K.1001 and subs.
12295	**Diaphragm Spring** (75575/10)	10–2	1	
11877	**Fulcrum Ring,** at each side of Diaphragm Spring (75574)	10–3	2	
12296	**Rivet** (Shouldered) locating Diaphragm Spring (75690/11)	10–4	9	
11879	**Release Plate** (76200)	10–5	1	
12297	**Sleeve,** retaining Release Plate to Diaphragm Spring (76085)	10–6	1	
11298	**Belleville Washer,** on Sleeve (75581)	10–7	1	
11882	**Pressure Plate** (76210)	10–8	1	
11883	**Clip,** retaining Pressure Plate to Diaphragm Spring (75284)	10–9	3	
11262	**Bolt,** securing Clips and Drive Straps to Pressure Plate (75387)	10–10	3	
11263	**Tab Washer,** locking Bolts (75104)	10–11	3	
11264	**Balance Weight** (.85 grammes) for Clutch Cover Assembly (75062)	10–12	As req'd	SELECTIVE WEIGHTS
11265	**Balance Weight** (2.13 grammes) for Clutch Cover Assembly (75063)	10–12	As req'd	
11266	**Balance Weight** (4.26 grammes) for Clutch Cover Assembly (75064)	10–12	As req'd	
11267	**Balance Weight** (5.38 grammes) for Clutch Cover Assembly (75065)	10–12	As req'd	
11268	**Balance Weight** (6.51 grammes) for Clutch Cover Assembly (75066)	10–12	As req'd	
11269	**Balance Weight** (7.56 grammes) for Clutch Cover Assembly (75067)	10–12	As req'd	
11270	**Balance Weight** (9.57 grammes) for Clutch Cover Assembly (75068)	10–12	As req'd	
11271	**Balance Weight** (11.57 grammes) for Clutch Cover Assembly (75069)	10–12	As req'd	
11272	**Pop Rivet,** securing Balance Weights to Clutch Cover	10–13	As req'd	
C.23251	**RELEASE BEARING AND CUP ASSEMBLY (76130)**	10–14	1	Fitted to Engine No. 7A.13231 and subs (Also to 7A.12185 and 7A.12186) 7K.1001 and subs
C.4163	**Retainer,** holding Cup to Operating Fork (41628)	10–15	2	
C.23252	**DRIVEN PLATE ASSEMBLY (51091/60)**	10–16	1	Fitted to Engine No. 7A.13231 and subs. (Also to 7A.12185 and 7A.12186) 7K.1001 and subs.
11885	**Facings Package** (comprising Clutch Linings and Rivets) (SSC.75959)		1	**FOR RENEWAL OF DRIVEN PLATES**
C.8631	**Dowel,** locating Clutch Cover to Flywheel	10–17	3	
UFS.131/6R	**Screw,** Set, securing Clutch Unit to Flywheel		6	
FG.105/X	**Washer,** Spring, on Setscrews		6	
C.27154/1	**Balance Weight** (.048" thick) for Clutch	10–18	As req'd	SELECTIVE WEIGHTS
C.27154/2	**Balance Weight** (.080" thick) for Clutch	10–18	As req'd	
C.27154/3	**Balance Weight** (.128" thick) for Clutch	10–18	As req'd	
C.27154/4	**Balance Weight** (.160" thick) for Clutch	10–18	As req'd	

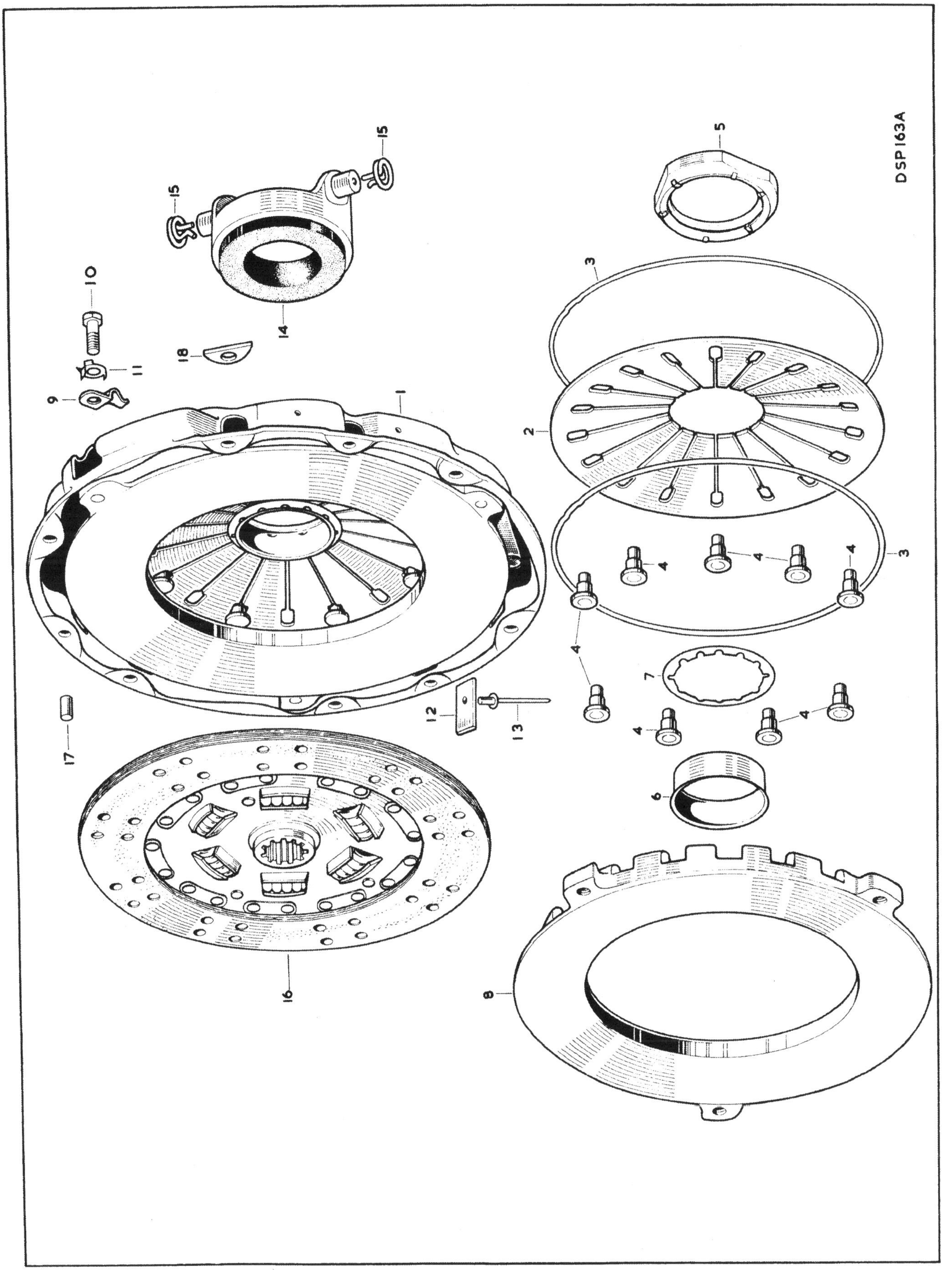

PLATE 10

CLUTCH CONTROLS

Part No.	Description	Plate No.	No. per Unit	Remarks
C.13246	**CLUTCH PEDAL (COMPLETE WITH BUSHES)**	11–1	1	**R.H. DRIVE CARS ONLY** Fitted to Chassis No. 1A.12581 and subs 1K.1001 and subs
C.13248	**CLUTCH PEDAL (COMPLETE WITH BUSHES)**		1	**L.H. DRIVE CARS ONLY** Fitted to Chassis No. 1A.20565 and subs 1K.30001 and subs
C.20955/4	**Bush**, in Clutch Pedal		2	
C.8966	**Return Spring**, for Clutch Pedal		1	
C.14852	**PAD (STEEL) ON CLUTCH PEDAL**	11–2	1	
C.8667/2	**Nut**, Self-locking, securing Pad to Pedal	11–3	1	
C.27521	**PAD (RUBBER) ON CLUTCH PEDAL**	11–4	1	
C.24014	**HOUSING, FOR BRAKE AND CLUTCH PEDAL**	11–5	1	See also Brake Controls
C.8954	**Stud**, mounting Housing to Scuttle	11–6	7	
UFN.131/L	**Nut**, on Studs	11–7	7	
C.725	**Washer**, Shakeproof, under Nuts	11–8	7	
C.8957	**Gasket**, between Housing and Scuttle	11–9	1	
C.8958	**Shaft**, mounting Clutch (and Brake) Pedals	11–10	1	
C.8962	**Washer** (Fibre) on Shaft, at each side of Clutch Pedal	11–11	2	
C.8963	**Return Spring**, on Shaft, for Clutch Pedal	11–12	1	
C.8967	**Bolt**, retaining Pedal Shaft	11–13	1	
C.8737/1	**Nut**, Self Locking, on Bolt	11–14	1	
FW.104/T	**Washer**, Plain, under Nut	11–15	1	
C.20526	**CLUTCH MASTER CYLINDER (91626)**	11–16	1	Fitted to Chassis No. 1A.12581 and subs. 1A.20565 and subs. also fitted to 1K.1001 and subs. 1K.30001 and subs.
9024	**Tank and Barrel Assembly** (91635)	11–17	1	
11447	**Filler Cap** (3847-218)	11–18	1	
6458	**Cork Seal**, for Filler Cap (86011)		1	
6459	**Return Spring** (85844)	11–19	1	
9025	**Retainer**, at bottom of Return Spring (92118)	11–20	1	
6449	**Seal** (85641)	11–21	1	
9026	**Disc**, between Seal and Piston (90897)	11–22	1	
6446	**Piston** (85486)	11–23	1	
6447	**Secondary Seal**, on Piston (37374)	11–24	1	
9027	**Push Rod Assembly** (91642)	11–25	1	
9028	**Circlip**, retaining Push Rod (K.19710)	11–26	1	
6445	**Rubber Boot**, over end of Push Rod (39994)	11–27	1	
C.8956	**Stud**, for Mounting of Clutch Master Cylinder to top of Housing	11–28	2	
UFN.131/L	**Nut**, on Studs		2	
C.725	**Washer**, Shakeproof, under Nuts		2	
C.28331	**Stop**, for Clutch Pedal		1	
C.28332	**Clevis Pin**, Master Cylinder Jaw to Clutch Pedal	11–29	1	
L.103/7U	**Split Pin**, retaining Clevis Pin, Master Cylinder Jaw to Pedal		1	
FW.105/T	**Washer**, Plain, behind Split Pin		1	
6593	**REPAIR KIT FOR SERVICING CLUTCH MASTER CYLINDER (KL.71534)**		1	
C.16343	**PIPE ASSEMBLY, BETWEEN FLEXIBLE HOSE AND CLUTCH MASTER CYLINDER (T.50/31B)**	11–30	1	**R.H. DRIVE CARS ONLY**
C.28235	**PIPE ASSEMBLY, BETWEEN FLEXIBLE HOSE AND CLUTCH MASTER CYLINDER (T.50/30B)**		1	**L.H. DRIVE CARS ONLY**

Part No.	Description	Plate No.	No. per Unit	Remarks
C.27504	**FLEXIBLE HOSE CONNECTING PIPES**	11–31	1	
UFN.237/L	**Locknut,** for Flexible Hose at bracket on Dash	11–32	1	
C.741	**Washer,** Shakeproof, under Nut	11–33	1	
C.27510	**BRACKET, SUPPORTING FLEXIBLE HOSE AND PIPE FOR SLAVE CYLINDER**	11–34	1	
UCS.143/5R	**Screw,** Set, securing Bracket to Clutch Housing		1	
FG.107/X	**Washer,** on Setscrew		1	
C.27505	**PIPE ASSEMBLY FROM FLEXIBLE HOSE TO SLAVE CYLINDER**	11–35	1	
C.27529	**Clip,** securing Pipe to Clutch Housing		1	
C.24145	**SLAVE CYLINDER FOR CLUTCH OPERATION (4252/374)**		1	Fitted from Engine No. 7K.1001 to 7K.3592
10944	**REPAIR KIT FOR SERVICING CLUTCH SLAVE CYLINDER (SSB.608)**		1	Fitted from Engine No. 7K.1001 to 7K.3592
C.29801	**SLAVE CYLINDER FOR CLUTCH OPERATION (4252/387)**	11–36	1	Fitted to Engine No. 7K.3593 and subs
C.24157	**Body, only** (3232/648)	11–37	1	
8336	**Spring** (89500)	11–38	1	
6139	**Cup Filler,** at top of Spring (87575)	11–39	1	
8591	**Seal,** behind Piston (27526)	11–40	1	
8592	**Piston** (103439)	11–41	1	
6142	**Rubber Boot,** over end of Slave Cylinder (33427)	11–42	1	
6096	**Bleed Screw,** in Slave Cylinder (28696)	11–43	1	
12287	**Circlip,** retaining Piston in Slave Cylinder Body	11–44	1	
C.9858	**Stud,** mounting Slave Cylinder to Clutch Housing		2	
UFN.137/L	**Nut,** on Studs		2	
FG.106/X	**Washer,** Spring, under Nuts		2	
8593	**REPAIR KIT FOR SERVICING CLUTCH SLAVE CYLINDER (KL.71590)**		1	
C.25881	**OPERATING ROD, BETWEEN SLAVE CYLINDER AND CLUTCH FORK**		1	Fitted from Engine No. 7K.1001 to 7K.3592
C.27326	**ADJUSTER ASSEMBLY FOR OPERATING ROD**		1	Fitted from Engine No. 7K.1001 to 7K.3592
C.20954/4	**Bush,** for Adjuster		1	Fitted from Engine No. 7K.1001 to 7K.3592
UFN.231/L	**Nut,** locking Adjuster to Operating Rod		1	Fitted from Engine No. 7K.1001 to 7K.3592
C.27327	**Pivot Pin,** for Operating Rod		1	Fitted from Engine No. 7K.1001 to 7K.3592
C.27328/5	**Safety Clip**		1	Fitted from Engine No. 7K.1001 to 7K.3592
C.29802	**OPERATING ROD, BETWEEN SLAVE CYLINDER AND CLUTCH FORK**	11–45	1	Fitted to Engine No. 7K.3593 and subs.
C.27326	**ADJUSTER ASSEMBLY FOR OPERATING ROD**	11–46	1	Fitted to Engine No. 7K.3593 and subs.
C.20954/4	**Bush,** for Adjuster		1	Fitted to Engine No. 7K.3593 and subs.
UFN.231/L	**Nut,** locking Adjuster to Operating Rod	11–47	1	Fitted to Engine No. 7K.3593 and subs.
C.29803	**Pivot Pin,** for Operating Rod	11–48	1	Fitted to Engine No. 7K.3593 and subs.
C.5120	**Return Spring,** for Clutch	11–49	1	Fitted to Engine No. 7K.3593 and subs.
C.5178	**Anchor Plate,** for Spring	11–50	1	Fitted to Engine No. 7K.3593 and subs.
C.27328/5	**Safety Clip,** retaining Pivot Pin	11–51	1	Fitted to Engine No. 7K.3593 and subs.

CLUTCH CONTROLS

Part No.	Description	Plate No.	No. per Unit	Remarks
C.27430	**FORK, OPERATING CLUTCH RELEASE BEARING**	11–52	1	
C.27431	**Shaft,** for Fork	11–53	1	
C.20725/14	**Roll Pin,** securing Shaft and Fork Assembly	11–54	1	
C.27428	**CLUTCH HOUSING (COMPLETE WITH BUSHES)**	11–55	1	
C.10025	**Bush,** for Release Fork Shaft		2	
UFB.143/13R	**Bolt,** securing Clutch Housing to Gearbox		6	
C.587	**Locking Plate,** for Bolt		1	
C.5073	**Locking Plate,** for Bolts		2	
C.20334	**Bolt,** securing Clutch Housing to Gearbox		1	
C.20335	**Screw,** Set, securing Clutch Housing to Gearbox		1	
FW.107/T	**Washer,** Plain, on Bolt and Setscrew		2	
C.3262	**Wire,** locking Bolt and Setscrew		As req'd	
C.11910	**Cover,** over Timing Aperture at bottom of Clutch Housing		1	
BD.524/2	**Rivet,** securing Cover to Clutch Housing		1	
420304	**Dowel,** locating Clutch Housing		1	
C.18739	**OIL SEAL, IN REAR OF CLUTCH HOUSING, FOR GEARBOX CONSTANT PINION SHAFT**	11–56	1	
C.27507	**BREATHER PIPE ASSEMBLY, FOR GEARBOX**	11–57	1	
C.20113	**Hose,** connecting Breather Assembly to Elbow on Gearbox Top Cover	11–58	1	
C.15886/5	**Clip,** securing Hose	11–59	2	
C.20218	**MID ENGINE MOUNTING, ON SCUTTLE**	11–60	1	
UFS.131/6R	**Screw,** Set, securing Mounting to Scuttle		2	
C.8667/2	**Nut,** Self-Locking, on Setscrews		2	
C.11688	**Bush,** Stepped, below Mounting	11–61	1	
C.11607	**Washer,** Stepped, above Mounting	11–62	1	
C.8667/3	**Nut,** Self-Locking, securing Link to Mounting	11–63	1	
C.21226	**LINK, FOR MID ENGINE MOUNTING**	11–64	1	
C.10940	**Bush** (Rubber/Steel) in Link	11–65	1	
UFB.143/15R	**Bolt,** securing Link to Mounting Bracket		1	
C.8737/4	**Nut,** Self-Locking on Bolt		1	
C.22112	**MOUNTING BRACKET ASSEMBLY FOR MID ENGINE MOUNTING LINK**	11–66	1	
UCS.131/6R	**Screw,** Set, securing Mounting Bracket to Clutch Housing		4	
FG.105/X	**Washer,** Spring, on Setscrews		4	

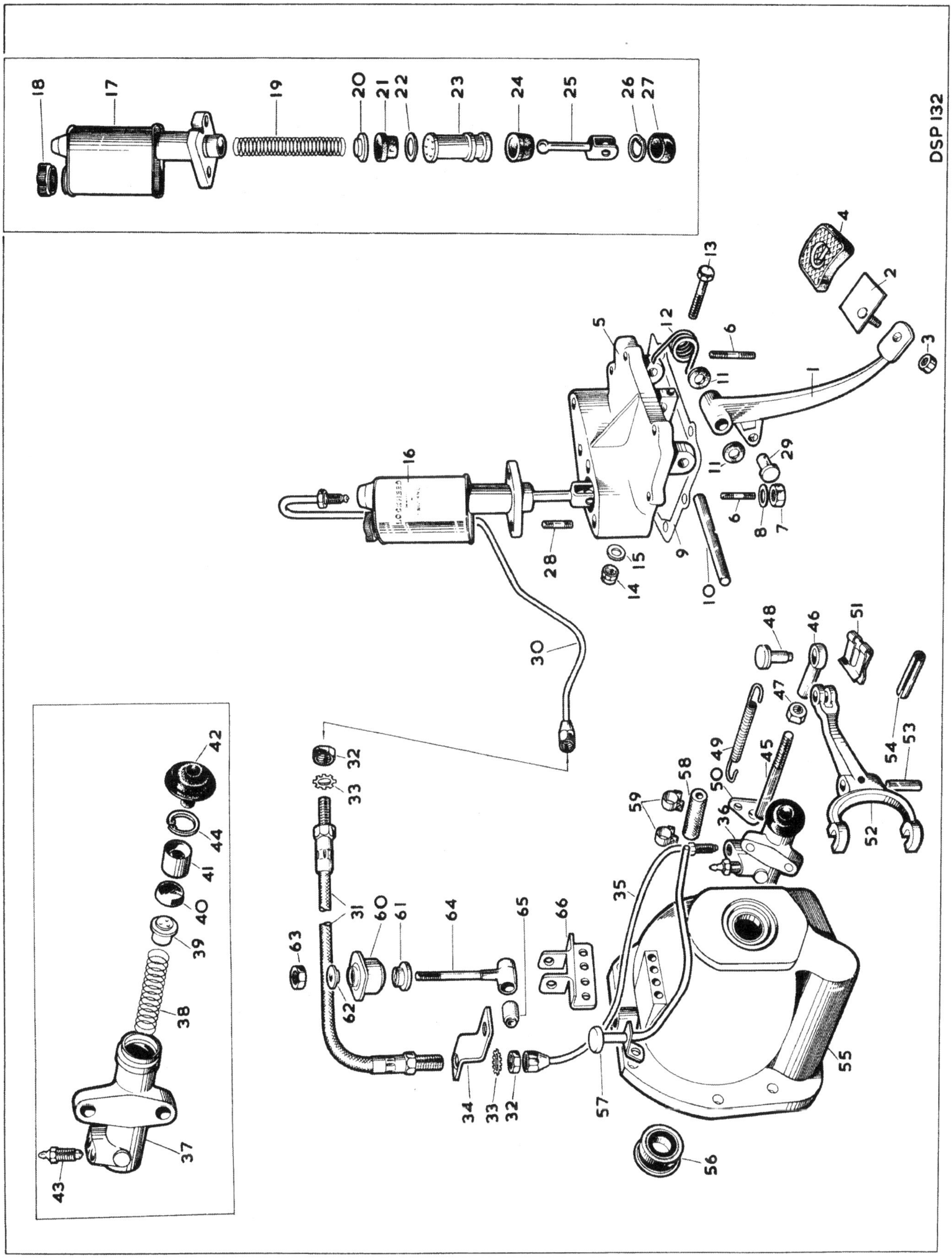

PLATE 11

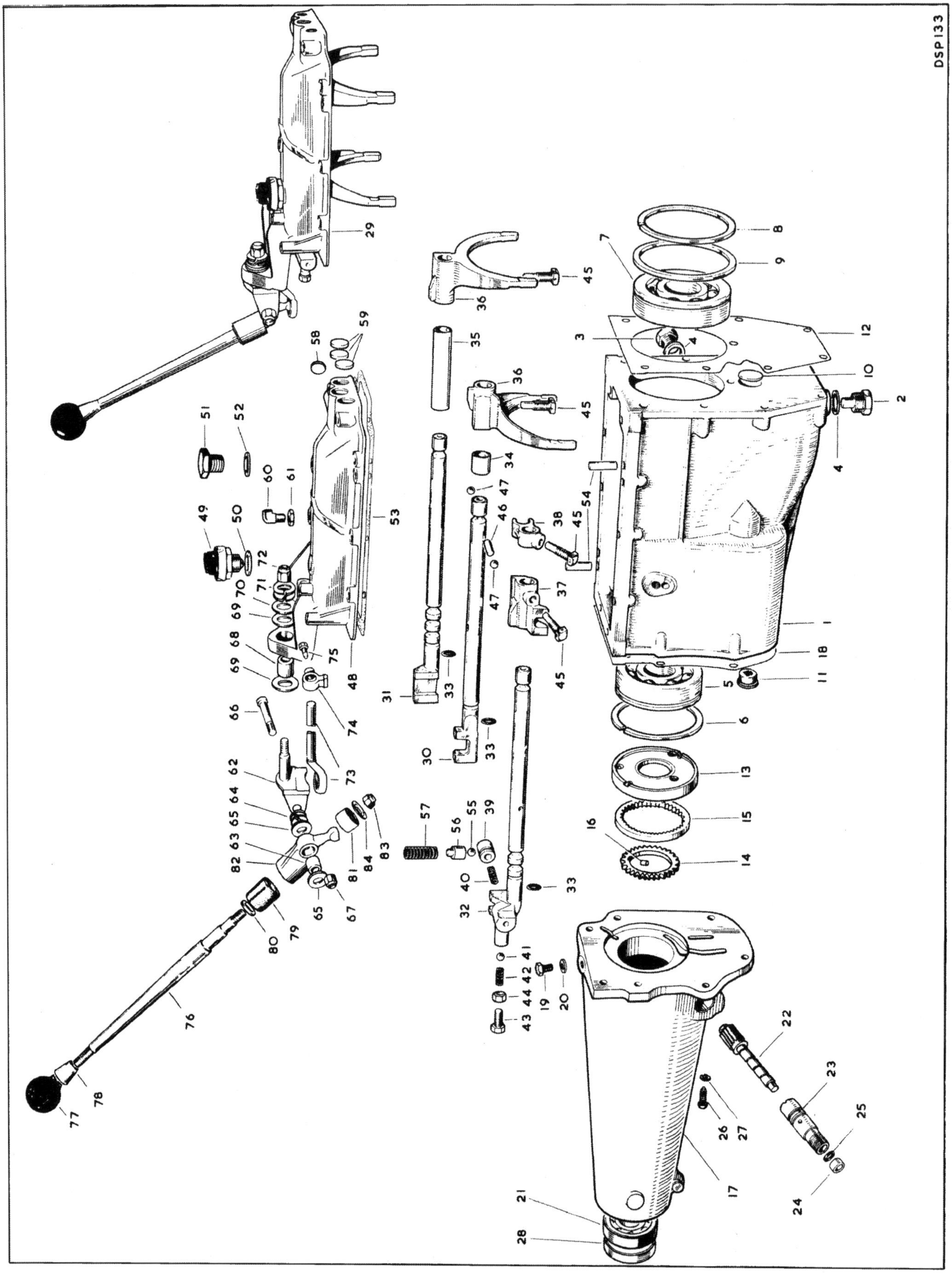

PLATE 12

Part No.	Description	Plate No.	No. per Unit	Remarks
C.28641	GEARBOX ASSEMBLY		1	Fitted to Engine No. 7A.13231 and subs. (Also to 7A.12185 and 7A.12186) 7K.1001 and subs.
C.20019	GEARBOX CASE	12–1	1	
C.19543	**Oil Drain Plug** (Magnetic) at bottom of Case	12–2	1	
C.799	**Oil Filter Plug**, at L.H. side of Case	12–3	1	
C.785	**Washer** (Fibre) on Plugs	12–4	2	
C.1845/1	**Ball Bearing**, for Mainshaft (MS.12½K)	12–5	1	
C.1855	**Circlip**, around Ball Bearing	12–6	1	
C.20491	**Ball Bearing**, for Constant Pinion Shaft (MS.13½K) (complete with Circlip C.20492)	12–7	1	
C.20492	**Circlip**, around Ball Bearing	12–8	1	
C.20494	**Collar**, between Circlip and Gearbox Case	12–9	1	
C.920	**Blanking Disc** (Fibre) in Case, at front of Countershaft	12–10	1	
C.22837	**Filter Gauze**, in rear face of Gearbox Case	12–11	1	
C.20658	**Gasket**, between Gearbox Case and Clutch Housing	12–12	1	
C.21306	OIL PUMP BODY AT REAR OF GEARBOX CASE	12–13	1	
C.21308	**Inner Gear**, for Oil Pump	12–14	1	
C.21307	**Outer Gear**, for Oil Pump	12–15	1	
C.21309	**Pin**, on Mainshaft, driving Oil Pump	12–16	1	
UCS.519/5H	**Screw**, Set, securing Oil Pump Body to Rear End Cover		3	
C.25763	REAR END COVER	12–17	1	
C.20020	**Gasket**, between Rear End Cover and Gearbox Case	12–18	1	
UFS.131/9R	**Screw**, Set, securing Rear End Cover to Gearbox Case		7	
FG.105/X	**Washer**, Spring, on Setscrews		7	
UCS.143/3R	**Plug**, at top of Rear End Cover	12–19	1	
FW.107/E	**Washer**, (Copper) on Plug	12–20	1	
C.18719	**Ball Bearing**, in Rear End Cover, supporting Mainshaft	12–21	1	
C.19118	**Location Pin**, for Rear Engine Mounting, in Gearbox Extension Case		1	
C.22646/1	**Spring Retainer**, on Location Pin		1	
UFN.156/L	**Nut**, on Location Pin, securing Spring Retainer		1	
C.21310	SPEEDOMETER DRIVEN GEAR	12–22	1	
C.21312	BEARING ASSEMBLY FOR SPEEDOMETER DRIVEN GEAR	12–23	1	
C.12377/1	**Oil Seal**, in Bearing, for Speedometer Driven Gear	12–24	1	
C.9177	**'O' Ring**, around Bearing	12–25	1	
C.22999	**Screw**, locking Bearing in Rear End Cover	12–26	1	
FG.105/X	**Washer**, Spring, on locking Screw	12–27	1	
C.23394	OIL SEAL AT END OF REAR COVER	12–28	1	
12306	REMOTE CONTROL UNIT COMPLETE (S.D.1168)	12–29	1	
	(COMPRISING TOP COVER, SELECTOR RODS, CHANGE-SPEED LEVER, FORKS, SELECTOR LEVER ETC.)			
	SELECTOR MECHANISM			
C.24915	**Selector Shaft**, for 1st and 2nd Gears	12–30	1	
C.24914	**Selector Shaft**, for 3rd and 4th Gears	12–31	1	
C.21478	**Selector Shaft**, for Reverse Gear	12–32	1	
C.20060	**'O' Ring**, on Selector Shafts	12–33	3	
C.20315	**Stop Sleeve**, on 1st/2nd Gear Selector Shaft	12–34	1	
C.20316	**Stop Sleeve**, on 3rd/4th Gear Selector Shaft	12–35	1	

GEARBOX
FOR STANDARD TRANSMISSION

Part No.	Description	Plate No.	No. per Unit	Remarks
	REMOTE CONTROL UNIT COMPLETE *(contd.)*			
C.21572	**Selector Fork**, on 1st/2nd and 3rd/4th Selector Shafts	12–36	2	Fitted to Engine No. 7A.13231 and subs. (Also to 7A.12185 and 7A.12186) 7K.1001 to 7K.2484
C.21572/1	**Selector Fork**, on 1st/2nd and 3rd/4th Selector Shafts		2	Fitted to Engine No. 7K.2485 and subs.
C.21616	**Selector Fork**, on Reverse Selector Shaft	12–37	1	
C.22925	**Locating Arm**, on Reverse Selector Shaft	12–38	1	
C.25468	**Plunger**, for Reverse Stop, in Selector Shaft	12–39	1	
C.9849	**Spring**, tensioning Plunger	12–40	1	
C.1319	**Ball**, locking Reverse Plunger	12–41	1	
C.20065	**Spring**, tensioning Ball	12–42	1	
UFS.131/5R	**Screw**, Set, adjusting tension on Spring	12–43	1	
UFN.231/L	**Locknut**, on Setscrew	12–44	1	
C.20314	**Dowel Screw**, securing Selector Forks and Locating Arm on Selector Shafts	12–45	4	
C.837	**Wire**, locking Dowel Screws		As req'd	
C.20066	**Roller**, in 1st/2nd Selector Shaft, for Interlock	12–46	1	
C.922	**Ball**, Between Selector Rods, for Interlock	12–47	2	
C.27840	**TOP COVER, FOR GEARBOX**	12–48	1	
C.1083	**Switch Unit**, in Top Cover, for Reverse Lamp	12–49	1	**See also Electrical Equipment**
C.4531	**Gasket**, for Switch (163682)	12–50	1	
C.7379	**Plug**, in Top Cover, sealing redundant hole for Top Gear Switch	12–51	1	
C.2296/1	**Washer**, (Copper) on Plug	12–52	1	
C.20057	**Gasket**, between Top Cover and Gearbox Case	12–53	1	
UFS.131/8R	**Screw**, Set, securing Top Cover to Gearbox		6	
UFB.131/13R	**Bolt**, securing Top Cover to Gearbox		2	
UFB.131/24R	**Bolt**, securing Top Cover to Gearbox		2	
C.16816	**Washer**, Spring, on Setscrews and Bolts		10	
OD.105/6H	**Dowel**, locating Top Cover on Gearbox	12–54	2	
C.1319	**Ball**, detent, for Reverse Striking Rod	12–55	1	
C.13334	**Plunger**, locating Detent Ball	12–56	1	
C.845	**Spring**, tensioning Plunger	12–57	1	
C.9840	**Welch Washer**, 7/16" dia, at L.H. side of Top Cover	12–58	1	
C.26113	**Welch Washer**, ¾" dia. at front of Top Cover (End of Striking Rods)	12–59	3	
C.7116	**Breather Elbow**, in Top Cover	12–60	1	
C.7125	**Nut**, for Breather Elbow	12–61	1	
C.27847	**SELECTOR PIVOT JAW ON TOP COVER, FOR SELECTOR LEVER**	12–62	1	
C.1906	**Bush**, in Selector Lever, for Pivot Pin	12–63	1	
C.1929	**Washer**, Spring (Double Coil) between Pivot Jaw and Tufnol Washer	12–64	1	
C.1908	**Washer**, (Tufnol) at each side of Selector Lever	12–65	2	
C.21110	**Pivot Pin**, mounting Selctor Lever in Pivot Jaw	12–66	1	
C.8737/4	**Nut**, Self-Locking, securing Pivot Pin	12–67	1	
C.1907	**Bush**, in Top Cover, for Pivot Jaw	12–68	1	
C.1894	**Washer** (Tufnol) at each side of Bush	12–69	2	
C.1914	**'D' Washer**, between Spring Washer and Fibre Washer	12–70	1	
C.1905	**Washer**, Spring (Double Coil) under Nut	12–71	1	
C.8737/4	**Nut**, Self-Locking, retaining Pivot Jaw to Top Cover	12–72	1	
C.28330	**Remote Control Shaft**, in Top Cover for Selector Lever	12–73	1	
C.27849	**Finger**, on Remote Control Shaft	12–74	1	
C.19772	**Screw**, Special, securing Finger to Remote Control Shaft	12–75	1	
C.837	**Locking Wire**, for Special Screw		As req'd	

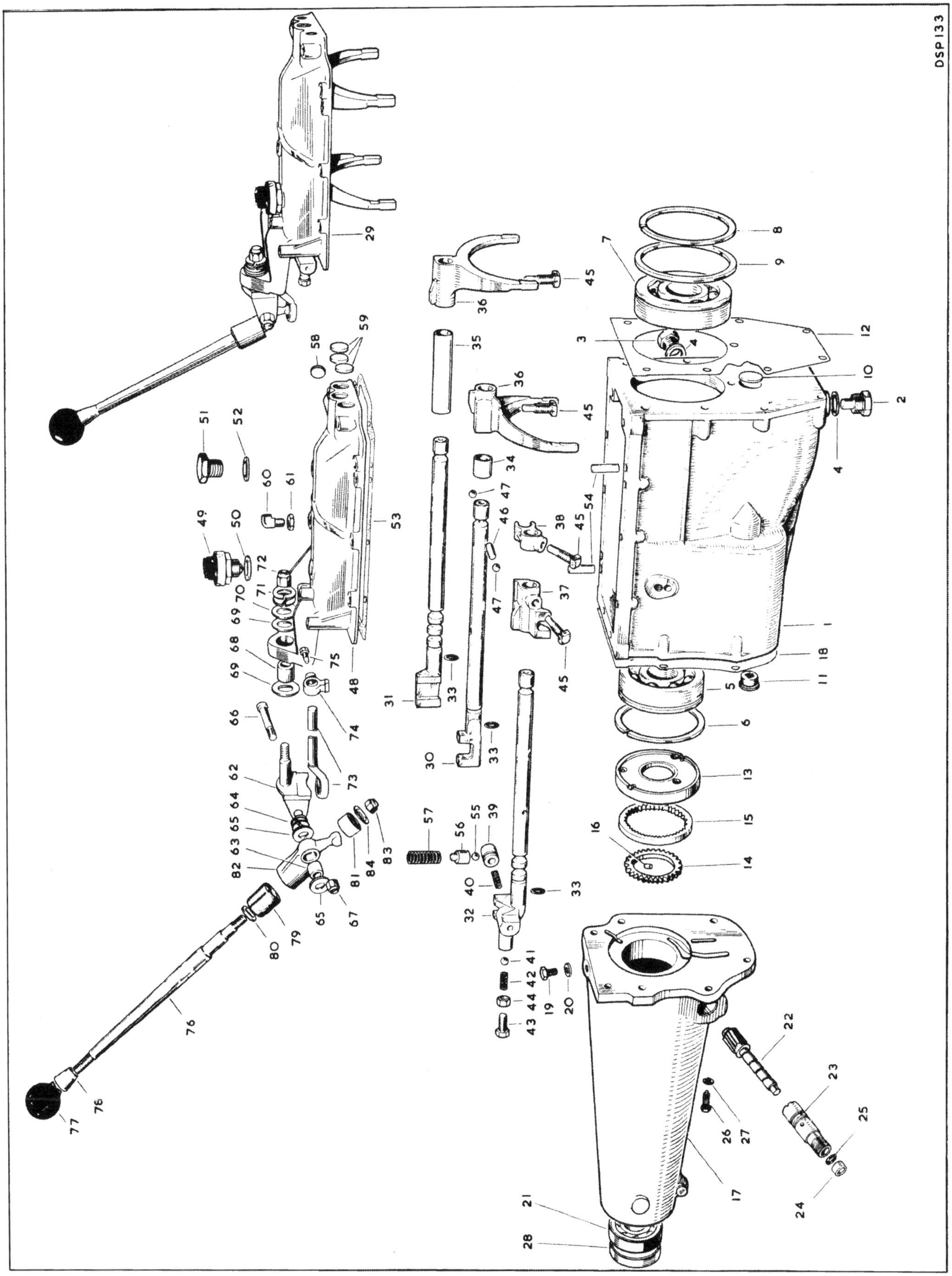

PLATE 12

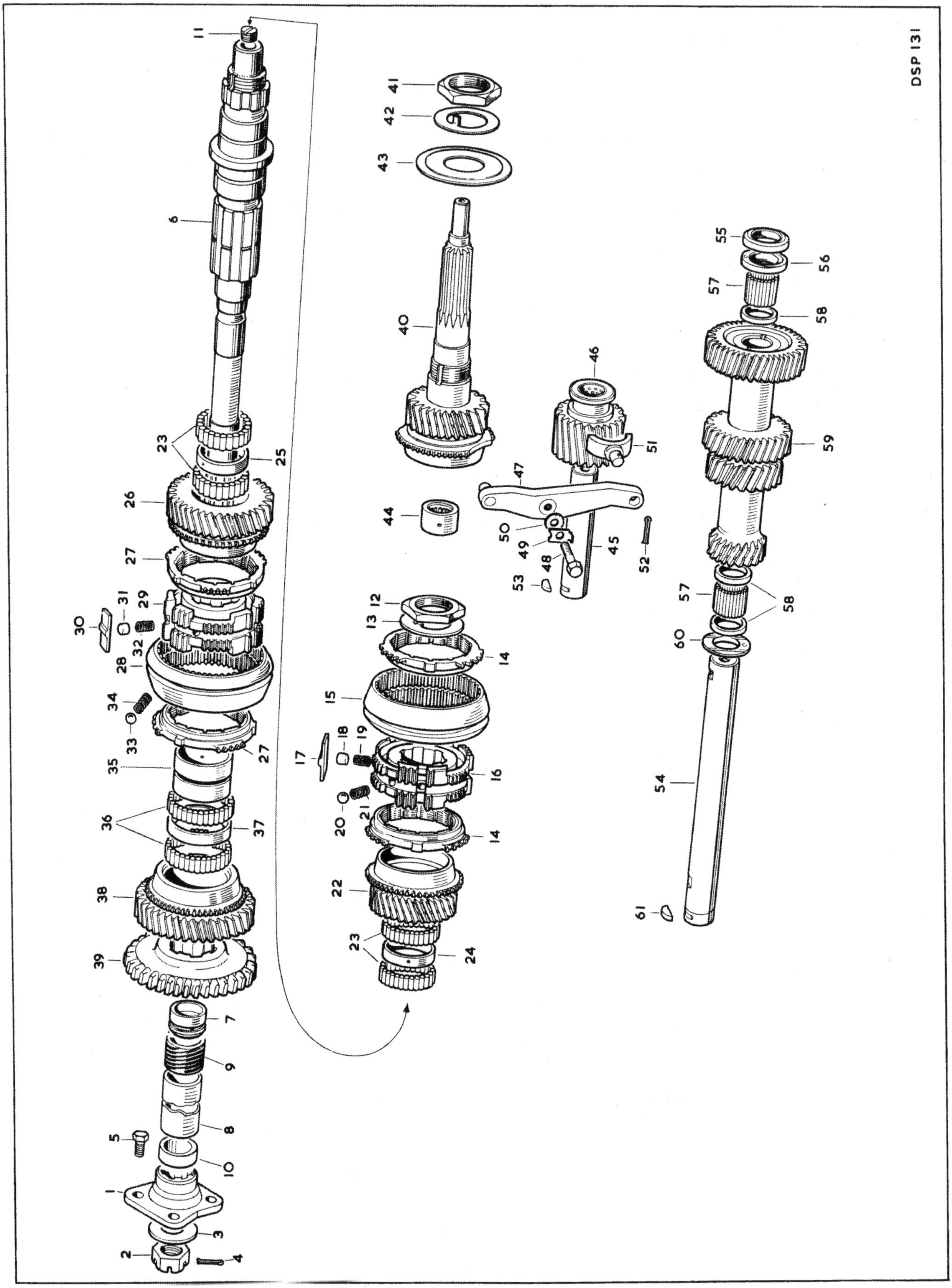

PLATE 13

Part No.	Description	Plate No.	No. per Unit	Remarks
C.27865	**CHANGE-SPEED LEVER**	12–76	1	
C.23007	**Knob,** on Change-Speed Lever	12–77	1	
C.23006	**Cone,** locking Knob to Lever	12–78	1	
C.22394	**Rubber Bush,** Upper, locating Change-Speed Lever in Selector Lever	12–79	1	
C.22395	**Washer,** Special, at top of Rubber Bush	12–80	1	
C.21921	**Bush Assembly** (Metalastic) lower, locating Change-Speed Lever in Selector Lever	12–81	1	
C.27848	**Selector Lever,** for Change-Speed Lever	12–82	1	
C.8737/2	**Nut,** Self-Locking, securing Change-Speed Lever Assembly to Selector Lever	12–83	1	
C.26595	**Washer,** Special, under Nut	12–84	1	
C.885	**FLANGE, ON MAINSHAFT, FOR UNIVERSAL JOINT**	13–1	1	
C.840	**Nut,** Slotted, securing Flange to Mainshaft	13–2	1	
C.23168	**Washer,** Special, under Slotted Nut	13–3	1	
L.104/12U	**Split Pin,** securing Slotted Nut	13–4	1	
C.19827	**Bolt,** connecting Flange to Propeller Shaft	13–5	4	**See also Propeller Shaft**
C.25821	**MAINSHAFT**	13–6	1	
C.18707	**Distance Piece,** between Universal Joint Flange and Bearing	13–7	1	
C.25824	**Distance Tube,** between Bearing and Speedometer Driving Gear	13–8	1	
C.25822	**Speedometer Driving Gear**	13–9	1	
C.21301	**Distance Piece,** between Mainshaft Bearing (in Case) and Propeller Shaft Flange	13–10	1	
C.20027	**Plug,** in end of Mainshaft	13–11	1	
C.21617	**Nut,** retaining 3rd Gear and 3rd/4th Synchro Hub on Mainshaft	13–12	1	
C.21618	**Tab Washer,** behind Nut	13–13	1	
C.22391	**Synchro Ring,** for 3rd and 4th speeds	13–14	2	
C.24640	**Operating Sleeve,** for 3rd and 4th speeds	13–15	1	Fitted to Engine No. 7A.13231 and subs. (Also to 7A.12185 and 7A.12186) 7K.1001 to 7K.2484
C.26439	**Synchro Hub,** for 3rd and 4th speeds	13–16	1	
C.28436	**Operating Sleeve,** for 3rd and 4th speeds		1	Fitted to Engine No. 7K.2485 and subs.
C.28432	**Synchro Hub,** for 3rd and 4th speeds		1	
C.24641	**Thrust Member,** between Sleeve and Synchro Hub	13–17	3	
C.21088	**Plunger,** under Thrust Member	13–18	3	
C.27491	**Spring,** under Plungers	13–19	3	
C.1319	**Detent Ball,** between Sleeve and Synchro Hub	13–20	3	
C.24181	**Spring,** tensioning Detent Balls	13–21	3	
C.21082	**3rd Speed Gear,** on Mainshaft	13–22	1	Fitted to Engine No. 7A.13231 and subs. (Also to 7A.12185 and 7A.12186) 7K.1001 to 7K.2484
C.28417	**3rd Speed Gear,** on Mainshaft		1	Fitted to Engine No. 7K.2485 and subs.
C.23109/1	**Needle Roller,** for 3rd and 2nd Speed Gears (3 mm + .0001" – .0000")	13–23	212	**ONLY USE ROLLERS OF ONE GRADE IN INDIVIDUAL GEAR ASSEMBLIES**
C.23109/2	**Needle Roller** for 3rd and 2nd Speed Gears (3 mm + .0000" – .0001")		212	
C.23109/3	**Needle Roller,** for 3rd and 2nd Speed Gears (3 mm – 0001" – .0002")		212	
C.23099	**Spacer,** for Needle Rollers in 3rd Speed Gear	13–24	1	
C.23098	**Spacer,** for Needle Rollers in 2nd Speed Gear	13–25	1	
C.21092	**2nd Speed Gear,** on Mainshaft	13–26	1	Fitted to Engine No. 7A.13231 and subs (Also to 7A.12185 and 7A.12186) 7K.1001 to 7K.2484
C.28430	**2nd Speed Gear,** on Mainshaft		1	Fitted to Engine No. 7K.2485 and subs.

GEARBOX
FOR STANDARD TRANSMISSION

Part No.	Description	Plate No.	No. per Unit	Remarks
	MAINSHAFT *(contd.)*			
C.22391	**Synchro Ring,** for 2nd and 1st speeds	13–27	2	
C.24640	**Operating Sleeve,** for 2nd and 1st speeds	13–28	1	Fitted to Engine No. 7A.13231 and subs. (Also to 7A.12185 and 7A.12186) 7K.1001 to 7K.2484
C.24638	**Sychro Hub,** for 2nd and 1st speeds	13–29	1	
C.28436	**Operating Sleeve,** for 2nd and 1st speeds		1	Fitted to Engine No. 7K.2485 and subs.
C.28432	**Synchro Hub,** for 2nd and 1st speeds		1	
C.24641	**Thrust Member,** between Sleeve and Synchro Hub	13–30	3	
C.21088	**Plunger,** under Thrust Member	13–31	3	
C.27491	**Spring,** under Plungers	13–32	3	
C.1319	**Detent Ball,** between Sleeve and Synchro Hub	13–33	3	
C.24181	**Spring,** tensioning Detent Balls	13–34	3	
C.21095	**Bearing Sleeve,** for 1st Speed Gear	13–35	1	
C.23109/1	**Needle Roller,** for 1st Speed Gear (3 mm + .0001" – 0000")	13–36	120	**ONLY USE ROLLERS OF ONE GRADE IN INDIVIDUAL GEAR ASSEMBLIES**
C.23109/2	**Needle roller,** for 1st Speed Gear (3 mm + .0000" – .0001")		120	
C.23109/3	**Needle Roller,** for 1st Speed Gear (3 mm – .0001" – .0002")		120	
C.23097	**Spacer,** for Needle Rollers	13–37	1	
C.21096	**1st Speed Gear,** on Mainshaft	13–38	1	Fitted to Engine No. 7A.13231 and subs. (Also to 7A.12185 and 7A.12186) 7K.1001 to 7K.2484
C.28429	**1st Speed Gear,** on Mainshaft		1	Fitted to Engine No. 7K.2485 and subs.
C.21098	**Reverse Gear,** on Mainshaft	13–39	1	
C.25436	**CONSTANT PINION SHAFT**		1	Fitted to Engine No. 7A.13231 and subs. (Also to 7A.12185 and 7A.12186) 7K.1001 to 7K.2484
C.28415	**CONSTANT PINION SHAFT**	13–40	1	Fitted to Engine No. 7K.2485 and subs.
C.21617	**Nut,** securing Constant Pinion Shaft in Bearing	13–41	1	
C.21618	**Tab Washer,** under Nut	13–42	1	
C.20493	**Oil Thrower,** on Constant Pinion Shaft	13–43	1	
C.25424	**Roller Bearing** (for Mainshaft) inside Constant Pinion Shaft	13–44	1	
C.20052	**REVERSE SPINDLE**	13–45	1	
C.20313	**Reverse Idler Gear,** complete with Bushes	13–46	1	
C.21198	**Lever Assembly,** operating Reverse Gear	13–47	1	
UFS.137/8R	**Screw,** Set, pivoting Lever	13–48	1	
C.21200	**Tab Washer,** locking Setscrew	13–49	1	
C.930	**Fibre Washer,** under Tab Washer	13–50	1	
C.866	**Reverse Slipper,** between Lever Assembly and Reverse Idler Gear	13–51	1	
L.103/8U	**Split Pin,** retaining Slipper in Lever	13–52	1	
C.753	**Woodruff Key,** locking Spindle in Gearbox Case	13–53	1	
C.25437	**COUNTERSHAFT**	13–54	1	
C.920	**Fibre Washer,** in Gear Case, at front of Counter-Shaft		1	
C.1862/1	**Thrust Washer,** for end float at front end of Countershaft (.159" thick)	13–55	1	**SELECTIVE SIZES. USE AS NECESSARY TO OBTAIN .004"/.006" END FLOAT**
C.1862/2	**Thrust Washer,** for end float at front end of Countershaft (.162" thick)		1	
C.1862/3	**Thrust Washer,** for end float at front end of Countershaft (.152" thick)		1	
C.1862/4	**Thrust Washer,** for end float at front end of Countershaft (.164" thick)		1	
C.1861	**Thrust Washer Assembly,** Inner, at front end of Countershaft	13–56	1	

Part No.	Description	Plate No.	No. per Unit	Remarks
	COUNTERSHAFT *(contd.)*			
C.918/1	**Needle Roller**, at each end of Gear Unit (Cluster) (3 mm + .0001" – .0000")	13–57	58	**ONLY USE ROLLERS OF ONE GRADE IN INDIVIDUAL GEAR UNIT ASSEMBLIES**
C.918/2	**Needle Roller**, at each end of Gear Unit (Cluster) (3 mm + .0000" – .0001")		58	
C.918/3	**Needle Roller**, at each end of Gear Unit (Cluster) (3 mm – .0001" – .0002")		58	
C.21099	**Spacer**, for Needle Rollers	13–58	3	
C.20712	**Gear Unit** (Cluster) on Countershaft	13–59	1	Fitted to Engine No. 7A.13231 and subs. (Also to 7A.12185 and 7A.12186) 7K.1001 to 7K.2484
C.28416	**Gear Unit** (Cluster) on Countershaft		1	Fitted to Engine No. 7K.2485 and subs.
C.21100	**Thrust Washer Assembly**, at rear of Gear Unit	13–60	1	
C.753	**Woodruff Key**, locking Countershaft in Gearbox Case	13–61	1	

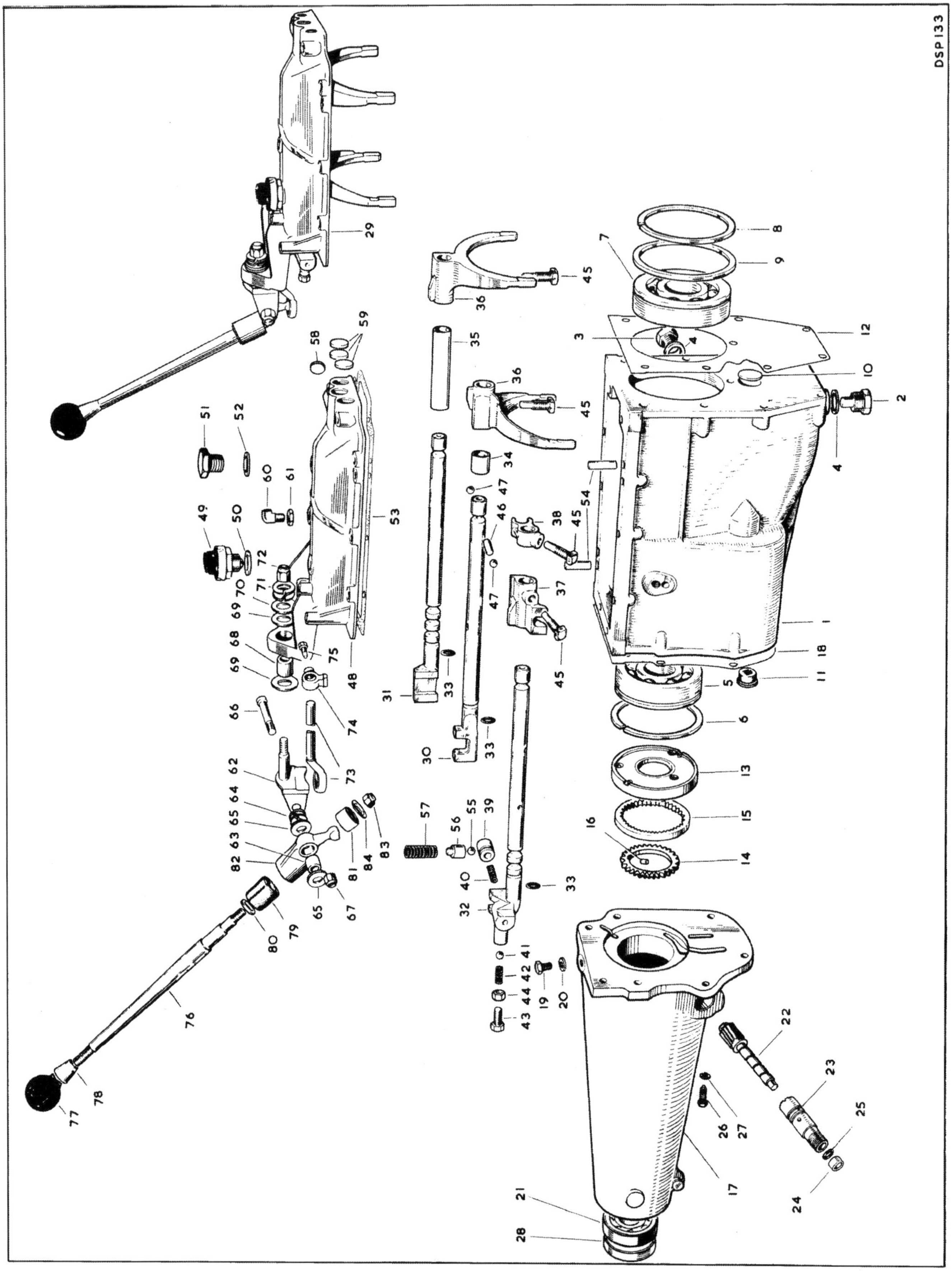

PLATE 12

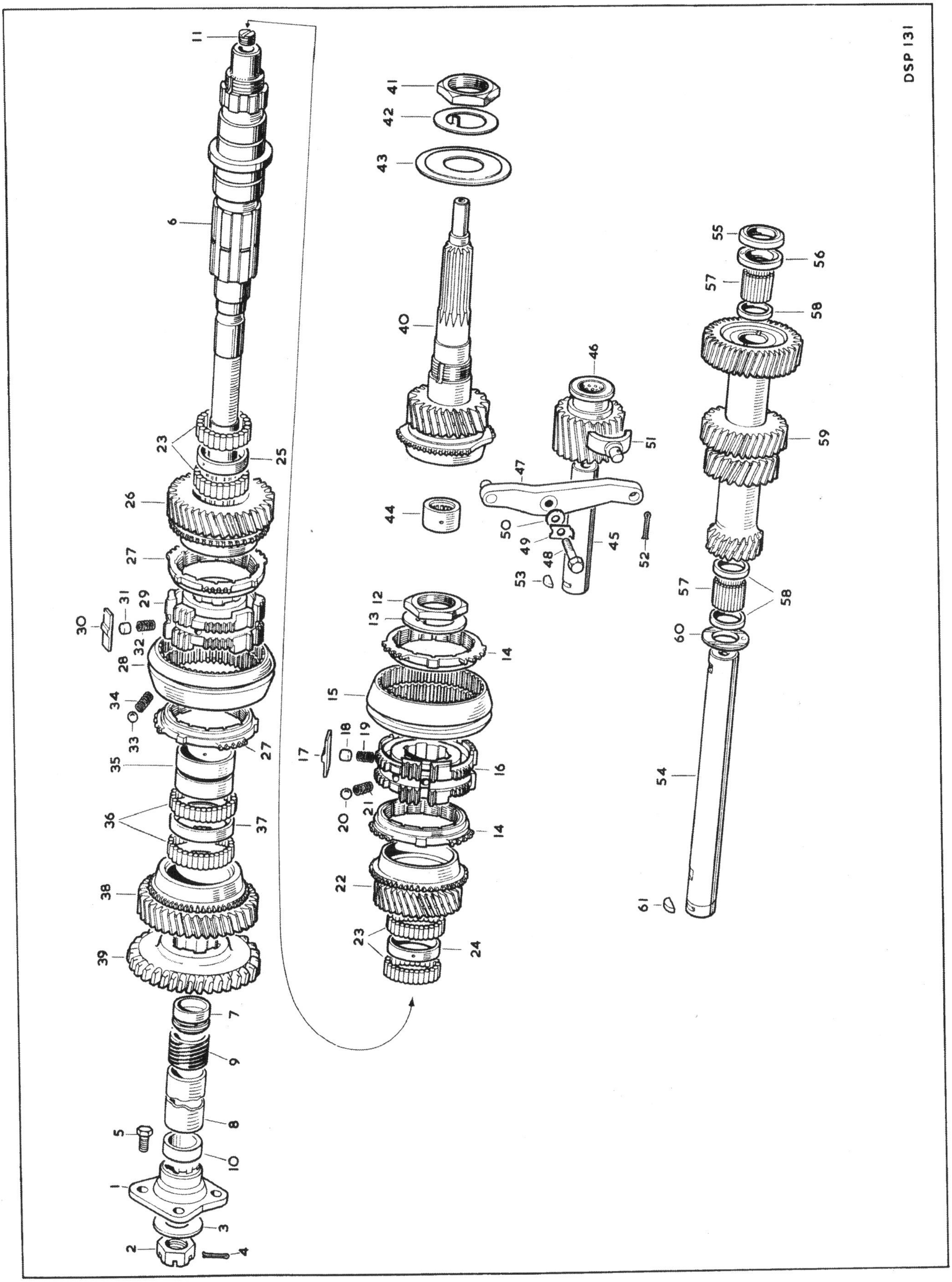
DSP 131
1
2
3
4
5
6
7
8
9
10
11
12
13
14
15
16
17
18
19
20
21
22
23
24
25
26
27
28
29
30
31
32
33
34
35
36
37
38
39
40
41
42
43
44
45
46
47
48
49
50
51
52
53
54
55
56
57
58
59
60
61

PLATE 13

GEARBOX

FOR OVERDRIVE TRANSMISSION

Part No.	Description	Plate No.	No. per Unit	Remarks
C.28645	**GEARBOX ASSEMBLY**		1	Fitted to Engine No. 7A.13231 and subs. (Also to 7A.12185 and 7A.12186) 7K.1001 and subs.
C.20019	**GEARBOX CASE**	14–1	1	
C.19543	**Oil Drain Plug** (Magnetic) at bottom of Case	14–2	1	
C.799	**Oil Filler Plug**, at L.H. side of Case	14–3	1	
C.785	**Washer**, (Fibre) on Plugs	14–4	2	
C.1845/1	**Ball Bearing**, for Mainshaft (MS.12½K)	14–5	1	
C.1855	**Circlip**, around Ball Bearing	14–6	1	
C.20491	**Ball Bearing**, for Constant Pinion Shaft (MS.13½K) (complete with Circlip C.20492)	14–7	1	
C.20492	**Circlip**, around Ball Bearing	14–8	1	
C.20494	**Collar**, between Circlip and Gearbox Case	14–9	1	
C.920	**Blanking Disc** (Fibre) in Case, at front of Countershaft	14–10	1	
C.20658	**Gasket**, between Gearbox Case and Clutch Housing	14–11	1	
C.22605	**Gasket**, between Gearbox Case and Overdrive Unit	14–12	1	
12307	**REMOTE CONTROL UNIT COMPLETE (SD.1169)**	14–13	1	
	(COMPRISING TOP COVER, SELECTOR RODS, CHANGE-SPEED LEVER, FORKS, SELECTOR LEVER, ETC.)			
	SELECTOR MECHANISM			
C.24915	**Selector Shaft**, for 1st and 2nd Gears	14–14	1	
C.24914	**Selector Shaft**, for 3rd and 4th Gears	14–15	1	
C.21478	**Selector Shaft**, for Reverse Gear	14–16	1	
C.20060	**'O' Ring**, on Selector Shafts	14–17	3	
C.20315	**Stop Sleeve**, on 1st/2nd Gear Selector Shaft	14–18	1	
C.20316	**Stop Sleeve**, on 3rd/4th Gear Selector Shaft	14–19	1	
C.21572	**Selector Fork**, on 1st/2nd and 3rd/4th Selector Shafts		2	Fitted to Engine No. 7A.13231 and subs. (Also to 7A.12185 and 7A.12186) 7K.1001 to 7K.2484
C.21572/1	**Selector Fork**, on 1st/2nd and 3rd/4th Selector Shafts	14–20	2	Fitted to Engine No. 7K.2485 and subs.
C.21616	**Selector Fork**, on Reverse Selector Shaft	14–21	1	
C.22925	**Locating Arm**, on Reverse Selector Shaft	14–22	1	
C.25468	**Plunger**, for Reverse Stop, in Selector Shaft	14–23	1	
C.9849	**Spring**, tensioning Plunger	14–24	1	
C.1319	**Ball**, locking Reverse Plunger	14–25	1	
C.20065	**Spring**, tensioning Ball	14–26	1	
UFS.131/5R	**Screw**, Set, adjusting tension on Spring	14–27	1	
UFN.231/L	**Locknut**, on Setscrew	14–28	1	
C.20314	**Dowel Screw**, securing Selector Fork and Locking Arm on Selector Shafts	14–29	4	
C.837	**Wire**, locking Dowel Screws		As req'd	
C.20066	**Roller**, on 1st/2nd Selector Shaft, for Interlock	14–30	1	
C.922	**Ball**, between Selector Rods, for Interlock	14–31	2	
C.27840	**TOP COVER, FOR GEARBOX**	14–32	1	
C.1083	**Switch Unit**, in Top Cover, for Reversing Lamp and Top Gear Circuit (31077/D-5510)	14–33	2	**See also Electrical Equipment**
C.4531	**Gasket**, for Reversing Lamp Switch (163682)	14–34	1	**See also Electrical Equipment**
C.4531	**Gasket**, for adjustment of Top Gear Switch (163682)	14–34	As req'd	**See also Electrical Equipment**
C.26377	**Gasket**, for adjustment of Top Gear Switch (.020" thick)		As req'd	**See also Electrical Equipment**
C.20057	**Gasket**, between Top Cover and Gearbox Case	14–35	1	
UFS.131/8R	**Screw**, Set, securing Top Cover to Gearbox		6	
UFB.131/13R	**Bolt**, securing Top Cover to Gearbox		2	
UFB.131/24R	**Bolt**, securing Top Cover to Gearbox		2	
C.16816	**Washer**, Spring, on Setscrews and Bolts		10	

Part No.	Description	Plate No.	No. per Unit	Remarks
	TOP COVER, FOR GEARBOX *(contd.)*			
OD.105/6H	**Dowel,** locating Top Cover on Gearbox	14–36	2	
C.1319	**Ball,** detent, for Reverse Striking Rod	14–37	1	
C.13334	**Plunger,** locating Detent Ball	14–38	1	
C.845	**Spring,** tensioning Plunger	14–39	1	
C.9840	**Welch Washer,** 7/16" dia. at L.H. side of Top Cover	14–40	1	
C.26113	**Welch Washer,** ¾" dia. at front of Top Cover (End of Striking Rods)	14–41	3	
C.7116	**Breather Elbow,** in Top Cover	14–42	1	
C.7125	**Nut,** for Breather Elbow	14–43	1	
C.27847	**SELECTOR PIVOT JAW ON TOP COVER, FOR SELECTOR LEVER**	14–44	1	
C.1906	**Bush,** in Selector Lever, for Pivot Pin	14–45	1	
C.1929	**Washer,** Spring (Double Coil) between Pivot Jaw and Tufnol Washer	14–46	1	
C.1908	**Washer** (Tufnol) at each side of Selector Lever	14–47	2	
C.21110	**Pivot Pin,** mounting Selector Lever in Pivot Jaw	14–48	1	
C.8737/4	**Nut,** Self-Locking, securing Pivot Pin	14–49	1	
C.1907	**Bush,** in Top Cover, for Pivot Jaw	14–50	1	
C.1894	**Washer** (Tufnol) at each side of Bush	14–51	2	
C.1914	**'D' Washer,** between Spring Washer and Fibre Washer	14–52	1	
C.1905	**Washer,** Spring (Double Coil) under Nut	14–53	1	
C.8737/4	**Nut,** Self-Locking, retaining Pivot Jaw to Top Cover	14–54	1	
C.28330	**Remote Control Shaft,** in Top Cover for Selector Lever	14–55	1	
C.27849	**Finger,** on Remote Control Shaft	14–56	1	
C.19772	**Screw,** Special, securing Finger to Remote Control Shaft	14–57	1	
C.837	**Locking Wire,** for Special Screw		As req'd	
C.27865	**CHANGE-SPEED LEVER**	14–58	1	
C.23007	**Knob,** on Change-Speed Lever	14–59	1	
C.23006	**Cone,** locking Knob to Lever	14–60	1	
C.22594	**Rubber Bush,** Upper, locating Change-Speed Lever in Selector Lever	14–61	1	
C.22395	**Washer,** Special, at top of Rubber Bush	14–62	1	
C.21921	**Bush Assembly,** (Metalastik) lower, locating Change-Speed Lever in Selector Lever	14–63	1	
C.27848	**Selector Lever,** for Change-Speed Lever	14–64	1	
C.8737/2	**Nut,** Self-Locking, securing Change-Speed Lever Assembly to Selector Lever	14–65	1	
C.26595	**Washer,** Special under Nut	14–66	1	
C.21939	**MAINSHAFT**	15–1	1	
C.20027	**Plug,** in end of Mainshaft	15–32	1	
C.20036	**Nut,** retaining 3rd Gear and 3rd/4th Synchro Hub on Mainshaft	15–2	1	
C.20037	**Tab Washer,** behind Nut	15–3	1	
C.22391	**Synchro Ring,** for 3rd and 4th speeds	15–29	2	
C.24640	**Operating Sleeve,** for 3rd and 4th speeds	15–23	1	Fitted to Engine No. 7A.13231 and subs. (Also to 7A.12185 and 7A.12186) 7K.1001 to 7K.2484
C.24639	**Sychro Hub,** for 3rd and 4th speeds	15–22	1	(as above)
C.28436	**Operating Sleeve,** for 3rd and 4th speeds	15–23	1	Fitted to Engine No. 7K.2485 and subs.
C.28432	**Synchro Hub,** for 3rd and 4th speeds	15–22	1	(as above)
C.24641	**Thrust Member,** between Sleeve and Synchro Hub	15–24	3	
C.21088	**Plunger,** under Thrust Member	15–25	3	
C.27491	**Spring,** under Plungers	15–26	3	
C.1319	**Detent Ball,** between Sleeve and Synchro Hub	15–27	3	
C.24181	**Spring,** tensioning Detent Balls	15–28	3	

GEARBOX

FOR OVERDRIVE TRANSMISSION

Part No.	Description	Plate No.	No. per Unit	Remarks
	MAINSHAFT *(contd.)*			
C.21082	**3rd Speed Gear,** on Mainshaft	15–18	1	Fitted to Engine No. 7A.13231 and subs. (Also to 7A.12185 and 7A.12186) 7K.1001 to 7K.2484
C.28417	**3rd Speed Gear,** on Mainshaft	15–18	1	Fitted to Engine No. 7K.2485 and subs.
C.21617	**Nut,** retaining 3rd Gear and 3rd/Top Synchro Hub on Mainshaft	15–30	1	
C.21618	**Tab Washer,** locking Nut	15–31	1	
C.23109/1	**Needle Roller,** for 3rd and 2nd Speed Gears (3 mm + .0001″ – .0000″)	15–19	212	**ONLY USE ROLLERS OF ONE GRADE IN INDIVIDUAL GEAR ASSEMBLIES**
C.23109/2	**Needle Roller,** for 3rd and 2nd Speed Gears 3 mm + .0000″ – .0001″)	15–19	212	
C.23109/3	**Needle Roller,** for 3rd and 2nd Speed Gears (3 mm – .0001″ – .0002″)	15–19	212	
C.23099	**Spacer,** for Needle Rollers in 3rd Speed Gear	15–21	1	
C.23098	**Spacer,** for Needle Rollers in 2nd Speed Gear	15–20	1	
C.21092	**2nd Speed Gear,** on Mainshaft	15–17	1	Fitted to Engine No. 7A.13231 and subs (Also to 7A.12185 and 7A.12186) 7K.1001 to 7K.2484
C.28430	**2nd Speed Gear,** on Mainshaft	15–17	1	Fitted to Engine No. 7K.2485 and subs.
C.22391	**Synchro Ring,** for 2nd and 1st speeds	15–16	2	
C.24640	**Operating Sleeve,** for 2nd and 1st speeds	15–10	1	Fitted to Engine No. 7A.13231 and subs. (Also to 7A.12185 and 7A.12186) 7K.1001 and 7K.2484
C.24638	**Synchro Hub,** for 2nd and 1st speeds	15–9	1	
C.28436	**Operating Sleeve,** for 2nd and 1st speeds	15–10	1	Fitted to Engine No. 7K.2485 and subs.
C.28432	**Synchro Hub,** for 2nd and 1st speeds	15–9	1	
C.24641	**Thrust Member,** between Sleeve and Synchro Hub	15–11	3	
C.21088	**Plunger,** under Thrust Member	15–12	3	
C.27491	**Spring,** under Plungers	15–13	3	
C.1319	**Detent Ball,** between Sleeve and Synchro Hub	15–14	3	
C.24181	**Spring,** tensioning Detent Balls	15–15	3	
C.21095	**Bearing Sleeve,** for 1st Speed Gear	15–6	1	
C.23109/1	**Needle Roller,** for 1st Speed Gear (3 mm + .0001″ – .0000″)	15–7	120	**ONLY USE ROLLERS OF ONE GRADE IN INDIVIDUAL GEAR ASSEMBLIES**
C.23109/2	**Needle Roller,** for 1st Speed Gear (3 mm + .0000″ – .0001″)	15–7	120	
C.23109/3	**Needle Roller,** 1st Speed Gear (3 mm – .0001″ – .0002″)	15–7	120	
C.23097	**Spacer,** for Needle Rollers	15–8	1	
C.21096	**1st Speed Gear,** on Mainshaft	15–5	1	Fitted to Engine No. 7A.13231 and subs. (Also to 7A.12185 and 7A.12186) 7K.1001 to 7K.2484
C.28429	**1st Speed Gear,** on Mainshaft	15–5	1	Fitted to Engine No. 7K.2485 and subs.
C.21098	**Reverse Gear,** on Mainshaft	15–4	1	
C.25436	CONSTANT PINION SHAFT	15–33	1	Fitted to Engine No. 7A.13231 and subs. (Also to 7A.12185 and 7A.12186) 7K.1001 to 7K.2484
C.28415	CONSTANT PINION SHAFT		1	Fitted to Engine No. 7K.2485 and subs.
C.21617	**Nut,** securing Constant Pinion Shaft in Bearing	15–36	1	
C.21618	**Tab Washer,** under Nut	15–37	1	
C.20493	**Oil Thrower,** on Constant Pinion Shaft	15–35	1	
C.25424	**Roller Bearing** (for Mainshaft) inside Constant Pinion Shaft	15–34	1	

Part No.	Description	Plate No.	No. per Unit	Remarks
C.20052	REVERSE SPINDLE	15–38	1	
C.20313	**Reverse Idler Gear,** complete with Bushes	15–40	1	
C.21198	**Lever Assembly,** operating Reverse Gear	15–41	1	
UFS.137/8R	**Screw,** Set, pivoting Lever	15–42	1	
C.21200	**Tab Washer,** locking Setscrew	15–44	1	
C.930	**Fibre Washer,** under Tab Washer	15–43	1	
C.866	**Reverse Slipper,** between Lever Assembly and Reverse Idler Gear	15–45	1	
L.103/8U	**Split Pin,** retaining Slipper in Lever	15–46	1	
C.753	**Woodruff Key,** locking Spindle in Gearbox Case	15–39	1	
C 25437	COUNTERSHAFT	15–47	1	
C.920	**Fibre Washer,** in Gear Case, at front of Countershaft		1	
C.1862/1	**Thrust Washer,** for end float at front end of Countershaft (.159" thick)	15–54	1	SELECTIVE SIZES. USE AS NECESSARY TO OBTAIN .004"/.006" END FLOAT
C.1862/2	**Thrust Washer,** for end float at front end of Countershaft (.162" thick)	15–54	1	
C.1862/3	**Thrust Washer,** for end float at front end of Countershaft (.152" thick)	15–54	1	
C.1862/4	**Thrust Washer,** for end float at front end of Countershaft (.164" thick)	15–54	1	
C.1861	**Thrust Washer** Assembly, Inner, at front end of Countershaft	15–53	1	
C.918/1	**Needle Roller,** at each end of Gear Unit (Cluster) (3 mm + .0001" – .0000")	15–50	58	ONLY USE ROLLERS OF ONE GRADE IN INDIVIDUAL GEAR UNIT ASSEMBLIES
C.918/2	**Needle Roller,** at each end of Gear Unit (Cluster) (3 mm + .0000" – .0001")	15–50	58	
C.918/3	**Needle Roller,** at each end of Gear Unit (Cluster) (3 mm – .0001" – .0002")	15–50	58	
C.21099	**Spacer,** for Needle Rollers	15–51	3	
C.20712	**Gear Unit** (Cluster) on Countershaft	15–49	1	Fitted to Engine No. 7A.13231 and subs. (Also to 7A.12185 and 7A.12186) 7K.1001 to 7K.2484
C.28416	**Gear Unit** (Cluster) on Countershaft	15–49	1	Fitted to Engine No. 7K.2485 and subs.
C.21100	**Thrust Washer Assembly,** at rear of Gear Unit	15–52	1	
C.753	**Woodruff Key,** locking Countershaft in Gearbox Case	15–48	1	

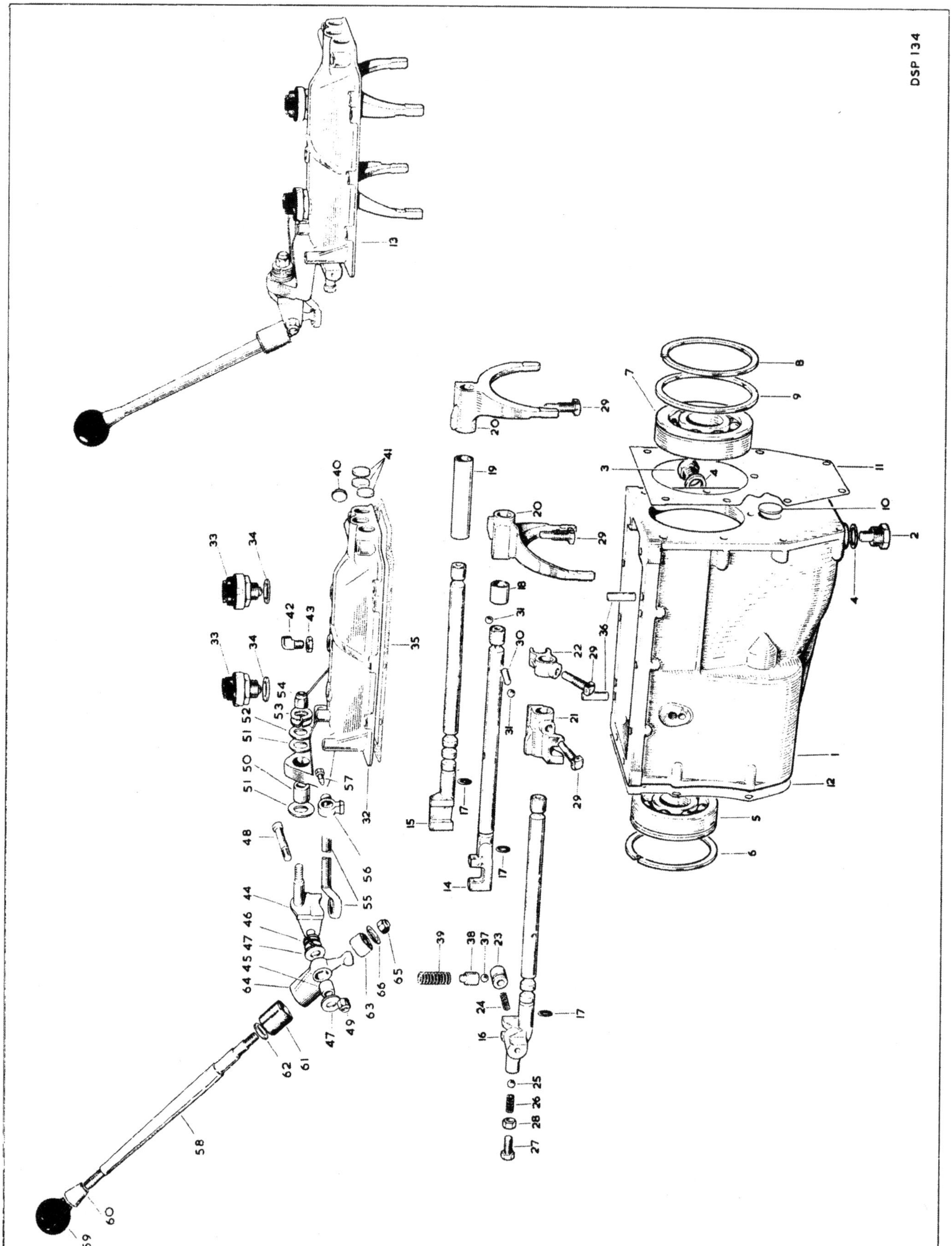

PLATE 14

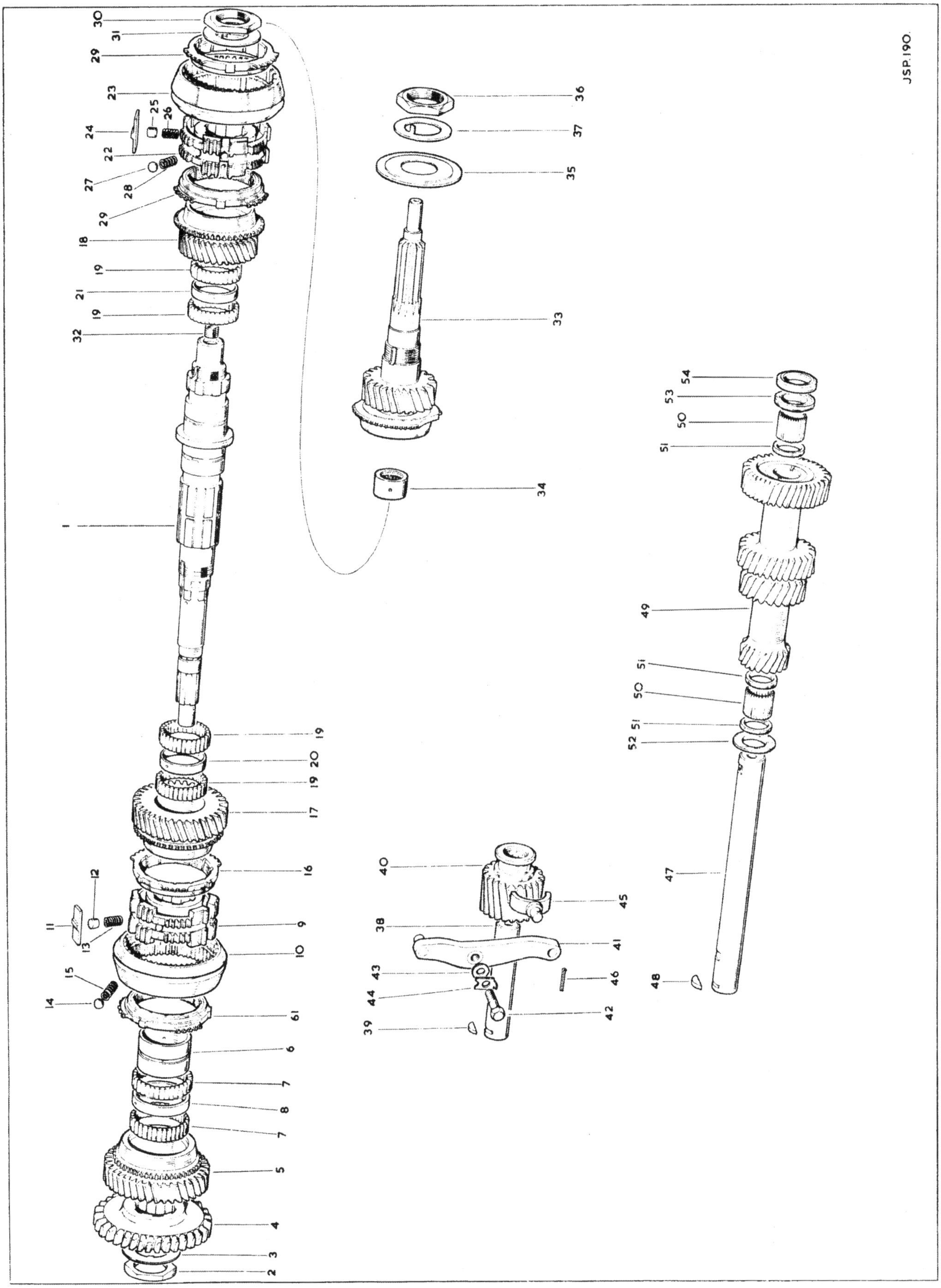

PLATE 15

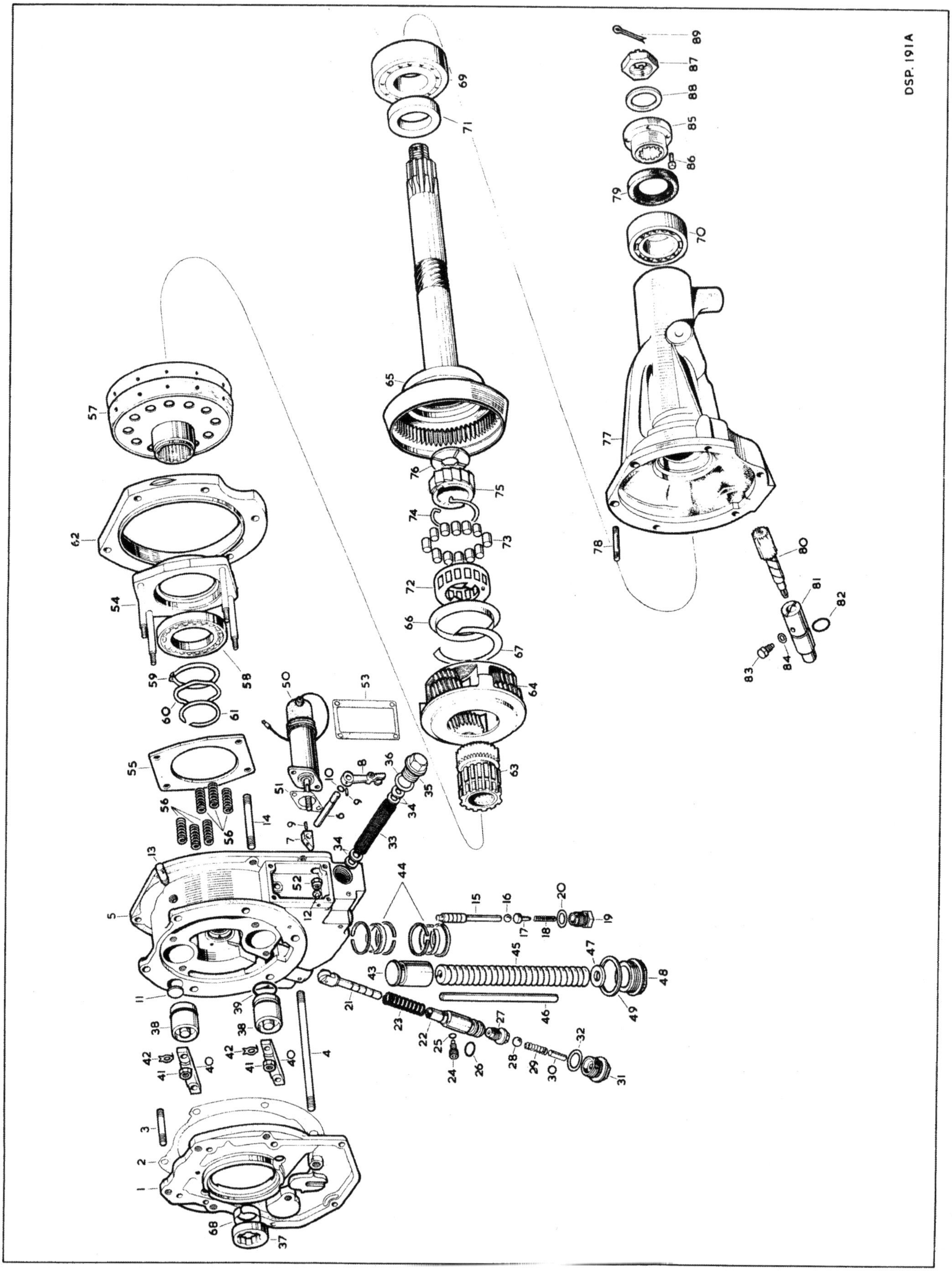

PLATE 16

Part No.	Description	Plate No.	No. per Unit	Remarks
C.25504	**OVERDRIVE UNIT, COMPLETE (28/61743)**		1	Fitted to Engine No. 7A.13231 and subs. (Also to 7A.12185 and 7A.12186) 7K.1001 and subs.
10811	**ADAPTOR PLATE FOR ATTACHMENT OF OVERDRIVE UNIT TO GEARBOX (60439)**	16–1	1	
10812	**Gasket,** between Adaptor Plate and Front Casing (60456)	16–2	1	
10813	**Stud,** fixing Adaptor Plate to Front Casing (63603/2)	16–3	4	
10814	**Stud,** fixing Adaptor Plate to Rear Casing (60395)	16–4	2	
UFN.131/L	**Nut,** on Studs (127/2/A)		6	
FG.105/X	**Washer,** Spring, under Nuts (1402)		6	
10815	**FRONT CASING ASSEMBLY (65904)**	16–5	1	
10816	**Shaft,** for Main Operating Valve (60388)	16–6	1	
10817	**Cam,** on Shaft (60595)	16–7	1	
10818	**Lever,** on Shaft (60387)	16–8	1	
10819	**Roll Pin,** securing Cam and Lever to Shaft (138/39)	16–9	2	
10378	**'O' Ring,** on Shaft between Lever and Casing (64486)	16–10	1	
10820	**Welch Washer,** sealing aperture in front of Casing (61068)	16–11	1	
C.5675	**Stop** (Rubber) for Solenoid (64283)	16–12	1	
C.5671	**Breather,** at top of Casing (64370)	16–13	1	
10822	**Stud,** fixing Front Casing to Rear Casing (60393)	16–14	2	
UFN.131/L	**Nut,** on Studs of Front Casing (127/2/A)		2	
FG.105/X	**Washer,** Spring, under Nuts (1402)		2	
10968	**MAIN OPERATIVE VALVE (60686)**	16–15	1	
C.5922	**Ball,** 5/16" diameter, at bottom of Valve (9809)	16–16	1	
C.5919	**Plunger,** below Ball (60036)	16–17	1	
7972	**Spring,** under Plunger (60440)	16–18	1	
10824	**Plug,** retaining Valve Parts (60385)	16–19	1	
C.5903	**Washer** (Copper) on Plug (60024)	16–20	1	
10825	**OIL PUMP PLUNGER ASSEMBLY IN FRONT CASING (65906)**	16–21	1	
10826	**Body,** for Oil Pump (60450)	16–22	1	
C.5938	**Spring,** tensioning Oil Pump Plunger (60536)	16–23	1	
10827	**Screw,** retaining Oil Pump Body in Front Casing (60399)	16–24	1	
10828	**Washer** (Fibre) on Screw (1678)	16–25	1	
10829	**'O' Ring,** around Oil Pump Body (131/3)	16–26	1	
10830	**NON-RETURN VALVE KIT (65851)**		1	
11090	**Body,** for Non-Return Valve (64548)	16–27	1	
C.9819	**Ball,** 7/32" diameter, for Non-Return Valve (9811)	16–28	1	
10831	**Spring,** for Non-Return Valve (60334)	16–29	1	
11089	**Support Rod,** for Spring (6662)	16–30	1	
10832	**Plug,** retaining Valve Parts (60452)	16–31	1	
10833	**Washer** (Copper) on Plug (60453)	16–32	1	
10834	**FILTER, IN FRONT CASING (60525)**	16–33	1	
10835	**MAGNETIC RING, AT EACH END OF FILTER (65902) (SET OF FOUR RINGS)**	16–34	1 set	
10836	**PLUG, RETAINING FILTER (60524)**	16–35	1	
10837	**Washer,** on Plug (60531)	16–36	1	
C.23406	**CAM, OPERATING OIL PUMP (60447)**	16–37	1	

OVERDRIVE UNIT

Part No.	Description	Plate No.	No. per Unit	Remarks
10838	**OPERATING PISTON ASSEMBLY IN FRONT CASING (65905)**	16–38	1	
7949	**'O' Ring**, on Operating Pistons (61647)	16–39	2	
10839	**Bridge Piece**, retaining Operating Pistons (60357)	16–40	2	
UFN.125/L	**Nut**, securing Bridge Pieces (65285)	16–41	4	
C.5965	**Tab Washer**, locking Nuts (65300)	16–42	4	
10840	**ACCUMULATOR PISTON ASSEMBLY COMPLETE WITH RINGS (65896)**	16–43	1	
C.5677	**Piston Ring Sets** (60814) comprising one wide and two narrow rings per set	16–44	2 sets	SUPPLIED ONLY IN SETS
11129	**SPRING, FOR ACCUMULATOR PISTON (61189)**	16–45	1	
10841	**Support Rod**, inside Accumulator Spring (60582)	16–46	1	
11334	**Packing Washer**, (.128" thick) at bottom of Accumulator Spring (64648/1)	16–47	As req'd	SELECTIVE SIZES
11335	**Packing Washer** (.056" thick) at bottom of Accumulator Spring (64648/2)	16–47	As req'd	SELECTIVE SIZES
11336	**Packing Washer** (.028" thick) at bottom of Accumulator Spring (64648/3)	16–47	As req'd	SELECTIVE SIZES
11337	**Packing Washer** (.104" thick) at bottom of Accumulator Spring (64648/4)	16–47	As req'd	SELECTIVE SIZES
11338	**Packing Washer** (.160" thick) at bottom of Accumulator Spring (64648/5)	16–47	As req'd	SELECTIVE SIZES
11339	**Packing Washer** (.048" thick) at bottom of Accumulator Spring (64648/6)	16–47	As req'd	SELECTIVE SIZES
10842	**Plug**, retaining Accumulator Spring (61345)	16–48	1	
11128	**Washer**, on Plug (61347)	16–49	1	
10002	**SOLENOID (76522/H-115) (65210)**	16–50	1	
9240	**Gasket**, between Solenoid and Front Casing (60121)	16–51	1	
UFS.119/4R	**Screw**, set, securing Solenoid to Front Casing		2	
AG.102/X	**Washer**, Spring, on Setscrews		2	
C.8737/1	**Nut**, for Solenoid Plunger Adjustment (5721)	16–52	1	
10844	**Gasket**, under Solenoid Inspection Cover (Name Plate) (60397)	16–53	1	
UFS.119/4R	**Screw**, Set, securing Inspection Cover to Front Casing		4	
C.723/A	**Washer**, Shakeproof, on Setscrews		4	
10845	**THRUST RING ASSEMBLY FOR CLUTCH (65899)**	16–54	1	
10846	**Plate**, retaining Thrust Ring to Bearing on Sliding Member (60372)	16–55	1	
11303	**SPRING, ON PINS OF THRUST RING (65966) (SUPPLIED ONLY IN SETS OF SIX SPRINGS)**	16–56	1 set	
10848	**SLIDING MEMBER ASSEMBLY, FOR CLUTCH (65920)**	16–57	1	
C.5968	**Ball Bearing**, on Sliding Member (65024)	16–58	1	
C.5977	**Circlip** (1 7/8" External) securing Ball Bearings (9003)	16–59	1	
9152	**Corrugated Washer**, on Sun Wheel, forward of Sliding Member (63616)	16–60	1	
10849	**Snap Ring**, retaining Sliding Member to Sun Wheel (60375)	16–61	1	
10850	**BRAKE RING (60438)**	16–62	1	
10851	**SUN WHEEL ASSEMBLY (65900)**	16–63	1	

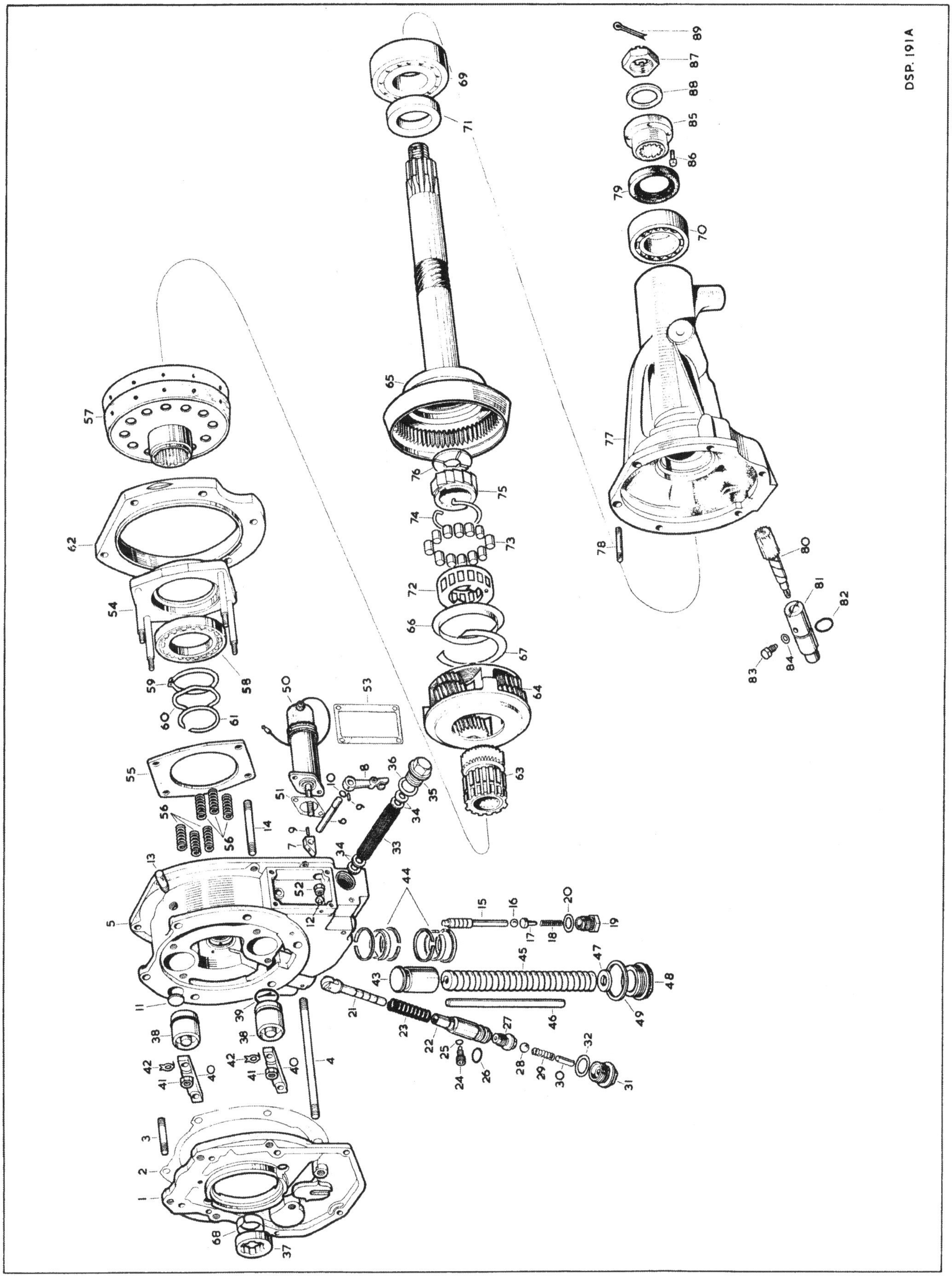

PLATE 16

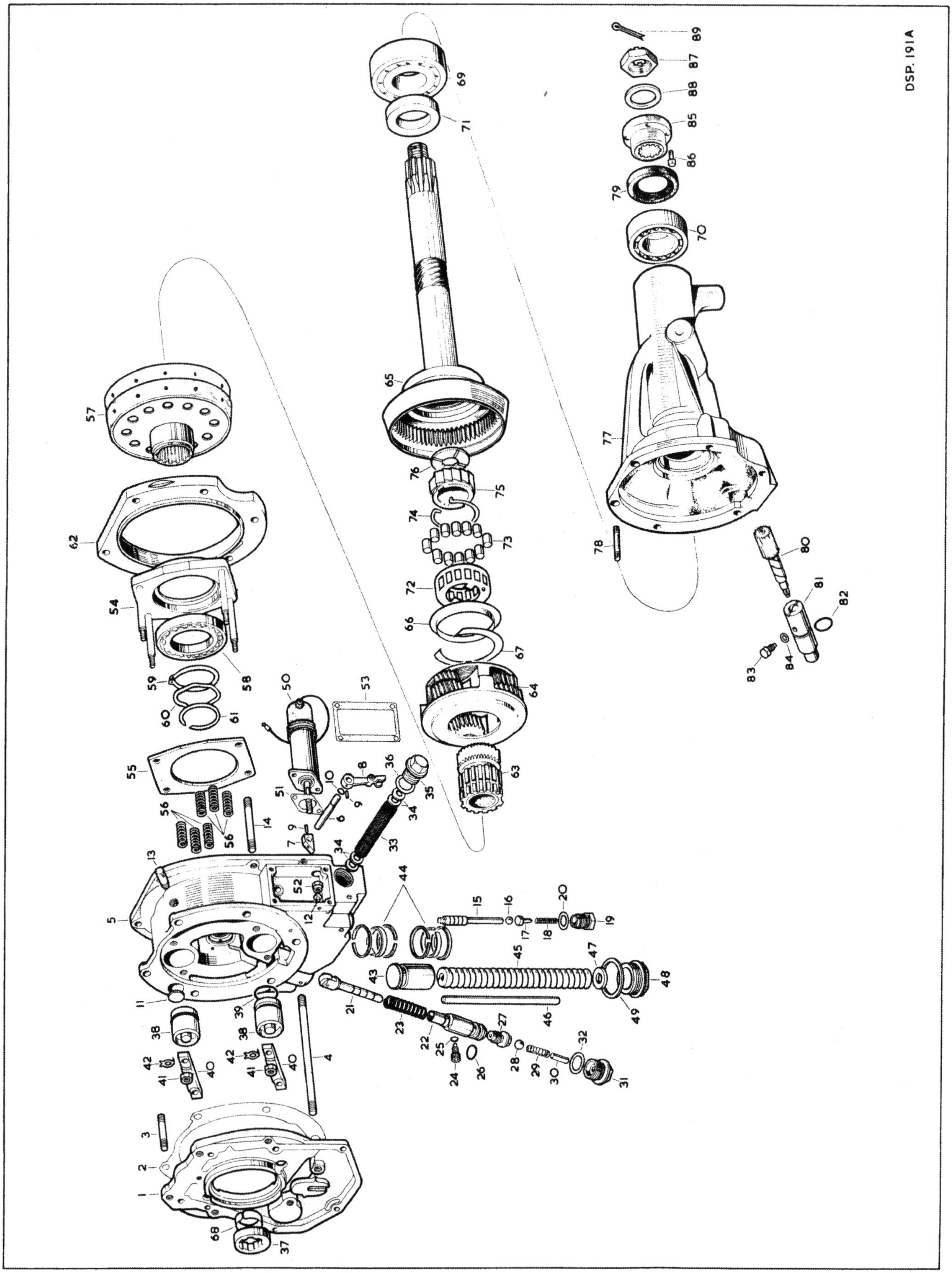

PLATE 16

Part No.	Description	Plate No.	No. per Unit	Remarks
10852	**PLANETARY CARRIER ASSEMBLY (65901)**	16–64	1	
10853	**Shaft,** for Planetary Bearings (60380)		3	
10854	**Mills Pin,** securing Shafts (6380)		3	
10855	**Bearing** (Torrington) inside Planetary Gears (62555)		6	
11125	**Spacing Collar,** between Bearings (63743)		3	
7979	**Thrust Washer,** behind Planetary Gears (65336)		3	
11124	**ANNULUS ASSEMBLY (65923)**	16–65	1	
10857	**Oil Thrower,** in Annulus (63714)	16–66	1	
9117	**Spring Ring,** securing Oil Thrower (61619)	16–67	1	
C.18957	**Spring Clip,** preventing forward movement of Annulus and Planetary Carrier	16–68	1	
C.5969	**BALL BEARING, ON SHAFT ADJACENT TO ANNULUS (1¼" x 2¾" x 11/16") (65025)**	16–69	1	
C.5687	**BALL BEARING, AT REAR END OF ANNULUS SHAFT (30 x 62 x 16 MM.) (65026)**	16–70	1	
C.5981	**Spacing Washer,** behind Bearing (.146" thick) (60657/E)	16–71	1	SELECTIVE SIZES
C.5694	**Spacing Washer,** behind Bearing (.151" thick) (60657/F)	16–71	1	SELECTIVE SIZES
C.5695	**Spacing Washer,** behind Bearing (.156" thick) (60657/G)	16–71	1	SELECTIVE SIZES
C.5696	**Spacing Washer,** behind Bearing (.161" thick) (60657/H)	16–71	1	SELECTIVE SIZES
C.5697	**Spacing Washer,** behing Bearing (.166" thick) (60657/J)	16–71	1	SELECTIVE SIZES
C.5956	**CAGE FOR UNI-DIRECTIONAL CLUTCH (64011)**	16–72	1	
C.5967	**Roller,** for Cage (12 rollers per set) (65534)	16–73	1 set	
9116	**Spring,** for Cage (60676)	16–74	1	
C.5957	**INNER MEMBER FOR UNI-DIRECTIONAL CLUTCH (FREE WHEEL) (64010)**	16–75	1	
C.5942	**Thrust Washer,** behind Inner Member (60634)	16–76	1	
11126	**REAR CASING ASSEMBLY (65924)**	16–77	1	
10861	**Stud,** fixing Rear Casing to Front Casing (60489)	16–78	2	
UFN.131/L	**Nut,** on Studs of Rear Casing (127/2/A)		2	
FG.105/X	**Washer,** Spring, under Nuts (1402)		2	
C.19118	**Location Pin,** for Rear Engine Mounting in Rear Casing		1	
C.22646/1	**Spring Retainer,** on Location Pin		1	
UFN.156/L	**Nut,** on Location Pin, securing Spring Retainer		1	
C.5686	**OIL SEAL, AT END OF REAR CASING (65357)**	16–79	1	
11130	**SPEEDOMETER DRIVEN GEAR (61191)**	16–80	1	
C.11927	**BEARING ASSEMBLY FOR SPEEDOMETER DRIVEN GEAR (65629)**	16–81	1	
C.11211	**'O' Ring,** around Bearing (64564)	16–82	1	
C.5680	**Screw,** retaining Bearing in Rear Casing (61082)	16–83	1	
7978	**Washer** (Copper) on Screw (64941)	16–84	1	

OVERDRIVE UNIT

Part No.	Description	Plate No.	No. per Unit	Remarks
11131	**FLANGE, CONNECTING OVERDRIVE UNIT TO PROPELLER SHAFT (61415)**	16–85	1	
C.19827	**Bolt,** in Flange, for connection of Propeller Shaft	16–86	4	
C.5681	**Nut,** Slotted, securing Flange to Annular Shaft (61085)	16–87	1	
11132	**Washer,** under Slotted Nut (65280)	16–88	1	
L.104/12U	**Split Pin,** locking Slotted Nut (6022)	16–89	1	

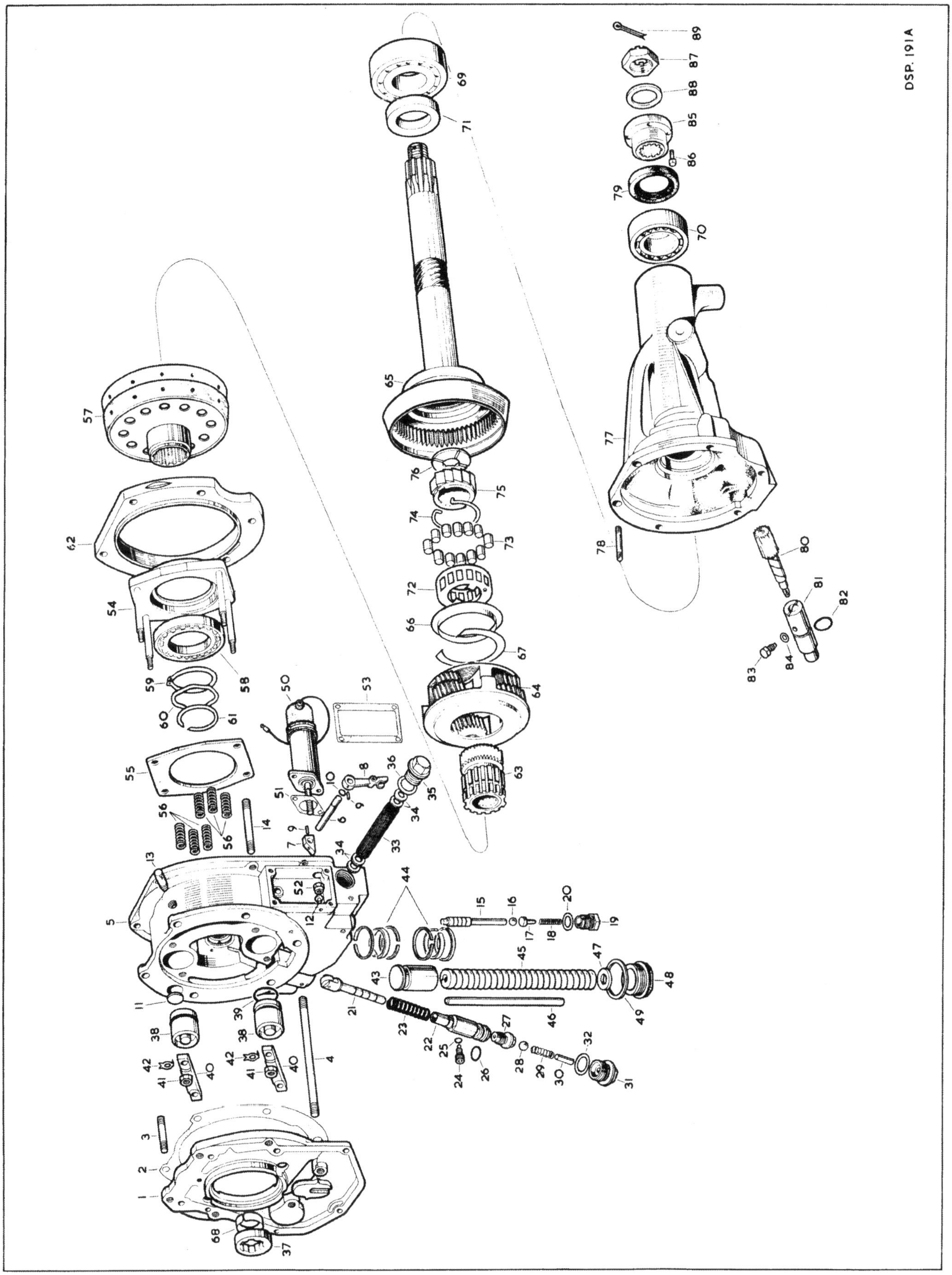

PLATE 16

AUTOMATIC TRANSMISSION UNIT
(SINGLE DRIVE RANGE)

Part No.	Description	Plate No.	No. per Unit	Remarks
C.21700	**TRANSMISSION UNIT ASSEMBLY, COMPLETE WITH CONVERTER (AS1-35EG) (35EG-509)**		1	Fitted from Engine No. 7A.1001 to 7A.3592
C.21004	**Converter Assembly**, only (35EG-1A368)	17–1	1	
C.21612	**Converter Housing** (35EG-1½C)		1	
C.21003	**Stoneguard**, over apertures in Converter Housing (35EA-393)		2	
C.21053	**Screw**, securing Stoneguards (EA.4626H)		4	
C.21054	**Nut**, on Screws (35EA-394)		4	
C.21055	**Screw and Spring Lockwasher**, securing Converter Housing to Transmission Case (443609)		6	
9768	**TRANSMISSION CASE ASSEMBLY (35EA-513/1)**	17–2	1	
9769	**Oil Seal**, for Manual Control Shaft (35EA-108)	17–3	2	
9869	**Plug**, for Pressure Take-Off Point (444688)	17–4	1	
C.21010	**Union**, for Oil Outlet and Return Pipes (35EA-293)	17–6	2	
C.21009	**Kickdown Cable Assembly** (35EG-A284A)	17–5	1	
11477	**Washer** (Copper) at fixing of Outer Cable to Transmission Case (EA.5022B)		1	
9870	**Screw**, for adjustment of Rear Servo (4626Y)	17–7	1	
9871	**Nut**, locking Adjusting Screw (220067)	17–8	1	
C.21007	**Adaptor**, for Dipstick Tube (35EA-385A)	17–9	1	
C.21008	**Switch**, for Starter Cut-Out and Reverse Lamp (35EA-A286C)	17–10	1	
9770	**Manual Control Shaft**, carrying Detent Lever (35EA-81A)	17–11	1	
9771	**Collar**, positioning Manual Shaft in Case (35EA-179A)	17–12	1	
9872	**Roll Pin**, securing Collar to Manual Shaft (273337)	17–13	1	
9772	**Lever Assembly**, on Shaft, for Manual Valve Detent (35EA-A78C)	17–14	1	
9873	**Roll Pin**, retaining Detent Lever (273436)	17–15	1	
9773	**Spring**, on Manual Control Shaft, between Detent Lever and Case (35EA-85)	17–16	1	
9874	**Ball**, (3/8" dia.) positioning Detent Lever (453593)	17–17	1	
9774	**Spring**, tensioning Ball (35EA-83)	17–18	1	
9775	**Link**, connecting Detent Lever to Torsion Lever (35EA-266A)	17–19	1	**May be replaced with 11787 and 11788** (9775, 9875, 9876)
9875	**Clip**, retaining Link (4857A)	17–20	2	
9876	**Torsion Lever**, for Parking Brake Linkage (3-261)	17–21	1	
11787	**Link**, connecting Detent Lever to Torsion Lever (35EA-266C)		1	Supersedes 9775, 9875, 9876 (11787, 11788)
11788	**Torsion Lever**, for Parking Brake Linkage (35EA-261)		1	
9877	**Spring**, providing tension between Torsion Lever and Forked Lever (3-264A)	17–22	1	
9878	**Forked Lever**, engaging Ball Pin in Toggle Lever (3-A262)	17–23	1	
9879	**Clip**, retaining Torsion Lever, Forked Lever and Spring (4743A)	17–24	1	
9890	**Washer**, Plain, under Clip (3-321)	17–25	1	
9880	**Toggle Lever**, for Parking Brake Pawl (3-272)	17–26	1	
9881	**Toggle Pin** (Shouldered) securing Toggle Lever (35-276)	17–27	1	
9882	**'O'Ring**, sealing Toggle Pin (4804P)	17–28	1	
9883	**Roll Pin**, retaining Toggle Pin in Case (EA.5004A)	17–29	1	
9884	**Ball Pin**, securing Link to Toggle Lever (3-274)	17–30	1	**May be replaced with 10577 and 10578** (9884 to 9887)
9885	**Collar**, on Ball Pin, carrying Pawl Tension Spring (3-270)	17–31	1	
9886	**Spring**, on Ball Pin, tensioning Parking Brake Pawl (3-271)	17–32	1	
9887	**Clip**, retaining Ball Pin (4743)	17–33	1	
10577	**Ball Pin**, securing Link to Toggle Lever (35EA-274)		1	
10578	**Spring**, on Ball Pin, tensioning Parking Brake Pawl (8-271)		1	
9888	**Link**, connecting Toggle Lever to Parking Brake Pawl (3-273)	17–34	1	
9776	**Pawl**, for Parking Brake (35-268)	17–35	1	
9777	**Anchor Pin**, securing Parking Brake Pawl (35-269)	17–36	1	
9889	**Pin**, securing Pawl to Link (3-275)	17–37	1	
C.21139	**SELECTOR LEVER ASSEMBLY ON TRANSMISSION UNIT**		1	**R.H. DRIVE CARS ONLY**

Part No.	Description	Plate No.	No. per Unit	Remarks
C.21142	**SELECTOR LEVER ASSEMBLY ON TRANSMISSION UNIT**		1	L.H. DRIVE CARS ONLY
C.8737/2	**Nut**, Self-Locking, securing Selector Lever to Manual Control Shaft		1	
FW.105/T	**Washer**, Plain, under Nut		1	
9778	**EXTENSION CASE ASSEMBLY (35EG-1A7A)**	17–38	1	
9891	**Oil Seal**, at rear end of Extension Case (R39-110)	17–39	1	
9779	**Gasket**, between Extension Case and Transmission Case (35-146)	17–40	1	
9892	**Bolt and Spring Lockwasher**, securing Extension Case to Transmission Case (431922)		4	
C.21296	**SPEEDOMETER DRIVEN GEAR**	17–41	1	
C.21297	**Bearing**, for Speedometer Driven Gear		1	
C.12377/1	**Oil Seal**, at bottom of Bearing		1	
C.21058	**'O' Ring**, around Speedometer Driven Gear Bearing	17–42	1	
C.21073	**Plate**, retaining Driven Gear in Extension Case	17–43	1	
UCS.125/4R	**Screw**, Set, securing Plate		2	
FG.104/X	**Washer**, Spring, on Setscrews		2	
9780	**FLANGE, COUPLING TRANSMISSION UNIT TO PROPELLER SHAFT (35EG-298)**	17–44	1	
9781	**Screw** (Wedglok) Socket Head, securing Flange to Transmission Drive Shaft (EG.5034)	17–45	1	
9782	**Washer**, Special, on Screw (EG.5003)	17–46	1	
9783	**FRONT PUMP ASSEMBLY (35EA-1A60B)**	17–47	1	
10427	**Oil Seal**, for Front Pump (35EA-62A)		1	
9785	**'O' Ring**, for Front Pump (EA4804D)		1	
9893	**Screw and Lockwasher**, for Assembly of Front Pump (443609)		5	
9894	**Screw and Lockwasher**, for Assembly of Front Pump (190601)		1	
11490	**Oil Inlet Tube**, for Front Pump (35-87A)		1	
9786	**'O' Ring**, for Front Pump Inlet Tube (4804Q)		1	
9787	**Gasket**, between Front Pump and Transmission Case (35-61)	17–49	1	
9893	**Screw and Lockwasher**, securing Front Pump Assembly to Transmission Case (443609)		6	
9788	**Thrust Washer**, between Front Pump and Input Shaft (35-15B)	17–50	1	SELECTIVE SIZES
9789	**Thrust Washer**, between Front Pump and Input Shaft (35-15C)		1	SELECTIVE SIZES
9855	**OIL TUBE, RETURNING OIL FROM CONVERTER TO OIL COOLER (35EA-297)**	17–48	1	
11893	**FRONT CLUTCH ASSEMBLY (35EE-1A70F)**		1	
9791	**Piston Assembly**, for Front Clutch (35EA-A35)	17–51	1	
9792	**Cylinder**, for Front Clutch (35FC-70)	17–52	1	
9793	**'O' Ring**, Inner, for Piston (EA-4804E)	17–53	1	
9896	**Sealing Ring**, Outer, for Piston (4805C)	17–54	1	
9794	**Split Ring**, seating Front Clutch Spring (35-33)	17–55	1	
11894	**Spring**, for Front Clutch (35-37)	17–56	1	
11895	**Snap Ring**, retaining Front Clutch Spring (4763F)	17–57	1	
N.S.S.	Pressure Plate, for Front Clutch (35EA-175B)	17–58	1	NOT SUPPLIED SEPARATELY USE SERVICE KIT 11908 SEE PAGE 73
N.S.S.	Clutch Plate, (Drive), for Front Clutch (35-65)	17–59	3	NOT SUPPLIED SEPARATELY USE SERVICE KIT 11908 SEE PAGE 73
N.S.S.	Clutch Plate, (Friction), for Front Clutch (35-A166)	17–60	4	NOT SUPPLIED SEPARATELY USE SERVICE KIT 11908 SEE PAGE 73
9799	**Hub**, carrying Front Clutch Plates (35-40)	17–61	1	
9800	**Thrust Washer**, between Hub and Input Shaft (35-17)	17–62	1	
9801	**Input Shaft**, (35-A16)	17–63	1	
9897	**Snap Ring**, retaining Input Shaft in Front Clutch Cylinder (4763D)	17–64	1	

AUTOMATIC TRANSMISSION UNIT

(SINGLE DRIVE RANGE)

Part No.	Description	Plate No.	No. per Unit	Remarks
9802	**REAR CLUTCH AND FRONT DRUM ASSEMBLY (35EG-2A50)**		1	
9803	**Front Drum Assembly** (35EG-514)	17–65	1	
9898	**Sealing Ring,** on Front Drum Assembly (4806M)	17–66	3	
9905	**Needle Bearing,** in Front Drum (4840L)	17–67	1	
9804	**Piston Assembly,** for Rear Clutch (35EP-A45)	17–68	1	
9899	**'O' Ring,** Inner, for Piston (4804S)	17–69	1	
9900	**Sealing Ring,** Outer, for Piston (4805E)	17–70	1	
9805	**Spring,** for Rear Clutch (35-47)	17–71	1	
9806	**Seat,** for Rear Clutch Spring (35-48)	17–72	1	
9901	**Snap Ring,** retaining Spring Seat (4653E)	17–73	1	
N.S.S.	Clutch Plate (Friction) for Rear Clutch (35-A166)	17–74	3	**NOT SUPPLIED SEPARATELY**
N.S.S.	Clutch Plate (Friction) for Rear Clutch (35EA-A166)	17–75	2	**USE SERVICE KIT 9865**
N.S.S.	Clutch Plate (Drive) for Rear Clutch (35-165)	17–76	5	**SEE PAGE 73**
N.S.S.	Pressure Plate, for Rear Clutch (35-67)	17–77	1	
9897	**Snap Ring,** retaining Rear Clutch Plates in Front Drum (4763D)	17–78	1	
9810	**Thrust Washer** (Bronze) between Front Clutch Cylinder and Front Drum (35-15)	17–79	1	
9811	**Thrust Washer** (Steel) at rear of Bronze Thrust Washer (35-21)	17–80	1	
9812	**BRAKE BAND ASSEMBLY FOR FRONT DRUM (35EG-1A155)**	17–81	1	
9813	**Strut** (Servo) operating Brake Band (35-156C)	17–82	1	
9814	**FRONT SERVO ASSEMBLY (35EE-A120)**		1	
9815	**Body,** for Front Servo (35EE-120)	17–83	1	
9816	**Lever,** operating Front Servo Strut (35EA-127)	17–84	1	
9817	**Pivot Pin,** for Operating Lever (35-118)	17–85	1	
9902	**Roll Pin,** securing Pivot Pin in Servo Body (3-138)	17–86	1	
9903	**Screw,** adjusting Front Servo (4626W)	17–87	1	
9904	**Nut,** locking Adjusting Screw (114503)	17–88	1	
9818	**Return Spring,** for Front Servo Piston (35EA-125)	17–89	1	
9819	**Piston Assembly,** for Front Servo (35EG-515)	17–90	1	
9820	**'O' Ring** (Small) on Piston (EA-4804C)	17–91	1	
9821	**'O' Ring** (Large) on Piston (EE-4804)	17–92	1	
9822	**Piston Sleeve,** for Front Servo (35EE-119)	17–93	1	
9823	**Sealing Ring,** for Piston Sleeve (EA-4805A)	17–94	1	
9824	**Snap Ring,** Retaining Piston and Sleeve in Body (EE.4763)	17–95	1	
9825	**Bolt and Spring Lockwasher,** securing Servo to Transmission Case (EA.4626F)		1	
9892	**Bolt and Spring Lockwasher,** securing Servo to Transmission Case (431922)		1	
11897	**FORWARD SUN GEAR ASSEMBLY (35EA-A104)**	17–96	1	
10431	**Sealing Ring,** on rear end of Shaft (4806P)	17–98	1	
9906	**Sealing Ring,** on front end of Shaft (4806L)	17–97	2	
9908	**Thrust Bearing** (Needle) between Front Drum and Sun Gear (4830B)	17–99	1	
9867	**CENTRE SUPPORT ASSEMBLY (35FC-A75)**	17–100	1	
9909	**Screw,** securing Centre Support in Transmission Case (4626X)	17–101	2	
9910	**Lockwasher,** on Screw (X.2978DR)		2	
11898	**PLANETARY GEARS AND REAR DRUM ASSEMBLY (35EA-1A150B)**	17–102	1	
9827	**Outer Race Assembly,** for One-Way Clutch (35-A148)	17–103	1	
9897	**Snap Ring,** retaining Outer Race (4763D)	17–104	1	
9808	**Thrust Bearing** (Needle) between Sun Gear and Planet Gear Cover (4830B)	17–105	1	
11508	**Plate,** for Thrust Bearing (35-20B)	17–106	1	

Part No.	Description	Plate No.	No. per Unit	Remarks
9829	**ONE-WAY CLUTCH ASSEMBLY IN REAR DRUM (35-A275)**	17–107	1	
9812	**BRAKE BAND FOR REAR DRUM (35EG-1A155)**	17–108	1	
9813	**Strut** (Servo) operating Brake Band (35-156C)	17–109	1	
9830	**REAR SERVO ASSEMBLY (35-1A130)**		1	
9831	**Body**, for Rear Servo (35-130)	17–110	1	
9832	**Lever**, operating Rear Servo Strut (35EA-137B)	17–111	1	
9817	**Pivot Pin**, for Operating Lever (35-118)	17–112	1	
9902	**Roll Pin**, securing Pivot Pin in Servo Body (3-138)	17–113	1	
9833	**Piston**, for Rear Servo (35-131)	17–114	1	
9911	**'O' Ring**, on Piston (4804U)	17–115	1	
9834	**Return Spring**, for Rear Servo Piston (35EA-135)	17–116	1	
9835	**Retainer Plate**, for Return Spring (35-134)	17–117	1	
9912	**Snap Ring**, securing Retainer Plate (4766B)	17–118	1	
9913	**Screw**, securing Rear Servo to Transmission Case (4626V)		1	
9914	**Lockwasher**, on Screw (103321)		1	
11515	**Bolt and Spring Lockwasher**, securing Rear Servo to Transmission Case (EA-5023B)		1	
10439	**Magnet**, on Rear Servo Fixing Bolt (8-199)		1	
9836	**RING GEAR, DRIVING MAINSHAFT (35-6)**	17–119	1	
11899	**MAINSHAFT ASSEMBLY (35EG-1A2B)**	17–120	1	
9916	**Snap Ring**, retaining Mainshaft in Ring Gear (4763A)	17–121	1	SELECTIVE SIZES USE SNAP RING OF DESIRED THICKNESS
9917	**Snap Ring**, retaining Mainshaft in Ring Gear (4763B)	17–121	1	
9918	**Snap Ring**, retaining Mainshaft in Ring Gear (4763C)	17–121	1	
9919	**Thrust Bearing** (Needle) between Planet Gear Cover and Mainshaft (4830C)	17–122	1	
9838	**Plate**, for Thrust Bearing (35-20A)	17–123	1	
9839	**Thrust Washer**, between Mainshaft and Transmission Case (35EA-159A)	17–124	1	
9920	**Sealing Ring**, on Mainshaft for Rear Pump Assembly (4806N)	17–125	3	
9840	**REAR PUMP ASSEMBLY (35EA-507)**	17–126	1	
9921	**Key**, on Mainshaft, driving Rear Pump (4897)	17–127	1	
9841	**Retaining Plate**, for Rear Pump Gears (35EA-162)	17–128	1	
9842	**Bolt and Lockwasher**, securing Rear Pump Assembly to Transmission Case (EA.4626K)		5	
9843	**Screw and Lockwasher**, securing Rear Pump Assembly to Transmission Case (EA.4626G)		1	FOR USE AS REQUIRED
9868	**GOVERNOR ASSEMBLY (35EG-510)**	17–129	1	
9922	**Ball** (Steel) locating Governor Assembly on Mainshaft (147485)	17–130	1	
9901	**Snap Ring**, retaining Governor Assembly on Mainshaft (4653E)	17–131	1	

AUTOMATIC TRANSMISSION UNIT
(SINGLE DRIVE RANGE)

Part No.	Description	Plate No.	No. per Unit	Remarks
C.21294	**DRIVE GEAR FOR SPEEDOMETER (35EG-169)**	17–132	1	
9764	**VALVE BODIES ASSEMBLY (35EG-1A200)**	18–1	1	
11524	**Spring**, for Primary Regulator Valve (35EG-247)		1	
11901	**Spring**, for Secondary Regulator Valve (35EA-246A)		1	
11526	**Spring**, for Throttle Valve (35-208)		1	
11527	**Return Spring**, for Throttle Valve (35-238)		1	
11778	**Stop Plate**, for Throttle (35EA-309)		1	
11779	**Stop**, for Throttle Valve (35EA-249)		1	
12298	**Spring**, for Servo Orifice Control Valve (35EE-296)		1	
11903	**Spring**, for Modulator Valve (35-212)		1	
11324	**Check Valve**, for Front Pump (35EA-129A)		1	
11529	**Check Valve**, for Rear Pump (35-229)		1	
11530	**Spring**, for Rear Pump Check Valve (3-252)		1	
9922	**Ball**, for Converter Check Valve (147485)		1	
10949	**Spring**, for Converter Check Valve (5-113)		1	
11531	**Spring**, for 1-2 Shift Valve (35EA-256A)		1	
11532	**Spring**, for 2-3 Shift Valve (35EA-223)		1	
9856	**Cam Assembly**, operating Downshift Valve (35EA-283D)	18–10	1	
11785	**Screw**, securing Cam Assembly to Valve Body (EA.5088E)		2	
9844	**Oil Filter**, on Valve Body, for Front Pump (35EA-A190)	18–2	1	
11780	**Screw**, securing Oil Filter (EA.5088)		4	
9845	**Oil Filter**, on Valve Body, for Rear Pump (35-A98)	18–3	1	
11784	**Screw**, securing Oil Filter (EA.5088D)		2	
12299	**Separating Plate** (35EG-204)		1	
11780	**Screw** (3/8" long) for assembly of Valve Bodies (EA.5088)		6	
11781	**Screw** (5/8" long) for assembly of Valve Bodies (EA.5088A)		11	
11782	**Screw** (15/16" long) for assembly of Valve Bodies (EA.5088B)		9	
11783	**Screw** (1 3/8" long) for assembly of Valve Bodies (EA.5088C)		1	
9848	**SPRING KIT FOR SERVICING VALVE BODIES (35EG-512)**		1	
	OIL TUBES			
9849	**Oil Tube**, connecting Valve Body to Front Servo (Apply) (35EA-185)	18–4	1	
9850	**Oil Tube**, connecting Valve Body to Front Servo (Release) (35EG-186)	18–5	1	
9851	**Oil Tube**, connecting Valve Body to Rear Clutch (35EA-86)	18–6	1	
9852	**Oil Tube**, connecting Valve Body to Rear Servo (35EA-91)	18–7	1	
9853	**Oil Tube**, connecting Valve Body to Front Pump (Outlet) (35-89)	18–8	1	
9854	**Oil Tube**, on Valve Body, supplying Converter (35-88)	18–9	1	
11490	**Oil Tube**, connecting Valve Body to Front Pump (Inlet) (35-87A)		1	
9858	**OIL PAN ASSEMBLY (35EU-508)**	18–11	1	
10345	**Drain Plug**, in Oil Pan (EA.5033A)		1	
9859	**Gasket**, between Oil Pan and Case (35EA-181)	18–12	1	
9895	**Bolt Assembly**, securing Oil Pan to Transmission Case (EA.5062)		15	
9860	**OIL SEAL AND 'O' RING SERVICE KIT (35EG-500)** **(Comprising all Oil Seals and 'O' Rings)**		1	**When supplies are exhausted use Service Kit Part No. 11906**

Part No.	Description	Plate No.	No. per Unit	Remarks
11906	OIL SEAL AND 'O' RING SERVICE KIT (35EG-500-1) (Comprising all Oil Seals and 'O' Rings)		1	
9861	GASKET SERVICE KIT (35EA-501) (Comprising all Gaskets)		1	
10435	SEALING RING AND PISTON RING SERVICE KIT (35EA-503-1) (Comprising all Sealing Rings and Piston Rings)		1	
9863	THRUST WASHER SERVICE KIT (35EA-504-1) (Comprising all Thrust Washers)		1	
9864	FRONT CLUTCH PLATE SERVICE KIT (35EA-505) (Comprising all Clutch Plates, 'O' Rings, and Sealing Rings, for Front Clutch)		1	
9865	REAR CLUTCH PLATE SERVICE KIT (35EG-506) (Comprising all Clutch Plates, 'O' Rings, and Sealing Rings, for Rear Clutch)		1	
10810	PLUG FOR PRESSURE TAKE-OFF POINT (COMPLETE WITH 'O' RING) (35EA-518)		1	FOR USE WHEN PRESSURE TAKE-OFF BOSS IS DAMAGED

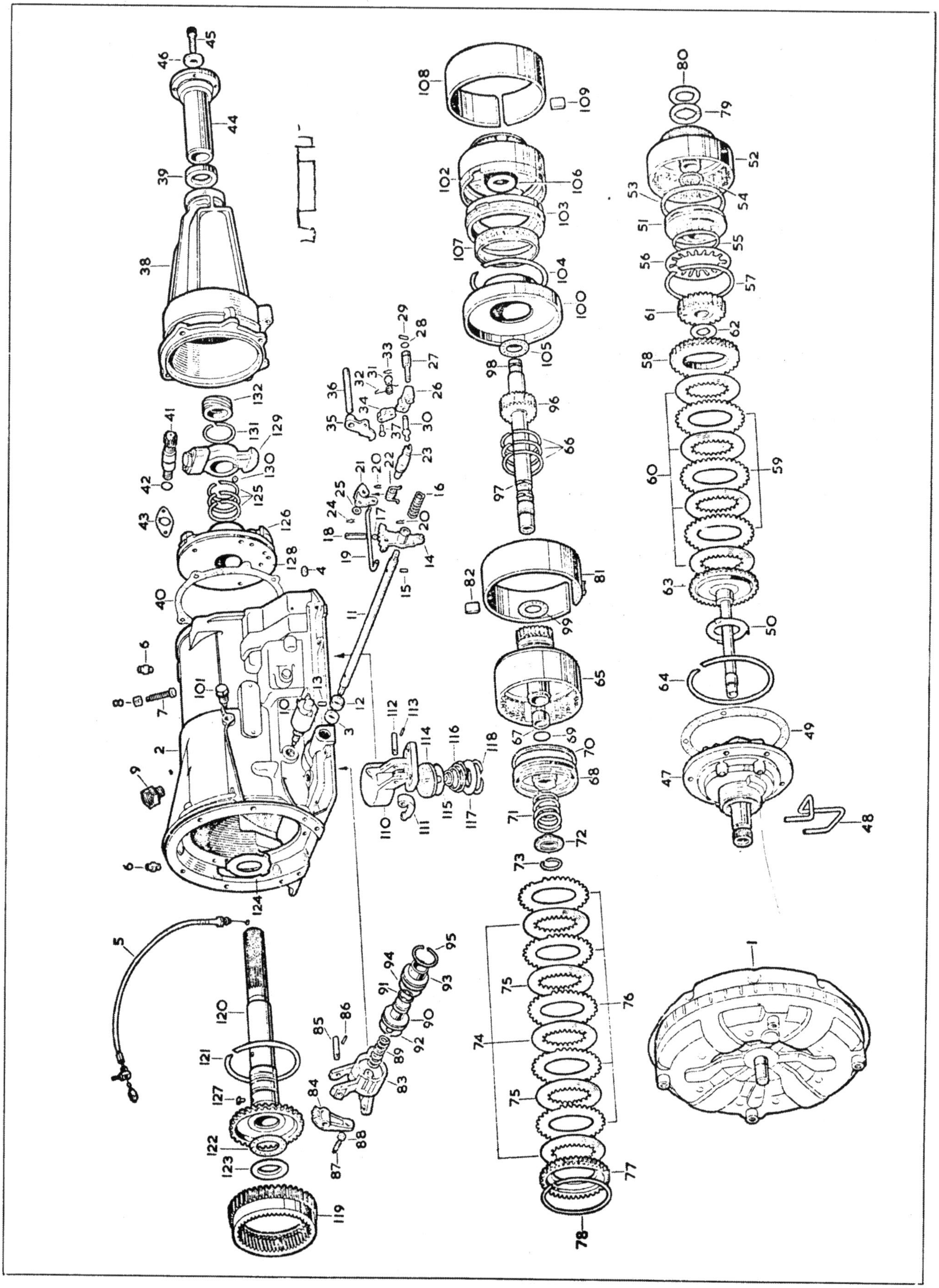

PLATE 17

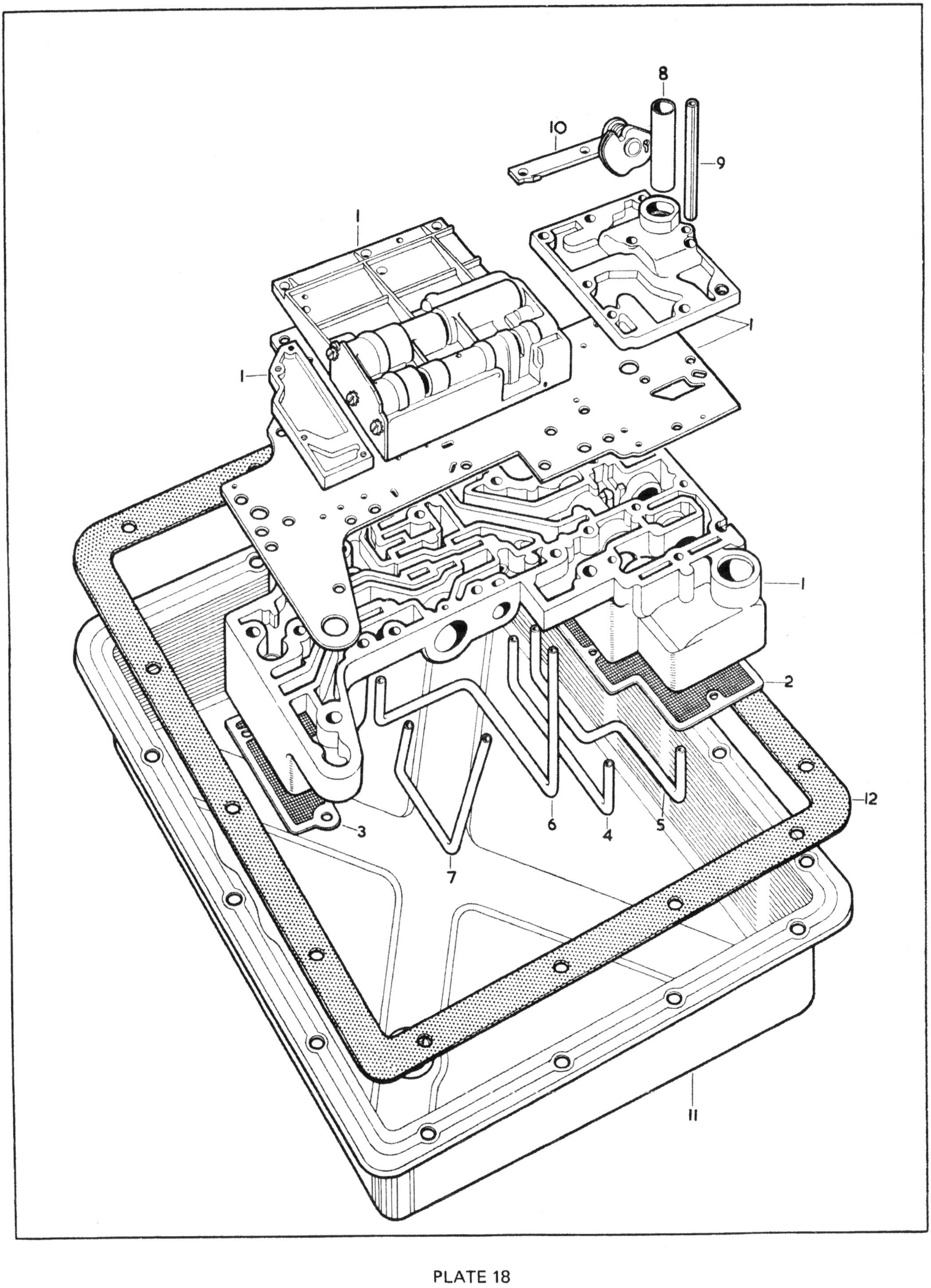

PLATE 18

AUTOMATIC TRANSMISSION UNIT
(DUAL DRIVE RANGE)

Part No.	Description	Plate No.	No. per Unit	Remarks
C.23185	**TRANSMISSION UNIT ASSEMBLY, COMPLETE WITH CONVERTER (AS3-35EG)**		1	Fitted from Engine No. 7A.3593 to 7A.9734
C.21004	**Converter Assembly only** (35EG-A1½C)		1	
C.21612	**Converter Housing** (35EG-1½C)		1	
C.21003	**Stoneguard,** over apertures in Converter Housing (35EA-393B)		2	
C.21053	**Screw,** securing Stoneguards (EA.4626H)		4	
C.21054	**Nut,** on Screws (35EA-394)		4	
C.21055	**Screw and Lockwasher,** securing Converter Housing to Transmission Case (443609)		6	
11288	**TRANSMISSION CASE ASSEMBLY (35EA-513-7)**		1	
9769	**Oil Seal,** for Manual Control Shaft (35EA-108)		2	
9869	**Plug,** for Pressure Take-Off Point (444688)		1	
C.21010	**Union,** for Oil Outlet and Return Pipe (35EA-293)		2	
C.21009	**Kickdown Cable Assembly** (35EG-A284A)		1	
11477	**Washer** (Copper) at fixing of Outer Cable to Transmission Case (EA.5022B)		1	
11399	**Screw,** for adjustment of Rear Servo (4626Z)		1	
11400	**Nut,** locking Adjustment Screw (4775R)		1	
C.21007	**Adaptor,** for Dipstick Tube (35EA-385A)		1	
C.21008	**Switch,** for Starter Cut-Out and Reverse Lamp (35EA-A286C)		1	
11890	**Manual Control Shaft,** carrying Detent Lever Assembly (35EA-81B)		1	
9771	**Collar,** positioning Manual Control Shaft in Case (35EA-179A)		1	
9872	**Roll Pin,** securing Collar to Shaft (273337)		1	
11786	**Lever Assembly,** on Shaft, for Manual Valve Detent (35EA-A78F)		1	
11481	**Pin,** retaining Lever (EA.5071)		1	
9773	**Spring,** on Manual Control Shaft, between Detent Lever and Case (35EA-85)		1	
9874	**Ball,** positioning Detent Lever (453593)		1	
9774	**Spring,** behind Ball (35EA-83)		1	
11787	**Link,** connecting Detent Lever to Torsion Lever (35EA-266C)		1	
11788	**Torsion Lever,** for Parking Brake Linkage (35EA-261)		1	
9877	**Spring,** providing tension between Torsion Lever and Forked Lever (3-264A)		1	
9878	**Forked Lever,** engaging Ball Pin in Toggle Lever (3-A262)		1	
9879	**Clip,** retaining Torsion Lever, Forked Lever, and Spring (4743A)		1	
9890	**Washer,** under Clip (3-321)		1	
9880	**Toggle Lever,** for Parking Brake Pawl (3-272)		1	
9881	**Toggle Pin** (Shouldered) securing Toggle Lever (35-276)		1	
9882	**'O' Ring,** sealing Toggle Pin (4804P)		1	
9883	**Roll Pin,** retaining Toggle Pin in Case (EA.5004A)		1	
10577	**Ball Pin,** securing Link to Toggle Lever (35EA-274)		1	
10578	**Spring,** on Ball Pin, tensioning Parking Brake Pawl (8-271)		1	
9888	**Link,** connecting Toggle Lever to Parking Brake Pawl (3-273)		1	
9776	**Pawl,** for Parking Brake (35-268)		1	
9777	**Anchor Pin,** securing Parking Brake Pawl (35-269)		1	
9889	**Pin,** securing Pawl to Link (3-275)		1	
C.23188	**SELECTOR LEVER ASSEMBLY ON TRANSMISSION UNIT**		1	**R.H. DRIVE CARS ONLY**
C.23177	**SELECTOR LEVER ASSEMBLY ON TRANSMISSION UNIT**		1	**L.H. DRIVE CARS ONLY**
C.8737/2	**Nut,** Self-Locking, securing Selector Lever to Manual Control Shaft		1	
FW.105/T	**Washer,** Plain, under Nut		1	

Part No.	Description	Plate No.	No. per Unit	Remarks
9778	**EXTENSION CASE ASSEMBLY (35EG-1A7A)**		1	
9891	**Oil Seal,** at rear end of Extension Case (R39-110)		1	
9779	**Gasket,** between Extension Case and Transmission Case (35-146)		1	
9892	**Bolt and Spring Lockwasher,** securing Extension Case to Transmission Case (431922)		4	
C.21296	**SPEEDOMETER DRIVEN GEAR**		1	
C.21297	**Bearing,** for Speedometer Driven Gear		1	
C.12377/1	**Oil Seal,** at bottom of Bearing		1	
C.21058	**'O' Ring,** around Speedometer Driven Gear Bearing		1	
C.21073	**Plate,** retaining Driven Gear in Extension Case		1	
UCS.125/4R	**Screw,** Set, securing Plate		2	
FG.104/X	**Washer,** Spring, on Setscrews		2	
9780	**FLANGE, COUPLING TRANSMISSION UNIT TO PROPELLER SHAFT (35EG-298)**		1	
9781	**Screw** (Wedglok) Socket Head, securing Flange to Transmission Drive Shaft (EG.5034)		1	
9782	**Washer,** Special, on Screw (FG.5003)		1	
9783	**FRONT PUMP ASSEMBLY (35EA-1A60B)**		1	
10427	**Oil Seal,** for Front Pump (35EA-62A)		1	
9785	**'O' Ring,** for Front Pump (EA.4804D)		1	
9893	**Screw and Lockwasher** (7/8" long) for assembly of Front Pump (443609)		5	
9894	**Screw and Lockwasher** (5/8" long) for assembly of Front Pump (190601)		1	
9787	**Gasket,** between Front Pump and Transmission Case (35-61)		1	
9893	**Screw and Lockwasher** (7/8" long) securing Front Pump Assembly to Transmission Case (443609)		1	
9788	**Thrust Washer,** between Front Pump and Input Shaft (35-15B)		1	SELECTIVE SIZES
9789	**Thrust Washer,** between Front Pump and Input Shaft (35-15C)		1	SELECTIVE SIZES
9855	**OIL TUBE, RETURNING OIL FROM CONVERTER TO OIL COOLER (35EA-297)**		1	
11893	**FRONT CLUTCH ASSEMBLY (35EE-1A70F)**		1	
9791	**Piston Assembly,** for Front Clutch (35EA-A35)		1	
9792	**Cylinder,** for Front Clutch (35FC-70)		1	
9793	**'O' Ring,** Inner, for Piston (EA.4804E)		1	
9896	**Sealing Ring,** Outer, for Piston (EA.4805C)		1	
9794	**Split Ring,** seating Front Clutch Spring (35-33)		1	
11894	**Spring,** for Front Clutch (35-37A)		1	
11895	**Snap Ring,** retaining Front Clutch Spring (4763F)		1	
N.S.S.	Pressure Plate, for Front Clutch (35EA-175B)		1	NOT SUPPLIED SEPARATELY USE SERVICE KIT 11908 SEE PAGE 81
N.S.S.	Clutch Plate (Drive) for Front Clutch (35-65)		3	NOT SUPPLIED SEPARATELY USE SERVICE KIT 11908 SEE PAGE 81
N.S.S.	Clutch Plate (Friction) for Front Clutch (35-A166)		4	NOT SUPPLIED SEPARATELY USE SERVICE KIT 11908 SEE PAGE 81
9799	**Hub,** carrying Front Clutch Plates (35-40)		1	
9800	**Thrust Washer,** between Hub and Input Shaft (35-17)		1	
9801	**Input Shaft** (35-A16)		1	
9897	**Snap Ring,** retaining Input Shaft in Front Clutch Cylinder (4763D)		1	

AUTOMATIC TRANSMISSION UNIT
(DUAL DRIVE RANGE)

Part No.	Description	Plate No.	No. per Unit	Remarks
10428	**REAR CLUTCH AND FRONT DRUM ASSEMBLY (35EN-2A50)**		1	
9803	**Front Drum Assembly** (35EG-514)		1	
9898	**Sealing Ring,** on Front Drum Assembly (4806M)		3	
9905	**Needle Bearing,** in Front Drum (4840L)		1	
9804	**Piston Assembly,** for Rear Clutch (35EP-A45)		1	
9899	**'O' Ring,** Inner, for Piston (4804S)		1	
9900	**Sealing Ring,** Outer, for Piston (4805E)		1	
9805	**Spring,** for Rear Clutch (35-47)		1	
9806	**Seat,** for Rear Clutch Spring (35-48)		1	
9901	**Snap Ring,** retaining Spring Seat (4653E)		1	
N.S.S.	Clutch Plate (Friction) for Rear Clutch (35-A166)		4	**NOT SUPPLIED SEPARATELY USE SERVICE KIT 10436 SEE PAGE 81**
N.S.S.	Clutch Plate (Drive) for Rear Clutch (35-165)		4	
N.S.S.	Pressure Plate, for Rear Clutch (35-67)		1	
9897	**Snap Ring,** retaining Rear Clutch Plates in Front Drum (4763D)		1	
9810	**Thrust Washer** (Bronze) between Front Clutch Cylinder and Front Drum (35-15)		1	
9811	**Thrust Washer** (Steel) at rear of Bronze Thrust Washer (35-21)		1	
9812	**BRAKE BAND ASSEMBLY, FOR FRONT DRUM (35EG-1A55)**		1	
9813	**Strut** (Servo) operating Brake Band (35-156C)		1	
9814	**FRONT SERVO ASSEMBLY (35EE-A120)**		1	
12300	**Body,** for Front Servo (35EE-120A)		1	
9816	**Lever,** operating Front Servo Strut (35EA-127)		1	
11498	**Pivot Pin,** for Operating Lever (35-118B)		1	
9903	**Screw,** adjusting Front Servo (4-626W)		1	
9904	**Nut,** locking Adjusting Screw (114503)		1	
9818	**Return Spring,** for Front Servo Piston (35EA-125)		1	
9819	**Piston Assembly,** for Front Servo (35EG-515)		1	
9820	**'O' Ring** (Small) on Piston (EA.4804C)		1	
9821	**'O' Ring** (Large) on Piston and on Sleeve (EE.4804)		2	
11892	**Piston Sleeve,** for Front Servo (35EE-119A)		1	
9824	**Snap Ring,** retaining Piston and Sleeve in Body (EE.4763)		1	
9825	**Bolt and Spring Lockwasher** (1 3/8" long) securing Servo to Transmission Case (EA.4626F)		1	
9892	**Bolt and Spring Lockwasher** (1" long) securing Servo to Transmission Case (431922)		1	
11897	**FORWARD SUN GEAR ASSEMBLY (35EA-A104)**		1	
10431	**Sealing Ring,** on rear end of Shaft (4806P)		1	
9906	**Sealing Ring,** on front end of Shaft (4806L)		2	
9908	**Thrust Bearing** (Needle) between Front Drum and Sun Gear (4830B)		1	
9867	**CENTRE SUPPORT ASSEMBLY (35FC-A75)**		1	
9909	**Screw,** securing Centre Support in Transmission Case (4626X)		2	
9910	**Lockwasher,** on Screw (X.2978DR)		2	
11898	**PLANETARY GEARS AND REAR DRUM ASSEMBLY (35EA-1A150B)**		1	
9827	**Outer Race Assembly,** for One-way Clutch (35-A148)		1	
9897	**Snap Ring,** retaining Outer Race (4763D)		1	
9908	**Thrust Bearing** (Needle) between Sun Gear and Planet Gear Cover (4830B)		1	
11508	**Plate,** for Thrust Bearing (35-20B)		1	

AUTOMATIC TRANSMISSION UNIT

(DUAL DRIVE RANGE)

Part No.	Description	Plate No.	No. per Unit	Remarks
9829	**ONE-WAY CLUTCH ASSEMBLY IN REAR DRUM (35-A275)**		1	
9812	**BRAKE BAND FOR REAR DRUM (35EG-1A155)**		1	
9813	**Strut** (Servo) operating Brake Band (35-156C)		1	
11510	**REAR SERVO ASSEMBLY (35EA-1A130A)**		1	
11511	**Body**, for Rear Servo (35EA-130A)		1	
9832	**Lever**, operating Rear Servo Strut (35EA-137B)		1	
11512	**Pivot Pin**, for Operating Lever (35-118A)		1	
11513	**Piston**, for Rear Servo (35-131A)		1	
9911	**'O' Ring**, on Piston (4804U)		1	
11514	**Return Spring**, for Rear Servo Piston (35-135A)		1	
9913	**Screw**, securing Rear Servo to Transmission Case (4626V)		1	
9914	**Lockwasher**, on Screw (103321)		1	
11515	**Bolt and Spring Lockwasher**, securing Rear Servo to Transmission Case (EA.5023B)		1	
10439	**Magnet**, on Rear Servo Fixing Bolt (8-199)		1	
9836	**RING GEAR, DRIVING MAINSHAFT (35-6)**		1	
11899	**MAINSHAFT ASSEMBLY (35EG-1A2B)**		1	
9916	**Snap Ring**, retaining Mainshaft in Ring Gear (4763A)		1	
9919	**Thrust Bearing** (Needle) between Planet Gear Cover and Mainshaft (4830C)		1	
9838	**Plate**, for Thrust Bearing (35-20A)		1	
9839	**Thrust Washer**, between Mainshaft and Transmission Case (35EA-159A)		1	
9920	**Sealing Ring**, on Mainshaft, for Rear Pump Assembly (4806N)		3	
9840	**REAR PUMP ASSEMBLY (35EA-507)**		1	
9921	**Key**, on Mainshaft, driving Rear Pump (4897)		1	
9841	**Retaining Plate**, for Rear Pump Gears (35EA-162)		1	
9842	**Bolt and Spring Lockwasher** (1 1/8" long) securing Rear Pump Assembly to Transmission Case (EA.4626K)		5	
11091	**GOVERNOR ASSEMBLY (35EG-A95)**		1	
9868	**Governor Service Kit** (35EG-510)		1	
9922	**Ball**, locating Governor Assembly on Mainshaft (147485)		1	
9901	**Snap Ring**, retaining Governor Assembly on Mainshaft (4653E)		1	
9868	**GOVERNOR SERVICE KIT (35EG-510)**		1	
11289	**Body only**, for Governor (35EG-95)		1	
11290	**Valve**, for Governor (35EG-96)		1	
11291	**Weight**, for Governor Valve (35ES-97A)		1	
11521	**Spring**, for Governor Valve (35EA-99A)		1	
11206	**Retainer**, for Spring (35EA-98)		1	
10703	**Plate**, covering Governor Valve (7B-93)		1	
10704	**Screw and Lockwasher**, securing Cover Plate (X2380BL)		2	
11522	**Screw and Lockwasher** (1 3/8" long) securing Governor Assembly to Sleeve (EA.4626A)		2	
C.21294	**DRIVE GEAR, FOR SPEEDOMETER (35EG-169)**		1	

AUTOMATIC TRANSMISSION UNIT
(DUAL DRIVE RANGE)

Part No.	Description	Plate No.	No. per Unit	Remarks
10432	**VALVE BODIES ASSEMBLY (35EG-1A200A)**		1	
11524	**Spring**, for Primary Regulator Valve (35EA-247)		1	
11901	**Spring**, for Secondary Regulator Valve (35EA-246A)		1	
11526	**Spring**, for Throttle Valve (35-208)		1	
11527	**Return Spring**, for Throttle Valve (35-238)		1	
11778	**Stop Plate**, for Throttle (35EA-309)		1	
11779	**Stop**, for Throttle Valve (35EA-249)		1	
11903	**Spring**, for Modulator Valve (35-212)		1	
11324	**Check Valve**, for Front Pump (35EA-129A)		1	
11529	**Check Valve**, for Rear Pump (35-229)		1	
11530	**Spring**, for Rear Pump Check Valve (3-252)		1	
9922	**Ball**, for Converter Check Valve (147485)		1	
10949	**Spring**, for Converter Check Valve (5-113)		1	
11531	**Spring**, for 1-2 Shift Valve (35EA-256A)		1	
11532	**Spring**, for 2-3 Shift Valve (35EA-223)		1	
9856	**Cam Assembly**, operating Downshift Valve (35EA-283D)		1	
11785	**Screw**, securing Cam Assembly to Valve Body (EA.5088E)		2	
9844	**Oil Filter**, on Valve Body, for Front Pump (35EA-A190)		1	
11780	**Screw**, securing Oil Filter (EA5088)		4	
9845	**Oil Filter**, on Valve Body, for Rear Pump (35-A98)		1	
11784	**Screw**, securing Oil Filter (EA.5088D)		2	
12301	**Separating Plate** (35EG-204-A)		1	
11780	**Screw** (3/8" long) for assembly of Valve Bodies (EA.5088)		6	
11781	**Screw** (5/8" long) for assembly of Valve Bodies (EA.5088A)		11	
11782	**Screw** (15/16" long) for assembly of Valve Bodies (EA.5088B)		9	
11783	**Screw** (1 3/8" long) for assembly of Valve Bodies (EA.5088C)		1	
9848	**SPRING KIT, FOR SERVICING VALVE BODIES (35EG-512)**		1	
	OIL TUBES			
9849	**Oil Tubes**, connecting Valve Body to Front Servo (Apply) (35EA-185)		1	
10434	**Oil Tube**, connecting Valve Body to Front Servo (Release) (35EA-186)		1	
9851	**Oil Tube**, connecting Valve Body to Rear Clutch (35EA-86)		1	
9852	**Oil Tube**, connecting Valve Body to Rear Servo (35EA-91)		1	
9853	**Oil Tube**, connecting Valve Body to Front Pump (Outlet) (35-89)		1	
9854	**Oil Tube**, on Valve Body, supplying Converter (35-88)		1	
11490	**Oil Tube**, connecting Valve Body to Front Pump (Inlet) (35-87A)		1	
11905	**OIL PAN ASSEMBLY (35EU-508-1)**		1	
10345	**Drain Plug**, in Oil Pan (EA-5033A)		1	
9859	**Gasket**, between Oil Pan and Transmission Case (35EA-181)		1	
9895	**Bolt and Spring Lockwasher**, securing Oil Pan to Transmission Case (EA.5062)		15	
11906	**OIL SEAL AND 'O' RING SERVICE KIT (35EG-500-1)** (Comprising all Oil Seals and 'O' Rings)		1	
9861	**GASKET SERVICE KIT (35EA-501)**		1	

Part No.	Description	Plate No.	No. per Unit	Remarks
10435	**SEALING RING AND PISTON RING SERVICE KIT (35EA-503-1)** (Comprising all Sealing Rings and Piston Rings)		1	
9863	**THRUST WASHER SERVICE KIT (35EA-504-1)** (Comprising all Thrust Washers)		1	
11908	**FRONT CLUTCH PLATE SERVICE KIT (35EA-505-3)** (Comprising all Clutch Plates, 'O' Rings and Sealing Rings, for Front Clutch)		1	
10436	**REAR CLUTCH PLATE SERVICE KIT (35EE-506-1)**		1	
10810	**PLUG FOR PRESSURE TAKE-OFF POINT (COMPLETE WITH 'O' RING) (35EA-518)**		1	FOR USE WHEN PRESSURE TAKE-OFF BOSS IS DAMAGED
C.21611	**CONVERTER HOUSING ASSEMBLY (35EG-A 1½C)**		1	
C.21055	**Screw and Spring Lockwasher**, securing Converter Housing to Transmission Case (443609)		6	
C.21003	**Stoneguard**, over apertures in Converter Housing (35EA-393B)		2	
UCB.137/12R	**Bolt**, securing Converter Housing to Gearbox Support Plate		2	
FW.106/T	**Washer**, Plain, on Bolts		2	
FG.106/X	**Washer**, Spring, on Bolts		2	
420304	**Dowel**, locating Converter Housing on Gearbox Support Plate		1	
11224	**CONVERSION KIT (35EG-520) REQUIRED WHEN CONVERTING AS3-35EG UNITS TO AS5-35EG CONDITION** (Comprising all parts necessary to convert earlier Dual Drive Range units to the later condition)		1	

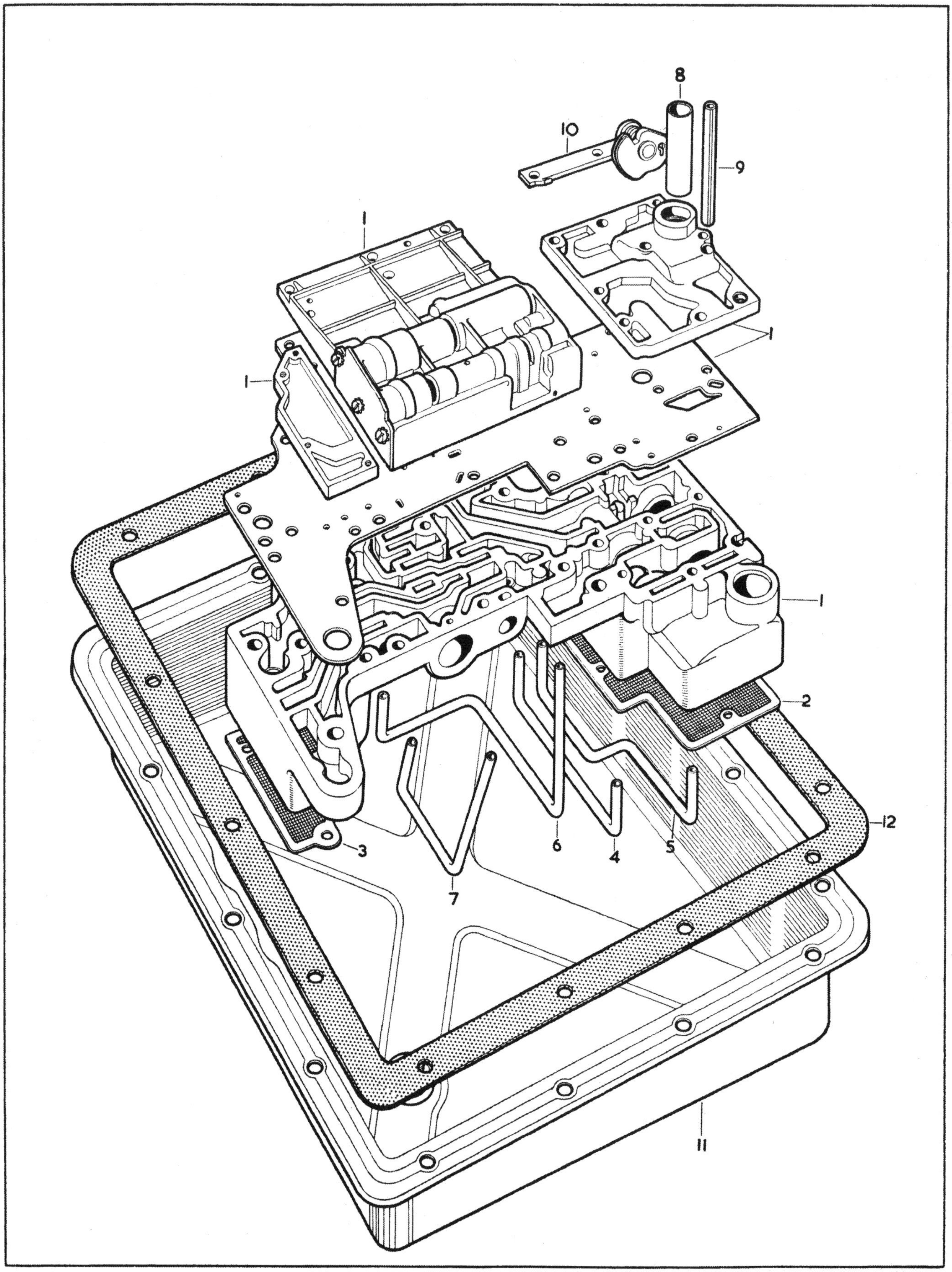

PLATE 18

AUTOMATIC TRANSMISSION UNIT

(DUAL DRIVE RANGE, RECALIBRATED)

Part No.	Description	Plate No.	No. per Unit	Remarks
C.26379	**TRANSMISSION UNIT ASSEMBLY, COMPLETE WITH CONVERTER (AS5-35EG) (35EG-509-2)**		1	Fitted to Engine No. 7A.9735 and subs. also from 7K.1001 to 7K.1997. Excepting 7K.1971, 7K.1973, 7K.1975 and 7K.1984 to 7K.1989 inclusive.
C.26380	**Converter Assembly only** (35EG-A1½E)		1	
C.26381	**Converter Housing** (35EG-1½E)		1	(Cast Iron)
C.21003	**Stoneguard,** over apertures in Converter Housing (35EA-393B)		2	
C.21053	**Screw,** securing Stoneguards		4	
C.21054	**Nut,** on Screws (35EA-394)		4	
C.21055	**Screw and Lockwasher,** securing Converter Housing to Transmission Case (443609)		6	
12288	**TRANSMISSION CASE ASSEMBLY (35EA-513-7)**		1	
9769	**Oil Seal,** for Manual Control Shaft (35EA-108)		2	
9869	**Plug,** for Pressure Take-Off Point (444-688)		1	
C.21010	**Union,** for Oil Outlet and Return Pipes (35EA-293)		2	
C.27146	**Kickdown Cable Assembly** (35EG-A284D)		1	
11477	**Washer** (Copper) at fixing of Outer Cable to Transmission Case (EA.5022B)		1	
11399	**Screw,** for adjustment of Rear Servo (4626Z)		1	
11400	**Nut,** locking Adjustment Screw (4775R)		1	
C.21007	**Adaptor,** for Dipstick Tube (35EA-385A)		1	
C.21008	**Switch,** for Starter Cut-Out and Reverse Lamp (35EA-A286C)		1	
11890	**Manual Control Shaft,** carrying Detent Lever (35EA-81B)		1	
9771	**Collar,** positioning Manual Shaft in Case (35EA-179A)		1	
9872	**Roll Pin,** securing Collar to Manual Shaft (273337)		1	
11786	**Lever Assembly,** on Shaft, for Manual Valve Detent (35EA-A78F)		1	
11481	**Pin,** retaining Detent Lever (EA.5071)		1	
9773	**Spring,** on Manual Control Shaft, between Detent Lever and Case (35EA-85)		1	
9874	**Ball** (3/8" dia) positioning Detent Lever (453593)		1	
9774	**Spring,** tensioning Ball (35EA-83)		1	
9876	**Torsion Lever,** for Parking Brake Linkage (3-261)		1	
11787	**Link,** connecting Detent Lever to Torsion Lever (35EA-266C)		1	
11788	**Torsion Lever,** for Parking Brake Linkage (35EA-261)		1	
9877	**Spring,** providing tension between Torsion Lever and Forked Lever (3-264A)		1	
9878	**Forked Lever,** engaging Ball Pin in Toggle Lever (3-A262)		1	
9879	**Clip,** retaining Torsion Lever, Forked Lever and Spring (4743A)		1	
9890	**Washer,** Plain, under Clip (3-321)		1	
9880	**Toggle Lever,** for Parking Brake Pawl (3-272)		1	
9881	**Toggle Pin,** (Shouldered) securing Toggle Lever (35-276)		1	
9882	**'O' Ring,** sealing Toggle Pin (4804P)		1	
9883	**Roll Pin,** retaining Toggle Pin in Case (EA.5004A)		1	
10577	**Ball Pin,** securing Link to Toggle Lever (35EA-274)		1	
10578	**Spring,** on Ball Pin, tensioning Parking Brake Pawl (8-271)		1	
9888	**Link,** connecting Toggle Lever to Parking Brake Pawl (3-273)		1	
9776	**Pawl,** for Parking Brake (35-268)		1	
9777	**Anchor Pin,** securing Parking Brake Pawl (35-269)		1	
9889	**Pin,** securing Pawl to Link (3-275)		1	
C.23188	**SELECTOR LEVER ASSEMBLY ON TRANSMISSION UNIT**		1	**R.H. DRIVE CARS ONLY**

AUTOMATIC TRANSMISSION UNIT
(DUAL DRIVE RANGE, RECALIBRATED)

Part No.	Description	Plate No.	No. per Unit	Remarks
C.23177	**SELECTOR LEVER ASSEMBLY ON TRANSMISSION UNIT**		1	**L.H. DRIVE CARS ONLY**
C.8737/2	**Nut**, Self-Locking, securing Selector Lever to Manual Control Shaft		1	
FW.105/T	**Washer**, Plain, under Nut		1	
9778	**EXTENSION CASE ASSEMBLY (35EG-1A7A)**		1	
9891	**Oil Seal**, at rear end of extension case (R39-110)		1	
9779	**Gasket**, between Extension Case and Transmission Case (35-146)		1	
9892	**Bolt and Spring Lockwasher**, securing Extension Case to Transmission Case (431922)		4	
C.21296	**SPEEDOMETER DRIVEN GEAR**		1	
C.21297	**Bearing**, for Speedometer Driven Gear		1	
C.12377/1	**Oil Seal**, at bottom of Bearing		1	
C.21058	**'O' Ring**, around Speedometer Driven Gear Bearing		1	
C.21073	**Plate**, retaining Driven Gear in Extension Case		1	
UCS.125/4R	**Screw**, Set, securing Plate		2	
FG.104/X	**Washer**, Spring, on Setscrews		2	
C.21294	**DRIVE GEAR, FOR SPEEDOMETER (35EG-169)**		1	
9780	**FLANGE, COUPLING TRANSMISSION UNIT TO PROPELLER SHAFT (35EG-298)**		1	
9781	**Screw** (Wedglok) Socket Head, securing Flange to Transmission Drive Shaft (EG.5034)		1	
9782	**Washer**, Special, on Screw (EG.5003)		1	
9783	**FRONT PUMP ASSEMBLY (35EA-1A60B)**		1	
	All component parts as listed for Transmission Unit C.23185 (AS3-35EG)			
9855	**OIL TUBE, RETURNING OIL FROM CONVERTER TO OIL COOLER (35EA-297)**		1	
11893	**FRONT CLUTCH ASSEMBLY (35EE-1A70F)**		1	
	All component parts as listed for Transmission Unit C.23185 (AS3-35EG)			
10428	**REAR CLUTCH AND FRONT DRUM ASSEMBLY (35EN-2A50)**		1	
	All component parts as listed for Transmission Unit C.23185 (AS3-35EG)			
9812	**BRAKE BAND ASSEMBLY, FOR FRONT DRUM (35EG-1A55)**		1	
9813	**Strut** (Servo) operating Brake Band (35-156C)		1	

AUTOMATIC TRANSMISSION UNIT
(DUAL DRIVE RANGE, RECALIBRATED)

Part No.	Description	Plate No.	No. per Unit	Remarks
11223	**FRONT SERVO ASSEMBLY (35EG-A120)**		1	
11225	**Body**, for Front Servo (35EG-120)		1	
9816	**Lever**, operating Front Servo Strut (35EA-127)		1	
11498	**Pivot Pin**, for Operating Lever (35-118B)		1	
9903	**Screw**, adjusting Front Servo (4626W)		1	
9904	**Nut**, locking Adjusting Screw (114503)		1	
9818	**Return Spring**, for Front Servo Piston (35EA-125)		1	
9819	**Piston Assembly**, for Front Servo (35EG-515)		1	
9820	**'O' Ring** (Small) on Piston (EA4804C)		1	
9821	**'O' Ring** (Large) on Piston (EE.4804)		2	
11892	**Piston Sleeve**, for Front Servo (35EE-119A)		1	
9824	**Snap Ring**, retaining Piston and Sleeve in Body (EE.4763)		1	
9825	**Bolt and Spring Lockwasher** (1 3/8" long) securing Servo to Transmission Case (EA.4626F)		1	
9892	**Bolt and Spring Lockwasher** (1" long) securing Servo to Transmission Case (431922)		1	
11897	**FORWARD SUNGEAR ASSEMBLY (35EA-A104)**		1	
10431	**Sealing Ring**, on rear end of Shaft (4806P)		1	
9906	**Sealing Ring**, on Front end of Shaft (4806L)		2	
9908	**Thrust Bearing** (Needle) between Front Drum and Sun Gear (4830B)		1	
9867	**CENTRE SUPPORT ASSEMBLY (35FC-A75)**		1	
9909	**Screw**, securing Centre Support in Transmission Case (4626X)		2	
9910	**Lockwasher**, on Screws (X.2978DR)		2	
11898	**PLANETARY GEARS AND REAR DRUM ASSEMBLY (35EA-1A150B)**		1	
9827	**Outer Race Assembly**, for One-Way Clutch (35-A148)		1	
9897	**Snap Ring**, Retaining Outer Race (4763D)		1	
9908	**Thrust Bearing** (Needle) between Sun Gear and Planet Gear Cover (4830B)		1	
11508	**Plate**, for Thrust Bearing (35-20B)		1	
9829	**ONE-WAY CLUTCH ASSEMBLY IN REAR DRUM (35-A275)**		1	
9812	**BRAKE BAND FOR REAR DRUM (35EG-1A155)**		1	
11510	**REAR SERVO ASSEMBLY (35EA-1A130A)**		1	
	All component parts as listed for Transmission Unit C.23185 (AS3-35EG)			
9836	**RING GEAR, DRIVING MAINSHAFT (35-6)**		1	
11899	**MAINSHAFT ASSEMBLY (35EG-1A2B)**		1	
9916	**Snap Ring**, retaining Mainshaft in Ring Gear (4763A)		1	
9919	**Thrust Bearing** (Needle) between Planet Gear Cover and Mainshaft (4830C)		1	
9838	**Plate**, for Thrust Bearing (35-20A)		1	
9839	**Thrust Washer**, between Mainshaft and Transmission Case (35EA-159A)		1	
9920	**Sealing Ring**, on Mainshaft, for Rear Pump Assembly (4806N)		3	

AUTOMATIC TRANSMISSION UNIT
(DUAL DRIVE RANGE, RECALIBRATED)

Part No.	Description	Plate No.	No. per Unit	Remarks
9840	**REAR PUMP ASSEMBLY (35EA-507)**		1	
9921	**Key**, on Mainshaft, driving Rear Pump (4897)		1	
9841	**Retaining Plate**, for Rear Pump Gear (35EA-162)		1	
9842	**Bolt and Spring Lockwasher** (1 1/8" long) securing Rear Pump Assembly to Transmission Case (EA.4626K)		5	
11393	**GOVERNOR ASSEMBLY (35EN-A95)**		1	
11150	**Governor Service Kit** (35EN-510)		1	
9922	**Ball**, locating Governor Assembly on Mainshaft (147485)		1	
9901	**Snap Ring**, retaining Governor Assembly on Mainshaft (4653E)		1	
11150	**GOVERNOR SERVICE KIT (35EN510)**		1	
12304	**Body only**, for Governor (35EN-95)		1	
12305	**Valve**, for Governor (35EN-96)		1	
11291	**Weight**, for Governor Valve (35ES-97A)		1	
11521	**Spring**, for Governor Valve (35EA-99A)		1	
11206	**Retainer**, for Spring (35EA-98)		1	
10703	**Plate**, covering Governor Valve (7B-93)		1	
10704	**Screw and Lockwasher**, securing Cover Plate (X2380BL)		2	
11522	**Screw and Lockwasher** (1 3/8" long) securing Governor Assembly to Sleeve (EA.4626A)		2	
C.21294	**DRIVE GEAR, FOR SPEEDOMETER (35EG-169)**		1	
11148	**VALVE BODIES ASSEMBLY (35EG-1A200C)**		1	
11149	**Cam Assembly**, operating Downshift Valve (35EB-A283)		1	
9857	**Screw and Lockwasher**, securing Cam to Valve Body (EA4626J)		2	
	All other components of the Valve Bodies Assembly are as for Valve Body Assembly (35EG-1A200A)			
12303	**SPRING KIT, FOR SERVICING VALVE BODIES (35EA-512)**		1	
	OIL TUBES			
9850	**Oil Tube**, connecting Valve Body to Front Servo (Release) (35EG-186)		1	
	All other Oil Tubes are as for Transmission Unit C.23185 (AS3-35EG)			
11905	**OIL PAN ASSEMBLY (35EU-508-1)**		1	
10345	**Drain Plug**, in Oil Pan (EA5033A)		1	
9859	**Gasket**, between Oil Pan and Transmission Case (35EA-181)		1	
9895	**Bolt and Spring Lockwasher**, securing Oil Pan to Transmission Case (EA.5062)		15	
11906	**OIL SEAL AND 'O' RING SERVICE KIT (35-EG-500-1)**		1	
	(Comprising all Oil Seals and 'O' Rings)			
9861	**GASKET SERVICE KIT (35EA-501)**		1	

Part No.	Description	Plate No.	No. per Unit	Remarks
10435	**SEALING RING AND PISTON RING SERVICE KIT (35EA-503-1)**		1	
	(Comprising all Sealing Rings and Piston Rings)			
9863	**THRUST WASHER SERVICE KIT (35EA-504-1)**		1	
	(Comprising all Thrust Washers)			
11908	**FRONT CLUTCH PLATE SERVICE KIT (35EA-505-3)**		1	
	(Comprising all Clutch Plates, 'O' Rings and Sealing Rings, for Front Clutch)			
10436	**REAR CLUTCH PLATE SERVICE KIT (35EE-506-1)**		1	
10810	**PLUG FOR PRESSURE TAKE-OFF POINT (COMPLETE WITH 'O' RING) (35EA-518)**		1	
C.21611	**CONVERTER HOUSING ASSEMBLY (935EG-A-1½C)**		1	
C.21055	**Screw and Spring Lockwasher,** securing Converter Housing to Transmission Case (443609)		6	
C.21003	**Stoneguard,** over apertures in Converter Housing (35EA-393B)		2	
UGB.137/12R	**Bolt,** securing Converter Housing to Gearbox Support Plate		2	
FW.106/T	**Washer,** Plain, on Bolts		2	
FG.106/X	**Washer,** Spring, on Bolts		2	
420304	**Dowel,** locating Converter Housing on Gearbox Support Plate		1	

AUTOMATIC TRANSMISSION UNIT

(DUAL DRIVE RANGE)

Part No.	Description	Plate No.	No. per Unit	Remarks
C.29250	**TRANSMISSION UNIT ASSEMBLY, COMPLETE WITH CONVERTER (AS11-35EG) (35EG-509-5)**		1	Fitted to Engine No. 7K.1984 and subs., also fitted to 7K.1971, 7K.1973, 7K.1975. Except 7K.1990 to 7K.1997 inclusive
C.26380	**Converter Assembly only** (35EG-A1½E)		1	
C.26381	**Converter Housing** (35EG-1½E)		1	(Cast Iron)
C.26382	**Stoneguard**, over aperture in Converter Housing (35EA.393A)		2	
C.26383	**Screw**, securing Stoneguards (EA.4626M)		4	
C.21055	**Screw and Lockwasher**, securing Converter Housing to Transmission Case (443609)		6	
12324	**TRANSMISSION CASE ASSEMBLY (35EZ-513)**		1	
9769	**Oil Seal**, for Manual Control Shaft (35EA-108)		2	
9869	**Plug**, for Pressure Take-Off Point (444688)		1	
C.21010	**Union**, for Oil Outlet and Return Pipes (35EA-293)		2	
C.27146	**Kickdown, Cable Assembly** (35EG-A284D)		1	
11477	**Washer** (Copper) at fixing of Outer Cable to Transmission Case (EA.5022B)		1	
11399	**Screw**, for adjustment of Rear Servo (4626Z)		1	
11400	**Nut**, locking Adjustment Screw (4775R)		1	
C.21007	**Adaptor**, for Dipstick Tube (35EA-385A)		1	
C.21008	**Switch**, for Starter Cut-Out and Reverse Lamp (35EA-A286C)		1	
11890	**Manual Control Shaft**, carrying Detent Lever (35EA-81B)		1	
9771	**Collar**, positioning Manual Shaft in Case (35EA-179A)		1	
9872	**Roll Pin**, securing Collar to Manual Shaft (273337)		1	
11786	**Lever Assembly**, on Shaft, for Manual Valve Detent (35EA-A78F)		1	
11481	**Pin**, retaining Detent Lever (EA.5071)		1	
9773	**Spring**, on Manual Control Shaft, between Detent Lever and Case (35EA-85)		1	
9874	**Ball**, (3/8" dia.) positioning Detent Lever (453593)		1	
9774	**Spring**, tensioning Ball (35EA-83)		1	
11788	**Torsion Lever**, for Parking Brake Linkage (35EA-261)		1	
11787	**Link**, connecting Detent Lever to Torsion Lever (35EA-266C)		1	
9877	**Spring**, providing tension between Torsion Lever and Forked Lever (3-264A)		1	
9878	**Forked Lever**, engaging Ball Pin in Toggle Lever (3-A262)		1	
9879	**Clip**, retaining Torsion Lever, Forked Lever and Spring (4743A)		1	
9890	**Washer**, Plain, under Clip (3-321)		1	
9880	**Toggle Lever**, for Parking Brake Pawl (3-272)		1	
9881	**Toggle Pin** (Shouldered) securing Toggle Lever (35-276)		1	
9882	**'O' Ring**, sealing Toggle Pin (4804P)		1	
9883	**Roll Pin**, retaining Toggle Pin in Case (EA.5004A)		1	
10577	**Ball Pin**, securing Link to Toggle Lever (35EA-274)		1	
10578	**Spring**, on Ball Pin, tensioning Parking Brake Pawl (8-271)		1	
9888	**Link**, connecting Toggle Lever to Parking Brake (3-273)		1	
9776	**Pawl**, for Parking Brake (35-268)		1	
9777	**Anchor Pin**, securing Parking Brake Pawl (35-269)		1	
9889	**Pin**, securing Pawl to Link (3-275)		1	
C.23188	**SELECTOR LEVER ASSEMBLY ON TRANSMISSION UNIT**		1	**R.H. DRIVE CARS ONLY**
C.23177	**SELECTOR LEVER ASSEMBLY ON TRANSMISSION UNIT**		1	**L.H. DRIVE CARS ONLY**
C.8737/2	**Nut**, Self-Locking, securing Selector Lever to Manual Control Shaft		1	
FW.105/T	**Washer**, Plain, under Nut		1	

Part No.	Description	Plate No.	No. per Unit	Remarks
9778	**EXTENSION CASE ASSEMBLY (35EG-1A7A)**		1	
9891	**Oil Seal**, at rear end of extension Case (R39-110)		1	
9779	**Gasket**, between Extension Case and Transmission Case (35-146)		1	
9892	**Bolt and Spring Lockwasher**, securing Extension Case to Transmission Case (431922)		4	
C.21296	**SPEEDOMETER DRIVEN GEAR**		1	
C.21297	**Bearing**, for Speedometer Driven Gear		1	
C.12377/1	**Oil Seal**, at bottom of Bearing		1	
C.21058	**'O' Ring**, aroung Speedometer Driven Gear Bearing		1	
C.21073	**Plate**, retaining Driven Gear in Extension Case		1	
UCS.125/4R	**Screw**, Set, securing Plate		2	
FG.104/X	**Washer**, Spring, on Setscrews		2	
C.21294	**DRIVE GEAR, FOR SPEEDOMETER (35EG-169)**		1	
9780	**FLANGE, COUPLING TRANSMISSION UNIT TO PROPELLER SHAFT (35EG-298)**		1	
9781	**Screw** (Wedglok) Socket Head, securing Flange to Transmission Drive Shaft (EG.5034)		1	
9782	**Washer**, Special, on Screw (EG.5003)		1	
9783	**FRONT PUMP ASSEMBLY (35EA-1A60B)**		1	
	All component parts as listed for Transmission Unit C.23185 (AS3-35EG)			
9855	**OIL TUBE, RETURNING OIL FROM CONVERTER TO OIL COOLER (35EA-297)**		1	
11893	**FRONT CLUTCH ASSEMBLY (35EE-1A70F)**		1	
	All component parts as listed for Transmission Unit C.23185 (AS3-35EG)			
10428	**REAR CLUTCH AND FRONT DRUM ASSEMBLY (35EN-2A50)**		1	
	All component parts as listed for Transmission Unit C.23185 (AS3-35EG)			
9812	**BRAKE BAND ASSEMBLY, FOR FRONT DRUM (35EG-1A155)**		1	
9813	**Strut** (Servo) operating Brake Band (35-156C)		1	
11223	**FRONT SERVO ASSEMBLY (35EG-120)**		1	
	All component parts as listed for Transmission Unit C.26379 (AS5-35EG)			
9825	**Bolt and Spring Lockwasher** (1 3/8" long) securing Servo to Transmission Case (EA.4626F)		1	
9892	**Bolt and Spring Lockwasher** (1" long) securing Servo to Transmission Case (431922)		1	

AUTOMATIC TRANSMISSION UNIT
(DUAL DRIVE RANGE)

Part No.	Description	Plate No.	No. per Unit	Remarks
11897	**FORWARD SUN GEAR ASSEMBLY (35EA-A104)**		1	
10431	**Sealing Ring**, on rear end of Shaft (4806P)		1	
9906	**Sealing Ring**, on front end of Shaft (4806L)		2	
9908	**Thrust Bearing** (Needle) between Front Drum and Sun Gear (4830B)		1	
9867	**CENTRE SUPPORT ASSEMBLY (35FC-A75)**		1	
9909	**Screw**, securing Centre Support in Transmission Case (4626X)		2	
9910	**Lockwasher**, on Screws (X2978DR)		2	
11898	**PLANETARY GEARS AND REAR DRUM ASSEMBLY (35EA-1A150B)**		1	
	All other items as listed for Transmission Unit C.26379 (AS5-35EG)			
9829	**ONE-WAY CLUTCH ASSEMBLY IN REAR DRUM (35-A275)**		1	
9812	**BRAKE BAND FOR REAR DRUM (35EG-1A155)**		1	
11510	**REAR SERVO ASSEMBLY (35EA-1A130A)**		1	
	All component parts as listed for Transmission Unit C.23185 (AS-35EG)			
9836	**RING GEAR, DRIVING MAINSHAFT (35-6)**		1	
11899	**MAINSHAFT ASSEMBLY (35EG-1A2B)**		1	
	All other parts as listed for Transmission Unit C.26379 (AS5-35EG)			
9840	**REAR PUMP ASSEMBLY (35EA-507)**		1	
	All other parts as listed for Transmission Unit C.26379 (AS5-35EG)			
11393	**GOVERNOR ASSEMBLY (35EN-A95)**		1	
11150	**Governor Service Kit** (35EN510)		1	
9922	**Ball**, locating Governor Assembly on Mainshaft (147485)		1	
9901	**Snap Ring**, retaining Governor Assembly on Mainshaft (4653E)		1	
11150	**GOVERNOR SERVICE KIT**		1	
	Contents as listed for Transmission Unit C.26379 (AS5-35EG)			
C.21294	**DRIVE GEAR, FOR SPEEDOMETER (35EG-169)**		1	
11900	**VALVE BODIES ASSEMBLY (35EG-1A200E)**		1	
11524	**Spring**, for Primary Regulator Valve (35EA-247)		1	
11901	**Spring**, for Secondary Regulator Valve (35EA-246A)		1	
11526	**Spring**, for Throttle Valve (35-208)		1	
11527	**Return Spring**, for Throttle Valve (35-238)		1	
11778	**Stop Plate**, for Throttle Valve (35EA-309)		1	
11779	**Stop**, for Throttle Valve (35EA-249)		1	

Part No.	Description	Plate No.	No. per Unit	Remarks
	VALVE BODIES ASSEMBLY (35EG-1A200E) *(contd.)*			
11902	**Spring,** for Servo Orifice Control Valve (35-296)		1	
11903	**Spring,** for Modulator Valve (35-212)		1	
11324	**Check Valve,** for Front Pump (35EA-129A)		1	
11529	**Check Valve,** for Rear Pump (35-229)		1	
11530	**Spring,** for Rear Pump Check Valve (3-252)		1	
9922	**Ball,** for Converter Check Valve (147485)		1	
10949	**Spring,** for Converter Check Valve (5-113)		1	
11531	**Spring,** for 1-2 Shift Valve (35EA-256A)		1	
11532	**Spring,** for 2-3 Shift Valve (35EA-223)		1	
11149	**Cam Assembly,** operating Downshift Valve (35EB-A288)		1	
11785	**Screw,** securing Cam Assembly to Valve Body (EA.5088E)		2	
9844	**Oil Filter,** on Valve Body, for Front Pump (35EA-A190)		1	
11780	**Screw,** securing Oil Filter (EA5088)		4	
9845	**Oil Filter,** on Valve Body, for Rear Pump (35-A98)		1	
11784	**Screw,** securing Oil Filter (EA5088D)		2	
12045	**Separating Plate** (35EU-204)		1	
11780	**Screw** (3/8" long) for Assembly of Valve Bodies (EA5088)		6	
11781	**Screw** (5/8" long) for assembly of Valve Bodies (EA5088A)		11	
11782	**Screw** (15/16" long) for assembly of Valve Bodies (EA5088B)		9	
11783	**Screw** (1 3/8" long) for assembly of Valve Bodies (EA5088C)		1	
9846	**Bolt and Spring Lockwasher** (1¼" long) securing Valve Bodies Assembly to Transmission Case (EA4626C)		1	
9847	**Bolt and Spring Lockwasher** (2" long) securing Valve Bodies Assembly to Transmission Case (EA4626D)		2	
12303	SPRING KIT, FOR SERVICING VALVE BODIES (35EA-512)		1	
	OIL TUBES			
9850	**Oil Tube,** connecting Valve Body to Front Servo (Release) (35EG-186)		1	
	All other Oil Tubes are as for Transmission Unit C.23185 (AS3-35EG)			
11905	OIL PAN ASSEMBLY (35EU-508-1)		1	
	All other parts as listed for Transmission Unit C.26379 (AS5-35EG)			
11906	OIL SEAL AND 'O' RING SERVICE KIT (35EG-500-1)		1	
9861	GASKET SERVICE KIT (35EA-501)		1	
10435	SEALING RING AND PISTON RING SERVICE KIT (35EA-503-1)		1	
9863	THRUST WASHER SERVICE KIT (35EA-504-1)		1	
11908	FRONT CLUTCH PLATE SERVICE KIT (35EA-505-3)		1	

AUTOMATIC TRANSMISSION UNIT
(DUAL DRIVE RANGE)

Part No.	Description	Plate No.	No. per Unit	Remarks
10436	**REAR CLUTCH PLATE SERVICE KIT (35EE-506-1)**		1	
10810	**PLUG FOR PRESSURE TAKE-OFF POINT (COMPLETE WITH 'O' RING)(35EA-518)**		1	
C.26380	**CONVERTER HOUSING ASSEMBLY (35EG-A1½E)**		1	
C.21055	**Screw and Spring Lockwasher**, securing Converter Housing to Transmission Case (443609)		6	
C.26382	**Stoneguard**, over aperture in Converter Housing (35EA-393A)		2	
C.26383	**Screw**, securing Stoneguards to Housing		4	
C.21055	**Screw and Lockwasher**, securing Converter Housing to Transmission Case (443609)		6	
UCB.137/12R	**Bolt**, securing Converter Housing to Gearbox Support Plate		2	
FW.106/T	**Washer**, Plain, on Bolts		2	
FG.106/X	**Washer**, Spring, on Bolts		2	
420304	**Dowel**, locating Converter Housing on Gearbox Support Plate		1	

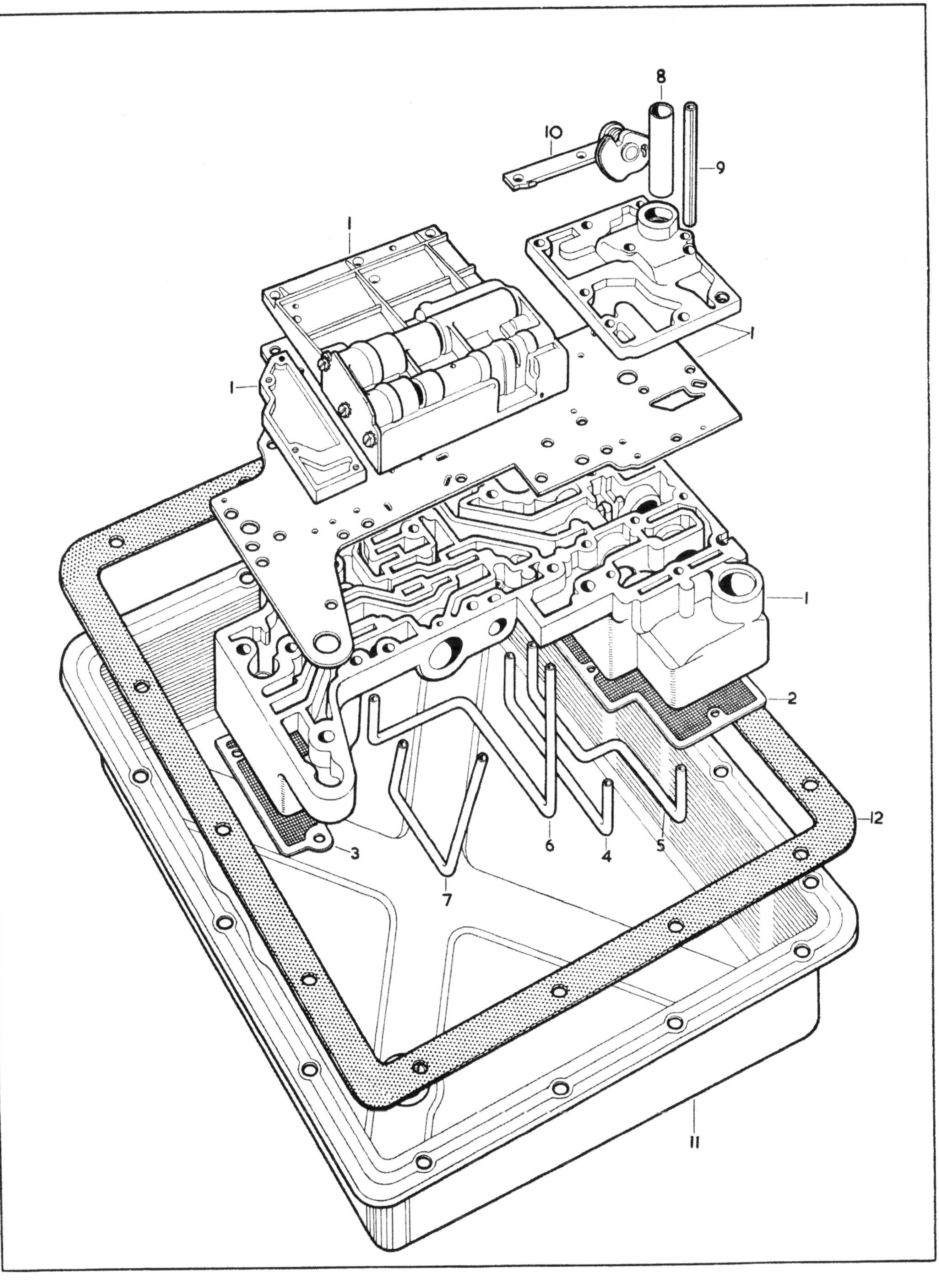

PLATE 18

AUTOMATIC TRANSMISSION MOUNTING DETAILS

Part No.	Description	Plate No.	No. per Unit	Remarks
C.21612	**CONVERTER HOUSING ASSEMBLY (35EG-A.1½C)**	19–1	1	
C.21055	**Screw and Washer Assembly**, securing Converter Housing to Transmission Case (443609)		6	
C.21003	**Stoneguard**, over apertures in Converter Housing (35EA-393)	19–2	2	
C.21053	**Screw**, securing Stoneguards (EA-4626H)		4	
C.21054	**Nut**, on Screws (35EA-394)		4	
UCB.137/12R	**Bolt**, securing Converter Housing to Gearbox Support Plate		2	
FW.106/T	**Washer**, Plain, on Bolts		2	
FG.106/X	**Washer**, Spring, on Bolts		2	
420304	**Dowel**, locating Converter Housing on Gearbox Support Plate	19–3	1	
C.22604	**Clip**, retaining Kickdown Cable to Converter Housing		1	Fitted from Engine No. 7A.3074 to 7A.12591
C.20218	**MID ENGINE MOUNTING, ON SCUTTLE**	19–4	1	
UFS.131/6R	**Screw**, Set, securing Mid-Engine Mounting to Scuttle		2	
C.8667/2	**Nut**, Self-Locking, on Setscrews		2	
C.11688	**Bush**, Stepped, below Mid-Engine Mounting	19–5	1	
C.11607	**Washer**, Stepped, above Mid-Engine Mounting	19–6	1	
C.8667/3	**Nut**, Self-Locking, securing Link to Mid-Engine Mounting		1	
C.21226	**LINK, FOR MID ENGINE MOUNTING**	19–7	1	
C.10940	**Bush** (Rubber/Steel) in Link	19–8	1	
UFB.143/15R	**Bolt**, securing Link to Mounting Bracket		1	
C.8737/4	**Nut**, Self-Locking on Bolt		1	
C.22112	**MOUNTING BRACKET ASSEMBLY FOR MID ENGINE MOUNTING LINK**	19–9	1	
UCS.131/6R	**Screw**, Set, securing Mounting Bracket to Converter Housing		4	
FG.105/X	**Washer**, Spring, on Setscrews		4	
C.22144	**DRIVE PLATE, ON CRANKSHAFT, FOR ATTACHMENT OF TORQUE CONVERTER**	19–10	1	
430239	**Bolt**, securing Drive Plate to Crankshaft Flange	19–11	6	
430237	**Lockplate**, on Bolts	19–12	3	
C.22147	**Spacing Washer**, for Drive Plate	19–13	2	
C.16682	**Screw**, Set, securing Torque Converter to Drive Plate		4	
C.21662	**Tab Washer**, securing Setscrews		4	
C.20871	**CHANNEL SUPPORT ASSEMBLY FOR REAR ENGINE MOUNTING (UNDER GEARBOX)**		1	
C.20872	**Channel Support only**	19–14	1	
C.12335	**Spring Seat** (Rubber) at bottom of Coil Spring	19–15	1	
C.12265	**Centre Bush** (Rubber)	19–16	1	
UFS.131/7R	**Screw**, Set, securing Channel Support to Body		6	
FG.105/X	**Washer**, Spring, on Setscrews		6	
FW.105/T	**Washer**, Plain, under Spring Washers		6	

NOTE: In all cases where only one Spring Seat is fitted and replacement is necessary, the new Coil Spring Part No. C.19988 must be fitted, with an additional Spring Seat at the top of the Spring.

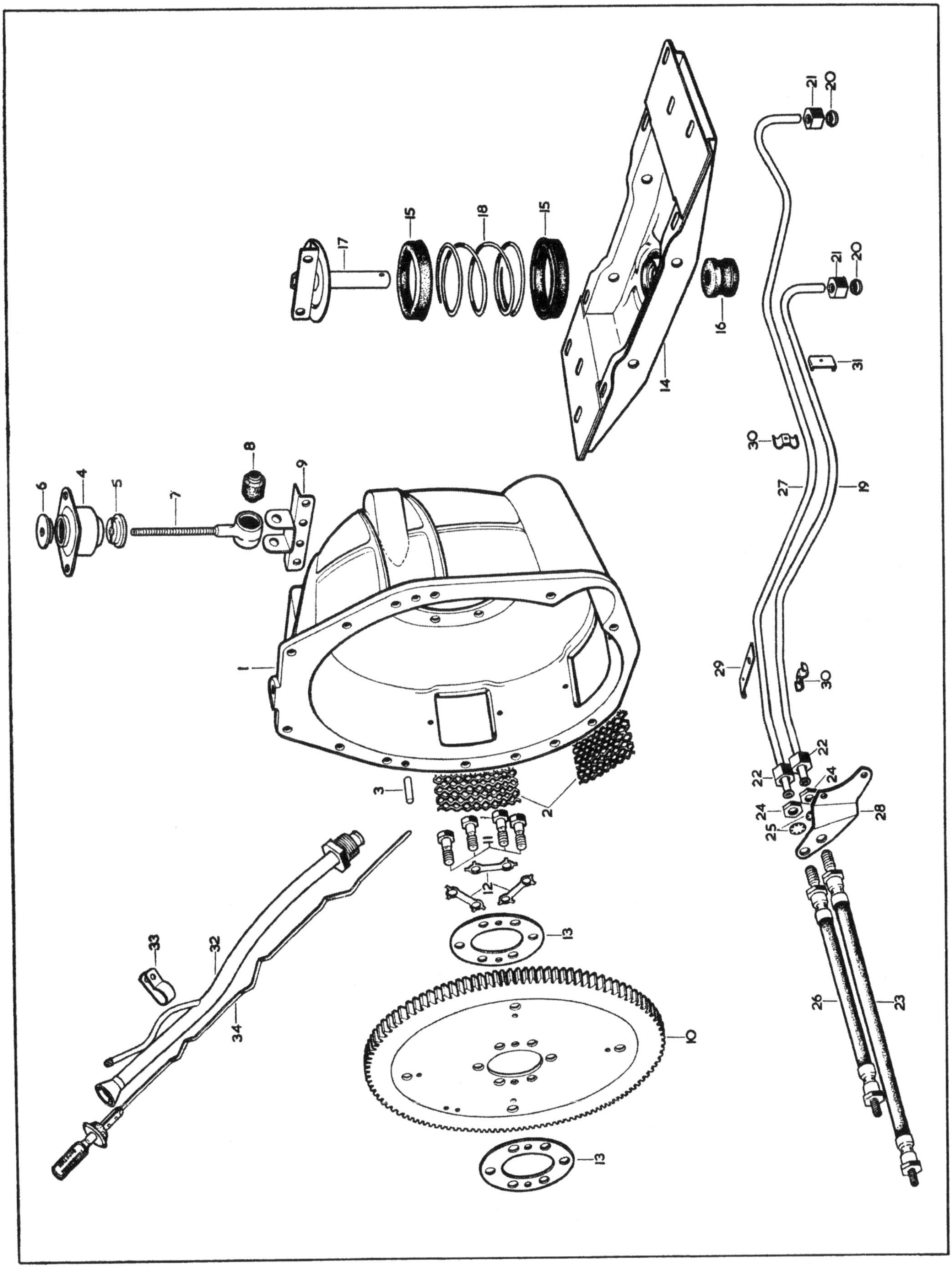
1
2
3
4
5
6
7
8
9
10
11
12
13
14
15
16
17
18
19
20
21
22
23
24
25
26
27
28
29
30
31
32
33
34

PLATE 19

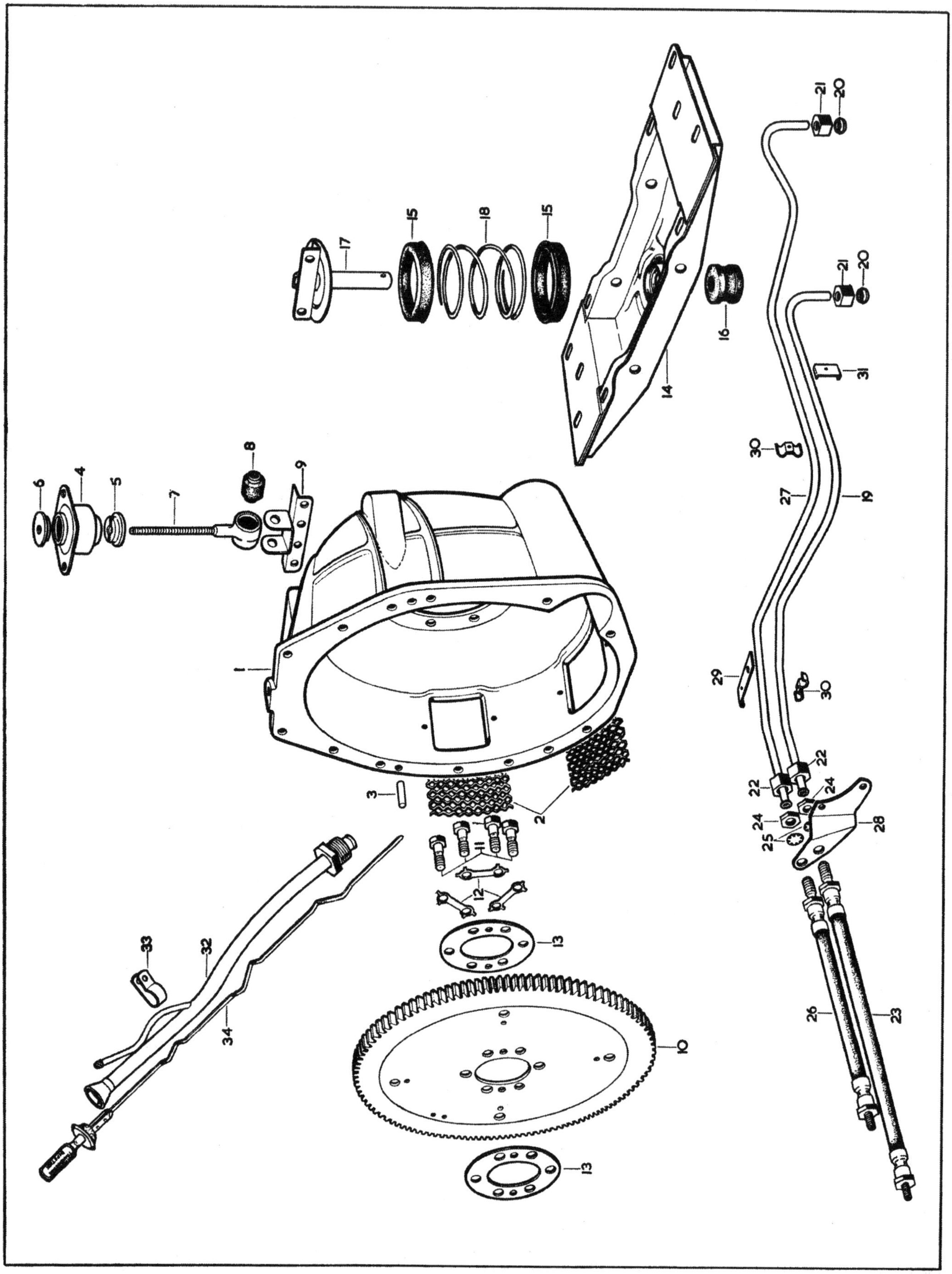
1
2
3
4
5
6
7
8
9
10
11
12
13
13
14
15
15
16
17
18
19
20
20
21
21
22
22
23
24
24
25
26
27
28
29
30
30
31
32
33
34

PLATE 19

AUTOMATIC TRANSMISSION MOUNTING DETAILS

Part No.	Description	Plate No.	No. per Unit	Remarks
C.20876	SPRING RETAINER ASSEMBLY FOR REAR ENGINE MOUNTING	19–17	1	Fitted from Chassis No. 1A.1001 to 1A.2835 1A.20001 to 1A.20248
C.22635	SPRING RETAINER ASSEMBLY FOR REAR ENGINE MOUNTING		1	Fitted to Chassis No. 1A.2836 and subs 1A.20249 and subs
C.19988	COIL SPRING, FOR REAR ENGINE MOUNTING	19–18	1	
C.12335	**Spring Seat** (Rubber) at Top of Coil Spring	19–15	1	
C.21777	OIL OUTLET PIPE, FROM TRANSMISSION UNIT TO FLEXIBLE HOSE	19–19	1	
C.21406	**Olive**, for Oil Outlet Pipe	19–20	1	
C.21407	**Union Nut**, securing Oil Outlet Pipe to Transmission Unit	19–21	1	
C.21779	**Union Nut**, securing Oil Outlet Pipe to Flexible Hose	19–22	1	
C.21420	FLEXIBLE HOSE ASSEMBLY (LONG) FROM OIL OUTLET PIPE TO RADIATOR	19–23	1	
C.21443	**Locking Nut**, on Hose	19–24	1	
C.728	**Washer**, Shakeproof, at fixing of Hose to Support Bracket	19–25	1	
C.21420/1	FLEXIBLE HOSE ASSEMBLY (SHORT) FROM RADIATOR TO OIL RETURN PIPE	19–26	1	
C.21443	**Locking Nut**, on Hose	19–24	1	
C.729	**Washer**, Shakeproof, at fixing of Hose to Support Bracket	19–25	1	
C.21778	OIL RETURN PIPE FROM FLEXIBLE HOSE TO TRANSMISSION UNIT	19–27	1	
C.21406	**Olive**, for Oil Return Pipe	19–20	1	
C.21407	**Union Nut**, securing Oil Return Pipe to Transmission Unit	19–21	1	
C.21779	**Union Nut**, securing Oil Return Pipe to Flexible Hose	19–22	1	
C.21500	BRACKET SUPPORTING HOSES AND PIPES AT FRONT OF ENGINE	19–28	1	
C.21391	SUPPORT PLATE, ON SUMP, FOR OIL PIPES	19–29	1	
C.24599	**Clip**, securing Oil Pipe to Support Plate	19–30	1	
UFS.119/3R	**Screw**, Set, securing Clip to Support Plate		1	
UFN.119/L	**Nut**, on Setscrew		1	
AG.102/X	**Washer**, Spring, under Nut		1	
C.21390	CLAMP, RECEIVING ANTI-RATTLE CLIP	19–31	1	
C.24599	**Clip**, Anti-Rattle, holding Oil Pipes apart	19–30	1	
UFS.119/4R	**Screw**, Set, securing Clip to Clamp		1	
UFN.119/L	**Nut**, on Setscrew		1	
FG.102/X	**Washer**, Spring, under Nut		1	

AUTOMATIC TRANSMISSION MOUNTING DETAILS

Part No.	Description	Plate No.	No. per Unit	Remarks
C.21246	**DIPSTICK TUBE ASSEMBLY**	19–32	1	
C.17617	**Clip,** securing Dipstick Tube to Support	19–33	1	
UFS.131/5R	**Screw,** Set, securing Clip		1	
C.8737/2	**Nut,** on Setscrew		1	
C.21197	**Bracket,** supporting Clip		1	
C.24149	**DIPSTICK ASSEMBLY**	19–34	1	
C.19361	**Felt Washer,** on Dipstick		1	
C.24153	**Felt Collar,** on Dipstick		3	
L.102/4U	**Split Pin,** retaining Felt Washer and Collar		1	
C.24152	**Washer,** Plain, behind Split Pin		1	
C.23285	**Hose,** on Dipstick Breather Tube		1	
C.2905/1	**Clip,** securing Hose to Breather Tube		1	
BD.26562	**Clip,** securing Hose to Wiring Clip		1	

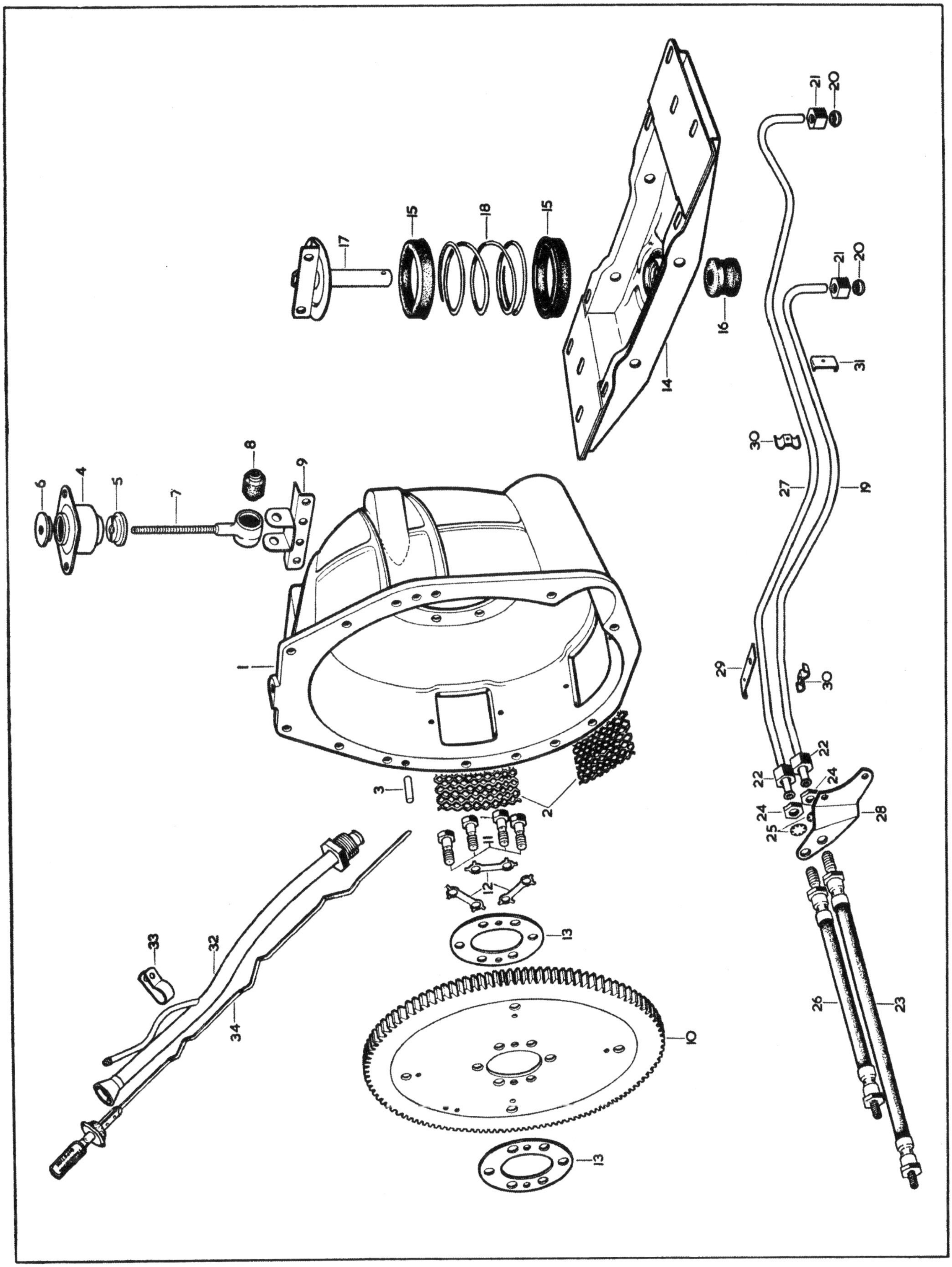
1
2
3
4
5
6
7
8
9
10
11
12
13
13
14
15
15
16
17
18
19
20
20
21
21
22
22
23
24
24
25
26
27
28
29
30
30
31
32
33
34

PLATE 19

PROPELLER SHAFT
STANDARD AND OVERDRIVE TRANSMISSION

Part No.	Description	Plate No.	No. per Unit	Remarks
C.25957	**PROPELLER SHAFT ASSEMBLY (66-0317)**		1	Fitted to Chassis No. 1A.12581 and subs 1A.20565 and subs 1K.1001 and subs 1K.30001 and subs
8404	**Flange Yoke** (K.2-2-329/C)	20–1	2	
9409	**Journal Assembly**, connecting Flange Yokes to Propeller Shaft (K.5-GB.162)	20–2	2	
9410	**Sleeve Yoke Assembly** (K.2-3GB.3758)	20–3	1	
9411	**Gaiter**, for Sliding Joint (94-GB.4007)	20–4	1	
9412	**Ring** (Rubber) retaining Gaiter to Shaft (K.3-86-GB.139)	20–5	2	
9413	**Ring** (Steel) retaining Gaiter to Sleeve Yoke (94-GB.3951)	20–6	1	
9409	**JOURNAL ASSEMBLY, CONNECTING FLANGE YOKES TO PROPELLER SHAFT (K.5-GB.162)**		2	
19	**Circlip**, retaining Journal Bearings (K.2-7-GB.19)	20–7	8	
C.19827	**Bolt**, securing Propeller Shaft to Flanges on Transmission and Final Drive Unit	20–8	8	
C.8737/3	**Nut**, Self-Locking, on Bolts	20–9	8	

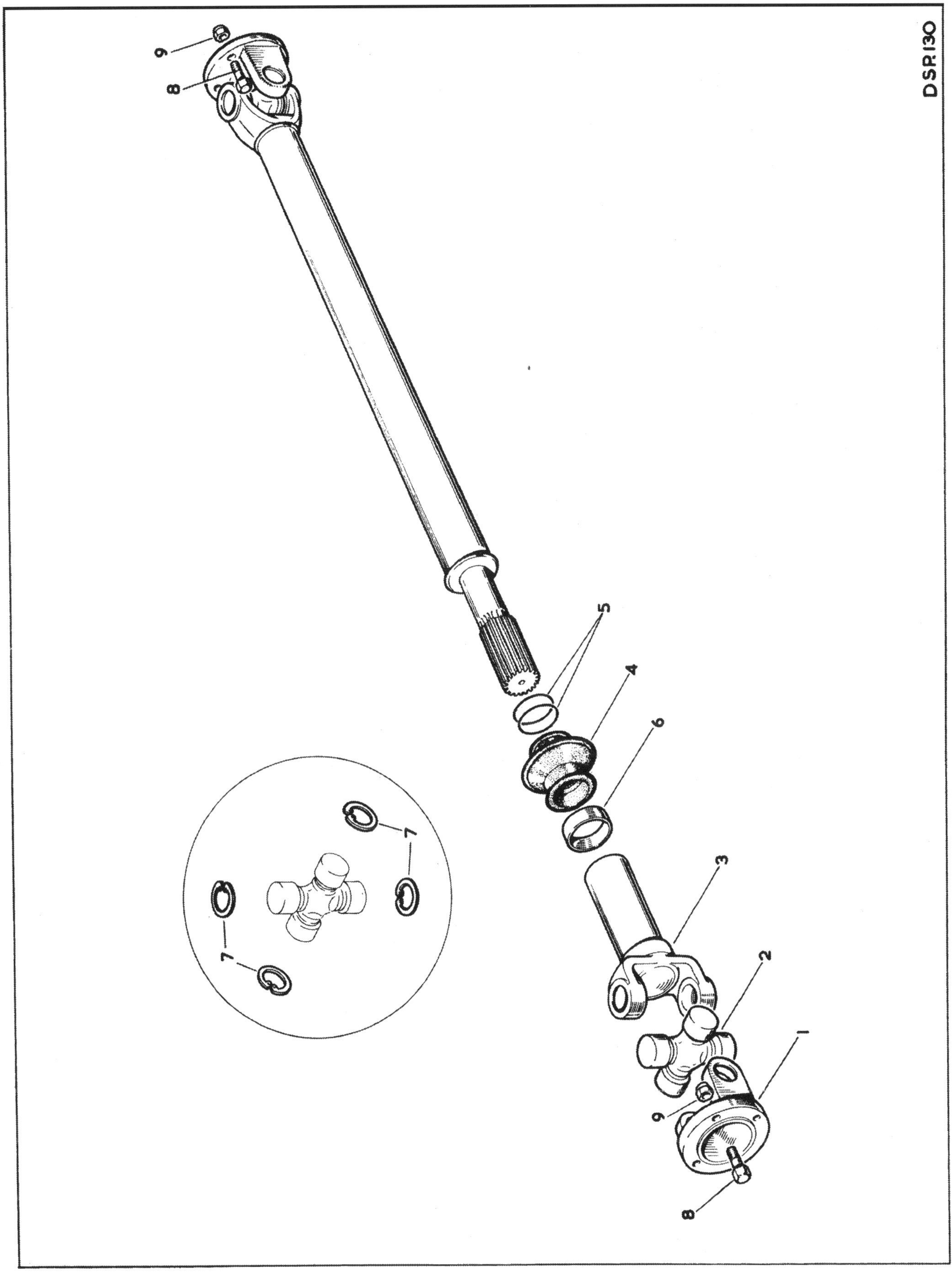

PLATE 20

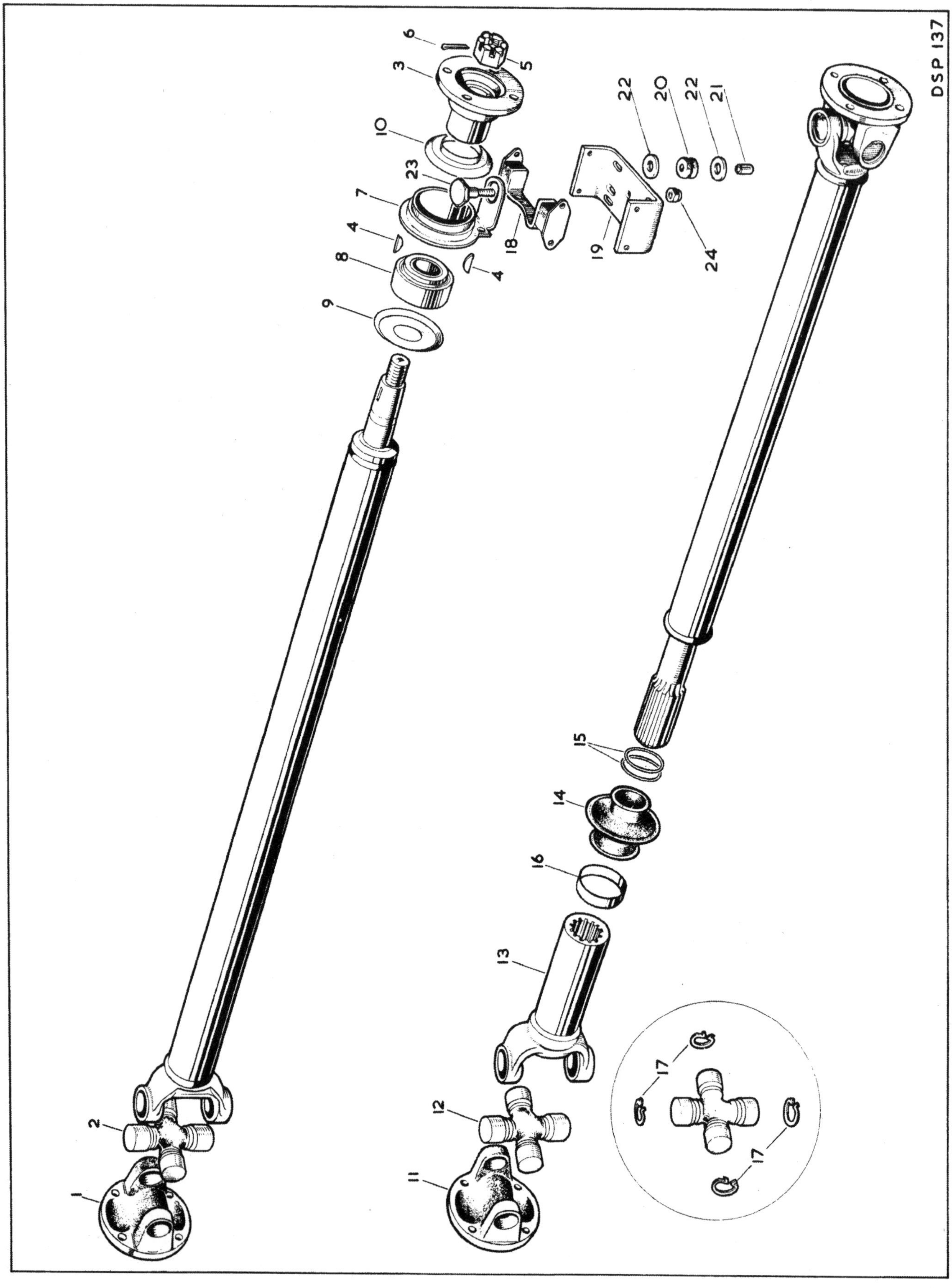

PLATE 21

Part No.	Description	Plate No.	No. per Unit	Remarks
C.26580	**FRONT PROPELLER SHAFT ASSEMBLY (COMPLETE WITH CENTRE BEARING)**		1	
8404	**Flange Yoke** (K/2/2/329/C)	21–1	1	
9409	**Journal Assembly,** connecting Flange Yoke to Propeller Shaft (K.5/GB.162)	21–2	1	
9607	**Centre Bearing Assembly** (SK.3555/F)		1	
9606	**Companion Flange,** at rear of Shaft (K.2/1/GB.2594)	21–3	1	
449	**Key,** locating Companion Flange on Shaft (94/GB.2254)	21–4	2	
113	**Nut,** securing Companion Flange (12/74/GB.161)	21–5	1	
73	**Split Pin,** retaining Nut (8/72/59)	21–6	1	
C.19827	**Bolt,** securing Front Propeller Shaft to Flange on Transmission Unit		4	
C.8737/3	**Nut,** Self-Locking, on Bolts		4	
9607	**CENTRE BEARING ASSEMBLY (SK.3555/F)**		1	
9635	**Housing,** for Centre Bearing (94/GB.4096)	21–7	1	
250	**Bearing** (94/GB.2147)	21–8	1	
249	**Shield,** at front of Bearing (94/GB.2146)	21–9	1	
278	**Shield,** at rear of Bearing (94/GB.2315)	21–10	1	
C.19827	**Bolt,** at joint of Flanges on Front and Rear Shafts		4	
C.8737/3	**Nut,** Self-Locking, on Bolts		4	
C.26579	**REAR PROPELLER SHAFT ASSEMBLY**		1	
8404	**Flange Yoke** (K.2/2/329/C)	21–11	2	
9409	**Journal Assembly,** connecting Yokes to Shaft (K.5/GB.162)	21–12	2	
9410	**Sleeve Yoke** (K.2/3/GB.3758)	21–13	1	
9411	**Gaiter,** for Sliding Joint (94/GB.4007)	21–14	1	
9412	**Ring** (Rubber) retaining Gaiter to Shaft (K.3/86/GB.139	21–15	2	
9413	**Ring** (Steel) retaining Gaiter to Sleeve Yoke (94/GB.3951)	21–16	1	
C.19827	**Bolt,** securing Rear Propeller Shaft to Flange on Rear Axle		4	
C.8737/3	**Nut,** Self-Locking, on Bolts		4	
9409	**JOURNAL ASSEMBLY, CONNECTING YOKES TO PROPELLER SHAFTS (K.5/GB.162)**	21–12	3	
19	**Circlip,** retaining Journal Bearings (K.2/7/GB.19)	21–17	12	
C.19509	**RUBBER MOUNTING, FIXING PROPELLER SHAFT CENTRE BEARING TO BODY**	21–18	1	
UFS.131/7R	**Screw,** Set, securing Centre Bearing to Rubber Mounting		2	
C.8667/2	**Nut,** Self-Locking, on Setscrews		2	
UFS.125/3R	**Screw,** Set, securing Rubber Mounting to Bracket		4	
C.724	**Washer,** Shakeproof, on Setscrews		4	
FW.104/T	**Washer,** Plain, under Shakeproof Washers		4	

PROPELLER SHAFT
AUTOMATIC TRANSMISSION

Part No.	Description	Plate No.	No. per Unit	Remarks
C.19510	**BRACKET, HOLDING RUBBER MOUNTING**	21–19	1	
C.19511	**Screw,** Set, securing Bracket to Body		2	Fitted from Chassis No. 1A.1001 to 1A.1093 1A.20001 to 1A.20017
FW.105/T	**Washer,** Plain, under head of Setscrews		2	
C.8737/2	**Nut,** Self-Locking, on Setscrews		2	
C.1904	**Special Washer,** under Self-Locking Nuts		2	
C.1904	**Special Washer** (Packing) between Bracket and Body		As req'd	
UFS.131/12R	**Screw,** Set, securing Bracket to Support Plate		2	Fitted to Chassis No. 1A.1094 and subs 1A.20018 and subs
FW.105/T	**Washer,** Plain, under Head of Setscrews		2	
C.10126	**Rubber Bush,** in Centre Bearing Bracket Support Plate	21–20	2	
C.22149	**Distance Tube,** in Rubber Bush	21–21	2	
BD.541/5	**Special Washer,** at each side of Rubber Bush	21–22	4	
C.8667/2	**Nut,** Self-Locking, on Setscrews		2	
C.19512	**Safety Bolt,** in Bracket	21–23	1	
C.8667/2	**Nut,** Self-Locking, on Safety Bolt	21–24	1	
C.21382	**SUPPORT PLATE FOR CENTRE BEARING BRACKET**		1	Fitted from Chassis No. 1A.1001 to 1A.2138 1A.20001 to 1A.20155
BD.9604/8	**Screw,** Set, (Pointed End) securing Support Plate to Body		4	
BD.541/2	**Distance Washer,** on Setscrews		As req'd	
FG.106/X	**Washer,** Spring, on Setscrews		4	
C.22233	**SUPPORT PLATE FOR CENTRE BEARING BRACKET**		1	Fitted to Chassis No. 1A.2139 and subs 1A.20156 and subs
BD.8291/7	**Screw,** Set, securing Support Plate to Body		4	
C.4156	**Distance Washer,** on Setscrews		As req'd	
FG.104/X	**Washer,** Spring, on Setscrews		4	
1861	**Washer,** Plain, under Spring Washers		4	

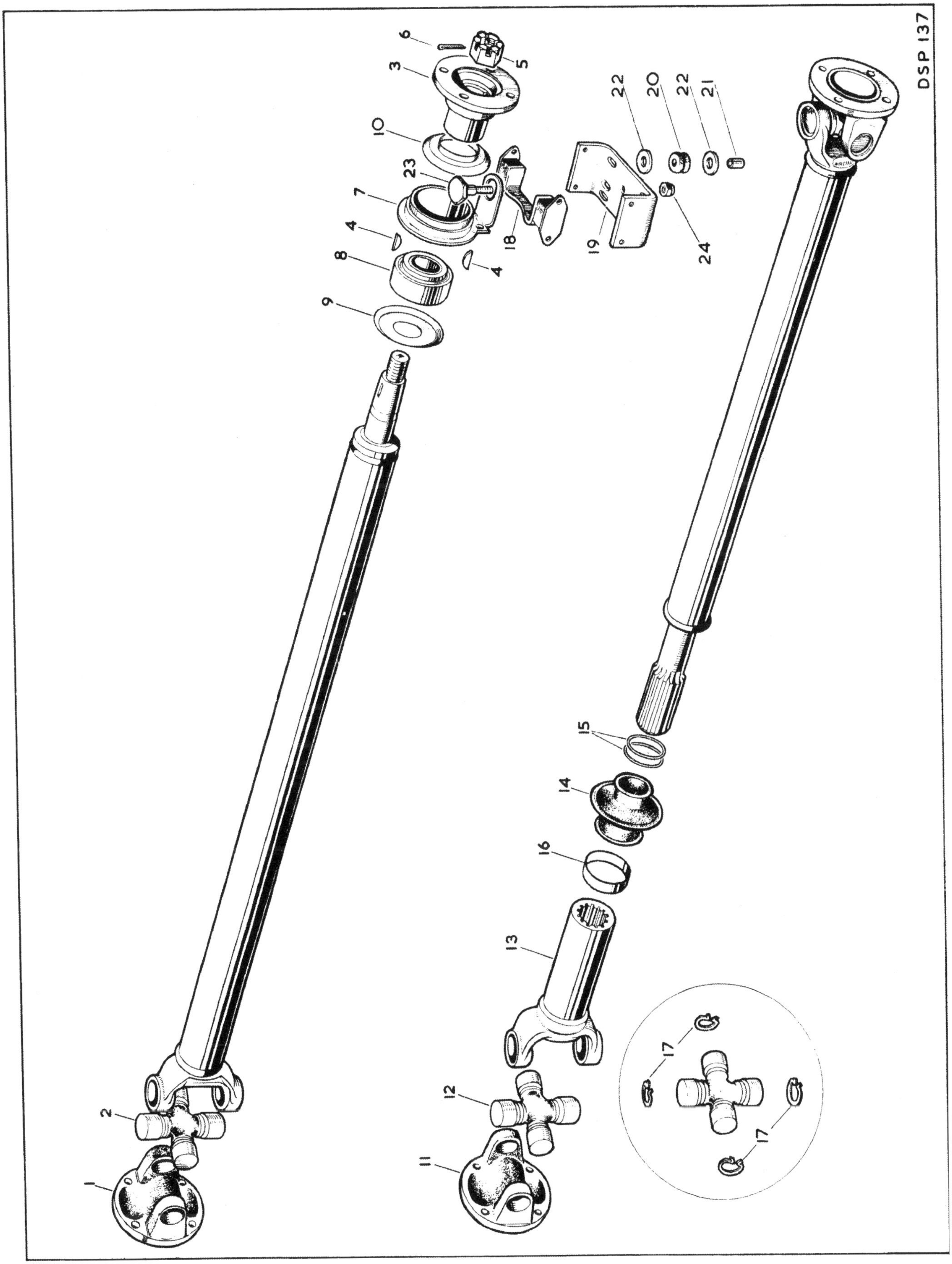

PLATE 21

REAR AXLE

Part No.	Description	Plate No.	No. per Unit	Remarks
C.17090	REAR AXLE, COMPLETE (RATIO 4.55 : 1) (4HA-001/60E)		1	FITTED TO CARS FOR ALL COUNTRIES EXCEPT ITALY, FRANCE, GERMANY, BELGIUM, AND HOLLAND From Chassis No. 1A.1001 to 1A.8639 1A.20001 to 1A.20403
7994	**Carrier and Tube Assembly** (4HA.101/40)	22–1	1	
74	**Screw,** Set, securing Differential Bearing Caps (8B-NC.36)	22–2	4	
C.728	**Washer,** Shakeproof, on Setscrews (8LW-14)	22–3	4	
3930	**Cover,** for Carrier (4HA-010)	22–4	1	
3823	**Plug,** for Filler and Drain Apertures (HA-059)	22–5	2	
3931	**Gasket,** for Cover (4HA-026)	22–6	1	
CS.131/5D	**Screw,** Set, securing Cover to Carrier (5B-NC.10)	22–7	10	
47	**Lock Washer,** on Setscrews (5LW-95)	22–8	10	
6312	**Differential Case Assembly** (4HA-082/7)		1	
3845	**Roller Bearing,** on Differential Case (Timken 24780-24721) (2HA-024)	22–9	2	
3935	**Shim,** adjusting Differential (.003" thick) (4HA-046)	22–10	2	USE SHIM OF DESIRED THICKNESS
3936	**Shim,** adjusting Differential (.005" thick) (4HA-047)	22–10	2	
3937	**Shim,** adjusting Differential (.010" thick) (4HA-048)	22–10	2	
3938	**Shim,** adjusting Differential (.030" thick) (4HA-049)	22–10	2	
6310	**Crown Wheel and Pinion** (Ratio 4.55 : 1) (4HA-105/14A)	22–11	1	
6301	**Screw,** Set, securing Crown Wheel to Differential Case (3HA-075/6)	22–12	8	
6316	**Lock Strap,** securing Crown Wheel Setscrews (4HA-074/2)	22–13	4	
3844	**Roller Bearing,** at inner end of Pinion (Timken 31593-31520) (2HA-023)	22–14	1	
3860	**Shim,** inner, adjusting Pinion (.003" thick) (2HA-043)	22–15	2	USE SHIM OF DESIRED THICKNESS
3861	**Shim,** inner, adjusting Pinion (.005" thick) (2HA-044)	22–15	2	
3862	**Shim,** inner, adjusting Pinion (.010" thick) (2HA-045)	22–15	2	
8029	**Distance Washer,** behind outer Pinion Shims (4HA-115)	22–15A	1	
3856	**Shim,** outer, adjusting Pinion (.003" thick) (2HA-039)	22–16	2	USE SHIM OF DESIRED THICKNESS
3857	**Shim,** outer, adjusting Pinion (.005" thick) (2HA-040)	22–16	2	
3858	**Shim,** outer, adjusting Pinion (.010" thick) (2HA-041)	22–16	2	
3859	**Shim,** outer, adjusting Pinion (.030" thick) (2HA-042)	22–16	2	
3843	**Roller Bearing,** at outer end of Pinion (Timken 02872-02820) (2HA-022)	22–17	1	
3934	**Oil Slinger,** on Pinion (4HA-036)	22–18	1	
3840	**Oil Seal,** on Pinion (2HA-019)	22–19	1	
3841	**Gasket,** for Pinion Oil Seal (2HA-020)	22–20	1	
4153	**Grease Nipple,** on Axle Tubes (Tec. NA.5799)	22–21	2	
3879	**Companion Flange Assembly** (2HA-083)	22–22	1	
115	**Nut,** securing Companion Flange to Pinion (12LN-NF.9)	22–23	1	
114	**Washer,** under Nut (12W-24)	22–24	1	
7995	**Shaft** (4HA-005/47)		2	
C.15303	REAR AXLE, COMPLETE (RATIO 4.27 : 1) (4HA-001/60C)		1	FITTED TO L.H. DRIVE CARS FOR ITALY, FRANCE, GERMANY, BELGIUM & HOLLAND. Up to and including Chassis No. 1A.20403 THEREAFTER FITTED TO ALL CARS IRRESPECTIVE OF COUNTRY
6309	**Crown Wheel and Pinion** (Ratio 4.27 : 1) (4HA-105/12A)		1	

For all other items, please refer to Rear Axle C.17090

NOTE: **Axles of either ratio can be fitted for any country to special order**

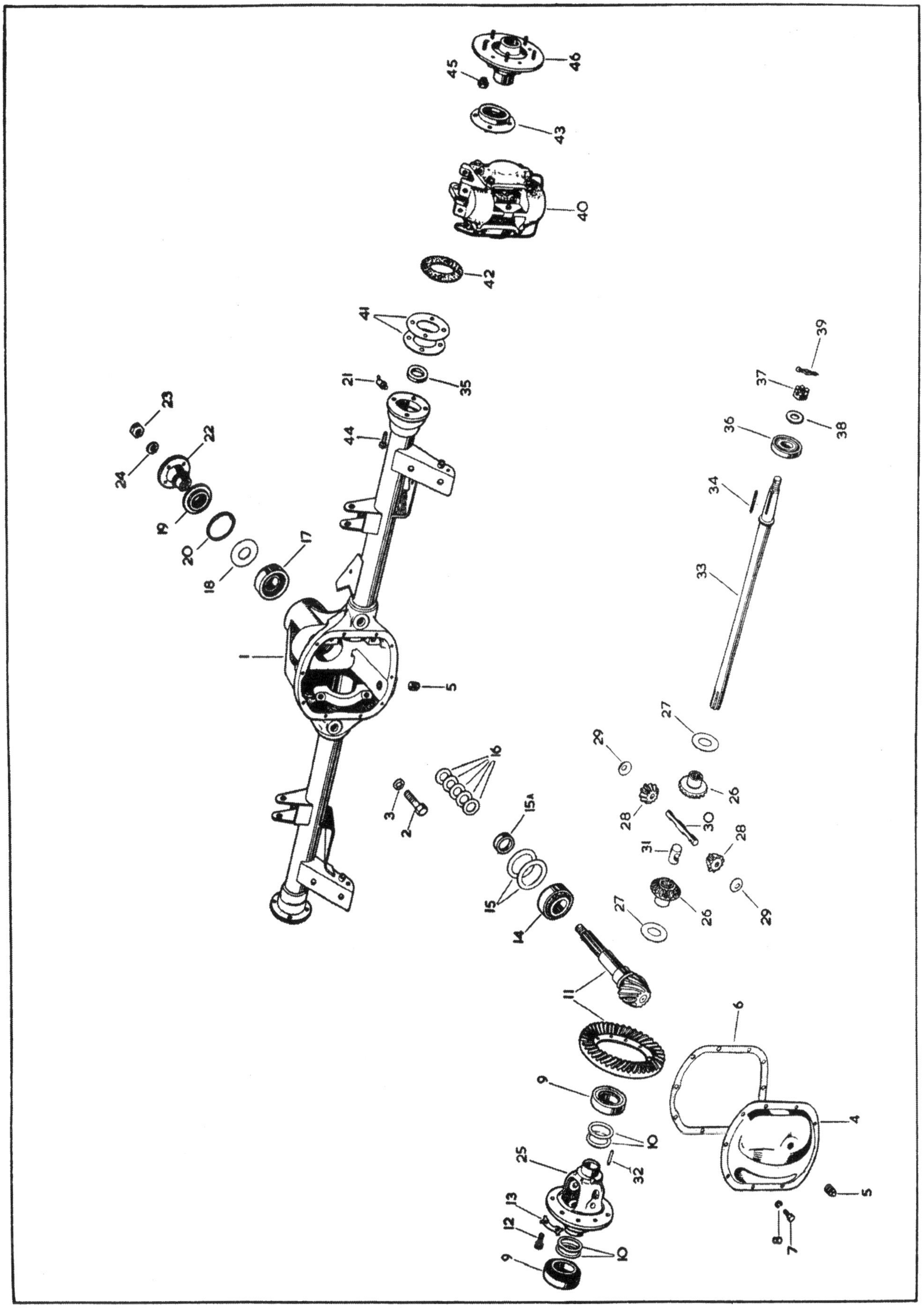
1
2
3
4
5
6
7
8
9
10
11
12
13
14
15
15A
16
17
18
19
20
21
22
23
24
25
26
27
28
29
30
31
32
33
34
35
36
37
38
39
40
41
42
43
44
45
46

PLATE 22

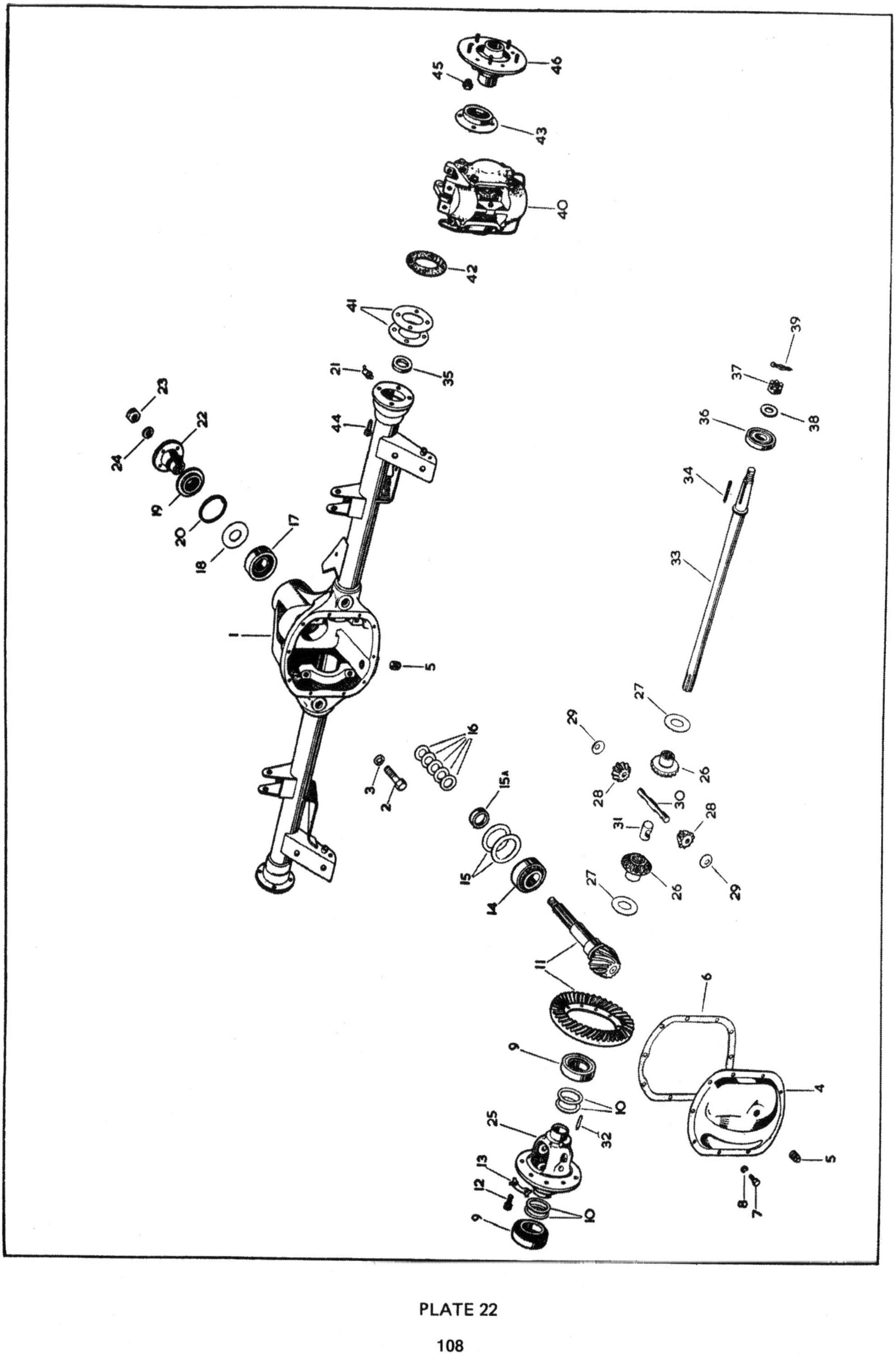

PLATE 22

Part No.	Description	Plate No.	No. per Unit	Remarks
6312	**DIFFERENTIAL CASE ASSEMBLY (4HA-082/7)**		1	
6314	**Case only**, for Differential (4HA-006/5)	22–25	1	
3929	**Gear**, Side, for Differential (4HA-007)	22–26	2	
3855	**Thrust Washer**, for Side Gears (2HA-038)	22–27	2	
3835	**Gear**, Differential Pinion Mate (2HA-008/1)	22–28	2	
3854	**Thrust Washer**, for Pinion Mate Gears (2HA-037)	22–29	2	
3837	**Shaft**, for Pinion Mate Gears (2HA-012)	22–30	1	
3932	**Spacer**, on Shaft (4HA-033)	22–31	1	
3838	**Pin**, locking Mate Gear Shaft in Differential (2HA-013)	22–32	1	
3879	**PINION COMPANION FLANGE ASSEMBLY (2HA-083)**	22–22	1	
3842	**Dust Shield**, for Companion Flange (2HA-021)		1	
7995	**REAR AXLE SHAFT (4HA-005/47)**	22–33	2	
C.11835	**Key**, securing Hubs to Axle Shafts (3HA-029/1)	22–34	2	
4144	**Oil Seal**, in ends of Carrier Tube for Axle Shafts (3HA-027/3)	22–35	2	
7993	**Roller Bearing**, for Hubs (Timken 14130X-14276) (7HA-025)	22–36	2	
4141	**Slotted Nut**, on Axle Shaft (14SN.-NF.10)	22–37	2	
4142	**Washer**, under Slotted Nut (14W.-26)	22–38	2	
L.104/14U	**Split Pin**, securing Nuts	22–39	2	
C.18613	**R.H. REAR BRAKE CALIPER ASSEMBLY (VB.1154A/N)**	22–40	1	Fitted from Chassis No. 1A.1001 to 1A.4415 1A.20001 to 1A.20282
C.18614	**L.H. REAR BRAKE CALIPER ASSEMBLY (VB.1154B/N)**		1	
C.24282	**R.H. REAR BRAKE CALIPER ASSEMBLY (VB.95059A/N)**		1	Fitted to Chassis No. 1A.4416 and subs 1A.20283 and subs
C.24283	**L.H. REAR BRAKE CALIPER ASSEMBLY (VB.95059B/N)**		1	
4145	**Shim**, adjusting Hub Bearing (.003" thick) (HA-050/2)	22–41	2	USE SHIM OF DESIRED THICKNESS
4146	**Shim**, adjusting Hub Bearing (.005" thick) (HA-051/2)	22–41	2	
4147	**Shim**, adjusting Hub Bearing (.010" thick) (HA-052/2)	22–41	2	
4148	**Shim**, adjusting Hub Bearing (.030" thick) (HA-053/2)	22–41	2	
4149	**Gasket**, behind Hub Bearing Shims (HA-031/4)	22–42	4	
7953	**Oil Seal Assembly**, for Hubs (3HA-057/4)	22–43	2	
UFB.137/8R	**Bolt**, securing Rear Brake Adaptor Plates to Axle Flanges	22–44	8	
C.8667/3	**Nut**, Self-Locking, on Bolts	22–45	8	
C.16287	**R.H. REAR CALIPER ADAPTOR PLATE**		1	
C.16288	**L.H. REAR CALIPER ADAPTOR PLATE**		1	
C.13198	**Shim** (.005" thick) between Adaptor Plates and Calipers		As req'd	SELECTIVE SIZES
C.13199	**Shim** (.020" thick) between Adaptor Plates and Calipers		As req'd	
C.13994	**Shim** (.040" thick) between Adaptor Plates and Calipers		As req'd	
C.16262	**Bolt**, securing Calipers to Adaptor Plates		4	
C.837	**Wire**, locking Bolts		As req'd	
C.13644	**REAR HUB ASSEMBLY**	22–46	2	FOR USE WHEN DISC TYPE ROAD WHEELS ARE FITTED
C.13365	**Stud**, in Hubs, fixing Road Wheels		10	

REAR AXLE

Part No.	Description	Plate No.	No. per Unit	Remarks
C.13094	R.H. REAR HUB ASSEMBLY		1	FOR USE WHEN WIRE SPOKE ROAD WHEELS ARE FITTED
C.13095	L.H. REAR HUB ASSEMBLY		1	
C.21270	R.H. HUB CAP		1	FOR USE WHEN WIRE SPOKE ROAD WHEELS ARE FITTED
C.21271	L.H. HUB CAP		1	
C.14927	TOOL FOR REMOVING AND FITTING HUB CAPS		1	
C.7860	CLIP, SECURING BRAKE PIPES TO REAR AXLE		2	3 required at Chassis No. 1A.4416 and subs 1A.20283 and subs
UFS.325/8H	**Screw**, Set, securing Clips to Axle Tubes		2	
UFN.125/L	**Nut**, on Setscrews		2	
C.724	**Washer**, Shakeproof, under Nuts		2	

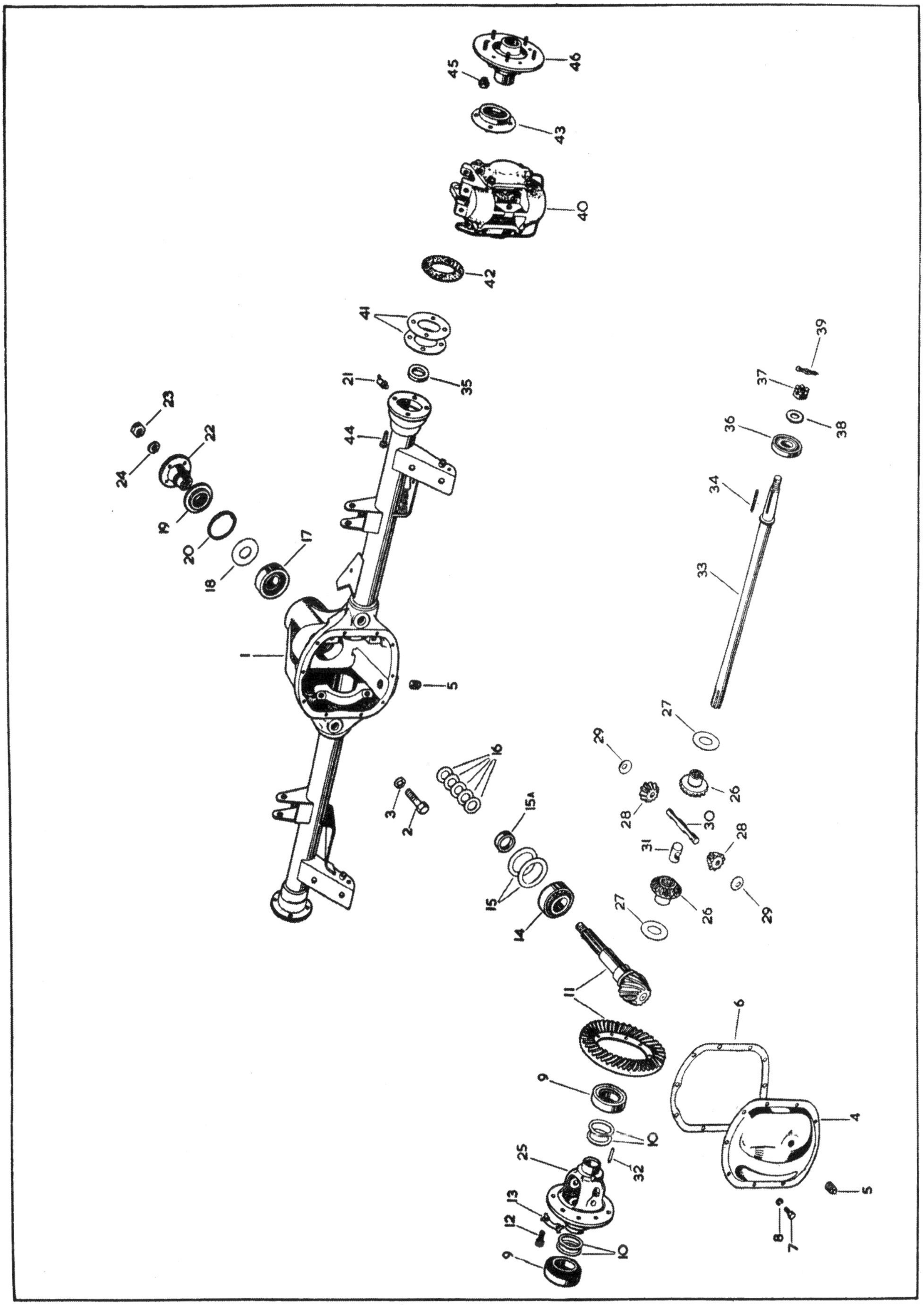

PLATE 22

REAR AXLE

WITH THORNTON "POWR-LOK" DIFFERENTIAL

Part No.	Description	Plate No.	No. per Unit	Remarks
C.17098	**REAR AXLE, COMPLETE (RATIO 4.55 : 1) (4HA)**			**SUPPLIED TO SPECIAL ORDER ONLY**
7994	**Carrier and Tube Assembly** (4HA-101/40)	23–1	1	
74	**Screw,** Set, securing Differential Bearing Caps (8B-NC.36)	23–2	4	
C.728	**Washer,** Shakeproof, on Setscrews (8LW-14)	23–3	4	
7745	**Cover,** for Carrier (4HA-064/3)	23–4	1	
3823	**Plug,** for Filler and Drain apertures (HA.059)	23–5	2	
3931	**Gasket,** for Cover (4HA-026)	23–6	1	
CS.131/5D	**Screw,** Set, securing Cover to Carrier (5B-NC.10)	23–7	10	
47	**Lock Washer,** on Setscrews (5LW-95)	23–8	10	
7380	**Differential Case Assembly** (4HA-082/7)		1	
3845	**Roller Bearing,** on Differential Case (Timken 24780-24721) (2HA-024)	23–9	2	
3935	**Shim,** adjusting Differential (.003" thick) (4HA-046)	23–10	2	**USE SHIM OF DESIRED THICKNESS**
3936	**Shim,** adjusting Differential (.005" thick) (4HA-047)	23–10	2	
3937	**Shim,** adjusting Differential (.010" thick) (4HA-048)	23–10	2	
3938	**Shim,** adjusting Differential (.030" thick) (4HA-049)	23–10	2	
6310	**Crown Wheel and Pinion** (Ratio 4.55 : 1) (4HA-105/14A)	23–11	1	
6301	**Screw,** Set, securing Crown Wheel to Differential Case (3HA-075/6)	23–12	10	
7384	**Lock Strap,** securing Crown Wheel Setscrews (4HA-074/4)	23–13	5	
3844	**Roller Bearing,** at inner end of Pinion (Timken 31593-31520) (2HA-023)	23–14	1	
3860	**Shim,** inner, adjusting Pinion (.003" thick) (2HA-043)	23–15	2	**USE SHIM OF DESIRED THICKNESS**
3861	**Shim,** inner, adjusting Pinion (.005" thick) (2HA-044)	23–15	2	
3862	**Shim,** inner, adjusting Pinion (.010" thick) (2HA-045)	23–15	2	
8029	**Distance Washer,** behind Outer Pinion Shims (4HA-115)	23–16	1	
3856	**Shim,** outer, adjusting Pinion (.003" thick) (2HA-039)	23–17	2	**USE SHIM OF DESIRED THICKNESS**
3857	**Shim,** outer, adjusting Pinion (.005" thick) (2HA-040)	23–17	2	
3858	**Shim,** outer, adjusting Pinion (.010" thick) (2HA-041)	23–17	2	
3859	**Shim,** outer, adjusting Pinion (.030" thick) (2HA-042)	23–17	2	
3843	**Roller Bearing,** at outer end of Pinion (Timken 02872-02820) (2HA-022)	23–18	1	
3934	**Oil Slinger,** on Pinion (4HA-036)	23–19	1	
3840	**Oil Seal,** on Pinion (2HA-019)	23–20	1	
3841	**Gasket,** for Pinion Oil Seal (2HA-020)	23–21	1	
3879	**Companion Flange Assembly** (2HA-083)	23–22	1	
115	**Nut,** securing Companion Flange to Pinion (12LN-NF.9)	23–23	1	
114	**Washer,** under Nut (12W-24)	23–24	1	
4153	**Grease Nipple,** on Axle Tubes (Tec.NA.5799)	23–25	2	
7995	**Shaft** (4HA-005/47)	23–26	2	
C.16204	**REAR AXLE, COMPLETE (RATIO 4.27 : 1) (4HA-001/66)**		1	**SUPPLIED TO SPECIAL ORDER ONLY**
6309	**Crown Wheel and Pinion** (Ratio 4.27 : 1) (4HA-105/12A)	23–11	1	
	All other items are as for Rear Axle C.17098			
7380	**DIFFERENTIAL CASE ASSEMBLY (4HA-082/11)**		1	
7888	**Case only,** for Differential (4HA-006/10)	23–27	2 halves	
7683	**Friction Plate** (flat) for Differential Clutch (4HA-111/3)	23–28	4	
8397	**Friction Plate** (dished) for Differential Clutch (4HA-111/4)	23–29	2	
7684	**Friction Disc,** for Differential Clutch (4HA-112)	23–30	4	
7682	**Ring,** for Side Gears (4HA-109/1)	23–31	2	
7681	**Gear,** Side, for Differential (4HA-007/1)	23–32	2	

Part No.	Description	Plate No.	No. per Unit	Remarks
	DIFFERENTIAL CASE ASSEMBLY *(contd.)*			
7685	**Gear,** Differential Pinion Mate (4HA-008)	23–33	4	
7686	**Shaft,** for Pinion Mate Gears (4HA-012/1)	23–34	2 halves	
7687	**Spacer,** on Shaft (4HA-033/1)	23–35	2	
7688	**Roll Pin,** for Spacer (RP.4-10)	23–36	1	
7680	**Bolt,** securing Differential Case (6B-NC.44)	23–37	10	
3879	**PINION COMPANION FLANGE ASSEMBLY (2HA-083)**	23–22	1	
3842	**Dust Shield,** for Companion Flange (2HA-021)		1	
7995	**REAR AXLE SHAFT (4HA-005/47)**	23–26	2	
C.11835	**Key,** securing Hubs to Axle Shaft (3HA-029/1)	23–38	2	
4144	**Oil Seal,** in end of Carrier Tube, for Axle Shafts (3HA-027/3)	23–39	2	
7993	**Roller Bearing,** for Hubs (Timken 14130X-14276) (7HA-025)	23–40	2	
4141	**Slotted Nut,** on Axle Shaft (14SN-NF.10)	23–41	2	
4142	**Washer,** under Slotted Nuts (14W.26)	23–42	2	
L.104/14U	**Split Pin,** securing Nuts	23–43	2	
C.18613	**R.H. REAR BRAKE CALIPER ASSEMBLY (VB.1154A/N)**	23–44	1	Fitted from Chassis No. 1A.1001 to 1A.4415 1A.20001 to 1A.20282
C.18614	**L.H. REAR BRAKE CALIPER ASSEMBLY (VB.1154B/N)**		1	
C.24282	**R.H. REAR BRAKE CALIPER ASSEMBLY (VB.95059A/N)**		1	Fitted to Chassis No. 1A.4416 and subs 1A.20283 and subs
C.24283	**L.H. REAR BRAKE CALIPER ASSEMBLY (VB.95059B/N)**		1	
4145	**Shim,** adjusting Hub Bearing (.003" thick) (HA-050/2)	23–45	2	USE SHIM OF DESIRED THICKNESS
4146	**Shim,** adjusting Hub Bearing (.005" thick) (HA-051/2)	23–45	2	
4147	**Shim,** adjusting Hub Bearing (.010" thick) (HA-052/2)	23–45	2	
4148	**Shim,** adjusting Hub Bearing (.030" thick) (HA-053/2)	23–45	2	
4149	**Gasket,** at side of Brake Adaptor Plates (HA.031/4)	23–46	2	
7953	**Oil Seal Assembly,** for Hubs (3HA-057/4)	23–47	2	
UFB.137/8R	**Bolt,** securing Rear Brake Adaptor Plates to Axle Flanges	23–48	8	
C.8667/3	**Nut,** Self Locking, on Bolts	23–49	8	
C.16287	**R.H. REAR ADAPTOR PLATE**		1	
C.16288	**L.H. REAR ADAPTOR PLATE**		1	
C.13198	**Shim** (.005" thick) between Adaptor Plates and Calipers (VBO.1333)		As req'd	SELECTIVE SIZES
C.13199	**Shim** (.020" thick) between Adaptor Plates and Calipers (VBO.1334)		As req'd	
C.13994	**Shim** (.040" thick) between Adaptor Plates and Calipers		As req'd	
C.16262	**Bolt,** securing Calipers to Adaptor Plates		4	
FG.107/X	**Washer,** Spring, on Bolts		4	
C.837	**Wire,** locking Bolts		As req'd	
C.13644	**REAR HUB ASSEMBLY**	23–50	2	FOR USE WITH DISC ROAD WHEELS
C.13265	**Stud,** in Hubs, fixing Road Wheels		10	
C.13094	**R.H. REAR HUB ASSEMBLY**		1	FOR USE WHEN WIRE SPOKE ROAD WHEELS ARE FITTED
C.13095	**L.H. REAR HUB ASSEMBLY**		1	

REAR AXLE
WITH THORNTON "POWR-LOK" DIFFERENTIAL

Part No.	Description	Plate No.	No. per Unit	Remarks
C.21270	R.H. HUB CAP		1	FOR USE WHEN WIRE SPOKE ROAD WHEELS ARE FITTED
C.21271	L.H. HUB CAP		1	
C.14927	TOOL FOR REMOVING AND FITTING HUB CAP		1	
C.7860	CLIP, RETAINING BRAKE PIPES	23–51	2	3 required at Chassis No. 1A.4416 and subs 1A.20283 and subs
UFS.325/8H	Screw, Set, securing Clips to Axle Casing		2	
UFN.125/L	Nut, on Setscrews		2	
C.724	Washer, Shakeproof, under Nuts		2	

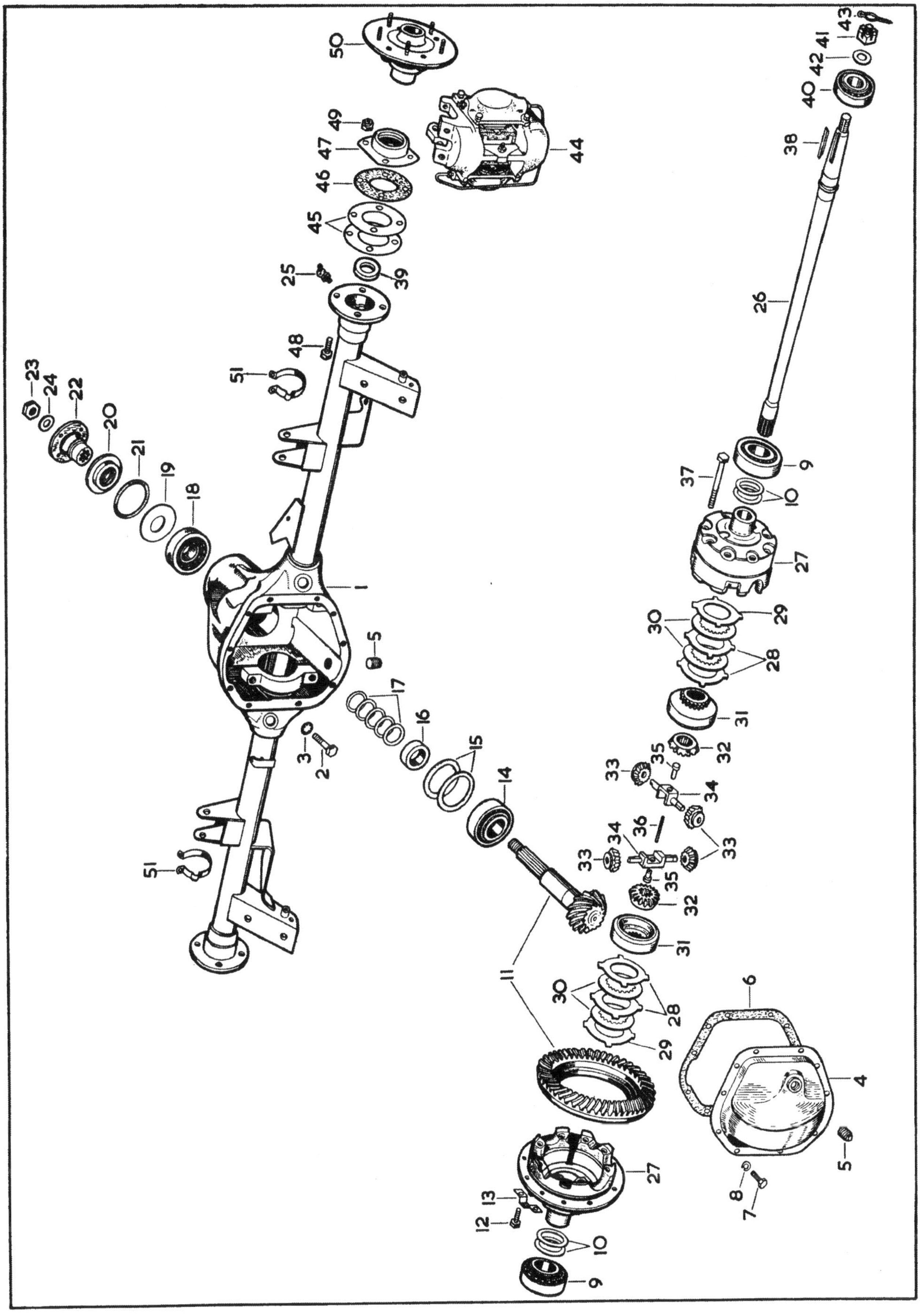
1
2
3
4
5
6
7
8
9
10
11
12
13
14
15
16
17
18
19
20
21
22
23
24
25
26
27
28
29
30
31
32
33
34
35
36
37
38
39
40
41
42
43
44
45
46
47
48
49
50
51

PLATE 23

FRONT SUSPENSION

Part No.	Description	Plate No.	No. per Unit	Remarks
C.19997	**R.H. FRONT SUSPENSION ASSEMBLY**		1	**FOR USE WITH DISC TYPE ROAD WHEELS from Chassis No. 1A.1001 to 1A.3941 1A.20001 to 1A.20269**
C.19998	**L.H. FRONT SUSPENSION ASSEMBLY**		1	
C.20013	**R.H. FRONT SUSPENSION ASSEMBLY**		1	**FOR USE WITH WIRE SPOKE ROAD WHEELS from Chassis No. 1A.1001 to 1A.3941 1A.20001 to 1A.20269**
C.20014	**L.H. FRONT SUSPENSION ASSEMBLY**	24–1	1	
C.23342	**R.H. FRONT SUSPENSION ASSEMBLY**		1	**FOR USE WITH DISC TYPE ROAD WHEELS from Chassis No. 1A.3942 to 1A.4415 1A.20270 to 1A.20282**
C.23343	**L.H. FRONT SUSPENSION ASSEMBLY**		1	
C.23346	**R.H. FRONT SUSPENSION ASSEMBLY**		1	**FOR USE WITH WIRE SPOKE ROAD WHEELS from Chassis No. 1A.3942 to 1A.4415 1A.20270 to 1A.20282**
C.23347	**L.H. FRONT SUSPENSION ASSEMBLY**		1	
C.24274	**R.H. FRONT SUSPENSION ASSEMBLY**		1	**FOR USE WITH DISC TYPE ROAD WHEELS at Chassis No. 1A.4416 and subs. 1A.20283 and subs.**
C.24275	**L.H. FRONT SUSPENSION ASSEMBLY**		1	
C.24276	**R.H. FRONT SUSPENSION ASSEMBLY**		1	**FOR USE WITH WIRE SPOKE ROAD WHEELS at Chassis No. 1A.4416 and subs. 1A.20283 and subs.**
C.24277	**L.H. FRONT SUSPENSION ASSEMBLY**		1	
C.15353	**FRONT SUSPENSION CROSS MEMBER**	24–2	1	Fitted to MANUAL Steering cars at Chassis No. 1A.1001 and subs, 1A.20001 and subs, also to POWER-ASSISTED Steering cars at Chassis No. 1A.1001 to 1A.13156 (excluding 1A.13003) 1A.20001 and subs 1K.30001 to 1K.30021
C.26599	**FRONT SUSPENSION CROSS MEMBER FOR R.H. DRIVE CARS FITTED WITH POWER ASSISTED STEERING UNIT C.26018**		1	Fitted to Chassis No. 1A.13157 and subs (also 1A.13003) 1K.1001 and subs
C.26600	**FRONT SUSPENSION CROSS MEMBER FOR L.H. DRIVE CARS FITTED WITH POWER ASSISTED STEERING UNIT C.26019**		1	Fitted to Chassis No. 1K.30022 and subs
BD.10341	**Rubber Plug**, sealing Tooling Holes in Top of Front Suspension Cross Member	24–3	2	
C.11089	**RUBBER MOUNTING AT FRONT OF SUSPENSION CROSS MEMBER**	24–4	2	Fitted from Chassis No. 1A.1001 to 1A.2200 1A.20001 to 1A.20166
C.23314	**RUBBER MOUNTING AT FRONT OF SUSPENSION CROSS MEMBER**		2	Fitted to Chassis No. 1A.2201 and subs 1A.20167 and subs
C.8737/4	**Nut**, Self-locking, securing Front Mountings to Cross Member		2	
C.3337	**Special Washer**, under Nuts		2	
UFS.137/7R	**Screw**, Set, securing Front Mountings to Body		4	
C.8737/3	**Nut**, Self-locking, on Setscrews		4	
FW.106/T	**Washer**, Plain, under Nuts		4	

Part No.	Description	Plate No.	No. per Unit	Remarks
C.13918	**BUMP STOP (RUBBER) ON TURRET OF FRONT SUSPENSION CROSS MEMBER**	24–5	2	
UFS.131/6R	**Screw,** Set, securing Bump Stops		4	
C.8667/2	**Nut,** Self-locking, on Setscrews		4	
C.9004	**RUBBER MOUNTING AT REAR OF FRONT SUSPENSION CROSS MEMBER**	24–6	2	
C.8737/3	**Nut,** Self-locking, securing Rear Mountings to Cross Member		4	
FW.106/T	**Washer,** Plain, under Nuts		4	
UFB.137/26R	**Bolt,** securing Rear Mountings to Body		2	
C.8737/3	**Nut,** Self-locking, on Bolts		2	
C.18573	**UPPER WISHBONE ASSEMBLY, R.H.**	24–7	1	Fitted from Chassis No. 1A.1001 to 1A.3941 1A.20001 to 1A.20269
C.21183	**UPPER WISHBONE ASSEMBLY, L.H.**		1	
C.17176	**Upper Wishbone Lever,** L.H. Rear and R.H. Front	24–8	2	
C.17177	**Upper Wishbone Lever,** R.H. Rear and L.H. Front	24–9	2	
C.15319	**Fulcrum Shaft,** for mounting of Upper Wishbone Levers	24–10	2	
C.8675	**Distance Washer,** on Fulcrum Shaft	24–11	4	
C.8672	**Bush** (Rubber) on Fulcrum Shaft	24–12	4	
NN.750/L	**Slotted Nut,** securing Upper Wishbone Levers to Fulcrum Shafts	24–13	4	
C.8676	**Special Washer,** under Nuts	24–14	4	
L.104/10U	**Split Pin,** retaining Slotted Nuts	24–15	4	
C.15890	**Ball Joint Assembly,** for Upper Wishbone Levers	24–16	2	
C.4740	**Distance Piece,** between Ball Joint and Inner Face of Upper Wishbone Lever	24–17	2	
C.15308	**Shim,** for adjustment of Castor Angle	24–18	8	
C.15903	**Bolt** (Short) securing Ball Joints to Upper Wishbone Levers		2	
C.17171	**Bolt** (Long) securing Ball Joints to Upper Wishbone Levers		2	
C.8667/3	**Nut,** Self-locking, on Bolts		4	
FW.106/T	**Washer,** Plain, under Nuts		4	
C.17172	**Rebound Stop** (Rubber) on Upper Wishbone Levers	24–19	2	
C.17170	**Bolt,** securing Rebound Stops		4	
FG.106/X	**Washer,** Spring, on Bolts		4	
C.24567	**UPPER WISHBONE ASSEMBLY, R.H.**		1	Fitted to Chassis No. 1A.3942 and subs 1A.20270 and subs
C.24568	**UPPER WISHBONE ASSEMBLY, L.H.**		1	
C.23034	**Ball Joint Assembly,** for Upper Wishbone Levers		2	
	All other items are as for Upper Wishbone Assemblies C.18573 and C.21183			
C.8918/1	**SHIM FOR ADJUSTMENT OF CAMBER ANGLE (.064" THICK)**	24–20	As req'd	**CHOOSE SHIMS OF DESIRED THICKNESS**
C.8918/2	**SHIM FOR ADJUSTMENT OF CAMBER ANGLE (.048" THICK)**	24–20	As req'd	
C.8918/3	**SHIM FOR ADJUSTMENT OF CAMBER ANGLE (.036" THICK)**	24–20	As req'd	
C.17078	**Bolt** (Short) securing Upper Fulcrum Shaft to Front Suspension Cross Member at top	24–21	4	
C.17079	**Bolt** (Long) securing Upper Fulcrum Shaft to Front Suspension Cross Member at bottom	24–22	4	
FG.106/X	**Washer,** Spring, on Bolts		8	

FRONT SUSPENSION

Part No.	Description	Plate No.	No. per Unit	Remarks
C.15890	**BALL JOINT ASSEMBLY FOR UPPER WISHBONE LEVERS**	24–16	2	Fitted from Chassis No. 1A.1001 to 1A.3941 1A.20001 to 1A.20269
C.3020	**Rubber Gaiter,** at bottom of Ball Joints		2	
C.9048	**Grease Nipple,** at top of Ball Joints		2	
C.8737/5	**Nut,** Self-locking, securing Ball Joints to Stub Axle Carriers		2	
C.791	**Special Washer,** under Nuts		2	
C.23024	**BALL JOINT ASSEMBLY FOR UPPER WISHBONE LEVERS**		2	Fitted to Chassis No. 1A.3942 and subs 1A.20270 and subs
C.22967	**Gaiter** (Rubber) at bottom of Ball Joints		2	
C.22970	**Insert** (Plastic) between Ball Joint		2	
C.22969	**Ring,** securing Gaiters to Plastic Insert		2	
C.9048	**Grease Nipple,** at top of Ball Joints (Tecalemit NA.6626)		2	
C.22971	**Washer** (Nylon) under Grease Nipples		2	
C.8737/5	**Nut,** Self-Locking, securing Ball Joints to Stub Axle Carriers		2	
FW.108/T	**Washer,** Plain, under Nuts		2	
C.15328	**LOWER WISHBONE LEVER**	24–23	2	
C.8673	**Bush,** in Wishbone Levers	24–24	4	
C.10778	**FULCRUM SHAFT FOR MOUNTING OF LOWER WISHBONE LEVERS TO FRONT SUSPENSION CROSS MEMBER**	24–25	2	
C.8693	**Special Washer,** on Fulcrum Shaft	24–26	4	
NN.756/L	**Slotted Nut,** securing Lower Wishbone Levers to Fulcrum Shaft		4	
L.104/10U	**Split Pin,** retaining Slotted Nuts		4	
C.15700	**LEFT-HAND SEAT ASSEMBLY FOR FRONT SUSPENSION COIL SPRING**	24–27	1	
C.15699	**RIGHT-HAND SEAT ASSEMBLY FOR FRONT SUSPENSION COIL SPRING**		1	
C.11917	**Screw,** Set, securing Coil Spring Seats to Lower Wishbone Levers		10	
C.11923	**Screw,** Set, securing Coil Spring Seats to Lower Wishbone Levers		2	
FG.106/X	**Washer,** Spring, on Setscrews		12	
C.16953	**FRONT SUSPENSION COIL SPRING**	24–28	2	**IDENTIFIED BY STRIP OF WHITE PAINT ALONG FULL LENGTH OF SPRING**
C.11874/1	**Packing Ring** (1/8" thick) at top of Spring	24–29	As req'd	
C.11874	**Packing Ring** (1/4" thick) at top of Spring	24–29	As req'd	

NOTE: Packing Rings are used in accordance with the Static Load Characteristics of the Springs,which are determined during manufacture. A patch of paint on the Centre Coils of the Spring indicates the packing required. A table showing the co-relation of Colour Code to Static Load Range is given below.
IT IS MOST IMPORTANT that Coil Springs of identical Static Load Range are fitted to an individual car.

Colour Code	Packing Required R.H. Spring	Packing Required L.H. Spring	Remarks
White	3/8"	¼"	**R.H. DRIVE CARS ONLY**
Blue	¼"	1/8"	
Green	1/8"	None	
White	¼"	¼"	**L.H. DRIVE CARS ONLY**
Blue	1/8"	1/8"	
Green	None	None	

Part No.	Description	Plate No.	No. per Unit	Remarks
C.8689	**BRACKET, UNDER SPRING SEAT, FOR MOUNTING OF FRONT SHOCK ABSORBERS**	24–30	2	
C.11916	**Screw,** Set, securing Shock Absorber Brackets to Spring Seats	24–31	8	
C.8690	**Tab Washer** (Long) retaining Setscrews	24–32	2	
C.8677	**Tab Washer** (Short) retaining Setscrews	24–33	2	
C.20989	**FRONT SHOCK ABSORBER (NFP.640-541-37E)**	24–34	2	
8299	**Rubber Buffer,** at top of Shock Absorbers (64539036)	24–35	4	
6744	**Inner Washer,** between Rubber Buffers (9046/934)	24–36	4	
8298	**Outer Washer,** at top and bottom of Rubber Buffers (64539035)	24–37	4	
7695	**Spacing Collar,** at top of Shock Absorbers (9046/1170)		2	
UFN.137/L	**Nut,** for Shock Absorbers (10203)	24–38	2	
UFN.237/L	**Locknut,** for Shock Absorbers (50062)	24–39	2	
C.386	**Washer,** Plain, at upper fixing		2	
C.8925	**Bolt,** securing Front Shock Absorbers through bottom Eye		2	
C.8737/4	**Nut,** Self-locking, on Bolts		2	
FW.107/T	**Washer,** Plain, under Nuts		2	
C.15337	**BALL PIN, IN STUB AXLE CARRIER, FOR ATTACHMENT OF LOWER WISHBONE LEVERS**	24–40	2	
C.15339	**Spigot,** over Ball Pins	24–41	2	Fitted from Chassis No. 1A.1001 to 1A.3941 1A.20001 to 1A.20269
C.15904	**Railko Socket,** supporting Lower Ball Pins	24–42	2	
C.22974	**Spigot,** over Ball Pins		2	Fitted to Chassis No. 1A.3942 and subs 1A.20270 and subs
C.23018	**Railko Socket,** supporting Lower Ball Pins		2	
C.15341	**Shim** (.002" thick) between Stub Axle and Cap	24–43	As req'd	**CHOOSE SHIMS OF DESIRED THICKNESS**
C.15342	**Shim** (.004" thick) between Stub Axle and Cap	24–43	As req'd	
C.17175	**Shim** (.010" thick) between Stub Axle and Cap	24–43	As req'd	
C.15338	**Cap,** under Lower Ball Pins		2	Fitted from Chassis No. 1A.1001 to 1A.3941 1A.20001 to 1A.20269
C.22973	**Cap,** under Lower Ball Pins		2	Fitted from Chassis No. 1A.3942 to 1A.6460 1A.20270 to 1A.20339
C.24222	**Cap,** under Lower Ball Pins		2	Fitted to Chassis No. 1A.6461 and subs 1A.20340 and subs
C.15344	**Bolt,** securing Caps to Stub Axle Carriers	24–45	8	
C.15343	**Tab Washer,** securing Bolts	24–46	4	
C.3044/1	**Grease Nipple,** in Cap, for lubrication of Lower Ball Pins	24–47	2	
C.24991	**Locating Washer,** under Grease Nipples		2	Fitted from Chassis No. 1A.3942 to 1A.6460 1A.20270 to 1A.20339
C.24990	**Washer** (Nylon) under Locating Washers		2	
C.22971	**Washer** (Nylon) under Grease Nipples		2	Fitted to Chassis No. 1A.6461 and subs 1A.20340 and subs
C.15346	**RUBBER GAITER FOR LOWER BALL PINS**	24–48	2	Fitted from Chassis No. 1A.1001 to 1A.3941 1A.20001 to 1A.20269
C.15345	**Retainer,** for Rubber Gaiters		2	
C.17174	**Clip,** securing Gaiters to Stub Axle Carriers	24–49	2	

FRONT SUSPENSION

Part No.	Description	Plate No.	No. per Unit	Remarks
C.22967	**RUBBER GAITER FOR LOWER BALL PINS**		2	Fitted to Chassis No. 1A.3942 and subs 1A.20270 and subs
C.22970	**Insert** (Plastic) on Ball Pin Spigots, seating Gaiters		2	
C.22969	**Ring,** securing Gaiters		2	
C.8737/6	**Nut,** Self-locking, securing Lower Ball Pins to Lower Wishbone Levers		2	
C.794	**Special Washer,** behind Nuts		2	
C.19224	**STUB AXLE CARRIER**	24–50	2	
C.19423	**STUB AXLE SHAFT**	24–51	2	
C.8667/7	**Nut,** Self-locking, securing Stub Axle Shafts and Tie Rod Levers to Carriers		2	
C.15350	**OIL SEAL FOR STUB AXLE SHAFTS**	24–52	2	
C.18842	**Water Deflector,** for Front Hub Oil Seals	24–53	2	
C.18843	**Water Shield,** for Front Hub Oil Seals	24–54	2	
C.15351	**INNER BEARING, ON STUB AXLE SHAFT, FOR HUBS (TIMKEN LM.67048–LM.67010)**	24–55	2	
C.15352	**OUTER BEARING, ON STUB AXLE SHAFT, FOR HUBS (TIMKEN LM.11949–LM.11910)**	24–56	2	
C.19229	**FRONT HUB ASSEMBLY**		2	FOR USE WHEN DISC ROAD WHEELS ARE FITTED from Chassis No. 1A.1001 to 1A.3941 1A.20001 to 1A.20269
C.13365	**Stud,** in Hubs, fixing Road Wheels		10	
C.23436	**FRONT HUB ASSEMBLY**		2	FOR USE WHEN DISC ROAD WHEELS ARE FITTED at Chassis No. 1A.3942 and subs 1A.20270 and subs
C.13365	**Stud,** in Hubs, fixing Road Wheels		10	
C.3044/1	**Grease Nipple,** in Hubs (Tec. NA.6625)		2	
C.19225	**R.H. FRONT HUB**		1	FOR USE WHEN WIRE SPOKE ROAD WHEELS ARE FITTED
C.19226	**L.H. FRONT HUB**	24–57	1	
C.3076/1	**Grease Nipple,** in Hubs (Tec. NA.6658)		2	
NN.756/L	**Slotted Nut,** securing Front Hubs to Stub Axle Shafts		2	
C.3400	**'D' Washer,** under Nuts		2	
L.104/10U	**Split Pin,** securing Nuts		2	
C.3039	**CAP IN END OF FRONT HUBS**		2	FOR USE ONLY WHEN DISC ROAD WHEELS ARE FITTED

Part No.	Description	Plate No.	No. per Unit	Remarks
C.21270	**R.H. HUB CAP**		1	
C.21271	**L.H. HUB CAP**	24–58	1	**FOR USE ONLY WHEN WIRE SPOKE ROAD WHEELS ARE FITTED**
C.14927	**TOOL, FOR REMOVING AND FITTING HUB CAPS**	24–59	1	
C.19962	**R.H. FRONT BRAKE CALIPER ASSEMBLY (VB.1194A/N)**		1	Fitted from Chasis No. 1A.1001 to 1A.4415 1A.20001 to 1A.20282
C.19963	**L.H. FRONT BRAKE CALIPER ASSEMBLY (VB.1194B/N)**		1	
C.24278	**R.H. FRONT BRAKE CALIPER ASSEMBLY (VB.95058A/N)**		1	Fitted to Chassis No. 1A.4416 and subs. 1A.20283 and subs.
C.24279	**L.H. FRONT BRAKE CALIPER ASSEMBLY (VB.95058B/N)**		1	
C.7569	**Shim** (.004" thick) between Brake Calipers and Stub Axle Carriers (8LO.2405)		As req'd	
C.11723	**Shim** (.010" thick) between Brake Calipers and Stub Axle Carriers (8LO.2411)		As req'd	**SELECTIVE SIZES**
C.11723/1	**Shim** (.030" thick) between Brake Calipers and Stub Axle Carriers		As req'd	
C.11723/2	**Shim** (.050" thick) between Brake Calipers and Stub Axle Carriers		As req'd	
C.15334	**DISC FOR FRONT BRAKE CALIPERS**		2	**SUPERSEDED BY C.23484**
C.23484	**DISC FOR FRONT BRAKE CALIPERS**		2	**SUPERSEDES C.15334**
C.13546	**Bolt,** securing Discs to Hubs		10	
C.24119/4	**Cleveloc Nut,** on Bolts		10	
C.23711	**SHIELD ASSEMBLY, UPPER, FOR R.H. FRONT BRAKE DISC**		1	Fitted from Chassis No. 1A.5800 to 1A.6150 1A.20327 to 1A.20335
C.23712	**SHIELD ASSEMBLY, UPPER, FOR L.H. FRONT BRAKE DISC**		1	
C.24715	**SHIELD ASSEMBLY, UPPER, FOR R.H. FRONT BRAKE DISC**		1	Fitted to Chassis No. 1A.6151 and subs. 1A.20336 and subs.
C.24716	**SHIELD ASSEMBLY, UPPER, FOR L.H. FRONT BRAKE DISC**		1	
UFS.150/6R	**Screw,** Set, securing Upper Disc Shields to Stub Axle Carriers		4	Fitted to Chassis No. 1A.5800 and subs. 1A.20327 and subs.
FG.108/X	**Washer,** Spring, on Setscrews		4	
C.23721	**SHIELD ASSEMBLY, LOWER, FOR R.H. FRONT BRAKE DISC**		1	
C.23722	**SHIELD ASSEMBLY, LOWER, FOR L.H. FRONT BRAKE DISC**		1	Fitted to Chassis No. 1A.5800 and subs. 1A.20327 and subs.
C.23723	**Clip,** retaining Lower Disc Shields to Stub Axle Carriers		2	
C.8737/1	**Nut,** Self-Locking, securing Clips to Lower Disc Shields and Lower Disc Shields to Upper Disc Shields		4	

FRONT SUSPENSION

Part No.	Description	Plate No.	No. per Unit	Remarks
C.15332	**L.H. TIE ROD LEVER**	24–60	1	
C.15336	**L.H. Front Hose Bracket**	24–61	1	Fitted from Chassis No. 1A.1001 to 1A.5799 1A.20001 to 1A.20326
C.23720	**L.H. Front Hose Bracket**		1	Fitted from Chassis No. 1A.5800 to 1A.6150 1A.20327 to 1A.20335
FW.108/T	**Washer,** Plain, on Setscrew securing Front Caliper and Upper Disc Shield to Stub Axle Carrier		1	Replaces Bracket C.23720 at Chassis No. 1A.6151 and subs. 1A.20336 and subs.
C.15331	**R.H. TIE ROD LEVER**		1	
C.15335	**R.H. Front Hose Bracket**		1	Fitted from Chassis No. 1A.1001 to 1A.5799 1A.20001 to 1A.20326
C.23719	**R.H. Front Hose Bracket**		1	Fitted from Chassis No. 1A.5800 to 1A.6150 1A.20327 to 1A. 20335
FW.108/T	**Washer,** Plain, on Setscrew securing Front Caliper and Upper Disc Shield to Stub Axle Carrier		1	Replaces Bracket C.23719 at Chassis No. 1A.6151 and subs 1A.20336 and subs
C.15333	**Screw,** Set, securing Tie Rod Levers, Front Calipers and Front Hose Brackets to Stub Axle Carriers		2	Fitted from Chassis No. 1A.1001 to 1A.5799 1A.20001 to 1A.20326
C.13547	**Screw,** Set, securing Front Calipers to Stub Axle Carriers		2	
C.15333	**Screw,** Set, securing Tie Rod Levers, Front Calipers, and Front Hose Brackets to Stub Axle Carriers		2	Fitted to Chassis No. 1A.5800 and subs. 1A.20327 and subs.
C.22770	**Screw,** Set, securing Front Calipers and Upper Disc Shields to Stub Axle Carriers		2	
FG.108/X	**Washer,** Spring, on Setscrews		4	
C.15897	**Shim** (.004" thick) between Tie Rod Levers and Stub Axle Carriers		As req'd	SELECTIVE SIZES
C.15898	**Shim** (.010" thick) between Tie Rod Levers and Stub Axle Carriers		As req'd	
C.8710	**L.H. OUTER TIE ROD ASSEMBLY**	24–62	1	Fitted from Chassis No. 1A.1001 to 1A.3941 1A.20001 to 1A.20269
C.9048	**Grease Nipple,** for lubrication of Ball Joints in Outer Tie Rods		2	
C.8717	**Rubber Gaiter,** on Ball Joints		2	
C.8718	**Retainer,** inside Gaiters		2	
C.8737/5	**Nut,** Self-Locking, securing Outer Tie Rods		2	
C.791	**Special Washer,** under Nuts		2	
C.8709	**R.H. OUTER TIE ROD ASSEMBLY**		1	
	Individual items are as for L.H. Outer Tie Rod C.8710			
C.22679	**L.H. OUTER TIE ROD ASSEMBLY**		1	Fitted to Chassis No. 1A.3942 and subs. 1A.20270 and subs.
C.9048	**Grease Nipple,** for lubrication of Ball Joint in Outer Tie Rod (Tecalemit NA.6626)		2	
C.22976	**Washer,** Plain, under Grease Nipples		2	
C.22975	**Washer,** (Rubber) under Plain Washers		2	
C.22683	**Gaiter** (Rubber) on Ball Joints		2	
C.22685	**Ring** (Polyurethane) securing Gaiters		2	
C.8737/5	**Nut,** Self-Locking, securing Outer Tie Rods		2	
C.791	**Special Washer,** under Nuts		2	
C.22678	**R.H. OUTER TIE ROD ASSEMBLY**		1	
	Individual items are as for L.H. Outer Tie Rod C.22679			

Part No.	Description	Plate No.	No. per Unit	Remarks
C.17179	**CENTRE TIE ROD ASSEMBLY, FOR ADJUSTMENT OF STEERING TRACK**		1	**NOT REQUIRED FOR CARS EQUIPPED WITH POWER-ASSISTED STEERING**
C.8720	**Tie Rod Tube**	24–63	1	
C.3165	**Clamp,** securing Tube to Tie Rod Ends	24–64	2	
UFB.131/11R	**Bolt,** securing Clamps		2	
UFN.131/L	**Nut,** on Bolts		2	
FG.105/X	**Washer,** Spring, under Nuts		2	
FW.105/T	**Washer,** Plain, under Spring Washers		2	
C.17100	**End Assembly,** at R.H. side of Centre Tie Rod		1	
C.17101	**End Assembly,** at L.H. side of Centre Tie Rod	24–65	1	
C.8737/5	**Nut,** Self-Locking, securing Tie Rod Ends to Drop Arm and Idler Lever	24–66	2	
C.791	**Special Washer,** under Self-Locking Nut	24–67	2	
C.17100	**END ASSEMBLY AT R.H. SIDE OF CENTRE TIE ROD**		1	**NOT REQUIRED FOR CARS EQUIPPED WITH POWER-ASSISTED STEERING**
C.17099	**Pin and Bush Assembly**		1	**FOR USE ONLY WITH C.17179**
C.17101	**END ASSEMBLY ATL.H. SIDE OF CENTRE TIE ROD**	24–65	1	**NOT REQUIRED FOR CARS EQUIPPED WITH POWER-ASSISTED STEERING**
C.17099	**Pin and Bush Assembly**		1	**FOR USE ONLY WITH C.17179**
C.29088	**CENTRE TIE ROD ASSEMBLY, FOR ADJUSTMENT OF STEERING TRACK**		1	**NOT REQUIRED FOR CARS EQUIPPEI WITH POWER-ASSISTED STEERING**
C.29089	**End Assembly,** at R.H. side of Centre Tie Rod		1	Supersedes C.17179 and is interchangeable as an assembly only
C.29090	**End Assembly,** at L.H. side of Centre Tie Rod		1	
	All other items are as for Centre Tie Rod Assembly C.17179			
C.19056	**CENTRE TIE ROD ASSEMBLY, FOR ADJUSTMENT OF STEERING TRACK**		1	**REQUIRED ONLY FOR CARS EQUIPPED WITH POWER-ASSISTED STEERING**
C.16021	**Tie Rod Tube**		1	
C.3165	**Clamp,** securing Tube to Tie Rod Ends		2	Fitted from Chassis No. 1A.1001 to 1A.4162 1A.20001 to 1A.20274
UFB.131/11R	**Bolt,** securing Clamps		2	
UFN.131/L	**Nut,** on Bolts		2	
FG.105/X	**Washer,** Spring, under Nuts		2	
FW.105/T	**Washer,** Plain, under Spring Washers		2	
C.19057	**End Assembly,** at R.H. side of Centre Tie Rod		1	
C.19059	**End Assembly,** at L.H. side of Centre Tie Rod		1	
C.8737/5	**Nut,** Self-Locking, securing Tie Rod Ends to Drop Arm and Idler Lever		2	
C.791	**Special Washer,** under Self-Locking Nuts		2	
C.23382	**CENTRE TIE ROD ASSEMBLY, FOR ADJUSTMENT OF STEERING TRACK**		1	**REQUIRED ONLY FOR CARS EQUIPPED WITH POWER-ASSISTED STEERING**
C.23329	**Tie Rod Tube**		1	Fitted from Chassis No. 1A.4163 to 1A.13157 1A.20275 and subs. 1K.1001 to 1K.3660 1K.30001 to 1K.30021
	All other items are as for Centre Tie Rod C.19056			
C.26393	**CENTRE TIE ROD ASSEMBLY**		1	**REQUIRED ONLY FOR CARS EQUIPPED WITH POWER-ASSISTED STEERING**
C.19057	**End Assembly,** at R.H. side of Centre Tie Rod		1	
C.19059	**End Assembly,** at L.H. side of Centre Tie Rod		1	
C.23329	**Tie Rod Tube**		1	
C.3165	**Clamp,** securing Tube to Tie Rod Ends		2	Fitted to Chassis No. 1A.13158 and subs. Also Fitted from Chassis No. 1K.30022 to 1K.30080
UFB.131/11R	**Bolt,** securing Clamps		2	
UFN.131/L	**Nut,** on Bolts		2	
FG.105/X	**Washer,** Spring, under Nuts		2	
FW.105/T	**Washer,** Plain, under Spring Washers		2	
C.8737/5	**Nut,** Self-Locking, securing Tie Rod Ends to Drop Arm and Idler Lever		2	
C.791	**Special Washer,** under Self-Locking Nuts		2	

FRONT SUSPENSION

Part No.	Description	Plate No.	No. per Unit	Remarks
C.29093	**CENTRE TIE ROD ASSEMBLY**		1	**REQUIRED ONLY FOR CARS EQUIPPED WITH POWER ASSISTED STEERING**
C.29094	**End Assembly,** at R.H. side of Centre Tie Rod		1	Fitted to Chassis No. 1K.3661 and subs. 1K.30081 and subs.
C.29095	**End Assembly,** at L.H. side of Centre Tie Rod		1	
	All other items are as for Centre Tie Rod Assembly C.26393			Supersedes C.26393 and is interchangeable as an assembly only
C.19057	**END ASSEMBLY AT R.H. SIDE OF CENTRE TIE ROD**		1	**REQUIRED ONLY FOR CARS EQUIPPED WITH POWER-ASSISTED STEERING FOR USE ONLY WITH C.19056**
C.17099	**Pin and Bush Assembly**		1	
C.19059	**END ASSEMBLY AT L.H. SIDE OF CENTRE TIE ROD**		1	
C.17099	**Pin and Bush Assembly**		1	
C.14035	**ANTI-ROLL BAR**	24–68	1	
C.20898	**Mounting Block** for Anti-Roll Bar		2	
C.3054	**Bracket,** supporting Anti-Roll Bar	24–69	2	
C.14983	**Rubber Bush,** in Brackets	24–70	2	
C.18303	**Keeper Plate,** for Rubber Bush	24–71	2	
UFB.137/13R	**Bolt,** securing Anti-Roll Bar Brackets		4	
FG.106/X	**Washer,** Spring, on Bolts		4	
C.20764	**LINK, BETWEEN ANTI-ROLL BAR AND COIL SPRING SEATS**	24–72	2	
C.10940	**Rubber Bush,** in Eye end of Links	24–73	2	
UFB.143/14R	**Bolt,** securing Links through Eye ends		2	
C.8737/4	**Nut,** Self-Locking, on Bolts		2	
C.10996	**RUBBER PAD AT UPPER END OF ANTI-ROLL BAR LINKS**	24–74	4	
C.11047	**Distance Tube,** through Rubber Pads	24–75	2	
C.11045	**Retaining Washer,** for Rubber Pads	24–76	4	
C.8737/3	**Nut,** Self-Locking, securing Links to Anti-Roll Bar		2	
C.14845	**STEERING BOX ASSEMBLY**	24–77	1	**STANDARD FITMENT ON R.H. DRIVE CARS**
C.14847	**Drop Arm,** on Steering Box		1	
C.14846	**STEERING BOX ASSEMBLY**		1	**STANDARD FITMENT ON L.H. DRIVE CARS**
C.14847	**Drop Arm,** on Steering Box		1	
C.8927	**HIGH RATIO STEERING BOX ASSEMBLY**	24–77	1	**SUPPLIED TO SPECIAL ORDER ON R.H. DRIVE CARS ONLY**
6429	**Drop Arm,** on Steering Box		1	
C.8928	**HIGH RATIO STEERING BOX ASSEMBLY**		1	**SUPPLIED TO SPECIAL ORDER ON L.H. DRIVE CARS ONLY**
6429	**Drop Arm,** on Steering Box		1	
C.17625	**POWER-ASSISTED STEERING BOX ASSEMBLY (PS.1-509)**		1	**R.H. DRIVE CARS ONLY** Fitted from Chassis No. 1A.1001 to 1A.1208
C.17626	**Drop Arm,** on Steering Box (S.9-171)		1	

Part No.	Description	Plate No.	No. per Unit	Remarks
C.14941	**POWER-ASSISTED STEERING BOX ASSEMBLY (PS.1-433)**		1	**L.H. DRIVE CARS ONLY** Fitted from Chassis No. 1A.20001 to 1A.20054
C.16045	**Drop Arm,** on Steering Box (S.9-154)		1	
C.21837	**POWER-ASSISTED STEERING BOX ASSEMBLY (PS.1-509)**		1	**R.H. DRIVE CARS ONLY** Fitted to Chassis No. 1A.1209 and subs.
C.17626	**Drop Arm,** on Steering Box (S.9-171)		1	
C.21836	**POWER-ASSISTED STEERING BOX ASSEMBLY (PS.1-433)**		1	**L.H. DRIVE CARS ONLY** Fitted to Chassis No. 1A.20055 and subs.
C.16045	**Drop Arm,** on Steering Box (S.9–154)		1	
C.17079	**Bolt,** securing Power-Assisted Steering Box to Front Suspension Cross Member		1	
FG.106/X	**Washer,** Spring, on Bolt		1	
C.25884	**Bolt,** securing Power-Assisted Steering Box to Cross Member and controlling Steering Lock		1	
UFN.137/L	**Nut,** on Bolt		1	
FG.106/X	**Washer,** Spring, under Nut		1	
C.16037	**Bolt,** securing Power-Assisted Steering Box to Front Suspension Cross Member		1	
C.8737/3	**Nut,** Self-Locking, on Bolt		1	
FW.106/T	**Washer,** Plain, under Nut		1	
C.8730	**STEERING IDLER ASSEMBLY**	24–78	1	**FOR USE WITH HIGH RATIO RATIO STEERING BOX ONLY. SUPPLIED TO SPECIAL ORDER** from Chassis No. 1A.1001 to 1A.8061 1A.20001 to 1A.20393
C.9052	**Bracket Assembly,** mounting Idler	24–79	1	
C.9048	**Grease Nipple,** for lubrication of Fulcrum Pin	24–80	1	
C.8733	**Fulcrum Pin**	24–81	1	
C.8734	**'O' Ring,** at bottom of Fulcrum Pin	24–82	1	
C.8735	**Idler Lever**	24–83	1	
C.8737/5	**Nut,** Self-Locking, securing Idler Lever to Fulcrum Pin	24–84	1	
C.791	**Special Washer,** under Nut	24–85	1	
UFS.137/9R	**Stop Screw,** controlling Steering Lock	24–86	2	
UFN.137/L	**Nut,** locking Stop Screws	24–87	2	
C.14887	**STEERING IDLER ASSEMBLY**		1	**FOR USE ONLY WITH STANDARD STEERING BOX** from Chassis No. 1A.1001 to 1A.4190 1A.20001 to 1A.20277
C.14848	**Idler Lever**		1	
	All other items as for Idler Assembly C.8730			**NOTE: C.14887 is no longer available. Use C.22440 for all replacements of C.14887**
C.22445	**STEERING IDLER ASSEMBLY**		1	**FOR USE WITH HIGH RATIO STEERING BOX ONLY SUPPLIED TO SPECIAL ORDER** at Chassis No. 1A.8062 and subs. 1A.20394 and subs.
C.22441	**Bracket,** mounting Idler		1	
C.19665	**Idler Spindle**		1	
UFN.150/L	**Nut,** at top of Spindle		1	
C.4490	**Tab Washer,** locking Nut		1	
C.16032	**'D' Washer,** under Tab Washer		1	
C.4219	**End Cap**		1	
C.16030	**Bearing,** at top of Idler Spindle (Timken 00050/00152X)		1	
C.16029	**Bearing,** at bottom of Idler Spindle (Timken 03062/03162)		1	
C.16035	**Seal** (Felt) on Spindle		1	
C.16034	**Retainer,** for Felt Seal		1	
C.16033	**Abutment Washer**		1	
C.16116	**Abutment Ring**		1	
C.22444	**Idler Lever**		1	
C.8737/4	**Nut,** Self-Locking, securing Idler Lever on Spindle		1	
C.789	**Washer,** Plain, under Nut		1	
UFS.137/9R	**Screw,** Set, controlling Steering Lock		2	
UFN.237/L	**Nut,** locking Setscrews		2	
FW.106/T	**Washer,** Plain, for Shimming Idler Bracket		As req'd	
C.24781	**Washer,** Plain, for Shimming Idler Bracket		As req'd	

FRONT SUSPENSION

Part No.	Description	Plate No.	No. per Unit	Remarks
C.22440	**STEERING IDLER ASSEMBLY**		1	FOR USE ONLY WITH STANDARD STEERING BOX at Chassis No. 1A.4191 and subs. 1A.20278 and subs.
C.22442	**Idler Lever**		1	
	All other items are as for Steering Idler Assembly C.22445			
C.11919	**Bolt** (Long) securing Steering Box and Steering Idler to Front Suspension Cross Member		4	NOT REQUIRED WHEN POWER ASSISTED STEERING IS FITTED
C.11920	**Bolt** (Short) securing Steering Box and Steering Idler to Front Suspension Cross Member		4	
C.11958	**Tab Washer,** retaining Bolts securing Steering Box		2	
FG.106/X	**Washer,** Spring, on Bolts securing Steering Idler		4	
C.16022	**STEERING IDLER ASSEMBLY (OAO.1814)**		1	FOR R.H. DRIVE CARS EQUIPPED WITH POWER-ASSISTED STEERING
C.16024	**Bracket,** mounting Idler (OLO.3937)		1	
C.16028	**Idler Spindle** (OLO.2729)		1	
C.16032	**'D' Washer,** on Spindle (8LO.2503)		1	
UFN.150/L	**Nut,** at Top of Spindle (7SO.1311)		1	
C.4490	**Tab Washer,** locking Nut (8LO.2908)		1	
C.4219	**End Cap** (7LO.1141)		1	
C.16030	**Bearing,** at top of Spindle (8LO.3081) (Timken 00050/00152X)		1	
C.16029	**Bearing,** at bottom of Spindle (8LO.3017) (Timken 03062/03162)		1	
C.16035	**Felt Seal,** on Spindle (8LO.1410)		1	
C.16034	**Retainer,** for Felt Seal (7LO.1185)		1	
C.16033	**Abutment Washer** (8LO.2731)		1	
C.16116	**Abutment Ring** (6LO.3504)		1	
C.16026	**Idler Lever** (OLO.3865)		1	
C.8737/5	**Nut,** Self-Locking, securing Idler Lever to Spindle (8SO.1501)		1	
C.789	**Washer,** Plain, under Nut (7SO.0801)		1	
C.16023	**STEERING IDLER ASSEMBLY (OAO.1815)**		1	FOR L.H. DRIVE CARS EQUIPPED WITH POWER-ASSISTED STEERING
C.16025	**Bracket,** mounting Idler (OLO.3938)		1	
C.16027	**Idler Lever** (OLO.3866)		1	
	All other items as for Idler Assembly C.16022			
C.11920	**Bolt,** securing Steering Idler to Front Suspension Cross Member		1	FOR USE ONLY WHEN POWER-ASSISTEI STEERING IS FITTED
FG.106/X	**Washer,** Spring, on Bolt		1	
C.25884	**Bolt,** securing Steering Idler and controlling Steering Lock		1	
UFN.137/L	**Nut,** on Bolt		1	
FG.106/X	**Washer,** Spring, under Nut		1	
C.16038	**Bolt,** securing Steering Idler to Cross Member		1	
C.8737/3	**Nut,** Self-Locking, on Bolt		1	
FW.106/T	**Washer,** Plain, under Nut		1	
C.11920	**Bolt,** securing Steering Idler to Front Suspension Cross Member		1	FOR USE WHEN POWER-ASSISTED STEERING ASSEMBLY C.26018 OR C.26019 IS FITTED
FG.106/X	**Washer,** Spring, on Bolt		4	
C.16038	**Bolt,** securing Steering Idler and controlling Steering Lock		1	
C.8737/3	**Nut,** Self-Locking, on Bolt		1	
FW.106/T	**Washer,** Plain, under Nut		1	
FW.106/T	**Washer,** Plain, for Shimming } To align Tie Rod Assembly		As req'd	
C.24781	**Washer,** Plain, for Shimming } To align Tie Rod Assembly		As req'd	
C.26406	**Mounting Plate,** for Pipe Clip		1	REQUIRED ON R.H.DRIVE CARS WHEN P.A.S. ASSEMBLY C.26018 IS FITTED
C.26407	**Clip,** for Oil Pipe		1	
UFS.119/5R	**Setscrew,** for Clip		1	

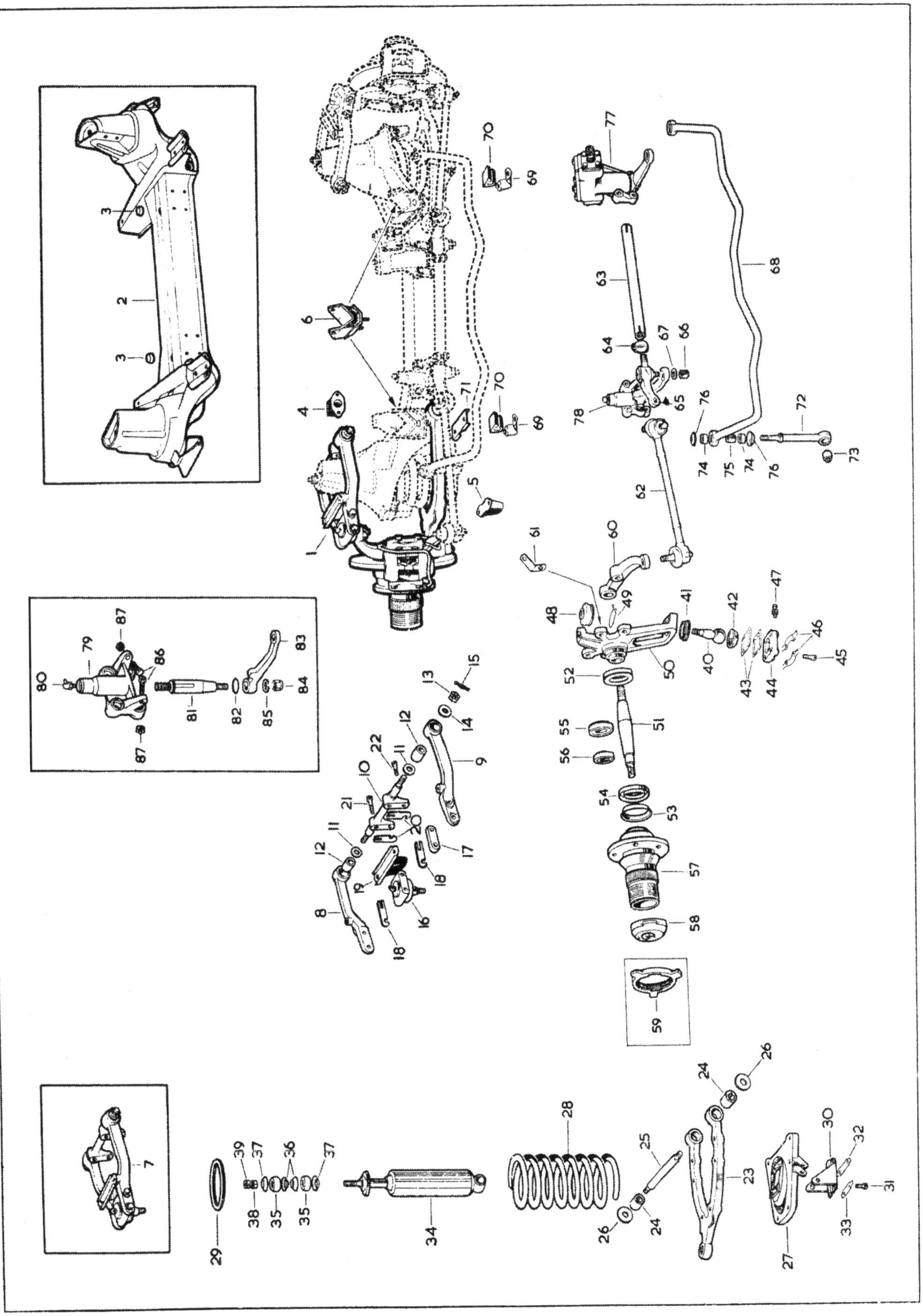

PLATE 24

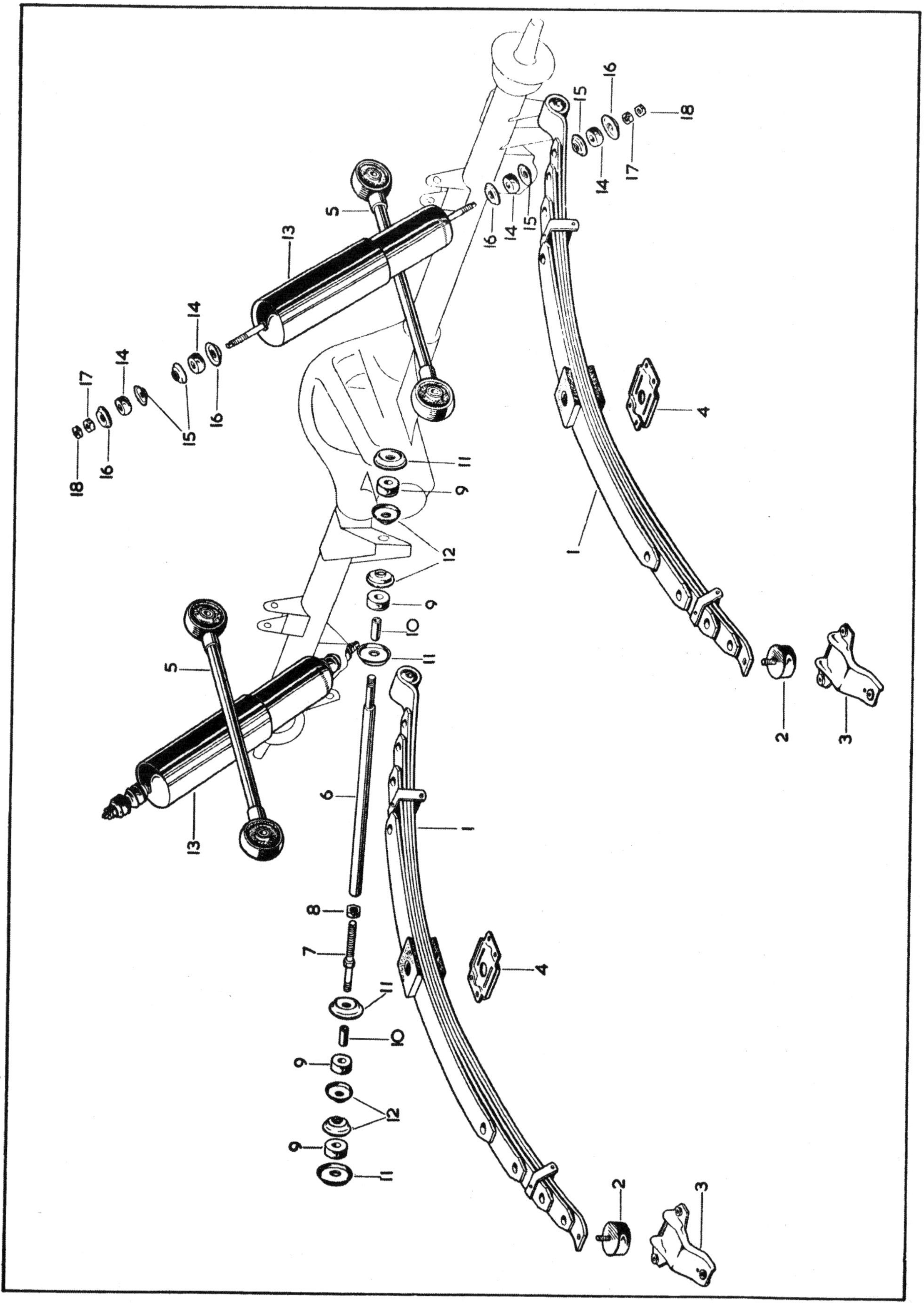

PLATE 25

Part No.	Description	Plate No.	No. per Unit	Remarks
C.10791/2	**REAR ROAD SPRING**	25–1	2	
C.10844	**Rubber Mounting,** at top and bottom of Road Springs		4	
C.8939	**Bush,** in eye of Road Springs		2	
C.12859	**Rubber Button,** between Spring Leaves		16	
C.17711	**Centre Bolt,** through Spring Leaves		2	
UFB.143/24R	**Bolt,** securing Road Springs through Eye		2	
C.8667/4	**Nut,** Self-Locking, on Bolts		2	
C.10843	**MOUNTING PAD AT FRONT END OF ROAD SPRINGS**	25–2	2	
C.8737/3	**Nut,** Self-Locking, securing Mounting Pads to Road Springs		2	
C.12343	**MOUNTING PLATE ASSEMBLY, AT FRONT END OF ROAD SPRINGS**	25–3	2	
UFS.131/7R	**Screw,** Set, securing Front Mounting Plate		2	
FG.105/X	**Washer,** Spring, on Setscrews		2	
UFS.131/9R	**Screw,** Set, securing Front Mounting Plate		4	
C.8667/2	**Nut,** Self-Locking, on Setscrews		4	
C.10845	**CLAMPING PLATE, SECURING ROAD SPRING AT CENTRE**	25–4	2	
UFS.131/7R	**Screw,** Set, securing Clamping Plates		8	
C.8667/2	**Nut,** Self-Locking, on Setscrews		8	
FW.105/T	**Washer,** Plain, under Nuts		8	
C.11091	**REAR TORQUE ARM ASSEMBLY**	25–5	2	
C.8949	**Bush,** in ends of Torque Arms		4	
UFB.143/20R	**Bolt,** securing Torque Arms		4	
FW.107/T	**Washer,** Plain, on Bolts		4	
C.8667/4	**Nut,** Self-Locking, on Bolts		4	
C.14140	**PANHARD ROD ASSEMBLY**	25–6	1	
C.11996	**ADJUSTING END FOR PANHARD ROD**	25–7	1	
UFN.256/L	**Nut,** locking Adjusting End	25–8	1	
C.10995	**RUBBER BUFFER FOR PANHARD ROD**	25–9	4	
C.11048	**Distance Tube,** between Rubber Buffers	25–10	2	
C.11042	**Outer Washer,** retaining Rubber Buffers on Panhard Rod	25–11	4	
C.11043	**Inner Washer,** retaining Rubber Buffers on Panhard Rod	25–12	4	
C.8737/5	**Nut,** Self-Locking, securing Panhard Rod to Brackets		2	
C.20990	**REAR SHOCK ABSORBER (NFP.640-539-15M)**	25–13	2	
8299	**Rubber Buffer,** at top and bottom of Rear Shock Absorbers (64539036)	25–14	8	
6744	**Inner Washer,** retaining Rubber Buffers (9046/934)	25–15	8	
8298	**Outer Washer,** retaining Rubber Buffers (64539035)	25–16	8	
UFN.137/L	**Nut,** securing Rear Shock Absorbers	25–17	4	
UFN.237/L	**Lock Nut,** for Rear Shock Absorbers	25–18	4	
C.386	**Plain Washer,** between Outer Washers and securing Nuts		4	

REAR SUSPENSION

Part No.	Description	Plate No.	No. per Unit	Remarks
C.21480	REAR SHOCK ABSORBER (NFP.640-541-38C)		2	**STIFFER FITTING— FITTED TO SPECIAL ORDER ONLY**
	Individual items are as for Rear Shock Absorber C.20990			

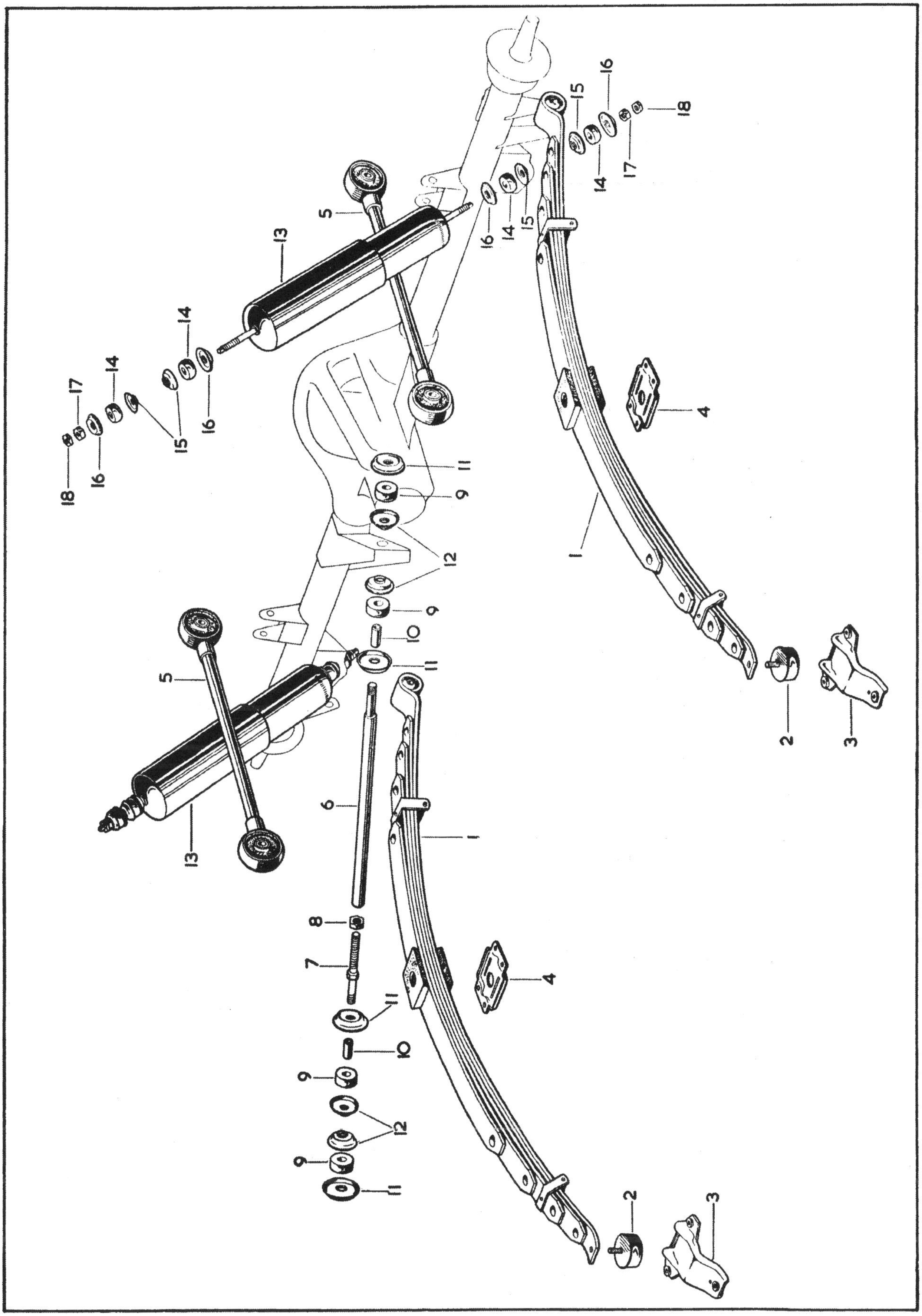

PLATE 25

ROAD WHEELS

Part No.	Description	Plate No.	No. per Unit	Remarks
C.16827	DISC TYPE ROAD WHEEL (5J. RIM x 15" DIAMETER)		5	Fitted from Chassis No. 1A.1001 to 1A.12434 1A.20001 to 1A.20555
C.27022	DISC TYPE ROAD WHEEL (5J. RIM x 15" DIAMETER)		5	Fitted from Chassis No. 1A.12435 and subs. 1A.20556 and subs.
C.3041	Nut, securing Disc type Road Wheels to Hubs		20	
C.14766	WIRE SPOKE ROAD WHEEL (5K. RIM x 15" DIAMETER) (PAINTED STOVED ALUMINIUM)		5	Fitted to Chassis No. 1A.1001 to 1A.12778 1A.20001 to 1A.20599
C.28043	WIRE SPOKE ROAD WHEELS (5K. RIM x 15" DIAMETER) (PAINTED STOVED ALUMINIUM)		5	Fitted to Chassis No. 1A.12779 and subs. 1A.20600 and subs.
7784	Long Spoke		120	SUPPLIED TO SPECIAL ORDER ONLY
7790	Short Spoke		240	
C.5604	Nipple, for Spokes		360	
C.14802	WIRE SPOKE ROAD WHEEL (5K. RIM x 15" DIAMETER) (CHROMIUM)		5	Fitted from Chassis No. 1A.1001 to 1A.12778 1A.20001 to 1A.20599
7796	Long Spoke (Chromium Plated)		120	SUPPLIED TO SPECIAL ORDER ONLY
7801	Short Spoke (Chromium Plated)		240	
C.9144	Nipple, for Spokes		360	
C.28044	WIRE SPOKE ROAD WHEEL (5K. RIM x 15" DIAMETER) (CHROMIUM)		5	Fitted to Chassis No. 1A.12779 and subs. 1A.20600 and subs. SUPPLIED TO SPECIAL ORDER ONLY
C.17422	TYRE FOR ROAD WHEELS ("ROAD SPEED" 6.40" x 15")		5	
C.22229	TYRE FOR ROAD WHEELS (WHITE WALL "ROAD SPEED" 6.40" x 15")		5	SUPPLIED TO SPECIAL ORDER ONLY
C.12567	TUBE FOR TYRES (6.40" x 15")		5	
C.21260	NAVE PLATE FOR ROAD WHEELS		4	Fitted from Chassis No. 1A.1001 to 1A.12580 1A.20001 to 1A.20564 FOR USE ONLY WITH DISC TYPE ROAD WHEELS
C.27134	NAVE PLATE FOR ROAD WHEELS		4	Fitted to Chassis No. 1A.12581 and subs. 1A.20565 and subs. FOR USE ONLY WITH DISC TYPE ROAD WHEELS
C.21256	BADGE "DAIMLER" FOR NAVE PLATE		4	FOR USE ONLY WITH DISC TYPE ROAD WHEELS
C.22583	Spire Nut (Push-on) securing Badge to Nave Plates		8	
	BALANCE WEIGHTS FOR ROAD WHEELS			
C.3985	Balance Weight (½ oz.)		As req'd	
C.3986	Balance Weight (1 oz.)		As req'd	
C.3987	Balance Weight (1½ ozs.)		As req'd	
C.3988	Balance Weight (2 ozs.)		As req'd	
C.3989	Balance Weight (2½ ozs.)		As req'd	
C.3990	Balance Weight (3 ozs.)		As req'd	

Part No.	Description	Plate No.	No. per Unit	Remarks
C.21983	RIMBELLISHER (FOR USE WITH 5J. ROAD WHEELS)		5	
C.17940	L.H. WHEEL DISC (FOR USE WITH 5J. ROAD WHEELS)		2	SUPPLIED TO SPECIAL ORDER ONLY
C.17940/1	R.H. WHEEL DISC (FOR USE WITH 5J ROAD WHEELS)		2	SUPPLIED TO SPECIAL ORDER ONLY
C.17923	RIMBELLISHER (FOR USE WITH 5J. ROAD WHEELS)		4	SUPPLIED TO SPECIAL ORDER ONLY
7486	Clip, securing Rimbellishers		12	

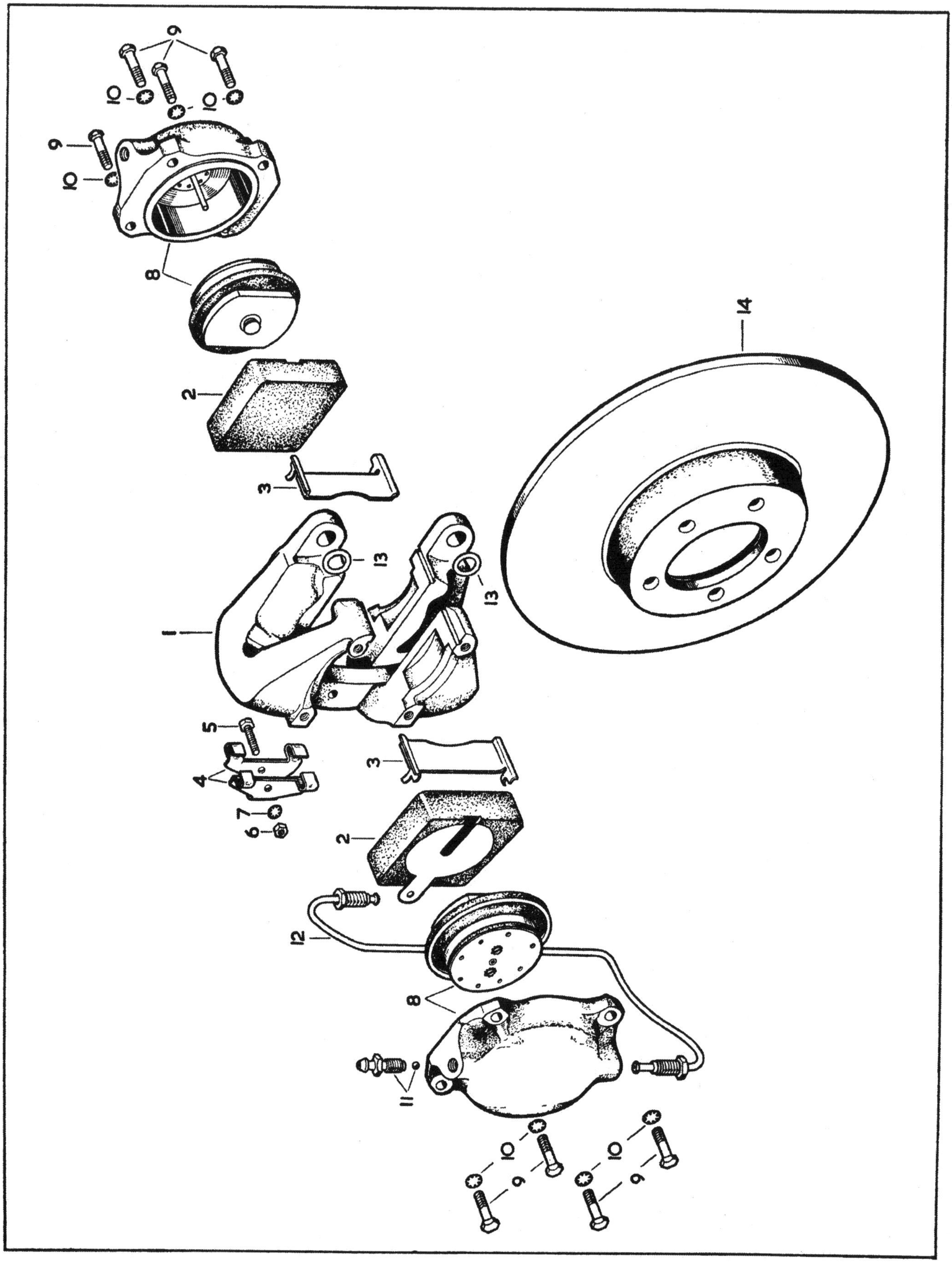

PLATE 26

Part No.	Description	Plate No.	No. per Unit	Remarks
C.19962	**R.H. FRONT CALIPER ASSEMBLY (VB.1194A/N)**		1	Fitted from Chassis No. 1A.1001 to 1A.4415 1A.20001 to 1A.20282
7715	**Front Caliper only** (VBM.1006)	26–1	1	
7937	**Friction Pad Assembly** (VBO.5138/N)	26–2	2	**NOT SUPPLIED SEPARATELY, SEE KIT**
7718	**Support Plate,** for Friction Pads (VBO.5133)	26–3	2	
7719	**Retaining Plate,** for Friction Pads (VBO.5123)	26–4	1	
UFB.125/7R	**Bolt,** securing Retaining Plate	26–5	1	
UFN.125/L	**Nut,** on Bolt	26–6	1	
C.724	**Washer,** Shakeproof, under Nut	26–7	1	
8777	**Piston and Cylinder Assembly** (VBO.5505)	26–8	2	
7720	**Bolt,** securing Cylinders to Front Caliper (VBO.5100)	26–9	8	
C.724	**Washer,** Shakeproof, on Bolts	26–10	8	
6911	**Bleed Screw and Ball Assembly** (VBO.5114)	26–11	1	
7783	**Bridge Pipe Assembly** (VBO.5203)	26–12	1	
C.24278	**R.H. FRONT CALIPER ASSEMBLY (VB.95058A/N)**		1	Fitted to Chassis No. 1A.4416 and subs. 1A.20283 and subs.
10474	**Bridge Pipe Assembly** (VB.20304)		1	
	All other items are as for R.H. Front Caliper C.19962			
C.19963	**L.H. FRONT CALIPER ASSEMBLY (VB.1194B/N)**		1	Fitted from Chassis No. 1A.1001 to 1A.4415 1A.20001 to 1A.20282
7822	**Bridge Pipe Assembly** (VBO.5204)		1	
	All other items are as for R.H. Front Caliper C.19962			
C.24279	**L.H. FRONT CALIPER ASSEMBLY (VB.95058B/N)**		1	Fitted to Chassis No. 1A.4416 and subs. 1A.20283 and subs.
10473	**Bridge Pipe Assembly** (VB.20303)		1	
	All other items are as for R.H. Front Caliper C.19962			
C.7569	**Shim** (.004" thick) for adjustment of Front Calipers (8LO.2405)	26–13	As req'd	SELECTIVE SIZES
C.11723	**Shim** (.010" thick) for adjustment of Front Calipers (8LO.2411)	26–13	As req'd	
C.13547	**Screw,** Set, securing Front Calipers to Stub Axle Carriers		2	Fitted from Chassis No. 1A.1001 to 1A.5799 1A.20001 to 1A.20326
C.15333	**Screw,** Set, securing Front Calipers, Tie Rod Levers, and Front Hose Brackets to Stub Axle Carriers		2	
C.22770	**Screw,** Set, securing Front Calipers and Upper Disc Shields to Stub Axle Carriers		2	Fitted to Chassis No. 1A.5800 and subs. 1A.20327 and subs.
C.15333	**Screw,** Set, securing Front Calipers, Tie Rod Levers and Front Hose Brackets to Stub Axle Carriers		2	
FG.108/X	**Washer,** Spring, on Setscrews		4	
C.837	**Lockwire,** for Caliper Setscrews		As req'd	
10821	**FRICTION PAD KIT (VB.20585/N)**		1	
	(Comprising Four Friction Pads)			
8820	**FRONT WHEEL CYLINDER REPAIR KIT (VBO.8210)**		4	
	(Comprising Piston Seal and Dust Seal)			
C.13686	**HOSE FOR FRONT CALIPERS (WDO.22211/15½)**		2	
C.3910	**Locknut,** securing Hose to Body Bracket		2	
C.3911	**Washer,** Shakeproof, under Nuts		2	
UFN.237/L	**Locknut,** securing Hose to Bracket on Front Suspension		2	
C.726	**Washer,** Shakeproof, under Locknuts		2	

FRONT BRAKES

Part No.	Description	Plate No.	No. per Unit	Remarks
C.15334	**DISC FOR FRONT CALIPERS**	26–14	2	SUPERSEDED BY C.23484
C.23484	**DISC FOR FRONT CALIPERS**		2	SUPERSEDES C.15334
C.13546	Bolt, securing Discs to Front Hubs		10	
C.24119/4	Cleveloc Nut, on Bolts		10	
C.19227	**FRONT HUB ASSEMBLY**		2	FOR USE WHEN DISC ROAD WHEELS ARE FITTED
C.13365	Stud, in Front Hubs, fixing Road Wheels		10	FOR USE WHEN DISC ROAD WHEELS ARE FITTED
C.19225	**R.H. FRONT HUB ASSEMBLY**		1	FOR USE WHEN WIRE SPOKE ROAD WHEELS ARE FITTED
C.19226	**L.H. FRONT HUB ASSEMBLY**		1	FOR USE WHEN WIRE SPOKE ROAD WHEELS ARE FITTED
C.18842	**WATERSHIELD FOR FRONT HUB**		2	

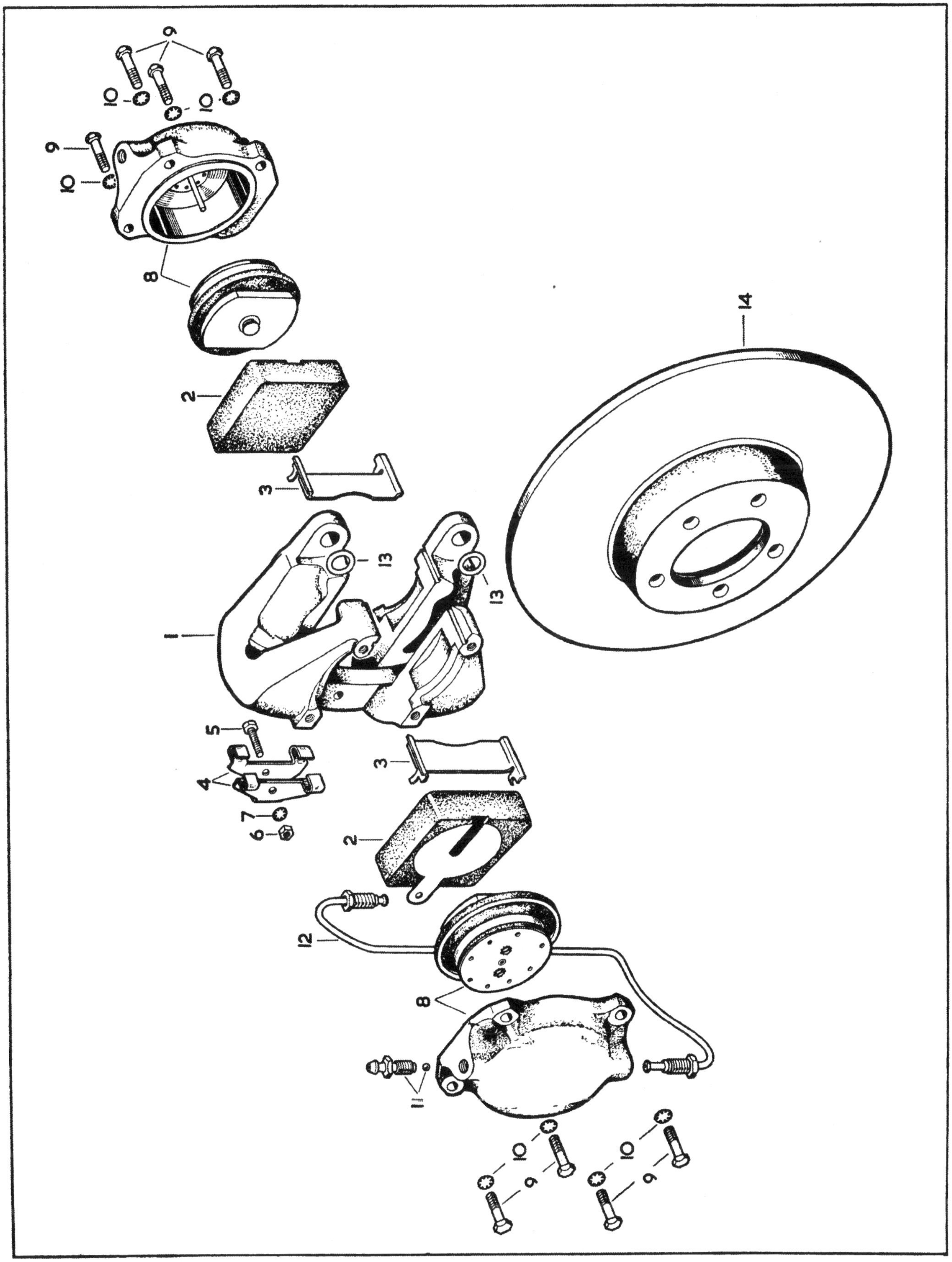

PLATE 26

REAR BRAKES

Part No.	Description	Plate No.	No. per Unit	Remarks
C.18613	**R.H. REAR CALIPER ASSEMBLY (VB.1154A/N)**		1	Fitted from Chassis No. 1A.1001 to 1A.4415 1A.20001 to 1A.20282
8009	**Rear Caliper only** (VBM.1080)	27–1	1	
7937	**Friction Pad** (VBO.5138N)	27–2	2	NOT SUPPLIED SEPARATELY. SEE KIT
7718	**Support Plate,** for Friction Pads (VBO.5133)	27–3	2	
7719	**Retaining Plate,** for Friction Pads (VBO.5123)	27–4	1	
7720	**Bolt,** securing Retaining Plate (VBO.5100)	27–5	1	
UFN.125/L	**Nut,** on Bolt	27–6	1	
C.724	**Washer,** Shakeproof, under Nut	27–7	1	
8816	**Piston and Cylinder Assembly** (VBO.5496)	27–8	2	
7720	**Bolt,** securing Cylinder to Rear Caliper (VBO.5100)	27–9	8	
C.724	**Washer,** Shakeproof, on Bolts	27–10	8	
6911	**Bleed Screw and Ball Assembly** (VBO.5114)	27–11	1	
7827	**Bridge Pipe Assembly** (VBO.3589)	27–12	1	
C.24282	**R.H. REAR CALIPER ASSEMBLY (VB.95059A/N)**		1	Fitted to Chassis No. 1A.4416 and subs. 1A.20283 and subs.
10476	**Bridge Pipe Assembly** (VB.20306)		1	
	All other items are as for R.H. Rear Caliper C.18613			
C.18614	**L.H. REAR CALIPER ASSEMBLY (VB.1154B/N)**		1	Fitted from Chassis No. 1A.1001 to 1A.4415 1A.20001 to 1A.20282
8010	**Rear Caliper only** (VBM.1081)		1	
7826	**Bridge Pipe Assembly** (VBO.3588)		1	
	All other items are as for R.H. Rear Caliper C.18613			
C.24283	**L.H. REAR CALIPER ASSEMBLY (VB.95059B/N)**		1	Fitted to Chassis No. 1A.4416 and subs. 1A.20283 and subs.
8010	**Rear Caliper only** (VBM.1081)		1	
10475	**Bridge Pipe Assembly** (VB.20305)		1	
	All other items are as for R.H. Rear Caliper C.18613			
C.13198	**Shim** (.005" thick) for adjustment of Rear Calipers (VBO.1333)	27–13	As req'd	SELECTIVE SIZES
C.13199	**Shim** (.020" thick) for adjustment of Rear Calipers (VBO.1334)	27–13	As req'd	SELECTIVE SIZES
C.13994	**Shim** (.040" thick) for adjustment of Rear Calipers	27–13	As req'd	SELECTIVE SIZES
C.16262	**Bolt,** securing Rear Calipers to Adaptor Plates		4	
C.837	**Wire,** locking Bolts		As req'd	
8821	**REAR WHEEL CYLINDER REPAIR KIT (VBO.8346)**		4	
	(Comprising Piston Seal and Dust Seal)			
10821	**FRICTION PAD KIT (VB.20585/N)**		1	
	(Comprising Four Friction Pads)			
C.19289	**HANDBRAKE MECHANISM ASSEMBLY R.H. (VBM.4212/A)**	27–14	1	
8829	**Pad Carrier Assembly,** R.H. Inner (VBM.4579/A)	27–15	1	
8831	**Pad Carrier Assembly,** R.H. Outer (VBO.8009/A)	27–16	1	
8833	**Anchor Pin,** in Inner Pad Carrier, retaining Return Springs (VBO.7330)	27–17	1	
8834	**Operating Lever Assembly** (VBO.7490)	27–18	1	
8835	**Return Spring,** between Operating Lever and Anchor Pin (VBO.7329)	27–19	1	
8836	**Pawl Assembly** (VBO.7995)	27–20	1	
8837	**Tension Spring,** for Pawl (VBO.7367)	27–21	1	
8838	**Anchor Pin,** retaining Tension Spring (VBO.7385)	27–22	1	
8839	**Adjusting Nut** (VBO.7483)	27–23	1	
8840	**Friction Spring,** on Adjusting Nut (VBO.7588)	27–24	1	

Part No.	Description	Plate No.	No. per Unit	Remarks
	HANDBRAKE MECHANISM ASSEMBLY R.H. (VBM.4212/A) *(contd.)*			
8841	**Hinge Pin,** securing Operating Lever (VBO.7289)	27–25	1	
L.102/3U	**Split Pin,** retaining Hinge Pin	27–26	1	
8842	**Protection Cover Assembly,** at rear of Operating Lever (VBO.8095)	27–27	1	
8843	**Protection Cover,** at front of Operating Lever (VBO.7939)	27–28	1	
8844	**Bolt,** securing Protection Covers (VBO.8090)	27–29	1	
8845	**Bolt,** securing Pad Carriers to Operating Lever (VBO.8024)	27–30	1	
L.102.5/8U	**Split Pin,** retaining Bolt	27–31	1	
C.19290	**HANDBRAKE MECHANISM ASSEMBLY L.H. (VBM.4212/B)**		1	
8830	**Pad Carrier Assembly,** L.H. Inner (VBM.4579/B)		1	
8832	**Pad Carrier Assembly,** L.H. Outer (VBO.8009/B)		1	
	All other items for L.H. Handbrake Mechanism as for Assembly C.19289			
8016	**Bolt,** securing Handbrake Mechanism to Rear Calipers (VBO.4190)	27–32	4	Fitted from Chassis No. 1A.1001 to 1A.9535 1A.20001 to 1A.20443
8846	**Retraction Plate** (VBO.4635)	27–33	2	
8014	**Tab Washer,** locking Bolts (VBO.4226)	27–34	2	
8016	**Bolt,** securing Handbrake Mechanism to Rear Calipers (VBO. 4190)		4	Fitted to Chassis No. 1A.9536 and subs. 1A.20444 and subs.
11115	**Retraction Plate,** for R.H. Handbrake Mechanism (VBO. 60248/A)		1	
11116	**Retraction Plate,** for L.H. Handbrake Mechanism (VB.60248/B)		1	
11133	**Retainer,** Top, for Retraction Plates (VB.20571)		2	
11134	**Retainer,** Bottom, for Retraction Plates (VB.20570)		2	
11117	**Tab Washer,** for Locking Bolts (VB.20357)		2	
	NOTE: TO ENABLE EARLIER CARS TO BE MODIFIED TO THE LATER RETRACTION CONDITION, THE NEW RETRACTION PLATES, RETAINERS, AND TAB WASHERS MAY BE OBTAINED AS A KIT TO PART NO.10792 (VB.20573)			
8021	**HANDBRAKE MECHANISM REPAIR KIT (JA.14)**		2	
	(Comprising Friction Pads and Fixings)			
C.16287	**ADAPTOR PLATE FOR R.H. REAR CALIPER**	27–35	1	
C.16288	**ADAPTOR PLATE FOR L.H. REAR CALIPER**		1	
C.13065	**DISC FOR REAR CALIPERS (VBM.720)**	27–36	2	
C.13546	**Bolt,** securing Discs to Rear Hubs		10	REQUIRED ONLY WHEN WIRE SPOKE ROAD WHEELS ARE FITTED
FG.107/X	**Washer,** Spring, on Bolts		10	
UFN.143/L	**Nut,** on Bolts		10	
C.23787	**SHIELD ASSEMBLY FOR R.H. REAR BRAKE DISC**		1	Fitted to Chassis No. 1A.5800 and subs. 1A.20327 and subs.
C.23786	**SHIELD ASSEMBLY FOR L.H. REAR BRAKE DISC**		1	

REAR BRAKES

Part No.	Description	Plate No.	No. per Unit	Remarks
C.22227	**HOSE FOR REAR CALIPERS (WDO.22671/15½)**		1	
C.3910	**Locknut,** securing Hose to Body Bracket (23897)		1	
C.3911	**Washer,** Shakeproof, under Nut (K.19411)		1	
C.11239	**Copper Gasket,** for Hose to 3-way Union (KL.44516)		1	
C.13685	**3-WAY UNION ON REAR AXLE (29587)**		1	
UFB.125/9R	**Bolt,** securing 3-way Union to Rear Axle		1	
FW.104/T	**Washer,** Plain, on Bolt		1	
UFN.125/L	**Nut,** on Bolt		1	
C.724	**Washer,** Shakeproof, under Nut		1	
C.13644	**REAR HUB ASSEMBLY**		2	FOR USE WHEN DISC ROAD WHEELS ARE FITTED
C.13365	**Stud,** in Rear Hubs, for Road Wheels		10	
C.13094	**R.H. REAR HUB ASSEMBLY**		1	FOR USE WHEN WIRE SPOKE ROAD WHEELS ARE FITTED
C.13095	**L.H. REAR HUB ASSEMBLY**		1	

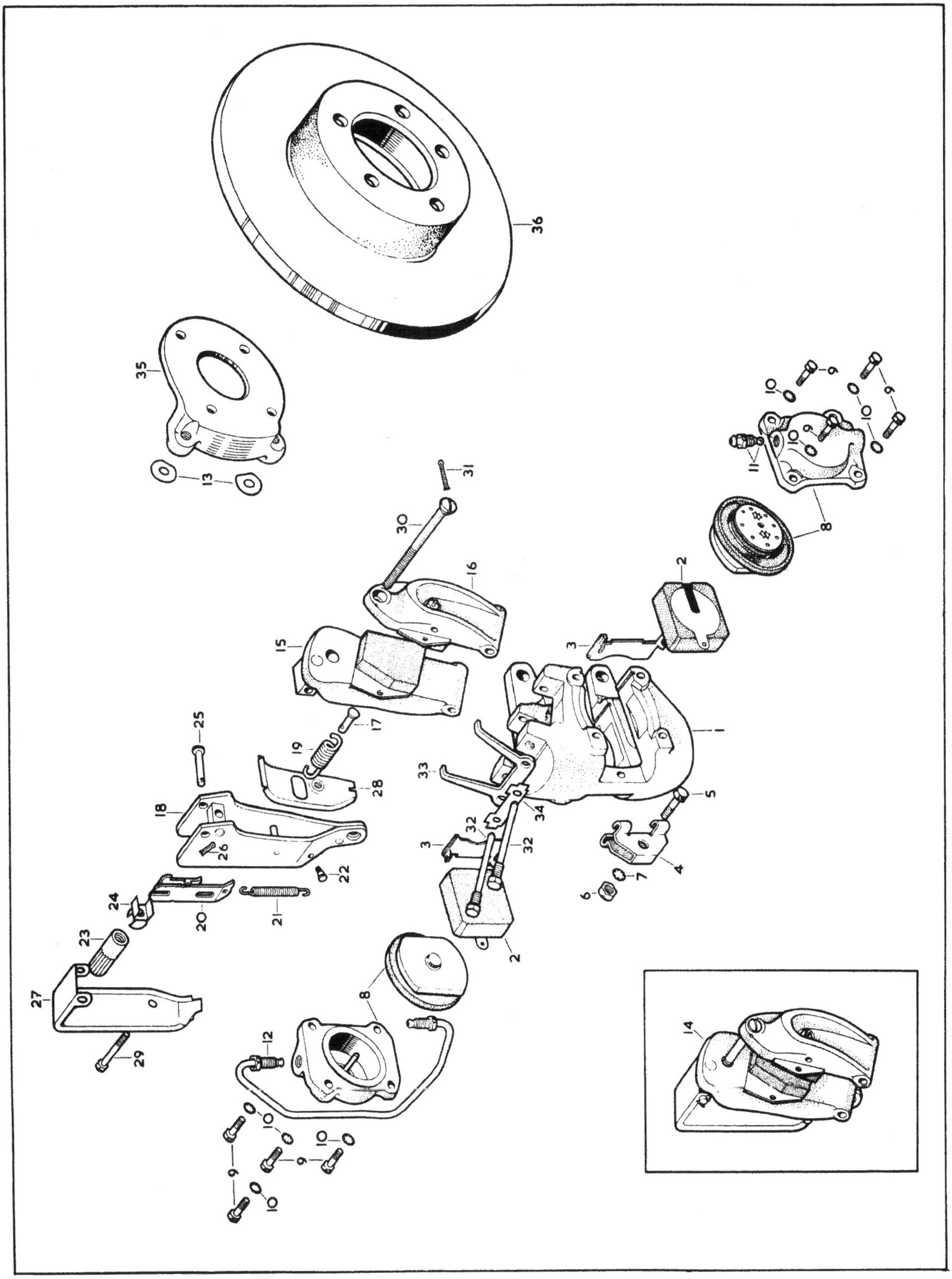

36
35
13
31
30
16
15
2
3
8
9
10
11
1
5
4
7
6
25
17
19
28
33
34
32
18
26
22
3
2
24
20
21
23
27
29
8
12
14

PLATE 27

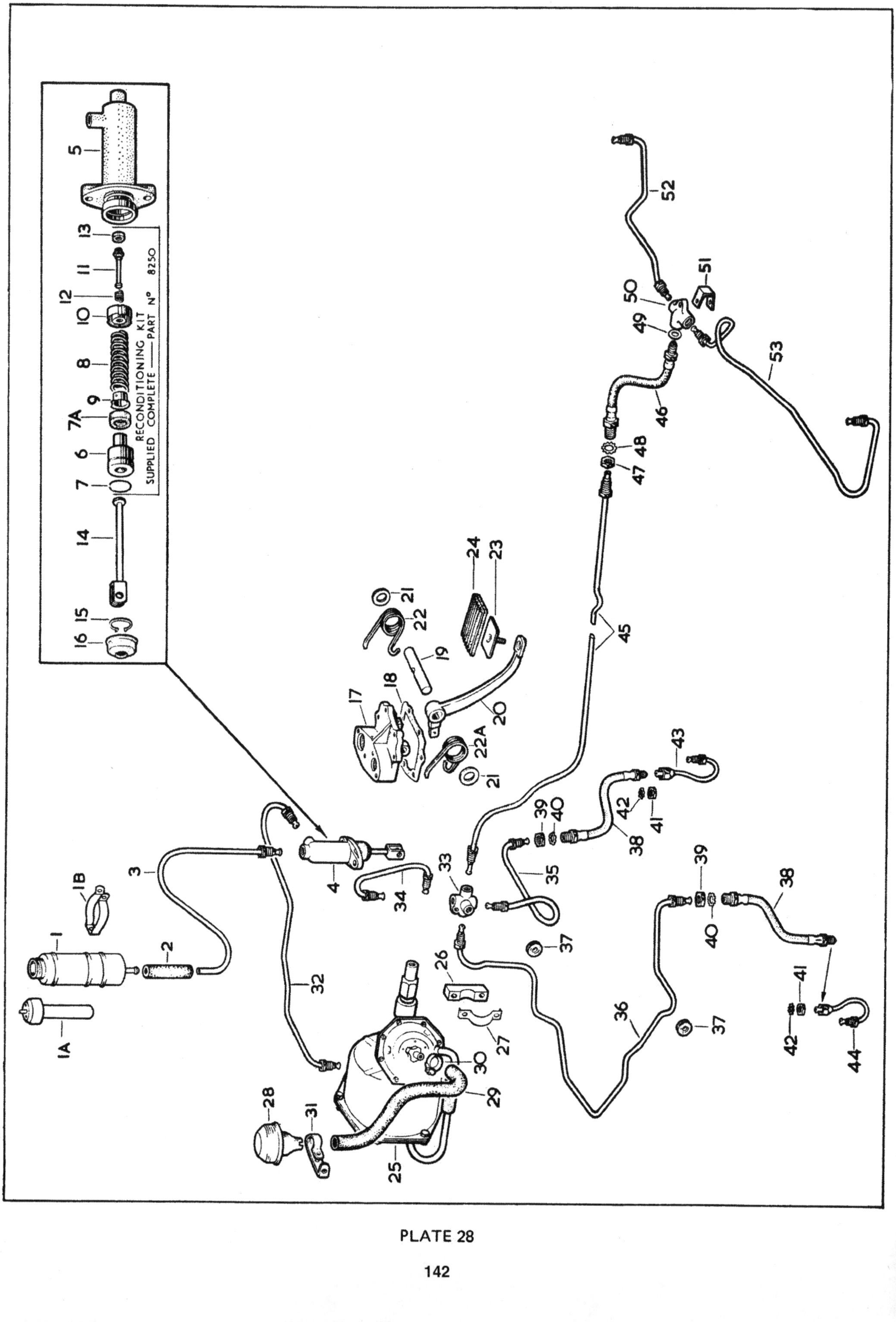
RECONDITIONING KIT
SUPPLIED COMPLETE —— PART No 8250
1
1A
1B
2
3
4
5
6
7
7A
8
9
10
11
12
13
14
15
16
17
18
19
20
21
22
22A
23
24
25
26
27
28
29
30
31
32
33
34
35
36
37
38
39
40
41
42
43
44
45
46
47
48
49
50
51
52
53

PLATE 28

Part No.	Description	Plate No.	No. per Unit	Remarks
C.22667	**BRAKE FLUID RESERVOIR**	28–1	1	Fitted from Chassis No. 1A.1001 to 1A.5651 1A.20001 to 1A.20323
C.16375	**Filler Cap**, complete with Level Indicator Switch	28–1A	1	
C.23638	**BRAKE FLUID RESERVOIR**		1	Fitted to Chassis No. 1A.5652 and subs. 1A.20324 and subs.
C.23632	**Filler Cap**, complete with Level Indicator Switch		1	
C.23627	**Cap**, protecting Level Indicator		1	
C.5173	**Clip**, around Fluid Reservoir	28–1B	1	
UFS.319/8H	**Screw**, Set, securing Clip to Reservoir		1	
UFN.119/L	**Nut**, on Setscrew		1	
C.723/A	**Washer**, Shakeproof, under Nut		1	
C.17597	**Distance Piece**, for mounting of Reservoir		1	
UFB.125/20R	**Bolt**, securing Reservoir to Valance		1	
UFN.125/L	**Nut**, on Bolt		1	
C.724	**Washer**, Shakeproof, under Nut		1	
FW.104/T	**Washer**, Plain, under Shakeproof Washer		1	
C.18717	**HOSE FROM RESERVOIR TO HYDRAULIC PIPE**	28–2	1	
C.15886/5	**Clip**, securing Hose to Reservoir		1	
C.15886/3	**Clip**, securing Hose to Hydraulic Pipe		1	
C.21427	**HYDRAULIC PIPE FROM HOSE TO MASTER CYLINDER (VBO.7563)**	28–3	1	
C.16469	**BRAKE MASTER CYLINDER ASSEMBLY (VBM.3280)**	28–4	1	Fitted to Chassis No. 1A.1001 to 1A.7355 1A.20001 to 1A.20366 **EXCHANGE SCHEME DOES NOT OPERATE ON BRAKE MASTER CYLINDER**
7474	**Body only** (VBM.3236)	28–5	1	
8291	**Piston** (VBO.4768)	28–6	1	
8294	**'O' Ring**, on Piston (VBO.4623)	28–7	1	
8293	**Cup Seal**, on Piston (VBO.4625/A)	28–7A	1	
8292	**Main Spring**, behind Piston (VBO.4767)	28–8	1	
6942	**Spring Support**, at forward end of Cylinder Body (VBO.2042)	28–9	1	
7675	**Spring Support**, at rear end of Cylinder Body (VBO.4141)	28–10	1	
7676	**Valve** (VBO.3539)	28–11	1	
6943	**Spring**, on Valve (VBO.1410)	28–12	1	
8296	**Seal**, at end of Valve (VBO.4842)	28–13	1	
8297	**Push Rod** (VBO.4238)	28–14	1	
6948	**Circlip**, retaining Push Rod in Cylinder Body (VBO.2048)	28–15	1	
6945	**Dust Excluder**, over end of Cylinder Body (VBO.1869)	28–16	1	

NOTE: Illustrated parts of Master Cylinder items 6 to 13 not supplied separately. Use Reconditioning Kit Part No. 8250 to service these items

Part No.	Description	Plate No.	No. per Unit	Remarks
C.25425	**BRAKE MASTER CYLINDER ASSEMBLY (VB.80029)**		1	Fitted to Chassis No. 1A.7356 and subs. 1A.20367 and subs. **EXCHANGE SCHEME DOES NOT OPERATE ON BRAKE MASTER CYLINDER**
11039	**Piston** (VB.20367)		1	
11040	**Main Spring**, behind Piston (VB.20368)		1	
9314	**Spring Support**, at forward end of Cylinder Body (VBO.8885)		1	
11041	**Spring Support**, at rear end of Cylinder Body (VB.20370)		1	

All other items are as Master Cylinder C.16469

BRAKE CONTROLS AND HYDRAULICS

Part No.	Description	Plate No.	No. per Unit	Remarks
C.27160	**BRAKE MASTER CYLINDER ASSEMBLY (64068590)**		1	**EXCHANGE SCHEME DOES NOT OPERATE ON BRAKE MASTER CYLINDER**
N.S.S.	Body only (64675906)		1	
N.S.S.	Seal on Valve (362802)		1	
7350	**Valve** (378641)		1	
11600	**Spring Retainer** (64673391)		1	Fitted to Chassis No. 1K.1001 and subs. 1K.30001 and subs.
C.27168	**Spacer** (318101)		1	
N.S.S.	Spring (64673425)		1	
N.S.S.	Seal, for Plunger (64673529)		1	
11605	**Plunger** (64675744)		1	
11606	**Push Rod Assembly** (64676262)		1	
N.S.S.	Circlip (378912)		1	
N.S.S.	Dust Cover (64673976)		1	
11607	**Retainer**, for Dust Cover (378070)		1	
N.S.S.	Spring Washer (378960)		1	
J.105/9.5S	**Joint Pin**, securing Master Cylinder to Pedal		1	
FW.105/T	**Washer**, Plain, on Joint Pin		2	**REQUIRED WHEN AUTOMATIC TRANSMISSION IS FITTED**
L.103/7U	**Split Pin**, securing Joint Pin		2	
8250	**BRAKE MASTER CYLINDER RECONDITIONING KIT (VBO.4843/A)**		1	**SUITABLE FOR MASTER CYLINDER C.16469 OR C.25425**
11608	**REPAIR KIT FOR SERVICING BRAKE MASTER CYLINDER (SP.2105)**		1	**SUITABLE FOR MASTER CYLINDER C.27160**
	Comprising:—			
	N.S.S. Seal, on Valve 1 off			
	N.S.S. Spring 1 off			
	N.S.S. Seal, for Plunger 1 off			
	N.S.S. Circlip 1 off			
	N.S.S. Dust Cover 1 off			
	N.S.S. Spring Washer 1 off			
C.13695	**HOUSING FOR BRAKE PEDAL AND MASTER CYLINDER**	28–17	1	**SUPERSEDED BY C.24014**
C.24014	**HOUSING FOR BRAKE AND CLUTCH PEDALS**		1	**SUPERSEDES C.13695**
C.8954	**Stud**, mounting Housing to Scuttle		7	
UFN.131/L	**Nut**, on Studs		7	
C.725	**Washer**, Shakeproof, under Nuts		7	
C.8957	**Gasket**, between Housing and Scuttle	28–18	1	**2 REQUIRED FOR L.H. DRIVE CARS**
C.27072	**Stud**, in Housing, mounting Brake Master Cylinder		2	
UFN.131/L	**Nut**, securing Brake Master Cylinder		2	
C.725	**Washer**, Shakeproof, under Nuts		2	
C.25337	**COVER PLATE ON PEDAL HOUSING**		1	**REQUIRED FOR CARS FITTED WITH AUTOMATIC TRANSMISSION ONLY**
UCS.131/4R	**Screw**, Set, securing Cover Plate to Housing		2	
C.725	**Washer**, Shakeproof, on Setscrews		2	
C.8958	**SHAFT, SUPPORTING BRAKE PEDAL IN HOUSING**	28–19	1	
C.8967	**Bolt**, securing Shaft in Housing		1	
C.8737/1	**Nut**, Self-Locking, on Bolt		1	
FW.104/T	**Washer**, Plain, under Nut		1	

Part No.	Description	Plate No.	No. per Unit	Remarks
C.8971	**COVER PLATE ASSEMBLY, SEALING PEDAL HOLE AT OPPOSITE HAND TO DRIVE**		1	
C.8957	**Gasket**, between Cover Plate and Dash		1	
UFS.125/5R	**Screw**, Set, securing Cover Plate to Dash		4	
UFN.125/L	**Nut**, on Setscrews		4	
C.739	**Washer**, Shakeproof, under Nuts		4	
C.13514	**BRAKE PEDAL (COMPLETE WITH BUSHES)**	28–20	1	R.H. DRIVE CARS FITTED WITH AUTOMATIC TRANSMISSION
C.20955/4	**Bush**, in Brake Pedal Boss		2	
C.13247	**BRAKE PEDAL (COMPLETE WITH BUSHES)**		1	R.H. DRIVE CARS FITTED WITH MANUAL TRANSMISSION ONLY
C.20955/4	**Bush**, in Brake Pedal Boss		2	
C.13249	**BRAKE PEDAL (COMPLETE WITH BUSHES)**		1	L.H. DRIVE CARS ONLY
C.20955/4	**Bush**, in Brake Pedal Boss		2	
C.13248	**BRAKE PEDAL (COMPLETE WITH BUSHES)**		1	L.H. DRIVE CARS WITH AUTOMATIC TRANSMISSION ONLY NOTE: This pedal operates on the normal Clutch Pedal Fulcrum and is joined to Pedal C.13249 by Steel Pad C.15326
C.20955/4	**Bush**, in Brake Pedal Boss		2	
C.8962	**Washer** (Fibre) at each side of Pedal Boss	28–21	2	4 REQUIRED FOR L.H. DRIVE CARS
C.8963	**RETURN SPRING, R.H., FOR BRAKE PEDAL**	28–22	1	
C.8966	**RETURN SPRING, L.H., FOR BRAKE PEDAL**	28–22A	1	
C.13066	**STEEL PAD FOR BRAKE PEDAL**	28–23	1	R.H. DRIVE CARS FITTED WITH AUTOMATIC TRANSMISSION
C.8667/2	**Nut**, securing Steel Pad to Brake Pedal		1	
C.15326	**STEEL PAD FOR BRAKE PEDAL**		1	L.H. DRIVE CARS FITTED WITH AUTOMATIC TRANSMISSION
C.8667/2	**Nut**, securing Steel Pad to Brake Pedals		2	
C.14852	**STEEL PAD FOR BRAKE PEDAL**		1	CARS FITTED WITH MANUAL TRANSMISSION ONLY
C.8667/2	**Nut**, securing Steel Pad to Brake Pedals		2	
C.6876	**RUBBER PAD FOR BRAKE PEDAL**	28–24	1	CARS FITTED WITH AUTOMATIC TRANSMISSION
C.27521	**RUBBER PAD FOR BRAKE PEDAL**		1	CARS FITTED WITH MANUAL TRANSMISSION
C.19612	**BRAKE VACUUM SERVO (4257–552)**	28–25	1	
BD.3210	**Grommet**, in Cowl mounting Brake Vacuum Servo		3	
C.11908	**Spacer**, inside Grommets		3	
UFN.125/L	**Nut**, securing Brake Vacuum Servo to Cowl		3	
C.724	**Washer**, Shakeproof, under Nuts		3	
FW.104/T	**Washer**, Plain, each side of Cowl		6	
C.11001	**Support Block**, for Clamp	28–26	1	
C.11002	**Clamp**, mounting Servo Unit to Front Wing Valance	28–27	1	
UFB.125/13R	**Bolt**, through Clamp		2	
BD.3210	**Grommet**, for mounting Support Block to Front Wing Valance		2	

Part No.	Description	Plate No.	No. per Unit	Remarks
	BRAKE VACUUM SERVO (4257–552) *(Contd.)*			
C.11908	**Spacer**, inside Grommets		2	
UFN.125/L	**Nut**, on Bolts		2	
FW.104/T	**Washer**, Plain, under Nuts		2	
C.724	**Washer**, Shakeproof, under Nuts		2	
BD.14155	**COWL, ON R.H. WING VALANCE, ENCLOSING SERVO UNIT**		1	
UFS.125/4R	**Screw**, Set, securing Cowl		4	Superseded by UFS.825/5H, C.724, DJZ.812/10C, BD.8536/3, and FW.104/T
UFS.125/5R	**Screw**, Set, securing Cowl		4	
FW.104/T	**Washer**, Plain, on Setscrews		8	
C.724	**Washer**, Shakeproof, on Setscrews		8	
UFS.825/5H	**Screw**, Set, securing Cowl		2	Supersedes UFS.125/4R, UFS.125/5R, FW.104/T, and C.724
C.724	**Washer**, Shakeproof, on Setscrews		2	
DJZ.812/10C	**Screw**, Self-Tapping, securing Cowl		5	
BD.8536/3	**Spire Nut**, receiving Self-Tapping Screws		5	
FW.104/T	**Washer**, Plain, under Shakeproof Washers and on Self-Tapping Screws		7	
C.13172	**AIR CLEANER FOR BRAKE VACUUM SERVO (7223289)**	28–28	1	
C.21351	**HOSE FROM BRAKE VACUUM SERVO TO AIR CLEANER**	28–29	1	
C.2905/1	**Clip**, securing Hose to Brake Vacuum Servo	28–30	1	
C.13882	**Support Clip**, fixing Hose to Air Cleaner and Body	28–31	1	
UFS.119/4R	**Screw**, Set, securing Support Clip		1	
UFN.119/L	**Nut**, on Setscrew		1	
C.723/A	**Washer**, Shakeproof, under Nut		1	
C.13706	**HYDRAULIC PIPE, FROM BRAKE VACUUM SERVO TO MASTER CYLINDER (VBO.3215)**	28–32	1	R.H. DRIVE CARS ONLY
C.16345	**HYDRAULIC PIPE, FROM BRAKE VACUUM SERVO TO MASTER CYLINDER (VBO.3220)**		1	L.H. DRIVE CARS ONLY
C.11008	**4-WAY ADAPTOR (80845)**	28–33	1	
UFB.125/9R	**Bolt**, securing 4-way Adaptor		1	
UFN.125/L	**Nut**, on Bolt		1	
FW.104/T	**Washer**, Plain, under Nut		1	
C.724	**Washer**, Shakeproof, under Nut		1	
C.13708	**HYDRAULIC PIPE, FROM 4-WAY ADAPTOR TO BRAKE VACUUM SERVO (T43/13½B)**	28–34	1	
C.13709	**HYDRAULIC PIPE, FROM 4-WAY ADAPTOR TO FLEXIBLE HOSE ON R.H. FRONT BRAKE (T43/22B)**	28–35	1	
C.12100	**HYDRAULIC PIPE, FROM 4-WAY ADAPTOR TO FLEXIBLE HOSE ON L.H. FRONT BRAKE (T43/66B)**	28–36	1	SUPERSEDED BY C.23288

Part No.	Description	Plate No.	No. per Unit	Remarks
C.23288	**HYDRAULIC PIPE, FROM 4-WAY ADAPTOR TO FLEXIBLE HOSE ON L.H. FRONT BRAKE (VBO.6075/72 3/8)**		1	**SUPERSEDES C.12100**
C.2706	**Grommet,** on Hydraulic Pipes (859059)	28–37	2	
C.13686	**FLEXIBLE HOSE TO FRONT BRAKES (WDO.22211/15½)**	28–38	2	
C.3910	**Locknut,** securing Hoses to Pipes (23897)	28–39	2	
C.3911	**Washer,** Shakeproof, under Locknuts (K.19411)	28–40	2	
UFN.237/L	**Locknut,** securing Flexible Hoses to Brackets on Front Suspension	28–41	2	
C.726	**Washer,** Shakeproof, under Locknuts	28–42	2	
C.15901	**HYDRAULIC PIPE FROM FLEXIBLE HOSE TO R.H. FRONT BRAKE CALIPER (VBO.3216)**	28–43	1	Fitted from Chassis No. 1A.1001 to 1A.4415 1A.20001 to 1A.20282
C.15902	**HYDRAULIC PIPE, FROM FLEXIBLE HOSE TO L.H. FRONT BRAKE CALIPER (VBO.3217)**	28–44	1	
C.24255	**HYDRAULIC PIPE, FROM FLEXIBLE HOSE TO R.H. FRONT BRAKE CALIPER (VBO.8480/10 3/8)**		1	Fitted to Chassis No. 1A.4416 and subs. 1A.20283 and subs.
C.24256	**HYDRAULIC PIPE, FROM FLEXIBLE HOSE TO L.H. FRONT BRAKE CALIPER (VBO.8480/10 3/8)**		1	
C.21426	**HYDRAULIC PIPE, FROM 4-WAY ADAPTOR TO REAR FLEXIBLE HOSE**	28–45	1	Fitted from Chassis No. 1A.1001 to 1A.13157
C.26112	**HYDRAULIC PIPE, FROM 4-WAY ADAPTOR TO REAR FLEXIBLE HOSE**		1	Fitted to Chassis No. 1A.13158 and subs. (also 1A.13003) on R.H. Drive Cars Fitted with C.26018 Power-Assisted Steering
C.28420	**HYDRAULIC PIPE, FROM 4-WAY ADAPTOR TO REAR FLEXIBLE HOSE**		1	**Supersedes C.26112**
C.22227	**FLEXIBLE HOSE, FROM PIPE TO 3-WAY ADAPTOR (WDO.22671/15½)**	28–46	1	
C.3910	**Locknut,** securing Hose to Body Bracket (23897)	28–47	1	
C.3911	**Washer,** Shakeproof, under Locknut (K.19411)	28–48	1	
C.11239	**Copper Gasket,** at joint of Hose to 3-way Adaptor (KL.44516)	28–49	1	
C.13685	**3-WAY ADAPTOR ON REAR AXLE (29587)**	28–50	1	
UFB.125/9R	**Bolt,** securing 3-way Adaptor to Bracket on Rear Axle		1	
FW.104/T	**Washer,** Plain, on Bolt		1	
UFN.125/L	**Nut,** on Bolt		1	
C.724	**Washer,** Shakeproof, under Nut		1	

BRAKE CONTROLS AND HYDRAULICS

Part No.	Description	Plate No.	No. per Unit	Remarks
C.22231	**BRACKET ASSEMBLY, MOUNTING 3-WAY ADAPTOR**	28–51	1	
UFN.125/L	**Nut,** securing Bracket to Rear Axle		1	
C.724	**Washer,** Shakeproof, under Nut		1	
FW.104/T	**Washer,** Plain, under Shakeproof Washer		1	
C.15556	**HYDRAULIC PIPE FROM 3-WAY ADAPTOR TO R.H. REAR BRAKE**	28–52	1	Fitted from Chassis No. 1A.1001 to 1A.4415 1A.20001 to 1A.20282
C.15557	**HYDRAULIC PIPE, FROM 3-WAY ADAPTOR TO L.H. REAR BRAKE**	28–53	1	
C.24253	**HYDRAULIC PIPE FROM 3-WAY ADAPTOR TO R.H. REAR BRAKE (VBO.6075/27¾)**		1	Fitted to Chassis No. 1A.4416 and subs. 1A.20283 and subs.
C.24254	**HYDRAULIC PIPE FROM 3-WAY ADAPTOR TO L.H. REAR BRAKE (VBO.6075/48¾)**		1	
C.4933	**Clip,** securing Hydraulic Brake Pipe to Body Underframe		1	
C.5023	**Clip,** securing Hydraulic Brake Pipes		9	**SUPERSEDED BY C.1040/29 AND DZZ.408/6C**
FW.104/T	**Washer,** Plain, for Clips		7	
BD.541/9	**Washer,** Special, for Clips		2	
C.1040/29	**Clip,** securing Hydraulic Brake Pipes to Body Underframe		8	**SUPERSEDES C.5023, FW.104/T, AND BD.541/9**
DZZ.408/6C	**Screw,** Self-Tapping, securing Clips		8	
C.7860	**Clip,** securing Hydraulic Brake Pipes to Rear Axle Tubes		2	3 required at Chassis No. 1A.4416 and subs. 1A.20283 and subs.
UFS.325/8H	**Screw,** Set, through Clips		2	
UFN.125/L	**Nut,** on Setscrews		2	
C.724	**Washer,** Shakeproof, under Nuts		2	
C.24272	**Clip,** on Rear Brake Caliper Fixing Bolts, securing Hydraulic Pipes		2	Fitted to Chassis No. 1A.4416 and subs. 1A.20283 and subs.
C.18414/2	**Washer,** Waved, at Clip Fixings		2	

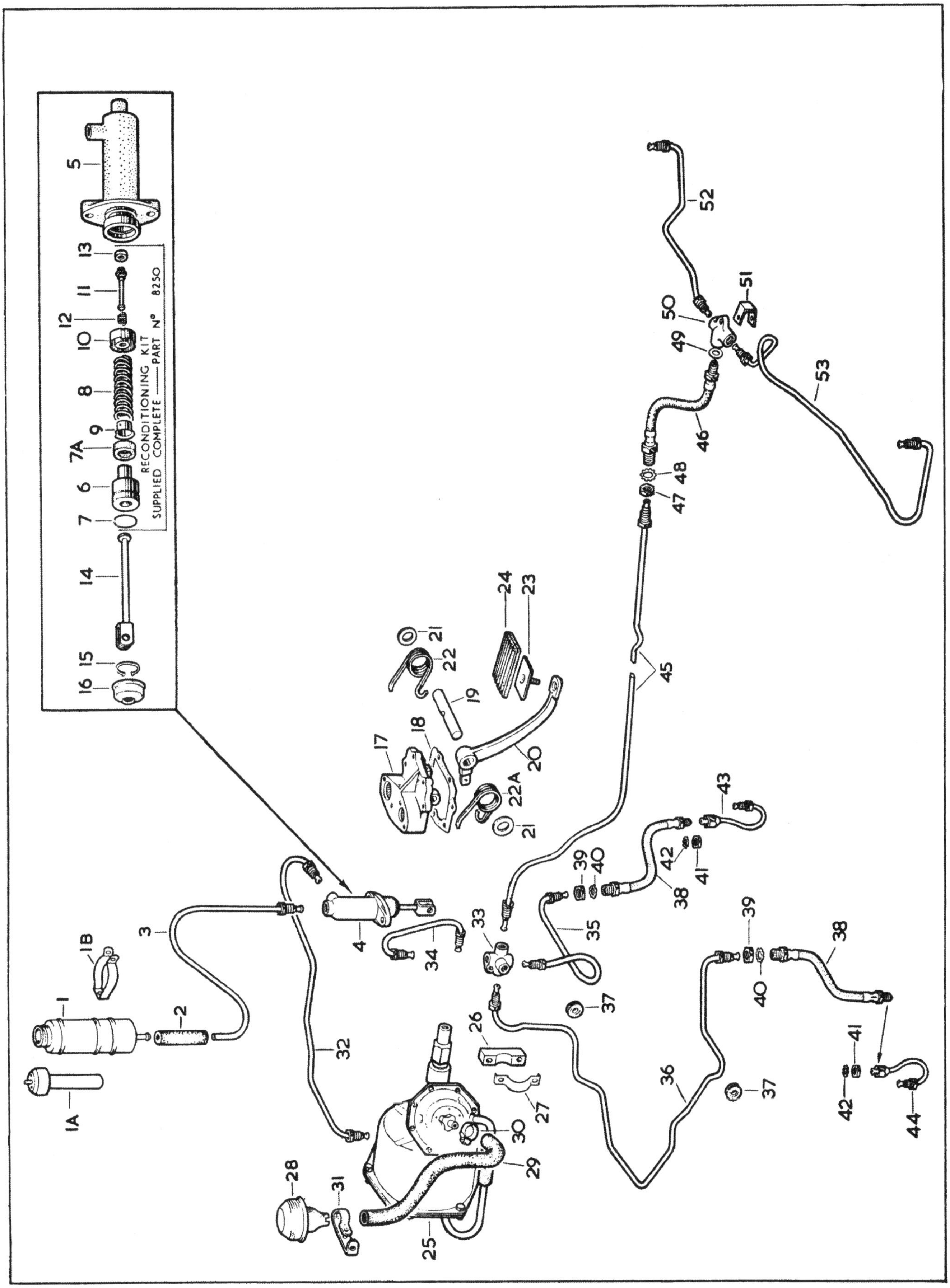

PLATE 28

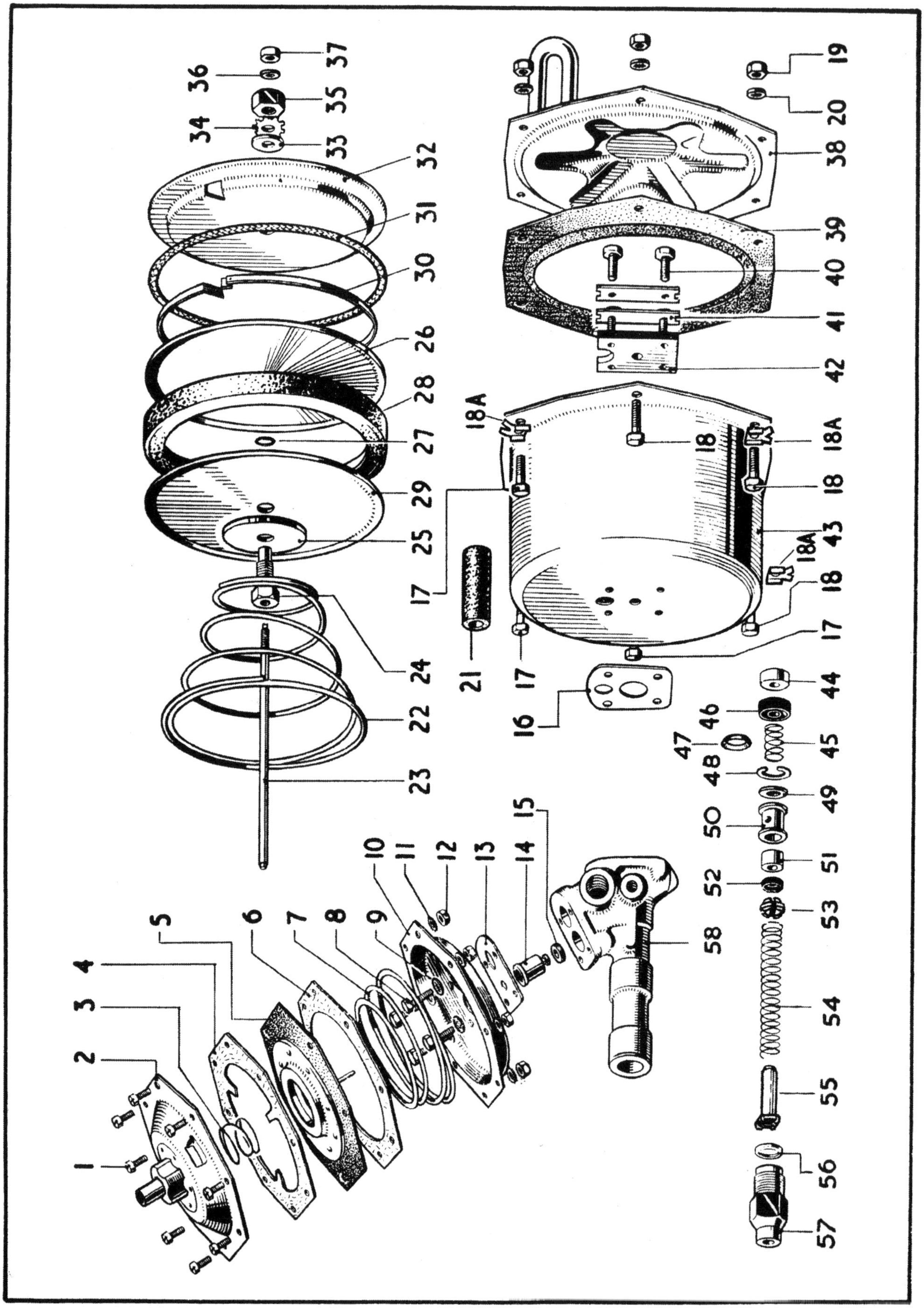
1
2
3
4
5
6
7
8
9
10
11
12
13
14
15
16
17
18
18A
19
20
21
22
23
24
25
26
27
28
29
30
31
32
33
34
35
36
37
38
39
40
41
42
43
44
45
46
47
48
49
50
51
52
53
54
55
56
57
58

PLATE 29

Part No.	Description	Plate No.	No. per Unit	Remarks
C.19612	**6 7/8" BRAKE VACUUM SERVO ASSEMBLY (4257–552)**		1	
AS.302/4H	**Screw**, securing Cover to Valve Housing (K.14303/RP)	29–1	8	
9205	**Cover and Valve Assembly** (106176)	29–2	1	
7163	**Return Spring**, for Diaphragm (89832)	29–3	1	
9206	**Gasket**, between Cover and Diaphragm (93737)	29–4	1	
9207	**Diaphragm** (93949)	29–5	1	
7160	**Gasket**, between Diaphragm and Valve Housing (88718)	29–6	1	
7167	**Balance Spring**, under Diaphragm (88734)	29–7	1	
UFS.125/4R	**Screw**, securing Valve Housing to Slave Cylinder Body (K.23502/RP)	29–8	4	
C.739	**Washer**, Shakeproof, on Bolts	29–9	4	
7169	**Valve Housing** (88805)	29–10	1	
C.721	**Washer**, Shakeproof, under Nuts	29–11	8	
AN.102/L	**Nut**, on Screws securing Valve Cover (K.18604)	29–12	8	
7168	**Gasket**, between Valve Housing and Slave Cylinder Body (88801)	29–13	1	
7161	**Valve Piston** (88719)	29–14	1	
8814	**Seal**, for Valve Piston (3862-423)	29–15	1	
6383	**Gasket**, between Slave Cylinder Body and Main Cylinder (32737)	29–16	1	
UFB.125/4R	**Bolt** (Short), securing Cover to Main Cylinder (K.23502/RP)	29–17	3	
UFB.125/9R	**Bolt** (Long), securing Cover to Main Cylinder and for Mounting Servo Unit (K.23508/RP)	29–18	3	
7754	**Tab Washer**, locking Long Bolts (99361)	29–18A	3	
UFN.125/L	**Nut**, on Bolts (K.24001/RP)	29–19	6	
C.724	**Washer**, Shakeproof, under Nuts	29–20	6	
7159	**Hose**, connecting Main Cylinder to Valve Cover (88199)	29–21	1	
7170	**Return Spring**, for Main Cylinder Piston (88806)	29–22	1	
7145	**Push Rod**, operating Main Cylinder Piston (86753)	29–23	1	MAY BE OBTAINED AS A COMPLETE ASSEMBLY TO PART No. 7146 (88799) (7145 to UFN.125/L)
7691	**Adjuster Nut**, on Push Rod (88023)	29–24	1	
7273	**Locating Washer**, for Adjuster Nut (88022)	29–25	1	
7692	**Outer Piston Plate**, in Main Cylinder (88731)	29–26	1	
7142	**'O' Ring**, between Piston Plates, on Adjuster Nut (25844)	29–27	1	
8385	**Rubber Cup**, for Main Cylinder Piston (107419)	29–28	1	
7693	**Inner Piston Plate**, in Main Cylinder (88732)	29–29	1	
7143	**Piston Ring**, in Main Cylinder (82138)	29–30	1	
7141	**Wick**, around Piston Ring (25835/21.25)	29–31	1	
7694	**End Stop**, for Main Cylinder Piston (88733)	29–32	1	
FW.108/T	**Plain Washer**, under Shakeproof Washer (K.20809)	29–33	1	
C.743	**Washer**, Shakeproof, under Nut	29–34	1	
UFN.150/L	**Nut**, securing Main Piston Assembly	29–35	1	
2538	**Copper Gasket**, under Nut (KL.44509)	29–36	1	
UFN.125/L	**Nut**, on Push Rod (K.24101)	29–37	1	
7140	**Cover and Pipe**, for Main Cylinder (88798)	29–38	1	
8414	**Gasket**, between Cover and Main Cylinder (106213)	29–39	1	
UFS.131/6R	**Screw**, securing Abutment Plate inside Main Cylinder (K.23603)	29–40	4	
6386	**Plate**, locking Bolts (82454)	29–41	2	
6385	**Abutment Plate**, inside Main Cylinder (32739)	29–42	1	
7137	**Main Vacuum Cylinder** (88807) or (114498)	29–43	1	
6382	**Spigot**, for Slave Cylinder Piston (32736)	29–44	1	
7164	**Spring** (Short) between Seal at rear of Distance Piece (92954)	29–45	1	
6402	**Seal**, between Spring and Spigot (85946)	29–46	1	
7896	**Spreader**, for Seal (90860)	29–47	1	
6444	**Circlip**, locating Short Spring (86158) or (113067)	29–48	1	
7171	**Washer**, behind Distance Piece (96462)	29–49	1	
7165	**Distance Piece**, between Piston and Washer (97266)	29–50	1	
6380	**Piston**, for Slave Cylinder (82601)	29–51	1	
6379	**Seal**, between Piston and Spring Support (92328)	29–52	1	
9209	**Spring Support**, adjacent to Piston (109579)	29–53	1	
7156	**Main Spring**, in Slave Cylinder (91861) or (109881)	29–54	1	
9210	**Spring Support**, adjacent to Adaptor (109580)	29–55	1	
C.11356	**Copper Washer**, on Adaptor in end of Slave Cylinder (KL.44526)	29–56	1	
7166	**Adaptor**, in end of Slave Cylinder (88728)	29–57	1	
7155	**Slave Cylinder Body** (88764)	29–58	1	

$6\frac{7}{8}$" BRAKE VACUUM SERVO

Part No.	Description	Plate No.	No. per Unit	Remarks
8055	REPAIR KIT FOR BRAKE VACUUM SERVO (KL.72548)		1	
8415	REPAIR KIT FOR MAIN CYLINDER PISTON (KL.72565)		1	
9211	REPAIR KIT FOR AIR CONTROL VALVE (KL.72551)		1	

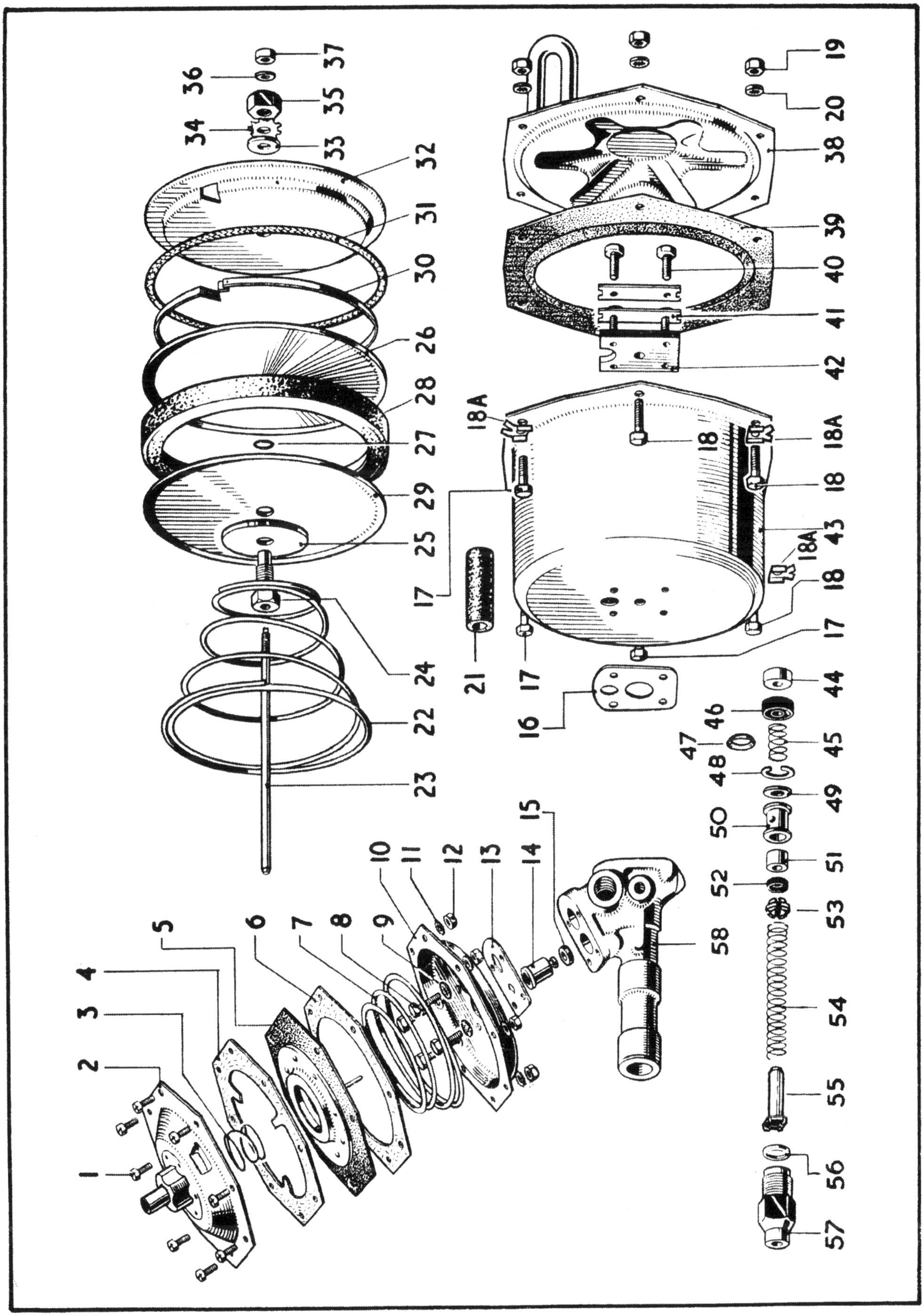
1
2
3
4
5
6
7
8
9
10
11
12
13
14
15
16
17
18
18A
19
20
21
22
23
24
25
26
27
28
29
30
31
32
33
34
35
36
37
38
39
40
41
42
43
44
45
46
47
48
49
50
51
52
53
54
55
56
57
58

PLATE 29

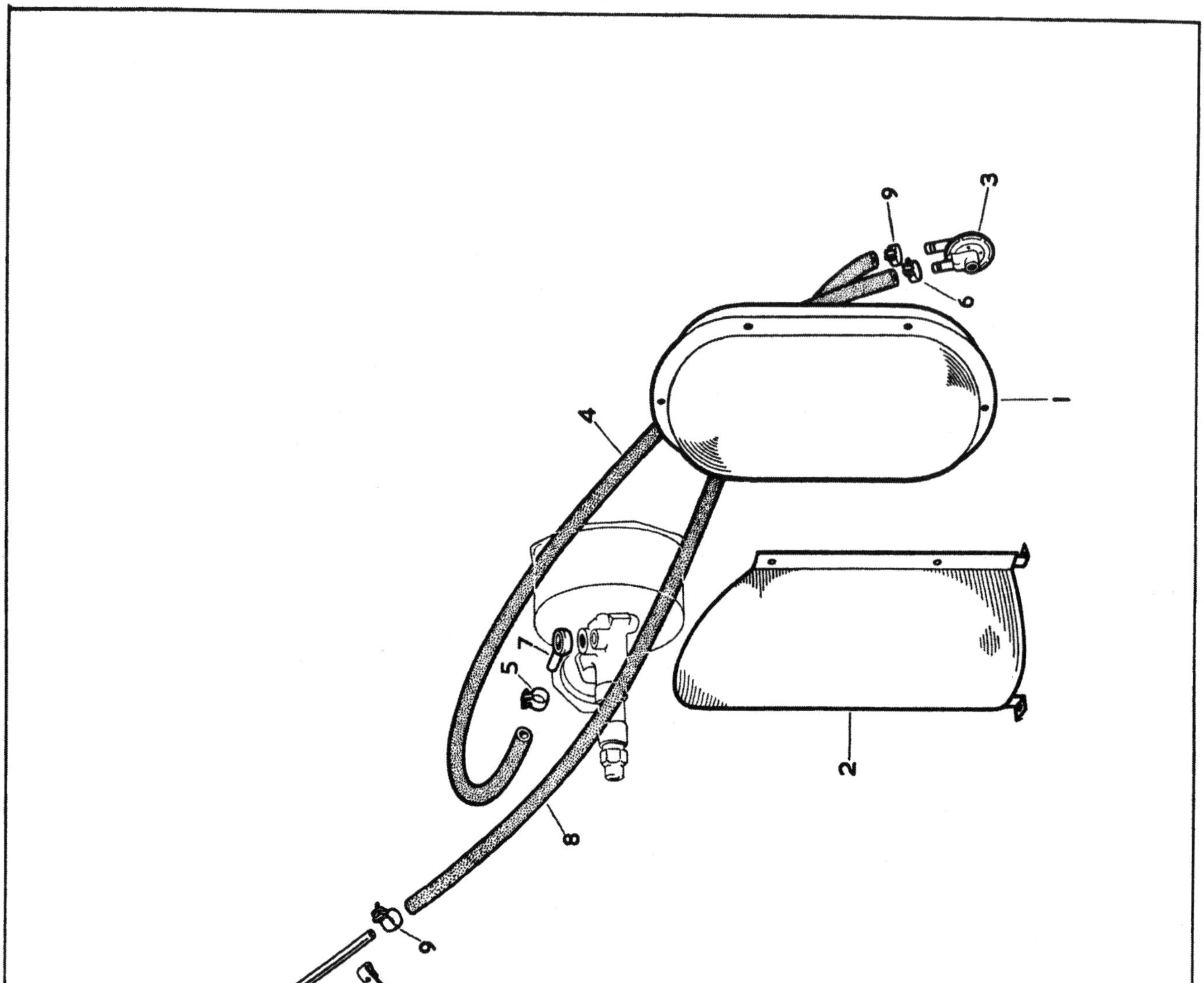

PLATE 30

RESERVAC TANK FOR BRAKE SERVO

Part No.	Description	Plate No.	No. per Unit	Remarks
C.14681	RESERVAC TANK FOR BRAKE SERVO	30–1	1	Fitted from Chassis No. 1A.1001 to 1A.6786 1A.20001 to 1A.20352
C.24910	RESERVAC TANK FOR BRAKE SERVO		1	Fitted to Chassis No. 1A.6787 and subs. 1A.20353 and subs.
C.14719	STONEGUARD ASSEMBLY FOR RESERVAC TANK	30–2	1	Fitted from Chassis No. 1A.1001 to 1A.2200 1A.20001 to 1A.20166
UFS.119/4R	**Screw,** Set, securing Stoneguard to Reservac Tank		4	
UFN,119/L	**Nut,** on Setscrews		4	
C.723/A	**Washer,** Shakeproof, under Nuts		4	
UFS.131/5R	**Screw,** Set, securing Reservac Tank and Stoneguard to Chassis Member under R.H. Front Wing		3	
UFN.131/L	**Nut,** on Setscrews		3	
FG.105/X	**Washer,** Spring, under Nuts		3	
C.22425	SUPPORT ROD ASSEMBLY FOR RESERVAC TANK		1	Fitted to Chassis No. 1A.2201 and subs. 1A.20167 and subs.
UFS.119/4R	**Screw,** Set, securing Support Rod to Brake Servo Cowl		1	
UFN.119/L	**Nut,** on Setscrew and Support Rod		2	
C.723/A	**Washer,** Shakeproof, under Nuts		2	
C.24913	STONEGUARD ASSEMBLY FOR RESERVAC TANK		1	
UFS.119/4R	**Screw,** Set, securing Stoneguard to Reservac Tank		3	
UFN.119/L	**Nut,** on Setscrews		3	
C.723/A	**Washer,** Shakeproof, under Nuts		3	
UFS.131/6R	**Screw,** Set, securing Reservac Tank and Stoneguard		1	
C.8737/2	**Nut,** Self-Locking, securing Reservac Tank and Stoneguard to Chassis Member under R.H. Front Wing		2	
FW.105/T	**Washer,** Plain, under Nuts		2	
C.27306	CHECK VALVE ASSEMBLY ON RESERVAC TANK	30–3	1	**Supersedes C.14693**
C.14963	HOSE FROM CHECK VALVE TO BRAKE SERVO	30–4	1	
C.15886/4	**Clip,** securing Hose to Banjo on Brake Servo	30–5	1	
C.15886/5	**Clip,** securing Hose to Check Valve	30–6	1	
C.14717	**Banjo,** for connection of Hose to Brake Servo	30–7	1	
C.11356	**Copper Washer,** at each side of Banjo (KL.44526)		2	
C.11421	**Bolt,** securing Banjo (30352)		1	
C.14963	HOSE FROM CHECK VALVE TO VACUUM PIPE	30–8	1	
C.15886/5	**Clip,** securing Hose to Check Valve and Vacuum Pipe	30–9	1	
C.21453	VACUUM PIPE ON CYLINDER BLOCK	30–10	1	
C.21541	**Clip,** securing Vacuum Pipe to Heater Water Pipe	30–11	1	
UFB.119/4R	**Bolt,** securing Clip		1	
AW.102/T	**Washer,** Plain, on Bolt		1	
UFN.119/L	**Nut,** on Bolt		1	
AG.102/X	**Washer,** Spring, under Nut		1	

RESERVAC TANK FOR BRAKE SERVO

Part No.	Description	Plate No.	No. per Unit	Remarks
C.21542	HOSE, FROM VACUUM PIPE TO SERVO ADAPTOR	30–12	1	
C.2905/1	**Clip,** securing Hose	30–13	2	
C.21452	ADAPTOR, AT REAR OF INDUCTION MANIFOLD, FOR VACUUM TAKE-OFF	30–14	1	
431365	**Copper Washer,** on Adaptor	30–15	1	

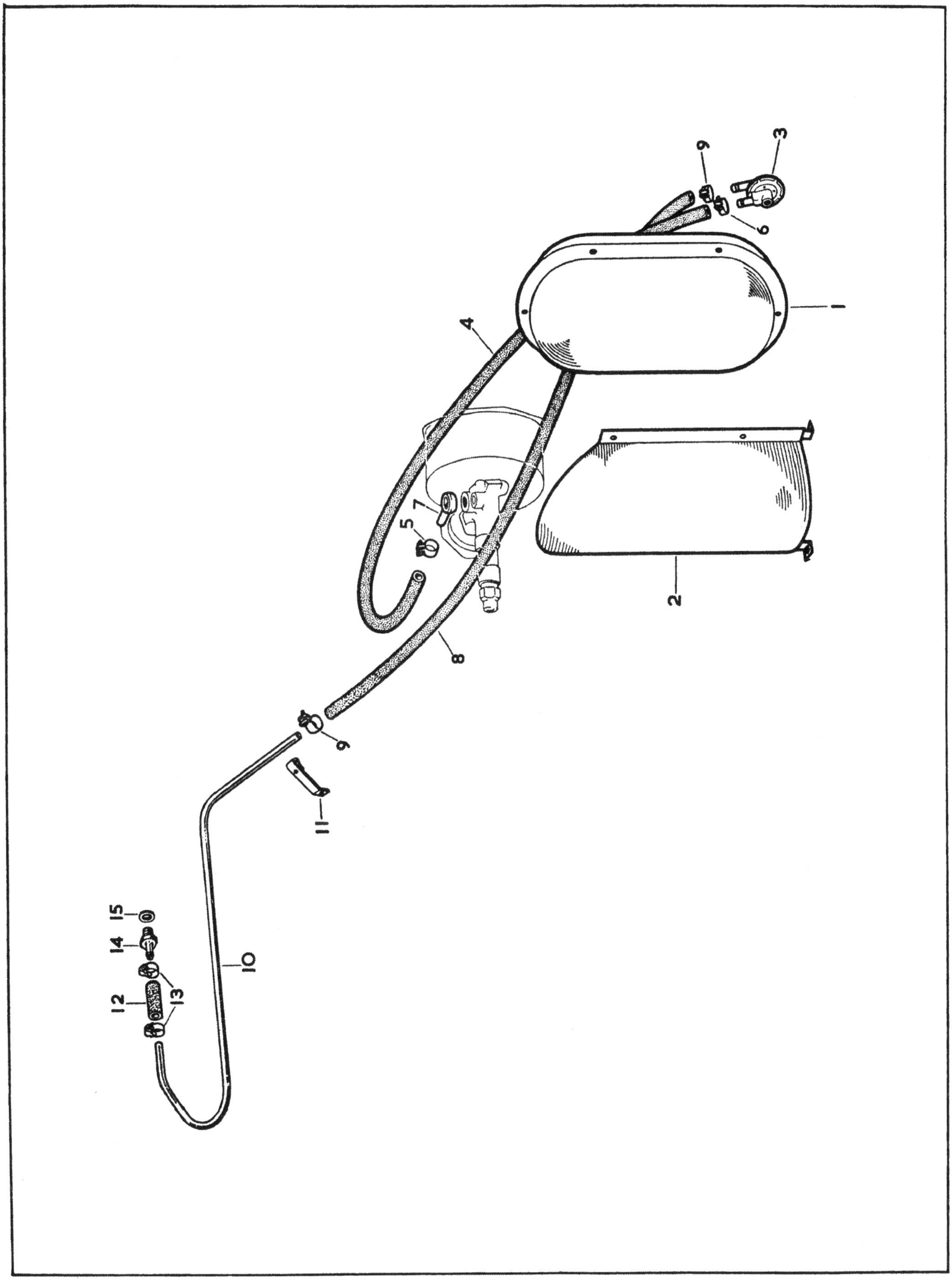

PLATE 30

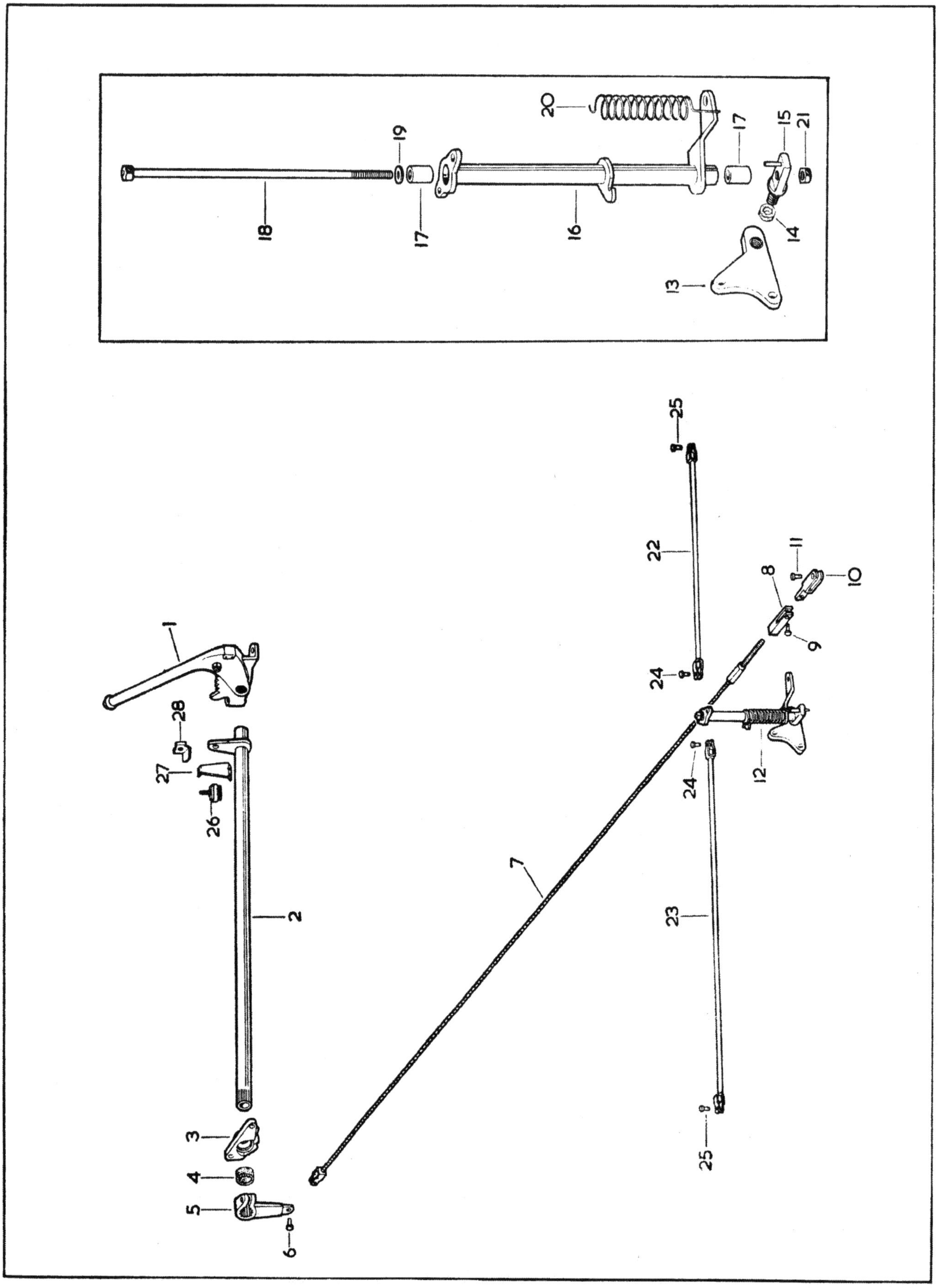

PLATE 31

Part No.	Description	Plate No.	No. per Unit	Remarks
C.15501	HANDBRAKE ASSEMBLY	31–1	1	R.H. DRIVE CARS ONLY
C.15502	HANDBRAKE ASSEMBLY		1	L.H. DRIVE CARS ONLY
C.17162	**Cover Plate,** exposing Handbrake Pawl		1	
UFS.137/7R	**Screw,** Set, securing Handbrake		2	
C.8667/3	**Nut,** on Setscrew, securing Handbrake to Floor		1	
FW.106/T	**Washer,** Plain, on Setscrew		1	
FG.106/X	**Washer,** Spring, under Nut		1	
C.15503	SHAFT ASSEMBLY FOR HANDBRAKE LEVER	31–2	1	
C.27855	**Circlip,** for Shaft		1	
C.8990	HOUSING FOR HANDBRAKE SHAFT	31–3	1	R.H. DRIVE CARS ONLY
C.8991	HOUSING FOR HANDBRAKE SHAFT		1	L.H. DRIVE CARS ONLY
C.8992	**Rubber Seal,** on Handbrake Shaft, for Housing	31–4	1	
UFB.125/7R	**Bolt,** securing Housing to Propeller Shaft Tunnel		2	
FG.104/X	**Washer,** Spring, on Bolts		2	
C.15511	HANDBRAKE LEVER FROM SHAFT TO PRIMARY CABLE	31–5	1	REQUIRED ONLY FOR R.H. DRIVE CARS
C.22240	HANDBRAKE LEVER FROM SHAFT TO PRIMARY CABLE		1	REQUIRED ONLY FOR L.H. DRIVE CARS
UFS.131/6R	**Screw,** Set, clamping Lever to Splined End of Shaft		1	
FG.105/X	**Washer,** Spring, on Setscrew		1	
J.105/11S	**Clevis Pin,** securing Cable to Lever	31–6	1	
L.103/U	**Split Pin,** retaining Clevis Pin		1	
FW.105/T	**Washer,** Plain, under Split Pin		1	
C.15512	PRIMARY CABLE, FROM LEVER TO COMPENSATOR	31–7	1	
C.14789	FORK END AT REAR OF PRIMARY HANDBRAKE CABLE	31–8	1	
UFN.125/L	**Nut,** securing Fork End to Primary Handbrake Cable		1	
J.105/12S	**Clevis Pin,** securing Fork End to Universal Jaw	31–9	1	
L.103/7U	**Split Pin,** retaining Clevis Pin		1	
FW.105/T	**Washer,** Plain, under Split Pin		1	
C.16214	UNIVERSAL JAW BETWEEN FORK END AND COMPENSATOR	31–10	1	
J.105/11S	**Clevis Pin,** securing Universal Jaw to Compensator	31–11	1	
L.103/7U	**Split Pin,** retaining Clevis Pin		1	
FW.105/T	**Washer,** Plain, under Split Pin		1	

HANDBRAKE CONTROL

Part No.	Description	Plate No.	No. per Unit	Remarks
C.20745	**HANDBRAKE COMPENSATOR ASSEMBLY**	31–12	1	
C.15524	**Bracket,** for mounting of Compensator to Rear Axle	31–13	1	
C.15532	**Felt Seal,** between Bracket and Fulcrum Pin	31–14	1	
C.20744	**Fulcrum Pin Assembly**	31–15	1	
C.15517	**Balance Lever,** complete with Bushes	31–16	1	
140	**Bush,** in Balance Lever Tube	31–17	2	
C.15523	**Spindle Assembly,** for Compensator Lever	31–18	1	
FW.106/T	**Washer,** Plain, at top of Spindle	31–19	1	
C.15899	**Spring,** for Compensator	31–20	1	
UFN.137/L	**Nut,** securing Spindle to Fulcrum Pin	31–21	1	
UCS.131/8R	**Screw,** Set, securing Compensator to Rear Axle		2	
C.22403	**HANDBRAKE CABLE SUPPORT BRACKET ASSEMBLY**		1	REQUIRED FOR L.H. DRIVE CARS ONLY
BD.1814/4	**Pop Rivet,** securing Handbrake Cable Support Bracket to Floor		2	REQUIRED FOR L.H. DRIVE CARS ONLY
C.19771	**Guide,** for Cable		1	REQUIRED FOR L.H. DRIVE CARS ONLY
UFS.119/7R	**Screw,** Set, securing Guide to Support Bracket		2	REQUIRED FOR L.H. DRIVE CARS ONLY
C.8667/17	**Nut,** Self-Locking, on Setscrews		2	REQUIRED FOR L.H. DRIVE CARS ONLY
AW.102/T	**Washer,** Plain, under Nuts		2	REQUIRED FOR L.H. DRIVE CARS ONLY
C.19495	**CABLE, FROM COMPENSATOR TO R.H. REAR CALIPER**	31–22	1	
C.19496	**CABLE, FROM COMPENSATOR TO L.H. REAR CALIPER**	31–23	1	
J.105/11S	**Clevis Pin,** securing Cables to Compensator	31–24	2	
J.105/12S	**Clevis Pin,** securing Cables to Handbrake Operating Levers	31–25	2	
FW.105/T	**Washer,** Plain, on Clevis Pins		4	
L.103/7U	**Split Pin,** retaining Clevis Pins		4	
C.11041	**Blanking Plate,** sealing redundant hole in Propeller Shaft Tunnel, for Handbrake Shaft		1	
C.10955	**Rubber Gasket,** under Blanking Plate		1	
UFS.125/5R	**Screw,** Set, securing Blanking Plate to Propeller Shaft Tunnel		2	
C.11046	**Rubber Plug,** sealing redundant hole in floor for Handbrake		1	
C.12833	**SWITCH, AT BASE OF HANDBRAKE LEVER OPERATING HANDBRAKE WARNING LIGHT (31504/D–12SA)**	31–26	1	**See also Electrical Equipment**
C.16180	**Bracket,** mounting Warning Light Switch	31–27	1	
UFS.119/4R	**Screw,** Set, securing Bracket		2	
UFN.119/L	**Nut,** on Setscrews		2	
C.723A	**Washer,** Shakeproof, under Nuts		2	
UFN.125/L	**Nut,** securing Switch to Bracket		2	
C.16185	**Bracket,** operating Handbrake Warning Light Switch	31–28	1	FITTED TO R.H. DRIVE CARS ONLY
C.16188	**Bracket,** operating Handbrake Warning Light Switch		1	FITTED TO L.H. DRIVE CARS ONLY

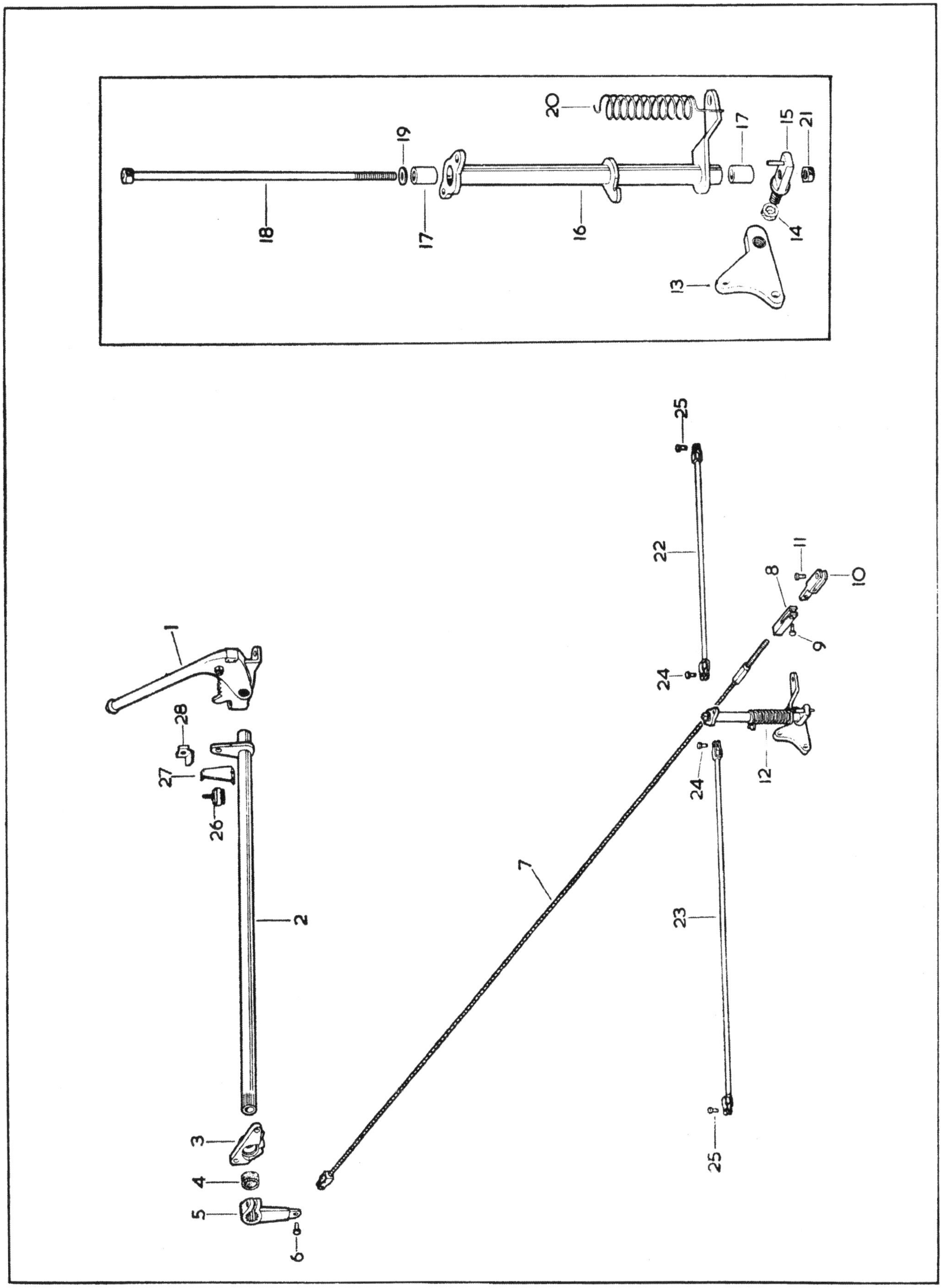
1
2
3
4
5
6
7
8
9
10
11
12
13
14
15
16
17
17
18
19
20
21
22
23
24
24
25
25
26
27
28

PLATE 31

STEERING UNIT

Part No.	Description	Plate No.	No. per Unit	Remarks
C.14845	**STEERING BOX ASSEMBLY, COMPLETE (SF.414) (RATIO 20.3:1)**		1	**STANDARD FITMENT TO R.H. DRIVE CARS ONLY**
7919	**Steering Box only** (S.1-135)	32–1	1	
6409	**Trunnion Bush,** at bottom of Steering Box (19-9-5)	32–2	1	
7921	**Inner Column Worm** (S.3-290)	32–3	1	
7922	**Main Nut** (complete with Transfer Tube but less Steel Balls (S.2-52)	32–4	1	
8753	**Roller,** on Spigot of Main Nut for Rockershaft traverse (24-5-4)	32–5	1	
136	**Steel Ball,** around Main Nut and Inner Column Worm (17-3-1)	32–6	51	
7925	**Ball Race** (Adjustable) at each end of Inner Column Worm (S.23-15)	32–7	2	
7926	**Distance Piece,** between Adjustable Ball Race and End Plates (12-9-32)	32–8	2	
6419	**End Plate,** at bottom of Steering Box (S.11-38)	32–9	1	
6414	**Gasket,** between Lower End Plate and Steering Box (S.10-15)	32–10	As req'd	
6416	**End Plate,** at upper end of Steering Box (adjacent to Splined end of Inner Column Worm) (S.11-43)	32–11	1	
7696	**Plate,** retaining Oil Seal in End Plate (S.10-99)	32–12	1	
6415	**Oil Seal,** in Upper End Plate, for Inner Column Worm (27-8-20)	32–13	1	
6417	**Gasket,** between Upper End Plate and Steering Box (S.10-42)	32–14	As req'd	
6418	**Shim,** under End Plates, for adjustment of End Float on Inner Column Worm (S.10-14)	32–15	As req'd	
UCS.131/7R	**Screw,** Set, securing End Plates to Steering Box	32–16	8	
FG.105/X	**Washer,** Spring, on Setscrews	32–17	8	
7927	**Rockershaft** (S.7-73)	32–18	1	
6427	**'O' Ring,** at bottom of Steering Box, for Rockershaft (27-9-5)	32–19	1	
6428	**Washer,** retaining 'O' Ring in Steering Box (12-9-2)	32–20	1	
C.14847	**Drop Arm,** on Rockershaft (S.9-149)	32–21	1	
UFN.275/L	**Nut,** securing Drop Arm to Rockershaft	32–22	1	
6430	**Spring Washer,** under Nut (12-8-22)	32–23	1	
7929	**Cover Plate,** at top of Steering Box (S.12-67)	32–24	1	
6423	**Gasket,** under Cover Plate (S.10-16)	32–25	1	
UCS.131/8R	**Screw,** Set, securing Cover Plate to Steering Box	32–26	4	
FG.105/X	**Washer,** Spring, on Setscrews	32–27	4	
7931	**Screw,** in Cover Plate, for Adjustment of Rockershaft (25-8-7)	32–28	4	
7932	**Nut,** locking Adjusting Screw (11-8-10)	32–29	1	
7597	**Spring,** inside Adjusting Screw (36-7-2)	32–30	1	
7933	**Bolt,** applying tension to Spring (10-7-3)	32–31	1	
144	**Oil Filler Plug,** in Cover Plate (21-7-2)	32–32	1	
C.2296/5	**Washer,** under Filler Plug	32–33	1	
C.14846	**STEERING BOX ASSEMBLY COMPLETE (SF.415) (RATIO 20.3:1)**		1	**STANDARD FITMENT TO L.H. DRIVE CARS ONLY**
7920	**Steering Box only** (S.1-134)		1	
7923	**Inner Column Worm** (S.3-292)		1	
7924	**Main Nut** (complete with Transfer Tube but less Steel Balls) (S.2-53)		1	
7930	**Cover Plate,** at top of Steering Box (S.12-68)		1	
	All other items are as for Steering Box C.14845			
C.11919	**Bolt** (Long), securing Steering Box to Front Suspension Cross Member	32–34	2	
C.11920	**Bolt** (Short), securing Steering Box to Front Suspension Cross Member	32–35	2	
C.11958	**Tab Washer,** retaining Bolts	32–36	2	

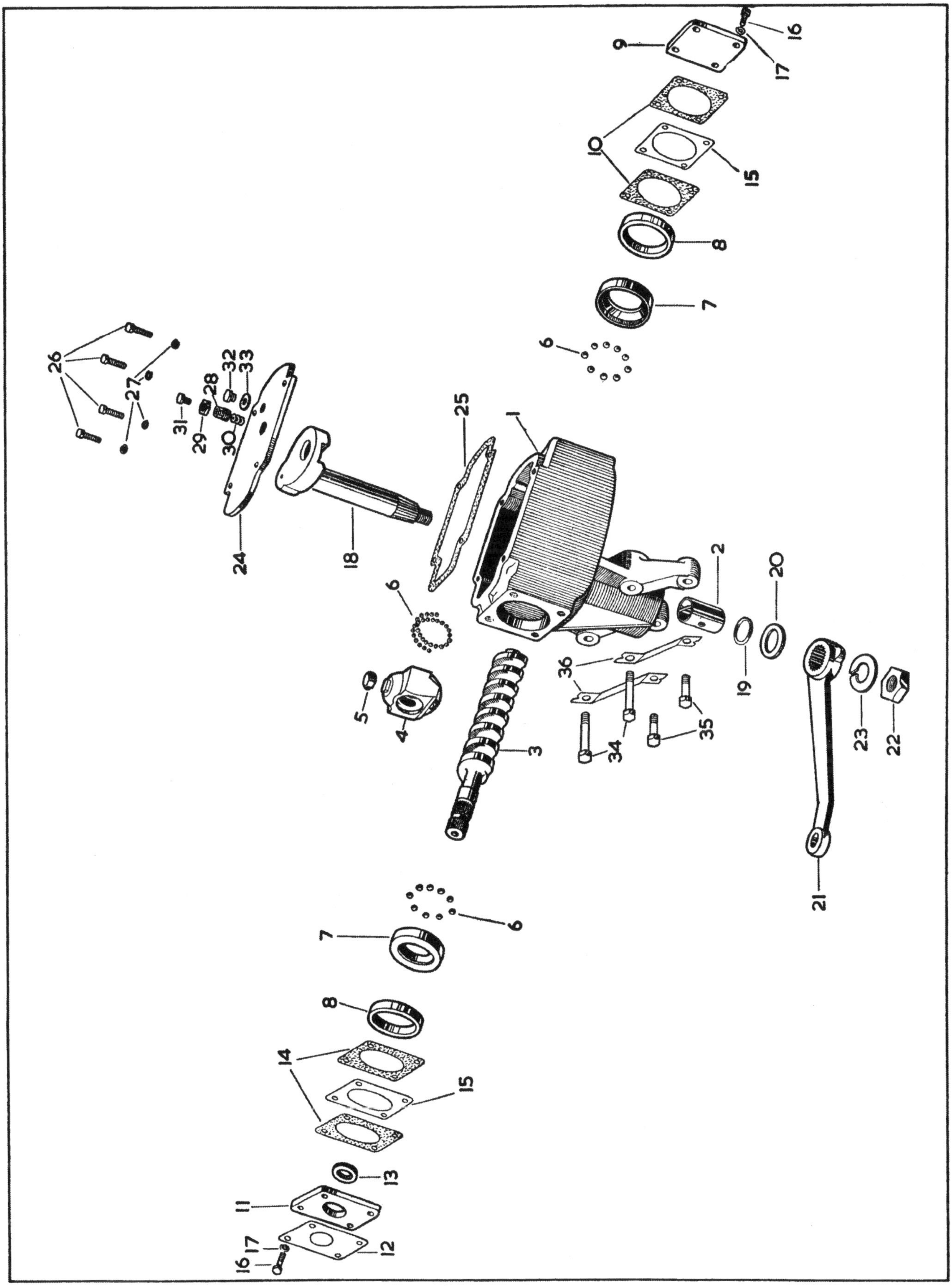

PLATE 32

STEERING UNIT

Part No.	Description	Plate No.	No. per Unit	Remarks
C.8927	**STEERING BOX ASSEMBLY, COMPLETE (SF.211) (RATIO 17.6:1)**		1	**SUPPLIED TO SPECIAL ORDER FOR R.H. DRIVE CARS ONLY**
6407	**Steering Box only** (S.1-80)	33–1	1	
6409	**Trunnion Bush,** at bottom of Steering Box (19-19-5)	33–2	1	
6412	**Inner Column Worm** (S.3-167)	33–3	1	
6420	**Main Nut** (complete with Transfer Tube but less Steel Balls)	33–4	1	
146	**Roller,** on Spigot of Main Nut, for Guide Slot in Cover Plate (24-5-1)	33–5	1	
137	**Steel Ball,** around Main Nut and Inner Column Worm (17-4-1)	33–6	30	
6411	**Washer,** for seating of Steel Balls, at each end of Inner Column Worm (12-8-32)	33–7	2	
6410	**Ball Race** (Adjustable) at each end of Inner Column Worm (5-23-10)	33–8	2	
6419	**End Plate,** at bottom of Steering Box (S.11-38)	33–9	1	
6414	**Gasket,** between Lower End Plate and Steering Box (S.10-15)	33–10	As req'd	
6416	**End Plate,** at upper end of Steering Box (adjacent to Splined end of Inner Column Worm) (S.11-43)	33–11	1	
7696	**Plate,** retaining Oil Seal in End Plate (S.10-99)	33–11A	1	
6415	**Oil Seal,** in Upper End Plate, for Inner Column Worm (27-8-20)	33–12	1	
6417	**Gasket,** between Upper End Plate and Steering Box (S.10-42)	33–13	As req'd	
6418	**Shim,** under End Plates, for adjustment of End Float on Inner Column Worm (S.10-14)	33–14	As req'd	
UCS.131/7R	**Screw,** Set, securing End Plates to Steering Box	33–15	8	
FG.105/X	**Washer,** Spring, on Setscrews	33–16	8	
6422	**Rockershaft** (S.7-44)	33–17	1	
6427	**'O' Ring,** at bottom of Steering Box, for Rockershaft (27-9-5)	33–18	1	
6428	**Washer,** retaining 'O' Ring in Steering Box (12-9-2)	33–19	1	
6429	**Drop Arm,** on Rockershaft (S.9-96)	33–20	1	
UFN.275/L	**Nut,** securing Drop Arm to Rockershaft	33–21	1	
6430	**Spring Washer,** under Nut (12-8-22)	33–22	1	
6424	**Cover Plate,** at top of Steering Box (S.12-12)	33–23	1	
6423	**Gasket,** under Cover Plate (S.10-16)	33–24	1	
UCS.131/8R	**Screw,** Set, securing Cover Plate to Steering Box	32–25	4	
FG.105/X	**Washer,** Spring, on Setscrews	33–26	4	
6426	**Screw,** in Cover Plate, for Adjustment of Rockershaft (25-7-2)	33–27	4	
UFN.262/L	**Nut,** locking Adjusting Screw	33–28	1	
144	**Oil Filler Plug,** in Cover Plate (21-7-2)	33–29	1	
C.2296/5	**Washer,** under Filler Plug	33–30	1	
C.8928	**STEERING BOX ASSEMBLY, COMPLETE (SF.212) (RATIO 17.6:1)**		1	**SUPPLIED TO SPECIAL ORDER FOR L.H. DRIVE CARS ONLY**
6408	**Steering Box only** (S.1-79)		1	
6413	**Inner Column Worm** (S.3-168)		1	
6421	**Main Nut** (complete with Transfer Tube but less Steel Balls)		1	
6425	**Cover Plate,** at top of Steering Box (S.12-13)		1	
	All other items are as for Steering Box C.8927			
C.11919	**Bolt** (Long) securing Steering Box to Front Suspension Cross Member	33–31	2	
C.11920	**Bolt** (Short) securing Steering Box to Front Suspension Cross Member	33–32	2	
C.11958	**Tab Washer,** retaining Bolts	33–33	2	

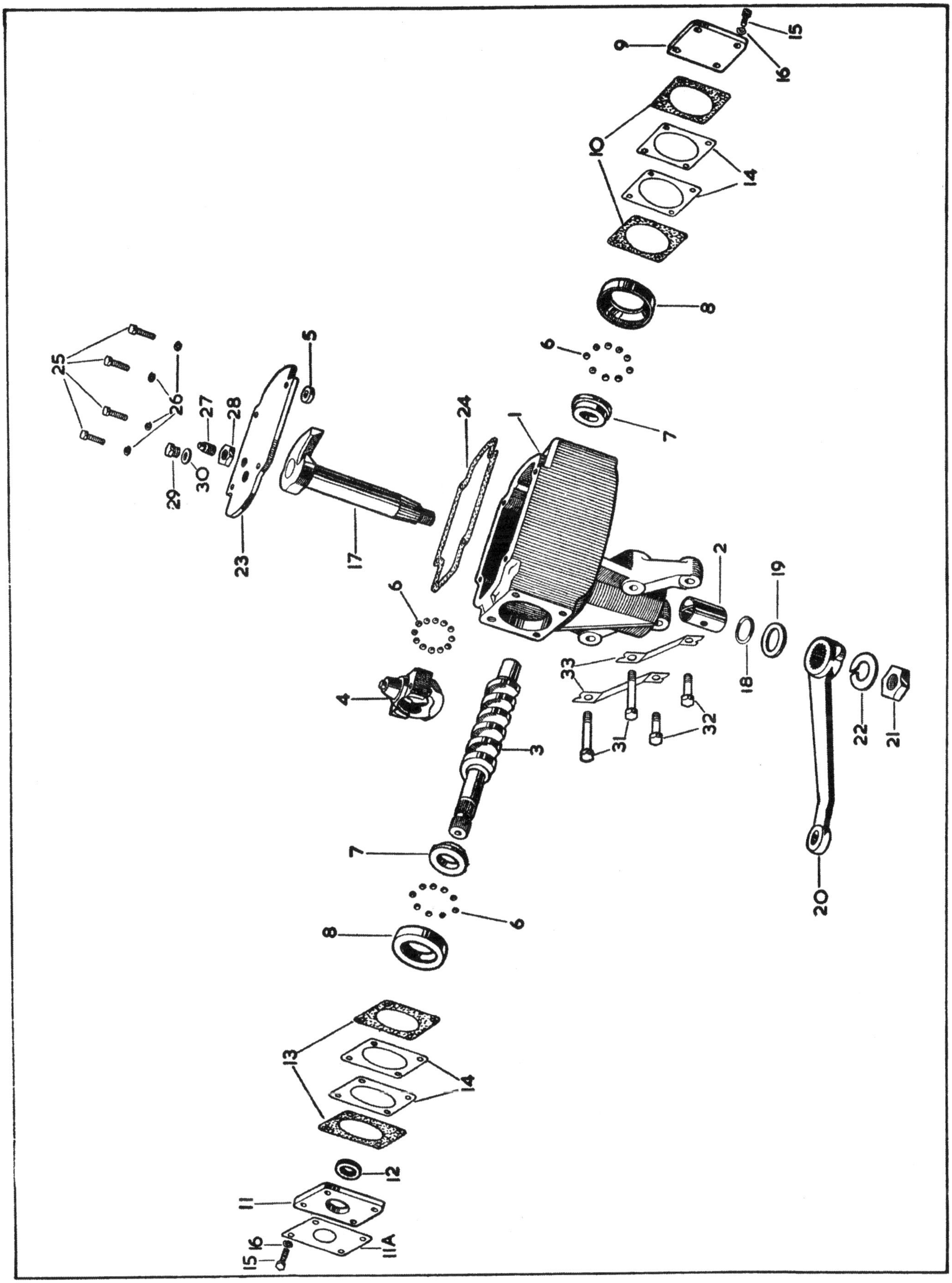
9
15
16
10
14
8
6
7
1
24
5
25
26
27
28
29
30
23
17
2
19
18
6
33
4
3
31
32
22
21
20
7
6
8
13
14
12
11
11A
16
15

PLATE 33

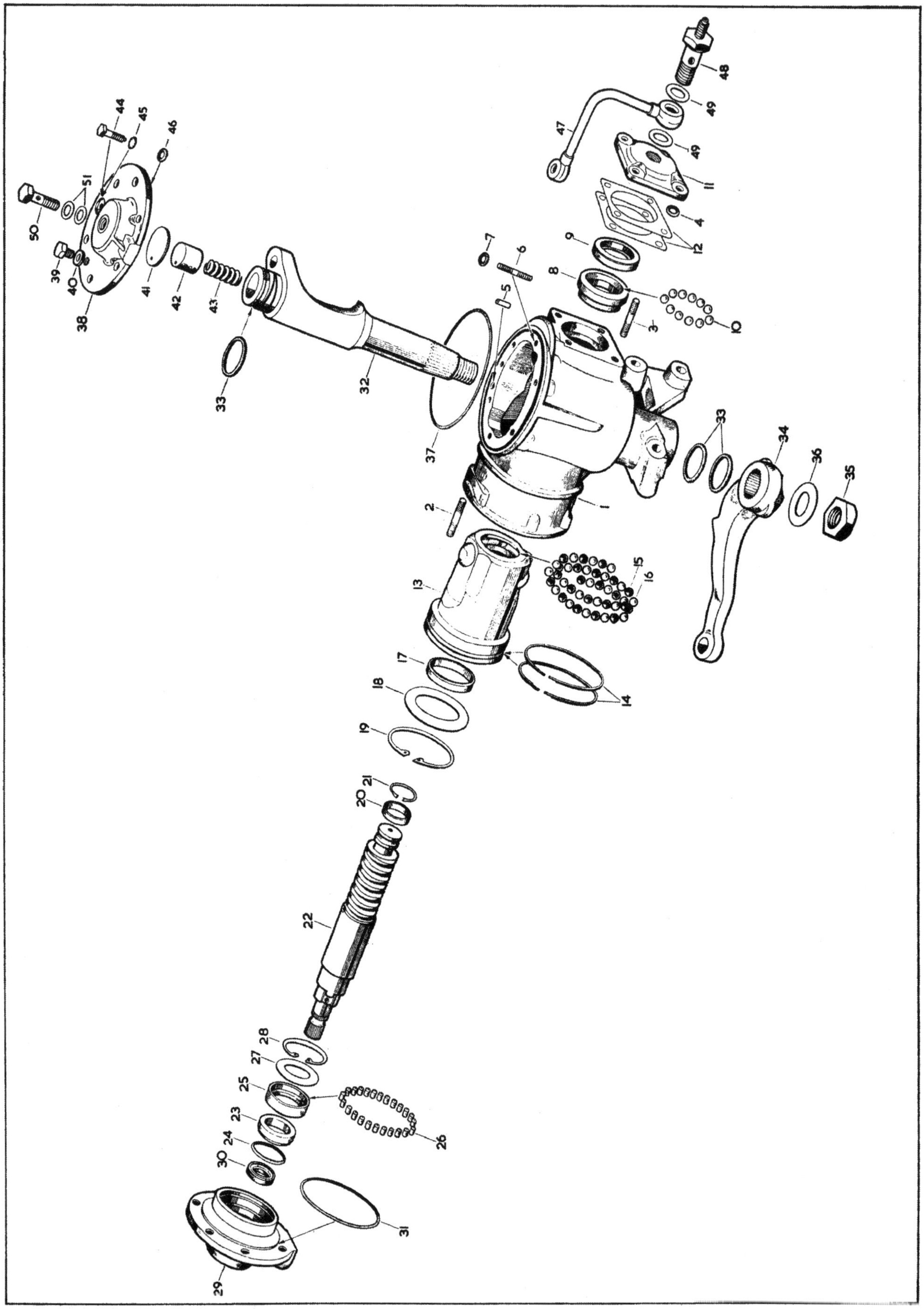

PLATE 34

POWER-ASSISTED STEERING UNIT

Part No.	Description	Plate No.	No. per Unit	Remarks
C.17625	**STEERING BOX ASSEMBLY (PS.1–509)**		1	**FOR R.H. DRIVE CARS ONLY** Fitted from Chassis No. 1A.1001 to 1A.1208
9095	**Steering Box** (complete with Trunnion Bush, Sleeve, and Oil Seal Housing) (S.1-152)		1	
8195	**Stud,** fixing top End Plate (13-2-23)		6	
8214	**Nut,** securing top End Plate (11-2-6)		6	
101	**Washer,** under Nuts (12-2-1)		6	
7608	**Stud,** fixing bottom End Plate (13-3-15)		4	
7570	**'O' Ring,** on Studs (27-3-1)		4	
8215	**Nut,** securing bottom End Plate (11-3-8)		4	
102	**Washer,** under Nuts (12-3-1)		4	
8197	**Dowel,** positioning Top Cover on Steering Box (35-2-14)		1	
8196	**Stud,** fixing Top Cover (13-2-22)		6	
7594	**'O' Ring,** on Studs for Top Cover (27-2-1)		6	
8222	**Adjustable Ball Race** (inner) (S.23-25)		1	
8223	**Adjustable Ball Race** (outer) (S.23-26)		1	
136	**Ball,** between Inner and Outer Adjustable Races (17-3-1)		11	
8212	**Bottom End Plate** (S.11-67)		1	
7609	**Shim,** for bottom End Plate (.0025" thick) (S.10-101)		As req'd	**CHOOSE SHIMS OF DESIRED THICKNESS**
8213	**Shim,** for bottom End Plate (.004" thick) (S.10-110)		As req'd	
7610	**Shim,** for bottom End Plate (.005" thick) (S.10-102)		As req'd	
7611	**Shim,** for bottom End Plate (.010" thick) (S.10-103)		As req'd	
8201	**Main Nut** (complete with Transfer Tube and Retainer) (S.2-57)		1	**Supplied complete with Inner Column to Part No. 8986 (3-2960)**
8221	**Piston Ring,** for Main Nut (27-12-15)		2	
8205	**Ball** (Black finish) in Main Nut (17-1-6)		22	
142	**Ball** (Bright finish) in Main Nut (17-1-1)		22	
8202	**Sleeve,** inside Main Nut (12-12-19)		1	
8203	**Washer,** retaining Sleeve (12-12-18)		1	
8204	**Circlip,** behind Washer (31-12-9)		1	
8224	**Sleeve,** at end of Inner Column, for Adjustable Ball Race (28-9-26)		1	
8225	**Circlip,** retaining Sleeve (31-8-10)		1	
8199	**Inner Column Assembly** (complete with Valve, Button, Ball, Plug, Tube and Spring) (S.3-303)		1	**Supplied complete with Main Nut to Part No. 8986 (3-2960)**
8200	**Packing Piece,** behind Roller Race (12-10-28)		1	
7572	**'O' Ring,** on Packing Piece (27-12-13)		1	
8228	**Race,** for Rollers on Inner Column (38-12-6)		1	
8227	**Rollers,** for Inner Column (18-1-3)		24	
8226	**Washer,** enclosing Rollers inside Race (12-11-19)		1	
8229	**Circlip,** retaining Roller Race in Top End Plate (31-12-8)		1	
8211	**Top End Plate** (S.11-66)		1	
8216	**Seal,** for Top End Plate (27-8-34)		1	
7582	**'O' Ring,** for Top End Plate (27-12-11)		1	
8206	**Rockershaft** (S.7-74)		1	
7607	**'O' Ring,** inside Trunnion of Steering Box and at top of Rockershaft (27-9-6)		3	
C.17626	**Drop Arm** (S.9-171)		1	
8209	**Nut,** securing Drop Arm to Rockershaft (11-8-13)		1	
8210	**Washer,** under Nut (12-8-56)		1	
8220	**'O' Ring,** in groove at top of Steering Box, sealing Cover Plate (27-12-14)		1	
8217	**Cover Plate,** at top of Steering Box (S.12-69)		1	
7600	**Thrust Pad,** in Cover Plate (.155" thick) (12-11-12)		1	**CHOOSE THRUST PAD OF DESIRED THICKNESS**
7661	**Thrust Pad,** in Cover Plate (.160" thick) (12-11-16)		1	
7601	**Thrust Pad,** in Cover Plate (.165" thick) (12-11-13)		1	
7662	**Thrust Pad,** in Cover Plate (.170" thick) (12-11-17)		1	
7602	**Thrust Pad,** in Cover Plate (.175" thick) (12-11-14)		1	
7663	**Thrust Pad,** in Cover Plate (.180" thick) (12-11-18)		1	
7603	**Thrust Pad,** in Cover Plate (.185" thick) (12-11-15)		1	
8207	**Plunger,** in recess at top of Rockershaft (S.24-15)		1	
8208	**Spring,** under Rockershaft Plunger (36-7-6)		1	
8219	**Special Bolt,** securing Cover Plate to Steering Box (10-2-16)		1	
8218	**Washer,** under head of Special Bolt (12-2-6)		1	
7594	**'O' Ring,** for Special Bolt, between Cover Plate and Steering Box (27-2-1)		1	
C.16103	**Feed Pipe**		1	
C.16102	**Connector,** securing Feed Pipe to bottom End Plate and for attachment of Hose from Steering Box to Pump		1	
C.2296/1	**Copper Washer,** for Connector		2	
C.13984	**Banjo Bolt,** securing Feed Pipe to Top Cover		1	
C.2296/4	**Copper Washer,** on Banjo Bolt		2	

POWER-ASSISTED STEERING UNIT

Part No.	Description	Plate No.	No. per Unit	Remarks
C.14941	**STEERING BOX ASSEMBLY (PS.1–433)**		1	**FOR L.H. DRIVE CARS ONLY** Fitted from Chassis No. 1A.20001 to 1A.20054
8231	**Steering Box** (complete with Trunnion Bush, Sleeve and Oil Seal Housing) (S.1-139)		1	
8331	**Main Nut** (complete with Transfer Tube and Retainer) (S.2-58)		1	**Supplied complete, to Part No. 8987 (3-2961)**
8232	**Inner Column Assembly** (complete with Valve, Button, Ball, Plug, Tube and Spring) (S.3-304)		1	
C.16045	**Drop Arm** (9-154)		1	
C.16104	**Feed Pipe**		1	
	All other items for Steering Box C.14941 are identical with Steering Box C.17625			
C.21837	**STEERING BOX ASSEMBLY (PS.1–509)**		1	**FOR R.H. DRIVE CARS ONLY** Fitted from Chassis No. 1A.1209 to 1A.11868
9095	**Steering Box** (complete with Trunnion Bush, Sleeve and Oil Seal Housing) (S.1-152)	34–1	1	
8195	**Stud,** fixing Top End Plate (13-2-23)	34–2	6	
8214	**Nut,** securing Top End Plate (11-2-6)		6	
101	**Washer,** under Nuts		6	
7608	**Stud,** fixing Bottom End Plate (13-3-5)	34–3	4	
7570	**'O' Ring,** on Studs (27-3-1)	34–4	4	
8215	**Nut,** securing Bottom End Plate (11-3-8)		4	
102	**Washer,** under Nuts (12-3-1)		4	
8197	**Dowel,** positioning Top Cover on Steering Box (35-2-14)	34–5	1	
8196	**Stud,** fixing Top Cover (13-2-22)	34–6	6	
7594	**'O' Ring,** on Studs for Top Cover (27-2-1)	34–7	6	
8222	**Adjustable Ball Race,** (Inner) (S.23-25)	34–8	1	
8223	**Adjustable Ball Race,** (Outer) (S.23-26)	34–9	1	
136	**Ball,** between Inner and Outer Adjustable Races (17-3-1)	34–10	11	
8212	**Bottom End Plate** (S.11-67)	34–11	1	
7609	**Shim,** for Bottom End Plate (.0025" thick) (S.10-101)	34–12	As req'd	**CHOOSE SHIMS OF DESIRED THICKNESS**
8213	**Shim,** for Bottom End Plate (.004" thick) (S.10-110)	34–12	As req'd	
7610	**Shim,** for Bottom End Plate (.005" thick) (S.10-102)	34–12	As req'd	
7611	**Shim,** for Bottom End Plate (.010" thick) (S.10-103)	34–12	As req'd	
8201	**Main Nut** (complete with Transfer Tube and Retainer) (S.2-57)	34–13	1	**Supplied complete with Inner Column, to Part No. 8986 (3-2960)**
8221	**Piston Ring,** for Main Nut (27-12-15)	34–14	2	
8205	**Ball** (Black Finish) in Main Nut (17-1-6)	34–15	22	
142	**Ball** (Bright Finish) in Main Nut (17-1-1)	34–16	22	
8202	**Sleeve,** inside Main Nut (12-12-19)	34–17	1	
8203	**Washer,** retaining Sleeve (12-12-18)	34–18	1	
8204	**Circlip,** behind Washer (31-12-9)	34–19	1	
8224	**Sleeve,** at end of Inner Column, for Adjustable Ball Race (28-9-26)	34–20	1	
8225	**Circlip,** retaining Sleeve (31-8-10)	34–21	1	
8199	**Inner Column Assembly** (complete with Valve, Button, Ball, Plug, Tube and Spring) (S.3-303)	34–22	1	**Supplied complete with Main Nut, to Part No. 8986 (3-2960)**
8200	**Packing Piece,** behind Roller Race (12-10-28)	34–23	1	
7572	**'O' Ring,** on Packing Piece (27-12-13)	34–24	1	
8228	**Race,** for Rollers on Inner Column (38-12-6)	34–25	1	
8227	**Roller,** for Inner Column (18-1-3)	34–26	24	
8226	**Washer,** enclosing Rollers inside Race (12-11-19)	34–27	1	
8229	**Circlip,** retaining Roller Race in Top End Plate (31-12-8)	34–28	1	
8211	**Top End Plate** (S.11-66)	34–29	1	
8216	**Seal,** for Top End Plate (27-8-34)	34–30	1	
7582	**'O' Ring,** for Top End Plate (27-12-11)	34–31	1	
8206	**Rockershaft** (S.7-74)	34–32	1	
7607	**'O' Ring,** inside Trunnion of Steering Box and at top of Rockershaft (27-9-6)	34–33	3	
C.17626	**Drop Arm** (S.9-171)	34–34	1	Fitted from Chassis No. 1A.1001 to 1A.11868
8209	**Nut,** securing Drop Arm to Rockershaft (11-8-13)	34–35	1	
8210	**Washer,** under Nut (12-8-56)	34–36	1	
8220	**'O' Ring,** in Groove at top of Steering Box, sealing Cover Plate (27-12-14)	34–37	1	
9966	**Cover Plate,** at top of Steering Box (S.12-113)	34–38	1	
9680	**Air Bleed Plug,** in Cover Plate (21-5-6)	34–39	1	
9681	**Washer,** on Plug (12-4-17)	34–40	1	

POWER-ASSISTED STEERING UNIT

Part No.	Description	Plate No.	No. per Unit	Remarks
	STEERING BOX ASSEMBLY (PS.1-509) *(contd.)*			
7600	**Thrust Pad**, in Cover Plate (.155" thick) (12-11-12)	34–41	1	CHOOSE THRUST PAD OF DESIRED THICKNESS
7661	**Thrust Pad**, in Cover Plate (.160" thick) (12-11-16)	34–41	1	
7601	**Thrust Pad**, in Cover Plate (.165" thick) (12-11-13)	34–41	1	
7662	**Thrust Pad**, in Cover Plate (.170" thick) (12-11-17)	34–41	1	
7602	**Thrust Pad**, in Cover Plate (.175" thick) (12-11-14)	34–41	1	
7663	**Thrust Pad**, in Cover Plate (.180" thick) (12-11-18)	34–41	1	
7603	**Thrust Pad**, in Cover Plate (.185" thick) (12-11-15)	34–41	1	
8207	**Plunger**, in recess at top of Rockershaft (S.24-15)	34–42	1	
8208	**Spring**, under Rockershaft Plunger (36-7-6)	34–43	1	
8219	**Special Bolt**, securing Cover Plate to Steering Box (10-2-16)	34–44	1	
8218	**Washer**, under head of Special Bolt (12-2-6)	34–45	1	
7594	**'O' Ring**, for Special Bolt, between Cover Plate and Steering Box (27-2-1)	34–46	1	
C.16103	**Feed Pipe Assembly**	34–47	1	
C.16102	**Connector**, securing Feed Pipe to Bottom End Plate and for attachment of Hose from Pump	34–48	1	
C.2296/1	**Copper Washer**, on Connector	34–49	2	
C.1398/4	**Banjo Bolt**, securing Feed Pipe to Top Cover	34–50	1	
C.2296/4	**Copper Washer**, on Banjo Bolt	34–51	2	
C.21836	**STEERING BOX ASSEMBLY (PS.1–433)**		1	**FOR L.H. DRIVE CARS ONLY** Fitted from Chassis No. 1A.20055 to 1A.20541
8231	**Steering Box** (complete with Trunnion Bush, Sleeve and Oil Seal Housing) (S.1-139)		1	
8331	**Main Nut** (complete with Transfer Tube and Retainer) (S.2-58)		1	**Supplied complete, to Part No. 8987 (3-2961)**
8232	**Inner Column Assembly** (complete with Valve, Button, Ball, Plug,Tube and Spring) (S.3-304)		1	
C.16045	**Drop Arm** (S.9-154)		1	
9965	**Cover Plate**, at top of Steering Box (S.12-111)		1	
C.16104	**Feed Pipe**		1	
	All other items are as for Steering Box C.21837			
C.11920	**Bolt**, securing Steering Box to Front Suspension Cross Member		1	
FG.106/X	**Washer**, Spring, on Bolt		1	
C.25884	**Bolt**, securing Steering Box to Front Suspension Cross Member and controlling Steering Lock		1	
UFN.137/L	**Nut**, on Bolt		1	
FG.106/X	**Washer**, Spring, under Nut		1	
C.16038	**Bolt**, securing Steering Box to Front Suspension Cross Member		1	
C.8737/3	**Nut**, Self-Locking, on Bolt		1	
FW.106/T	**Washer**, Plain, under Nut		1	
8316	**SEALS KIT FOR POWER-ASSISTED STEERING BOX**		1	
C.25376	**STEERING BOX ASSEMBLY (SP.1–903)**		1	**FOR R.H. DRIVE CARS ONLY**
10494	**Steering Box** (complete with Top Cover, Dowels, Trunnion Bush, Sleeve and Oil Seal Housing) (4-2379)		1	Fitted from Chassis No. 1A.11869 to 1A.13157
10162	**Bolt**, securing Top End Plate (10-2-6)		2	
10163	**Bolt**, securing Top End Plate (10-2-9)		6	
8218	**Washer**, Spring, on Bolts (12-2-6)		8	
10164	**Stud**, fixing Bottom End Plate (13-3-24)		4	
7570	**'O' Ring**, on Studs (27-3-1)		4	
8215	**Nut**, securing Bottom End Plate (11-3-8)		4	
102	**Washer**, spring, under Nuts (12-3-1)		4	
7570	**'O' Ring**, on Top Cover Dowels (27-3-1)		2	
10180	**Stud**, fixing Top Cover (13-3-27)		6	
7570	**'O' Ring**, on Studs (27-3-1)		6	
8215	**Nut**, securing Top Cover (11-3-8)		6	
102	**Washer**, Spring, under Nuts (12-3-1)		6	
9081	**Race**, for Rollers, at bottom of Inner Column (38-10-5)		1	
8227	**Rollers**, at bottom of Inner Column (18-1-3)		21	

POWER-ASSISTED STEERING UNIT

Part No.	Description	Plate No.	No. per Unit	Remarks
	STEERING BOX ASSEMBLY (SP.1-903) *(contd.)*			
8223	**Ball Race** (Adjustable) below Inner Column Rollers (S.23-26)		2	
136	**Ball,** between Adjustable Races		11	
9078	**Bottom End Plate** (S.11-78)		1	
9077	**Sealing Sleeve,** in Bottom End Plate (27-7-12)		1	
7609	**Shim,** for Bottom End Plate (.0025" thick) (S.10-101)		As req'd	SELECTIVE SIZES
8213	**Shim,** for Bottom End Plate (.004" thick) (S.10-110)		As req'd	SELECTIVE SIZES
7610	**Shim,** for Bottom End Plate (.005" thick) (S.10-102)		As req'd	SELECTIVE SIZES
7611	**Shim,** for Bottom End Plate (.010" thick) (S.10-103)		As req'd	SELECTIVE SIZES
11238	**Inner Column and Main Nut Assembly** (complete with Torsion Bar) (3-4167)		1	
10167	**Piston Ring,** for Main Nut (27-12-22)		2	
9080	**Balls** (Black Finish) in Main Nut (17-3-3)		18	
136	**Balls** (Bright Finish) in Main Nut (17-3-1)		19	
8200	**Packing Piece,** behind Roller Race at top of Inner Column (12-10-28)		1	
7572	**'O' Ring,** on Packing Piece (27-12-13)		1	
8228	**Race,** for Rollers, at top of Inner Column (28-12-6)		1	
8227	**Rollers,** at top of Inner Column (18-1-3)		24	
8226	**Washer,** retaining Rollers in Race (12-11-19)		1	
8229	**Circlip,** retaining Race in Top End Plate (31-12-8)		1	
10168	**Top End Plate** (S.11-82)		1	
8216	**Seal,** in Top End Plate, for Inner Column (27-8-34)		1	
7582	**'O' Ring,** on Top End Plate (27-12-11)		1	
10495	**Rockershaft** (S.7-110)		1	
10170	**'O' Ring,** inside Trunnion of Steering Box (27-10-2)		2	
C.22923	**Drop Arm** (S.9-226)		1	
10171	**Nut,** securing Drop Arm to Rockershaft (11-8-14)		1	
10172	**Washer,** under Nut (12-8-36)		1	
8220	**'O' Ring,** between Steering Box and Top Cover (27-12-14)		1	
9085	**'O' Ring,** at top of Rockershaft (27-8-37)		1	
10497	**Adjusting Screw,** in Top Cover, for Rockershaft (25-8-9)		1	
10498	**Spring,** for Adjusting Screw (36-7-9)		1	
10499	**Shim** (Copper) on Adjusting Screw (12-8-79)		2	
10502	**Shim** (Steel) on Adjusting Screw (.005" thick) (12-8-80)		As req'd	SELECTIVE SIZES
10503	**Shim** (Steel) on Adjusting Screw (.010" thick) (12-8-81)		As req'd	SELECTIVE SIZES
10504	**Shim** (Steel) on Adjusting Screw (.015" thick) (12-8-82)		As req'd	SELECTIVE SIZES
9680	**Air Bleed Plug,** in Top Cover (21-5-6)		1	
9681	**Washer,** on Plug (12-4-17)		1	
C.23372	**Feed Pipe Assembly**		1	
C.16102	**Connector,** securing Feed Pipe to Bottom End Plate and for attachment of Hose from Pump		1	
C.2296/1	**Washer** (Copper) on Connector		2	
C.13984	**Banjo Bolt,** securing Feed Pipe to Top Cover		1	
C.2296/4	**Washer** (Copper) on Banjo Bolt		2	
C.25377	**STEERING BOX ASSEMBLY (SP.1–904)**		1	**FOR L.H. DRIVE CARS ONLY** Fitted to Chassis No. 1A.20542 and subs. 1K.30001 to 1K.30021
10500	**Steering Box** (complete with Top Cover, Dowels, Trunnion Bush, Sleeve, and Oil Seal Housing) (4-2380)		1	
11239	**Inner Column and Main Nut Assembly** (Complete with Torsion Bar) (3-4168)		1	
C.22924	**Drop Arm** (S.9-227)		1	
C.23373	**Feed Pipe Assembly**		1	
	All other items for Steering Box C.25377 are identical with Steering Box C.25376			
10178	**SEALS KIT FOR POWER ASSISTED STEERING BOX (SD.1130)**		1	
C.19125	**DUST SEAL AT TOP OF POWER ASSISTED STEERING BOX**		1	

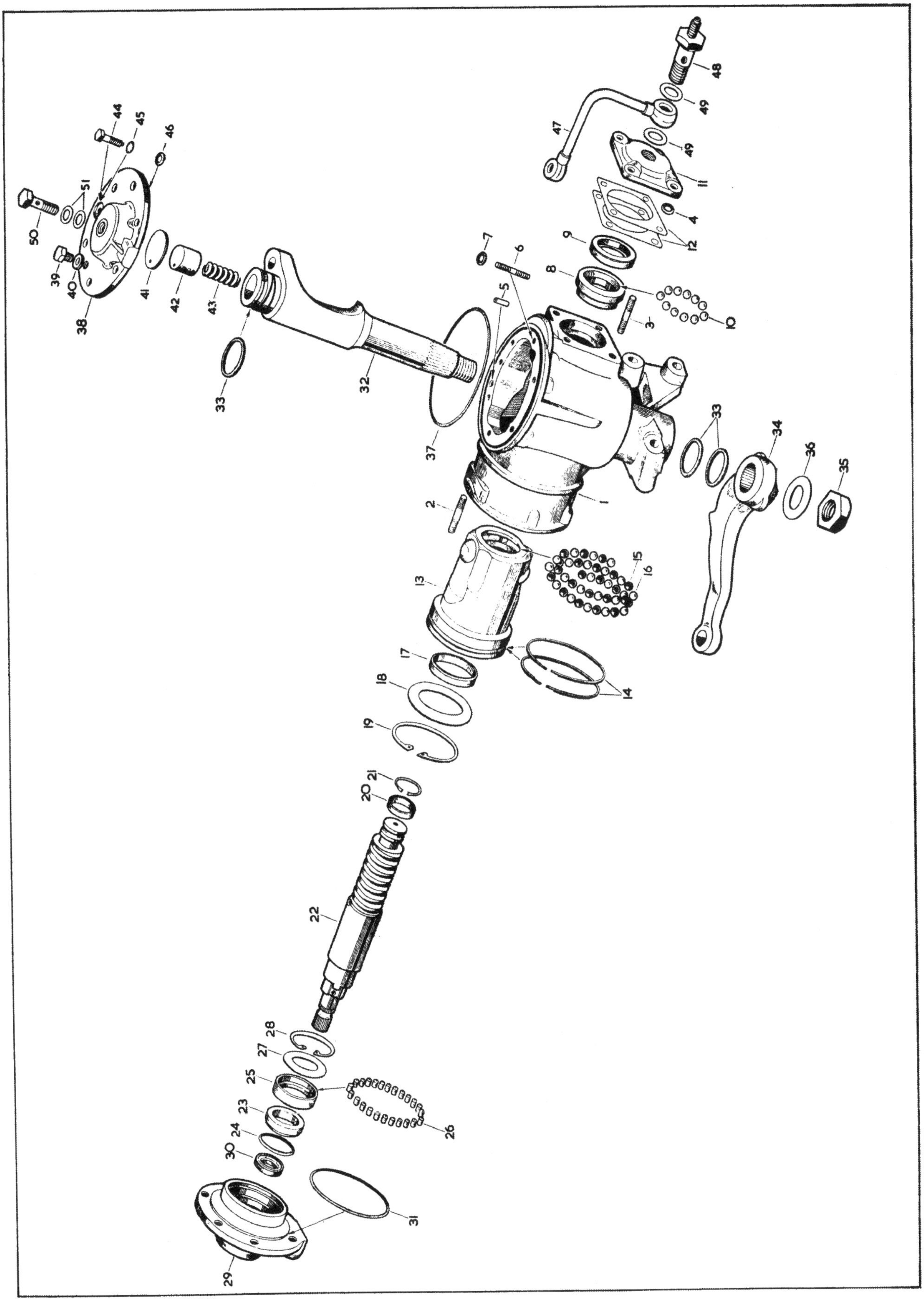

PLATE 34

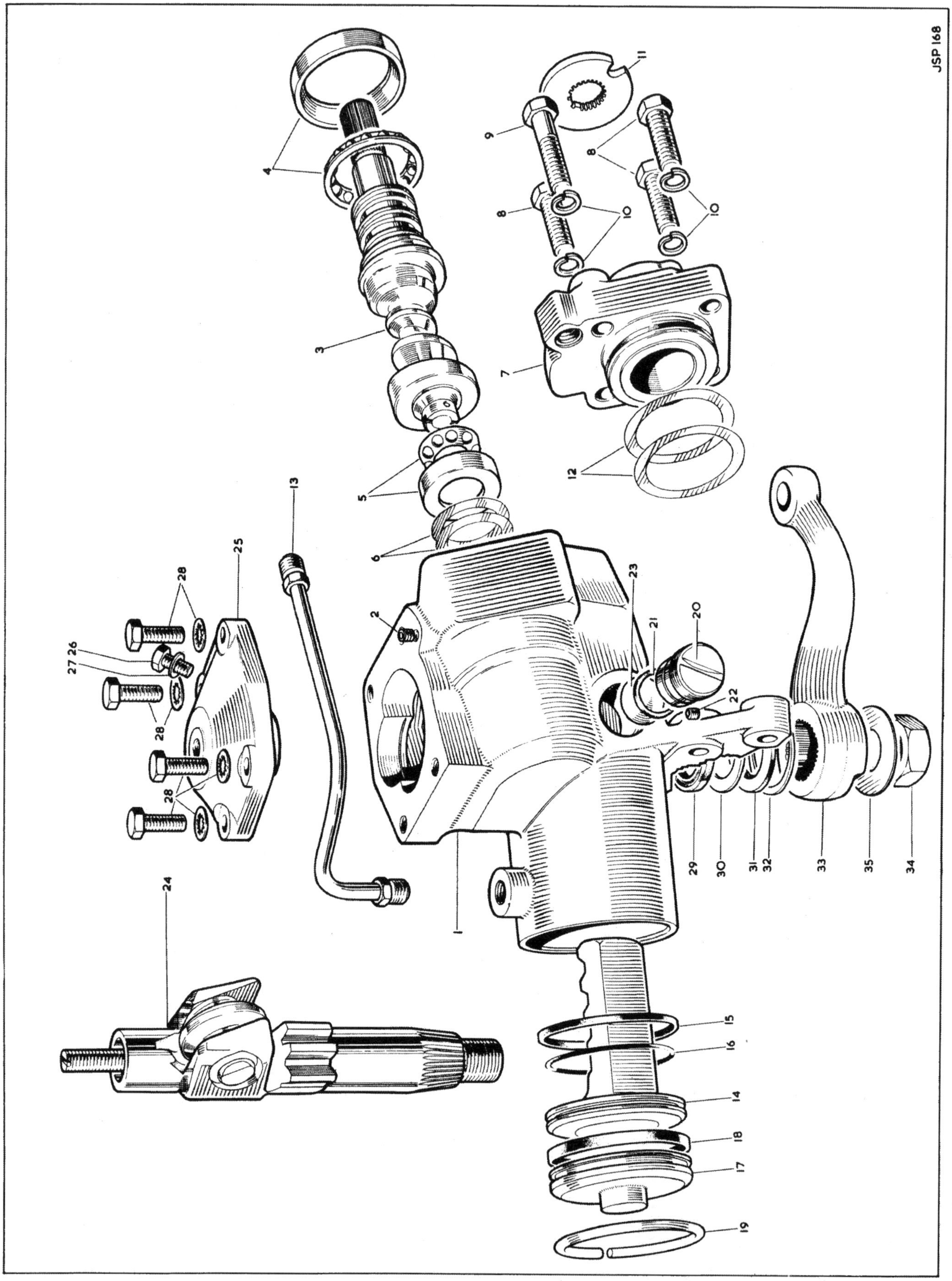
JSP 168
1
2
3
4
5
6
7
8
9
10
11
12
13
14
15
16
17
18
19
20
21
22
23
24
25
27 26
28
29
30
31
32
33
34
35

PLATE 35

POWER-ASSISTED STEERING UNIT

Part No.	Description	Plate No.	No. per Unit	Remarks
C.26018	**STEERING BOX ASSEMBLY (HBE.1500)**		1	**FOR R.H. DRIVE CARS ONLY** Fitted to Chassis No. 1A.13158 and subs. (also 1A.13003) 1K.1001 and subs.
11277	**Steering Box**, complete with Sector Shaft, Bush and Feed Pipe Seat (HBD.1510)	35-1	1	
10996	**Screw**, sealing Valve Adjustment Aperture at top of Steering Box (HBA.1028)	35-2	1	
11278	**Valve and Worm Assembly** (HBB.1530)	35-3	1	
10998	**Ball Bearing Assembly**, supporting Worm at top (HBA.1165)	35-4	1	
10999	**Ball Bearing Assembly**, supporting Worm at bottom (HBA.1155)	35–5	1	
11000/1	**Shim** (.002" thick) between Bottom Bearing and Steering Box (HBB.1038)	35–6	As req'd	SELECTIVE SIZES
11000/2	**Shim** (.003" thick) between Bottom Bearing and Steering Box (HBB.1038)	35–6	As req'd	SELECTIVE SIZES
11000/3	**Shim** (.010" thick) between Bottom Bearing and Steering Box (HBB.1038)	35–6	As req'd	SELECTIVE SIZES
11279	**Valve Housing Assembly** (HBC.1520)	35–7	1	
11003	**Bolt** (Short) securing Valve Housing to Steering Box (HB.1032)	35–8	3	
11004	**Bolt** (Long) securing Valve Housing to Steering Box (HB.1033)	35–9	1	
FG.107/X	**Washer**, Spring, on Bolts (S.9030)	35–10	4	
11280	**Alignment Washer**, on Worm Shaft, above Valve Housing (HBB.1508)	35–11	1	
11005/1	**Shim** (.001" thick) between Valve Housing and Top Bearing (HBA.1018)	35–12	As req'd	SELECTIVE SIZES
11005/2	**Shim** (.002" thick) between Valve Housing and Top Bearing (HBA.1018)	35–12	As req'd	SELECTIVE SIZES
11005/3	**Shim** (.010" thick) between Valve Housing and Top Bearing (HBA.1018)	35–12	As req'd	SELECTIVE SIZES
11281	**Feed Pipe Assembly** (HBC.1509)	35–13	1	
11007	**Piston and Rack** (HBC.1002)	35–14	1	
N.S.S.	Sealing Ring, ("Teflon") on Piston (HBB.1012)	35–15	1	**NOT SUPPLIED SEPARATELY. USE SEALS KIT 11027**
N.S.S.	'O' Ring, inside Sealing Ring (HBB.1014)	35–16	1	**NOT SUPPLIED SEPARATELY. USE SEALS KIT 11027**
11008	**End Plate**, for Cylinder (HBB.1004)	35–17	1	
N.S.S.	Seal, around End Plate (HB.1013)	35–18	1	**NOT SUPPLIED SEPARATELY USE SEALS KIT 11026**
11009	**Clip**, securing End Plate (HBB.1011)	35–19	1	
11010	**Plug**, adjusting Rack (HBB.1007)	35–20	1	
N.S.S.	Seal, on Adjusting Plug (HB.1016)	35–21	1	**NOT SUPPLIED SEPARATELY. USE SEALS KIT 11026**
11011	**Screw**, locking Adjusting Plug (HBA.1022)	35–22	1	
11012	**Thrust Pad**, between Plug and Rack (HBA.1006)	35–23	1	
11013	**Sector Shaft and Worm Follower Assembly** (HBA.1195)	35–24	1	
11014	**Top Cover Assembly** (HBA.1175)	35–25	1	
11015	**Bleed Screw**, in Top Cover (HBB.1074)	35–26	1	
N.S.S.	Washer, on Bleed Screw (S.9742)	35–27	1	**NOT SUPPLIED SEPARATELY. USE SEALS KIT 11026**
11016	**Screw and Lockwasher Assembly**, securing Top Cover to Steering Box (HBB.1023)	35–28	4	
N.S.S.	Seal, Inner, at bottom of Sector Shaft (HBB.1021)	35–29	1	**NOT SUPPLIED SEPARATELY. USE SEALS KIT 11026**
N.S.S.	Washer, between Inner and Outer Seals (HBB.1039)	35–30	1	**NOT SUPPLIED SEPARATELY. USE SEALS KIT 11026**
N.S.S.	Seal, Outer, at bottom of Sector Shaft (HBB.1019)	35–31	1	**NOT SUPPLIED SEPARATELY. USE SEALS KIT 11026**
11017	**Circlip**, retaining Sector Shaft Seals (HB.1027)	35–32	1	
C.25973	**Drop Arm**, on Sector Shaft (HBC.1504)	35–33	1	
11018	**Nut**, securing Drop Arm (S.9309)	35–34	1	
11019	**Tab Washer**, locking Nut (P.4101)	35–35	1	
C.26019	**STEERING BOX ASSEMBLY (HBE.1600)**		1	**FOR L.H. DRIVE CARS ONLY** Fitted to Chassis No. 1K.30022 and subs.
11282	**Steering Box**, complete with Sector Shaft, Bush and Feed Pipe (HBD.1610)		1	
11283	**Valve and Worm Assembly** (HBB.1630)		1	
11284	**Valve Housing Assembly** (HBB.1620)		1	
11285	**Feed Pipe Assembly** (HBC.1609)		1	
11024	**Sector Shaft and Worm Follower Assembly** (HBA.1205)		1	
11025	**Top Cover Assembly** (HBA.1215)		1	
C.25974	**Drop Arm**, on Sector Shaft (HBC.1604)		1	

All other items are as for Steering Box C.26018

POWER-ASSISTED STEERING UNIT

Part No.	Description	Plate No.	No. per Unit	Remarks
	STEERING BOX ASSEMBLY (HBE.1600) *(contd.)*			
C.26604	**Bolt,** securing Steering Box to Front Suspension Cross Member		1	
C.26605	**Bolt,** securing Steering Box to Front Suspension Cross Member		2	
FG.106/X	**Washer,** Spring, on Bolts		3	
11026	**SEALS KIT (EXTERNAL SEALS) FOR STEERING BOX (HBA.1185)**		1	
	Comprising:—			
	Seat, for Feed Pipe (HBB.1129)		2	
	Seat, for High Pressure Pipe (HBC.1132)		1	
	Seat, for Low Pressure Pipe (HBB.1124)		1	
	Rotor Seal, Inner (HBA.1126)		1	
	Rotor Seal, Outer (HBA.1127)		1	
	Seal, under Top Cover (HBA.1031)		1	
	Nut, Self-Locking, for Sector Shaft Adjustment (HBA.1029)		1	
	Washer, on Bleed Screw (S.9742)		1	
	Seal, on Rack Adjusting Plug (HB.1016)		1	
	Seal, around Cylinder End Plate (HB.1013)		1	
	Sector Shaft Seal, Inner (HBB.1021)		1	
	Sector Shaft Seal, Outer (HBB.1019)		1	
	Washer, between Sector Shaft Seals (HBB.1039)		1	
11027	**SEALS KIT (INTERNAL SEALS) FOR STEERING BOX (HBA.1190)**		1	
	Comprising:—			
	'O' Ring, on Valve Housing (HB.1131)		1	
	Sealing Ring ("Teflon") on Valve Sleeve (HBA.1107)		3 plus 1 Spare	
	Sealing Ring ("Teflon") on Piston (HBB.1012)		1	
	'O' Ring, inside "Teflon" Ring (HBB.1014)		1	

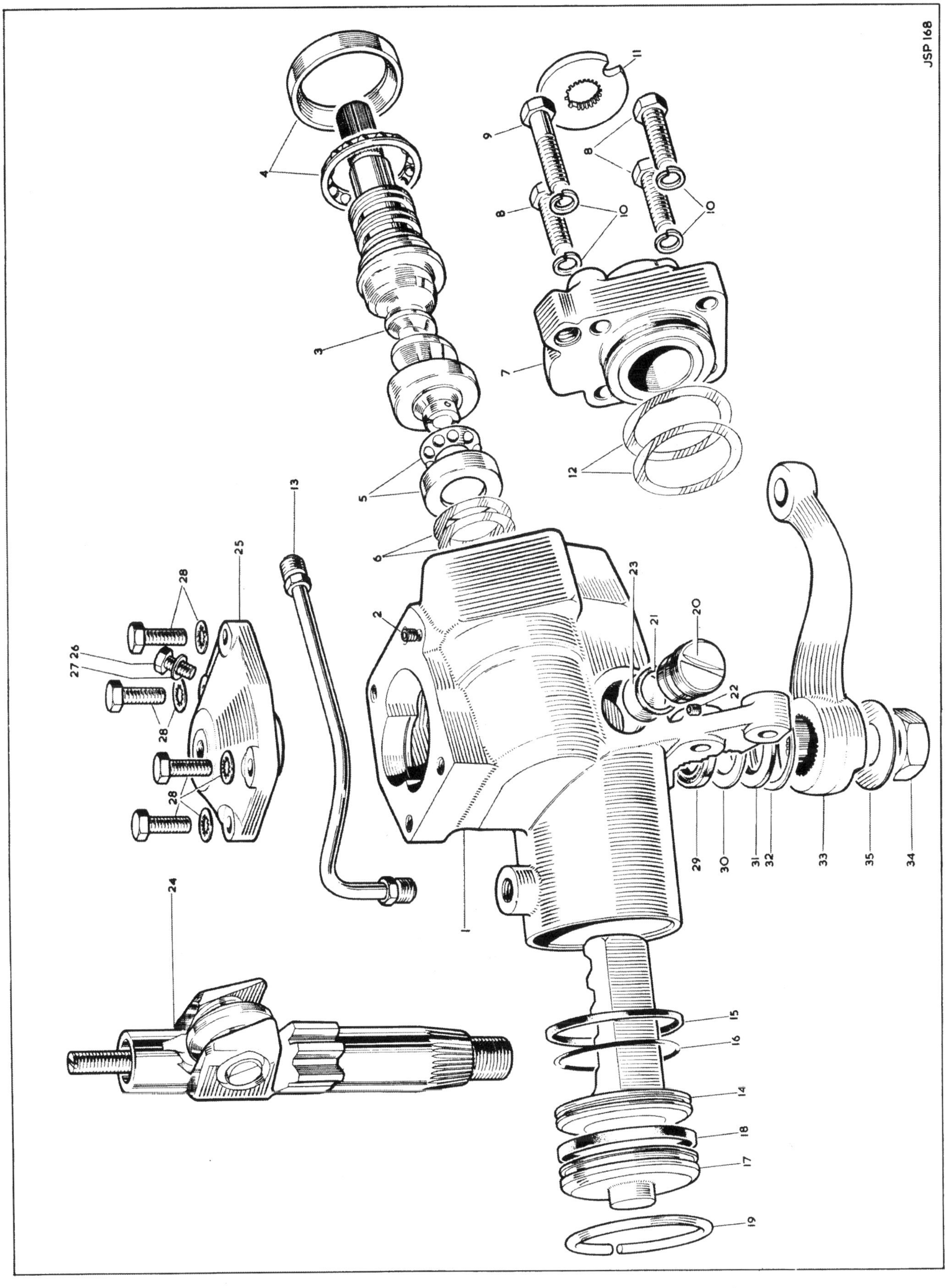
JSP 168
1
2
3
4
5
6
7
8
9
10
11
12
13
14
15
16
17
18
19
20
21
22
23
24
25
26
27
28
29
30
31
32
33
34
35

PLATE 35

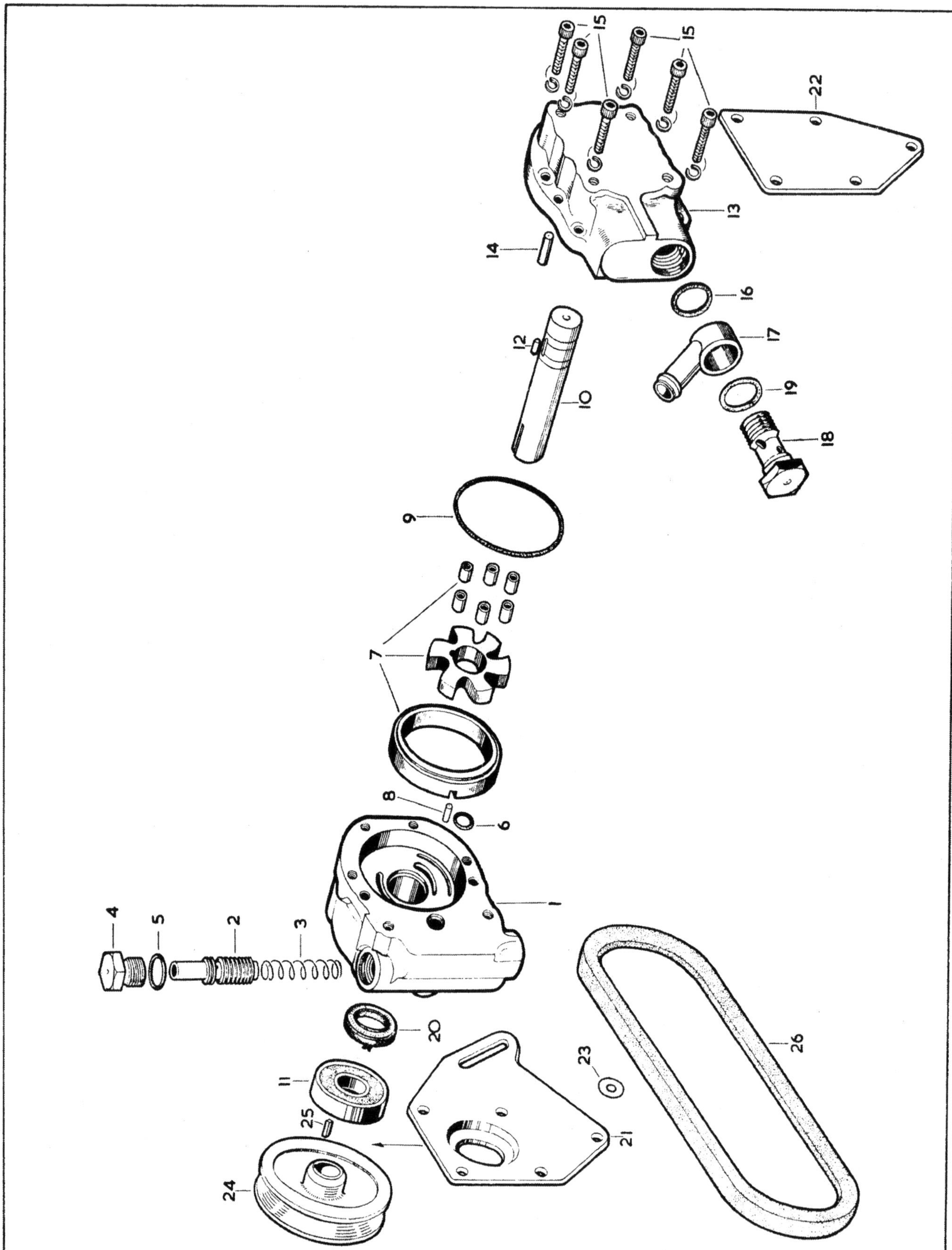

PLATE 36

OIL PUMP FOR POWER-ASSISTED STEERING UNIT

Part No.	Description	Plate No.	No. per Unit	Remarks
C.21113	**OIL PUMP ASSEMBLY (HE.5503/8518/11)**		1	SUPERSEDED BY C.24218
9728	**Body Assembly** (HE.5884/18)	36–1	1	
8792	**Flow Control Valve Assembly** (HE.3092/85)	36–2	1	
8791	**Return Spring**, for Flow Control Valve (HE.2836)	36–3	1	
9729	**Cap**, securing Flow Control Valve in Body (HE.5732)	36–4	1	
9730	**'O' Ring**, for Cap (HE.5734)	36–5	1	
8794	**'O' Ring**, (Small) for By-Pass Aperture (HE.3546)	36–6	1	
9731	**Cam Set** (HE.5322/35) (comprising Cam Insert, Carrier and Rollers)	36–7	1	
8796	**Peg**, locking Cam Insert in Body (HE.2845)	36–8	1	
8797	**'O' Ring** (Large) between Body and Cover (HE.3631)	36–9	1	
9732	**Shaft**, operating Pump (HE.3534)	36–10	1	
9733	**Bearing**, on Shaft (HE.1417)	36–11	1	
9734	**Drive Pin**, between Shaft and Roller Carrier (HE.3539)	36–12	1	
9735	**Cover Assembly** (HE.5407/A)	36–13	1	
8788	**Dowel**, in Cover (HE.3538)	36–14	2	
8803	**Screw**, securing Cover to Body (HE.3544/1)	36–15	6	
FG.104/X	**Washer**, Spring, on Screws		6	
7545	**Gasket**, for Venturi Tube (HE.1322)	36–16	1	
8804	**Adaptor**, at Top of Body (HE.5007)	36–17	1	
8805	**Banjo Bolt**, securing Adaptor (HE.3069)	36–18	1	
8806	**Fibre Washer**, under Head of Banjo Bolt (HE.3213)	36–19	1	
9728	**BODY ASSEMBLY (HE.5884/18)**	36–1	1	
7560	**Oil Seal**, for Drive Shaft (HE.3545)	36–20	1	
9739	**REPAIR KIT FOR OIL PUMP (HE.5321/11)** Comprising:— 7560. Oil Seal; 8794. 'O' Ring; 7545. Gasket; 9730. 'O' Ring; 8797. 'O' Ring; 8806. Washer		1	REQUIRED FOR OIL PUMP C.21113
C.24218	**OIL PUMP ASSEMBLY (HE.5503/8518/14)**		1	SUPERSEDES C.21113
10450	**Body Assembly** (HE.6884/18)		1	
10235	**'O' Ring** (Small) for By-Pass aperture (HE.6252)		1	
	All other items are as for Oil Pump C.21113			
10450	**BODY ASSEMBLY (HE.6884/18)**		1	
7560	**Oil Seal**, for Drive Shaft (HE.3545)		1	
10236	**REPAIR KIT FOR OIL PUMP (HE.5321/17)** Comprising:— 7560. Oil Seal; 10235. 'O' Ring; 7545. Gasket; 9730. 'O' Ring; 8797. 'O' Ring; 8806. Washer		1	REQUIRED FOR OIL PUMP C.24218
C.21114	**FRONT MOUNTING PLATE ASSEMBLY FOR OIL PUMP**	36–21	1	

OIL PUMP FOR POWER-ASSISTED STEERING UNIT

Part No.	Description	Plate No.	No. per Unit	Remarks
C.21106	**REAR MOUNTING PLATE FOR OIL PUMP**	36–22	1	
UFS.125/6R	**Screw,** Set, securing Plates to Pump		8	
FG.104/X	**Washers,** Spring, on Setscrews		8	
UFS.131/8R	**Screw,** Set, securing Oil Pump Assembly to Engine Mounting Bracket		3	
C.21105/2	**Shim** (.003" thick) between Mounting Plates and Engine Mounting Bracket	36–23	As req'd	
C.21105/1	**Shim** (.010" thick) between Mounting Plates and Engine Mounting Bracket	36–23	As req'd	
401883	**Washer** (Serrated), on Clamping Screw		1	
FG.105/X	**Washer,** Spring, on Setscrews		3	
C.27391	**OIL PUMP ASSEMBLY (HE.5503/10022/14)**		1	SUPERSEDES C.24218. REQUIRED FOR CARS FITTED WITH P.A.S. UNITS C.26018 AND C.26019 (C.27391 to 7560)
12313	**Body Assembly** (HE.6884/22)		1	
9412	**Flow Control Valve Assembly** (HE.3092/100)		1	
8791	**Return Spring,** for Flow Control Valve (HE.2836)		1	
9729	**Cap,** securing Flow Control Valve in Body (HE.5732)		1	
9730	**'O' Ring,** for Cap (HE.5734)		1	
10235	**'O' Ring,** (Small) for By-Pass Aperture (HE.6252)		1	
9731	**Cam Set** (HE.5322/35) (comprising Cam Insert, Carrier and Rollers)		1	
8796	**Peg,** locking Cam Insert in Body (HE.2845)		1	
8797	**'O' Ring,** (Large) between Body and Cover (HE.3631)		1	
9732	**Shaft,** operating Pump (HE.3534)		1	
9733	**Bearing,** on Shaft (HE.1417)		1	
9734	**Drive Pin,** between Shaft and Roller Carrier (HE.3539)		1	
9735	**Cover Assembly** (HE.5407/A)		1	
8788	**Dowel,** in Cover (HE.3538)		2	
8803	**Screw,** securing Cover to Body (HE.3544/1)		6	
FG.104/X	**Washer,** Spring, on Screws		6	
7545	**Gasket,** for Venturi Tube (HE.1322)		1	
8804	**Adaptor,** at Top of Body (HE.5007)		1	
8805	**Banjo Bolt,** securing Adaptor (HE.3069)		1	
8806	**Fibre Washer,** under Head of Banjo Bolt (HE.3213)		1	
12313	**BODY ASSEMBLY (HE.6884/22)**		1	
7560	**Oil Seal,** for Drive Shaft (HE.3545)		1	
10236	**REPAIR KIT FOR OIL PUMP (HE.5321/17)** Comprising:– **7560. Oil Seal; 10235. 'O' Ring; 7545. Gasket; 9730. 'O' Ring; 8797. 'O' Ring; 8806. Washer**		1	REQUIRED FOR OIL PUMP C.27391
C.21108	**PULLEY ON OIL PUMP SHAFT**	36–24	1	
C.21111	**Key,** locating Pulley on Shaft	36–25	1	
UCS.131/6R	**Screw,** Set, securing Pulley to Shaft		1	
432429	**Washer,** Plain, on Setscrew		1	
FG.105/X	**Washer,** Spring, on Setscrew		1	
C.21109	**BELT, DRIVING OIL PUMP**	36–26	1	

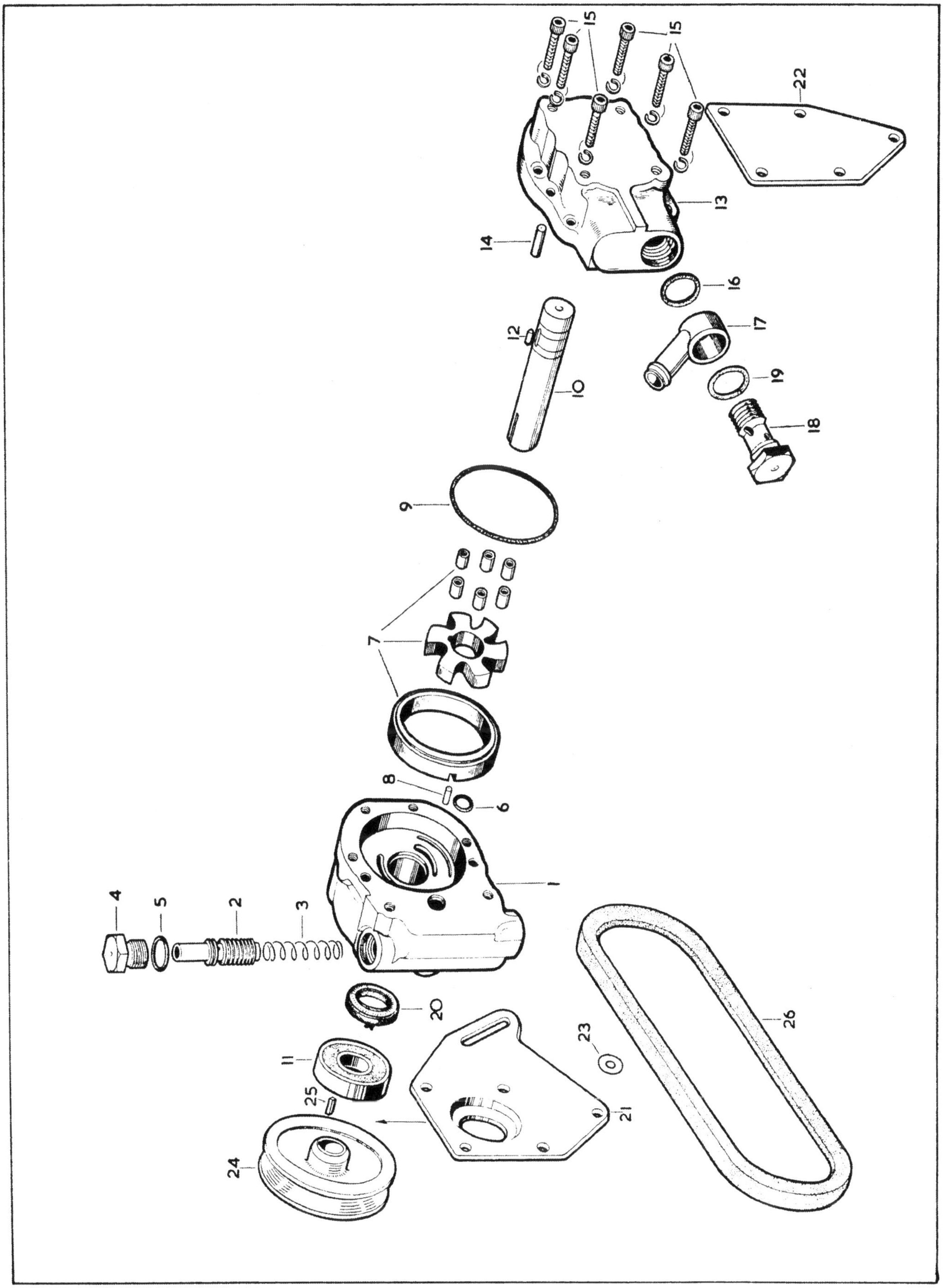

PLATE 36

HYDRAULICS FOR POWER-ASSISTED STEERING

Part No.	Description	Plate No.	No. per Unit	Remarks
C.16189/1	**RESERVOIR ASSEMBLY, FOR OIL FEED TO STEERING BOX (HE.4142/2)**	37–1	1	
8281	**Filler Cap,** complete with Level Indicator (HE.2960)	37–2	1	
7753	**Gasket,** on Cover Bolt (HE.1324)	37–3	1	
7751	**Gasket,** under Cover (HE.1306)	37–4	1	
7752	**Filter** (Purolator MFH.117) (HE.1307)	37–5	1	
7545	**'O' Ring,** between Reservoir Body and Mounting Bracket (HE.1322)	37–6	2	
9595	**Adaptor,** at bottom of Reservoir for Oil Feed Hose to Pump (HE.5379)	37–7	1	
9597	**'O' Ring,** on Adaptor (HE.5391)	37–8	1	
9596	**Adaptor,** at bottom of Reservoir, for Oil Return Hose (HE.5380)	37–9	1	
9345	**'O' Ring,** on Adaptor (HE.5390)	37–10	1	
UFB.125/8R	**Bolt,** securing Reservoir to Mounting Bracket		3	
C.8737/1	**Nut,** Self-locking, on Bolts		3	
FW.104/T	**Washer,** Plain, under Nuts		3	
C.21387	**MOUNTING BRACKET FOR RESERVOIR**	37–11	1	
UFS.119/6R	**Screw,** Set, securing Bracket to Front Wing Valance		3	
C.18111	**Washer,** Plain, on Setscrews		3	
C.723A	**Washer,** Spring, on Setscrews		3	
C.18110	**Grommet,** on Bracket	37–12	3	
C.21422	**HOSE FROM RESERVOIR TO PUMP**	37–13	1	
C.2905/2	**Clip,** securing Hose	37–14	2	
C.21422	**HOSE FROM RESERVOIR TO PUMP**		1	**R.H. DRIVE CARS ONLY, FITTED WITH P.A.S. UNIT C.26018**
C.28409	**HOSE FROM RESERVOIR TO PUMP**		1	**L.H. DRIVE CARS ONLY, FITTED WITH P.A.S. UNIT C.26019**
C.2905/2	**Clip,** securing Hose		2	
C.18415	**HOSE FROM PUMP TO STEERING BOX**	37–15	1	**R.H. DRIVE CARS ONLY** Fitted from Chassis No. 1A.1001 to 1A.8856
C.1040/21	**Clip,** securing Hose to Engine Mounting Bracket		1	
C.25563	**HOSE FROM PUMP TO STEERING BOX**		1	**R.H. DRIVE CARS ONLY** Fitted to Chassis No. 1A.8857 and subs.
C.1040/31	**Clip,** securing Hose to Engine Mounting Bracket		1	
C.21740	**HOSE FROM PUMP TO STEERING BOX**		1	**L.H. DRIVE CARS ONLY**
C.28286	**HOSE FROM PUMP TO STEERING BOX**		1	**R.H. DRIVE CARS ONLY, FITTED WITH P.A.S. UNIT C.26018**
C.28301	**HOSE FROM PUMP TO STEERING BOX**		1	**L.H. DRIVE CARS ONLY, FITTED WITH P.A.S. UNIT C.26019**
C.28287	**HOSE FROM STEERING BOX TO RESERVOIR**		1	**R.H. DRIVE CARS FITTED WITH P.A.S. UNIT C.26018**
C.1040/21	**Clip,** securing Hose to Engine Mounting Bracket		1	
C.28302	**HOSE FROM STEERING BOX TO RESERVOIR**		1	**L.H. DRIVE CARS FITTED WITH P.A.S. UNIT C.26019**
C.2905/3	**Clip,** securing Hose to Reservoir		1	
C.13857	**BANJO, ON STEERING BOX END PLATE, FOR HOSE FROM RESERVOIR**	37–16	1	
C.15273	**Banjo Bolt,** securing Banjo to End Plate	37–17	1	
C.2296/1	**Copper Washer,** at each side of Banjo	37–18	2	
C.24503	**HOSE, FROM BANJO ON END PLATE TO RESERVOIR**	37–19	1	**R.H. DRIVE CARS ONLY**
C.21739	**HOSE, FROM BANJO ON END PLATE TO RESERVOIR**		1	**L.H. DRIVE CARS ONLY**
C.2905/2	**Clip,** securing Hose	37–20	2	
C.1040/25	**Clip,** securing Hose to Engine Bracket		1	**R.H. DRIVE CARS ONLY**
C.1040/16	**Clip,** securing Hose to Cross Member		1	

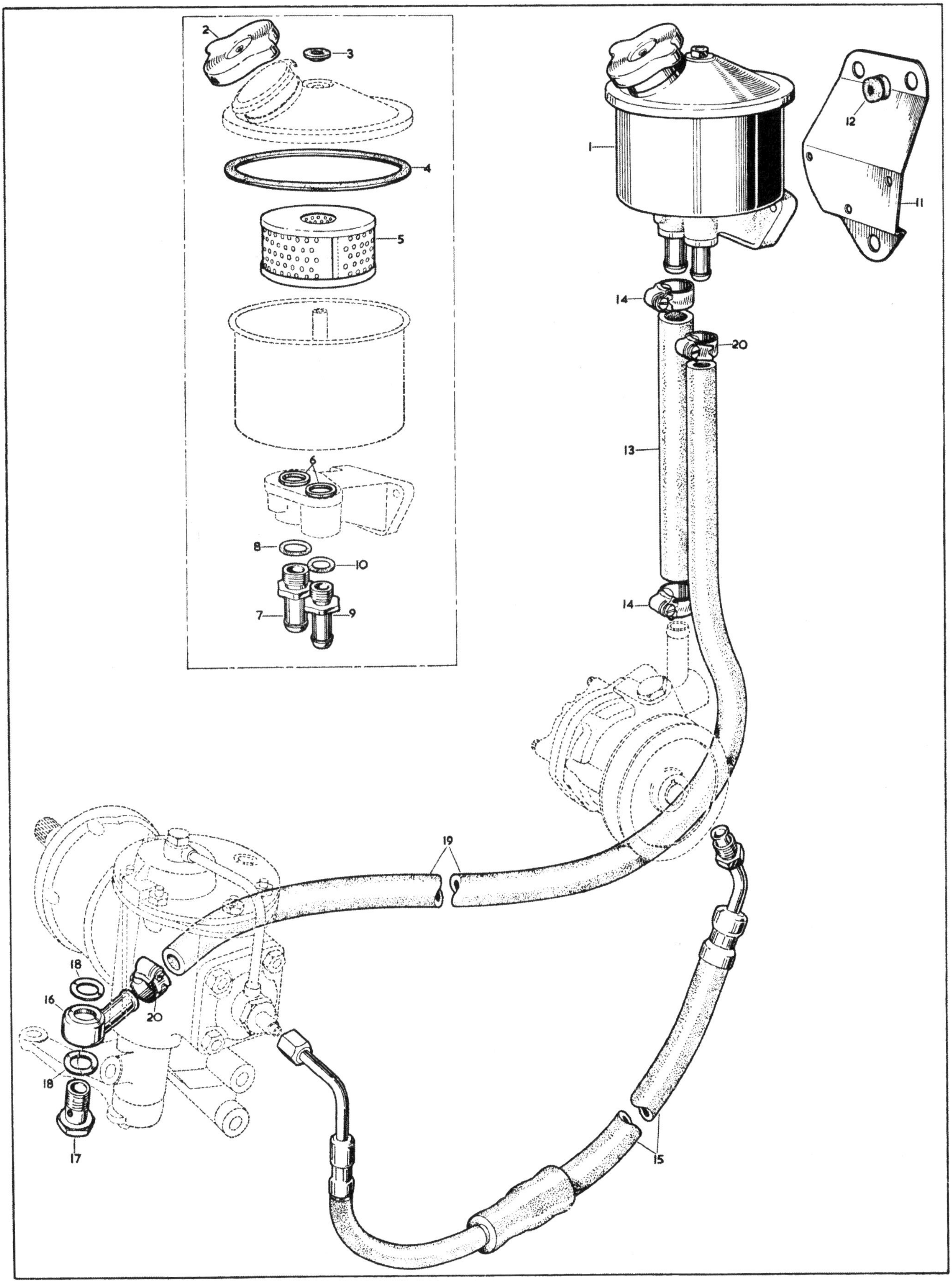

PLATE 37

STEERING COLUMN

Part No.	Description	Plate No.	No. per Unit	Remarks
C.20899	**LOWER STEERING COLUMN ASSEMBLY (01/500865)**		1	**NOT REQUIRED WITH POWER-ASSISTED STEERING** Fitted from Chassis No. 1A.1001 to 1A.1234 1A.20001 to 1A.20034
9612	**End Yoke** (01/206012)		1	
5426	**Journal Assembly,** connecting End Yoke to Steering Box (01/500582 or KO.5-GB.1)		1	
9613	**Socket,** at Upper End of Column (01/245383)		1	
9614	**Nylon Roller,** for Universal Joint of Socket (01/245387)		2	
9615	**Circlip,** retaining Rollers to Pivot Pin (01/245439)		2	
9616	**Gaiter,** at upper end of Column (01/245437)		1	
9617	**Clip,** securing Gaiter to Socket (245459/16)		1	
9618	**Clip,** securing Gaiter to Column (245459/17)		1	
C.20918	**LOWER STEERING COLUMN ASSEMBLY (01/500866)**		1	**FOR USE ONLY WITH POWER-ASSISTED STEERING** Fitted from Chassis No. 1A.1001 to 1A.1234 1A.20001 to 1A.20034
	Individual items are as for Lower Steering Column C.20899			
C.22355/2	**LOWER STEERING COLUMN ASSEMBLY (01/501285)**	38–1	1	**NOT REQUIRED WITH POWER-ASSISTED STEERING** Fitted from Chassis No. 1A.1235 to 1A.5854 1A.20035 to 1A.20328
9612	**End Yoke** (01/206012)	38–2	1	
5426	**Journal Assembly,** connecting End Yoke to Lower Steering Column (01/500582 or KO.5-GB.1)	38–3	1	
9980	**Socket,** at upper end of Column (01/245586)	38–4	1	
9614	**Nylon Roller,** for Universal Joint of Socket (01/245387)	38–5	2	
9615	**Circlip,** retaining Rollers to Pivot Pin (01/245439)	38–6	2	
9616	**Gaiter,** at upper end of Column (01/245437)	38–7	1	
9617	**Clip,** securing Gaiter to Socket (245459/16)	38–8	1	
9618	**Clip,** securing Gaiter to Column (245459/17)	38–9	1	
C.22355/3	**LOWER STEERING COLUMN ASSEMBLY (01/501286)**		1	**FOR USE ONLY WITH POWER-ASSISTED STEERING** Fitted from Chassis No. 1A.1235 to 1A.5919 1A.20035 to 1A.20331
	Individual items are as for Lower Steering Column C.22355/2			
C.23605/2	**LOWER STEERING COLUMN ASSEMBLY (01/501450)**		1	**NOT REQUIRED WITH POWER-ASSISTED STEERING** Fitted to Chassis No. 1A.5855 and subs. 1A.20329 and subs.
9612	**End Yoke** (01/206012)		1	
5426	**Journal Assembly,** connecting End Yoke to Steering Column (01/500582 or KO.5-GB.1)		1	
10756	**Socket,** at upper end of Column (01/245647)		1	
9614	**Roller** (Nylon) for Universal Joint of Socket (01/245387)		2	
9615	**Circlip,** retaining Rollers to Pivot Pin (01/245439)		2	
10757	**End Plate,** for Socket (01/245648)		1	
10758	**Screw** (Allen Head) securing End Plate to Socket (245817)		4	
10759	**Washer,** Spring, on Screws (245677)		4	
9616	**Gaiter,** at upper end of Column (01/245437)		1	
10760	**Clip,** securing Gaiter to Socket (245625)		1	
10761	**Clip,** securing Gaiter to Column (245626)		1	
C.23605/3	**LOWER STEERING COLUMN ASSEMBLY (01/501449)**		1	**FOR USE ONLY WITH POWER-ASSISTED STEERING** Fitted to Chassis No. 1A.5920 and subs. 1A.20332 and subs.
10762	**End Yoke** (01/206175)		1	
	All other items are as for Lower Steering Column C.23605/2			

Part No.	Description	Plate No.	No. per Unit	Remarks
UFB.131/12R	**Bolt**, securing Lower Steering Column Socket to Upper Steering Column		1	
C.22283	**Distance Washer**, between jaws of Socket	38–10	1	FOR USE ONLY WITH LOWER STEERING COLUMNS C.20899, C.20918, C.22355/2, AND C.22355/3
UFB.131/14R	**Bolt**, securing Lower Steering Column End Yoke to Steering Box		1	
C.8737/2	**Nut**, Self-locking, on Bolts		2	
C.13739	STONEGUARD FOR LOWER STEERING COLUMN		1	L.H. DRIVE CARS ONLY. NOT REQUIRED WITH POWER-ASSISTED STEERING
C.13740	STONEGUARD FOR LOWER STEERING COLUMN	38–11	1	R.H. DRIVE CARS ONLY. NOT REQUIRED WITH POWER-ASSISTED STEERING
FW.105/T	**Washer**, Plain, for Stoneguard		1	
C.25575	UPPER STEERING COLUMN ASSEMBLY		1	FOR CARS FITTED WITH STANDARD OR OVERDRIVE TRANSMISSION
C.25057	**Outer Tube Assembly**		1	
C.23593	**Bearing** (Plastic) at top of Outer Tube		1	
C.23592	**Bearing** (Plastic) at bottom of Outer Tube		1	
C.25044	**Inner Column**		1	
C.20163	**Shaft**, inside Inner Column, for attachment of Steering Wheel		1	
C.16676	**Stop Button**, restricting movement of Shaft		1	
C.15419	**Locknut**, clamping Shaft to Inner Column		1	
8671	**Split Collet**, inside Locknut (complete with Circlip C.18235)		1	
C.18235	**Circlip** (supplied with Split Collet)		1	
C.20928	UPPER STEERING COLUMN ASSEMBLY		1	Fitted from Chassis No. 1A.1001 to 1A.9210 1A.20001 to 1A.20430
C.20923	**Outer Tube Assembly**	38–12	1	
C.23593	**Bearing** (Plastic) at top of Outer Tube	38–13	1	
C.23592	**Bearing** (Plastic) at bottom of Outer Tube	38–14	1	
C.20880	**Inner Column**	38–15	1	
C.20163	**Shaft**, inside Inner Column for attachment of Steering Wheel	38–16	1	
C.16676	**Stop Button**, restricting movement of Shaft	38–17	1	
C.15419	**Locknut**, clamping Shaft to Inner Column	38–18	1	
8671	**Split Collet**, inside Locknut	38–19	1	SUPPLIED COMPLETE WITH CIRCLIP C.18235
C.18235	**Circlip**, for Split Collet	38–20	1	
C.26043	UPPER STEERING COLUMN ASSEMBLY		1	Fitted from Chassis No. 1A.9211 to 1A.10326 1A.20431 to 1A.20479
C.26040	**Outer Tube Assembly**		1	
	All other items are as for Upper Steering Column Assembly C.20928			
C.26108	UPPER STEERING COLUMN ASSEMBLY		1	FOR CARS FITTED WITH AUTOMATIC TRANSMISSION Fitted to Chassis No. 1A.10327 and subs. 1A.20480 and subs.
C.26109	**Outer Tube Assembly**		1	
C.25044	**Inner Column Assembly**		1	
	All other items as for Upper Steering Column Assembly C.20928			

STEERING COLUMN

Part No.	Description	Plate No.	No. per Unit	Remarks
C.26025	**CLAMP, ON DIRECTION INDICATOR SWITCH SCREWS, ADJUSTING UPPER BEARING**		1	Fitted to Chassis No. 1A.9211 and subs. 1A.20431 and subs.
UFN.119/L	**Nut,** locking Clamp		1	
C.26023/1	**Spacer,** at lower end of Clamp (.188" thick)		1	SELECTIVE SIZES Fitted to Chassis No. 1A.9211 and subs. 1A.20431 and subs.
C.26023/2	**Spacer,** at lower end of Clamp (.160" thick)		1	
C.23762/1	**Thrust Bearing,** retaining Inner Column in Outer Tube (.312"/.314" thick)		1	SELECTIVE SIZES
C.23762/2	**Thrust Bearing,** retaining Inner Column in Outer Tube (.314"/316" thick)		1	
C.20126	**Retaining Ring,** for Thrust Bearing	38–22	1	
C.20127	**Spring Clip,** securing Retaining Ring	38–23	1	
C.17754	**Screw Plate Assembly,** mounting Upper Steering Column to Scuttle	38–24	1	
FW.105/T	**Washer,** Plain, on Screws		2	
FG.105/X	**Washer,** Spring, on Screws		2	
UFN.131/L	**Nut,** on Screws		2	
C.17753	**Spacer,** for Upper Column Bracket (10 S.W.G. Aluminium)	38–25	As req'd	
C.16370	**Shim,** for adjustment of Upper Steering Column at Upper Bracket	38–26	As req'd	
C.27046	**LOCK AND IGNITION SWITCH ASSEMBLY ON STEERING COLUMN**		1	STANDARD FITMENT ON CARS EXPORTED TO GERMANY AND SWEDEN. MAY ALSO BE FITTED TO SPECIAL ORDER ON ANY OTHER CAR
C.26851	**Cover** (Rubber) for Switch Terminals		1	
11242	**Switch Assembly,** only for Lock and Ignition Switch Assembly		1	
11243	**Screw,** securing Switch to Lock		3	
BD.12010	**BRACKET ASSEMBLY, MOUNTING UPPER STEERING COLUMN AT BOTTOM**	38–27	1	R.H. DRIVE CARS ONLY
BD.27497	**Gasket,** under Bracket	38–28	1	
BD.18587	**BRACKET ASSEMBLY, MOUNTING UPPER STEERING COLUMN AT BOTTOM**		1	L.H. DRIVE CARS WITH AUTOMATIC TRANSMISSION ONLY
BD.27497	**Gasket,** under Bracket		1	
UFN.125/L	**Nut,** on Screw of Lower Bracket		1	
C.724	**Washer,** Shakeproof, under Nut		1	
FW.104/T	**Washer,** Plain, under Shakeproof Washer		1	
UFS.125/4R	**Screw,** Set, securing Lower Bracket to Dash		3	
C.724	**Washer,** Shakeproof, on Setscrews		3	
C.2905/4	**Clip,** securing Upper Column in Bracket	38–29	1	
C.7878	**SPLIT CONE, ON INNER COLUMN SHAFT, SEATING STEERING WHEEL**	38–30	2 halves	
C.15285/1	**STEERING WHEEL (17" DIAMETER)** **(NOT REQUIRED FOR CANADA, DENMARK, FINLAND, GERMANY, ICELAND, NEWFOUNDLAND, NORWAY, SWEDEN, OR U.S.A.)**	38–31	1	Fitted from Chassis No. 1A.1001 to 1A.2529 1A.20001 to 1A.20219
C.15285/2	**STEERING WHEEL (17" DIAMETER)** **(FOR CARS EXPORTED TO CANADA, DENMARK, FINLAND, GERMANY, ICELAND, NEWFOUNDLAND, NORWAY, SWEDEN, OR U.S.A.)**		1	

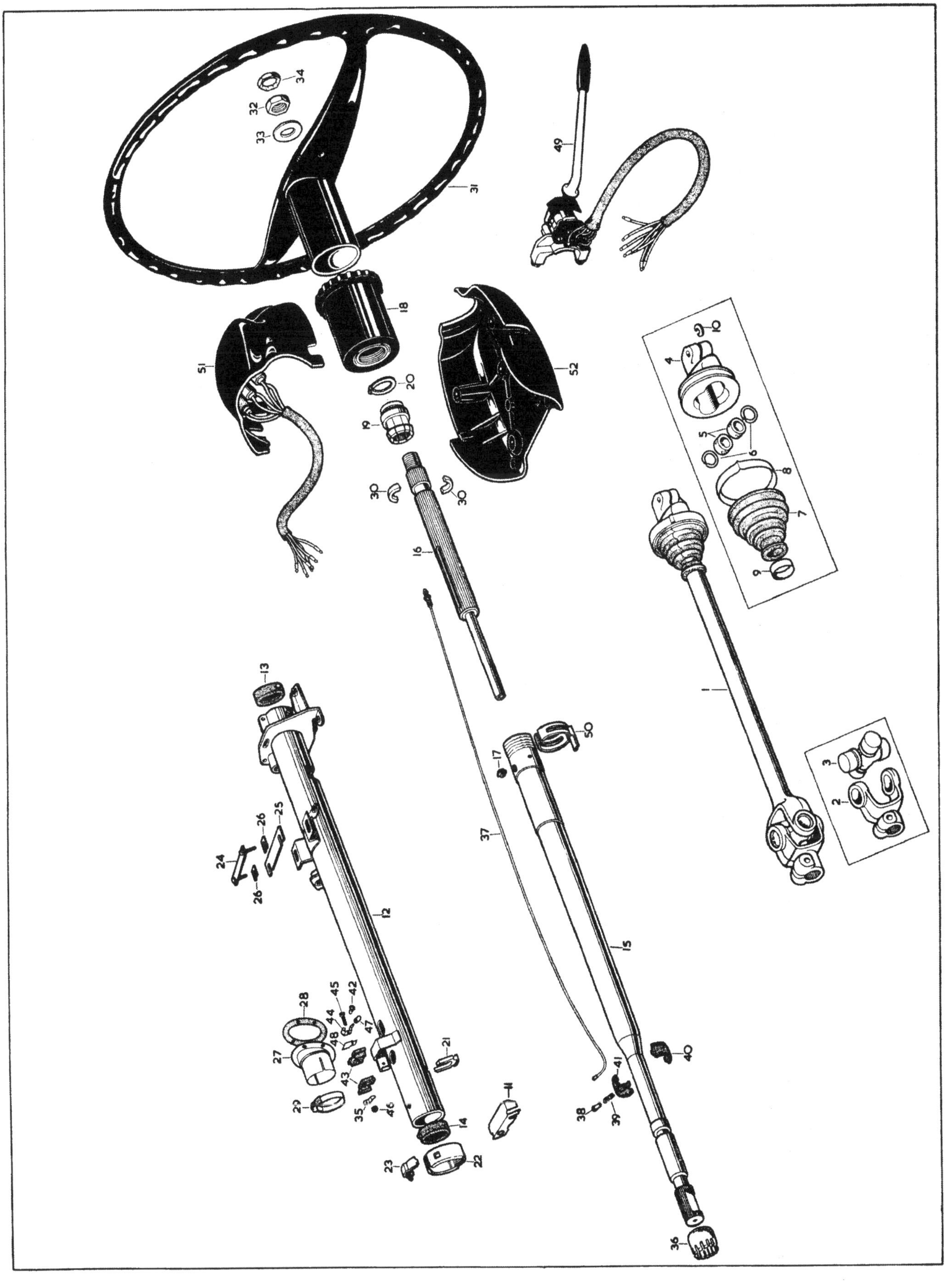

PLATE 38

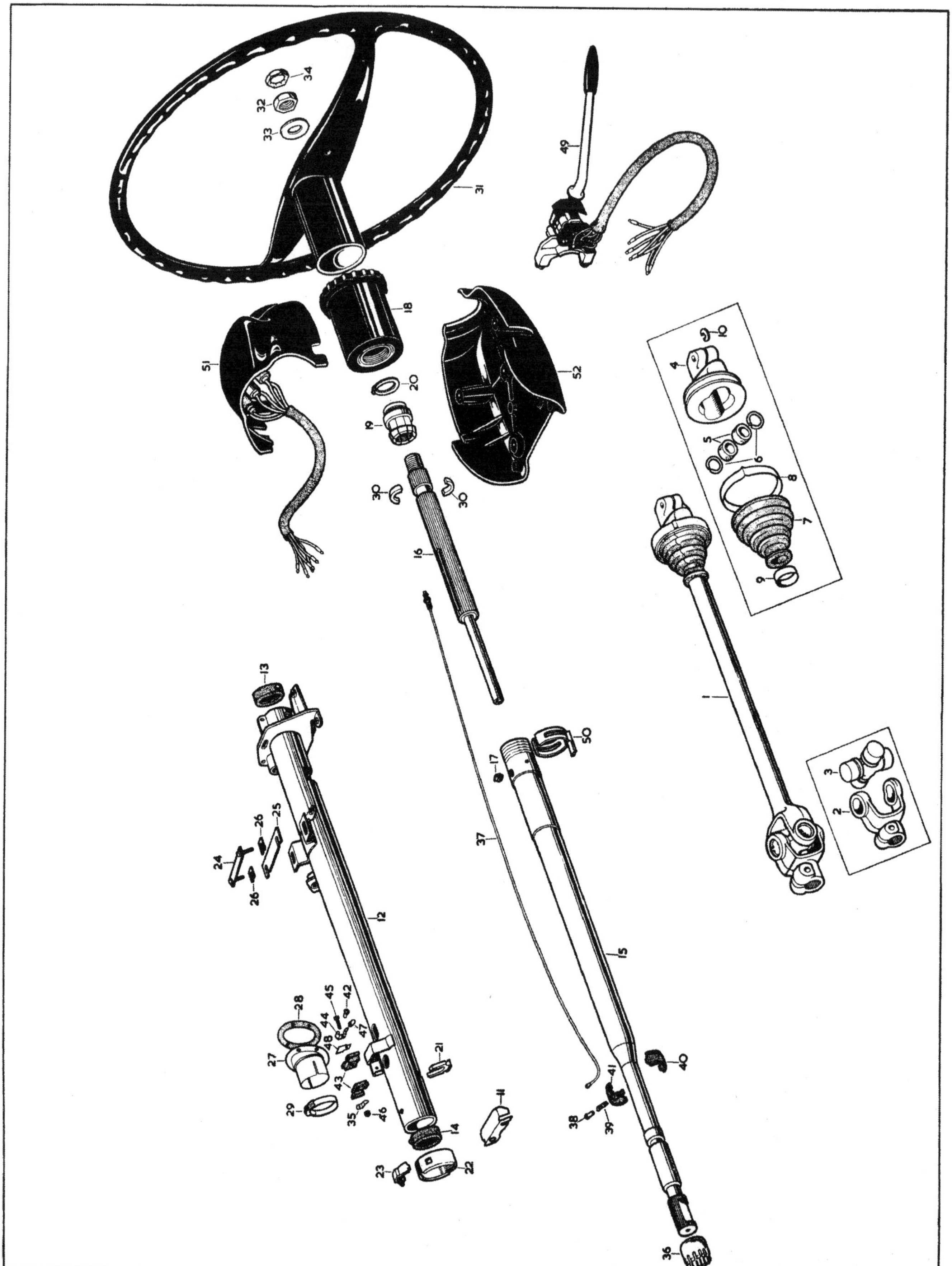

31
32
33
34
49
18
51
52
20
19
10
4
5
6
8
7
9
30
30
16
13
1
50
17
37
3
2
24
26
25
26
12
15
28
45
42
44
47
48
27
43
21
40
29
35
46
41
14
11
38
39
23
22
36

PLATE 38

Part No.	Description	Plate No.	No. per Unit	Remarks
C.18758	**STEERING WHEEL (17" DIAMETER)**		1	Fitted to Chassis No. 1A.2530 and subs. 1A.20220 and subs.
C.7879	**Nut,** securing Steering Wheel to Inner Column Shaft	38–32	1	
C.15885	**Washer,** Special, under Nut	38–33	1	
C.11820/7	**Palnut,** locking Nut	38–34	1	
C.11746	**EARTH CONTACT FOR OPERATION OF HORN**	38–35	1	
UCS.313/3R	**Screw,** Set, securing earth contact to bracket on Upper Column Outer Tube		1	
UCN.113/L	**Nut,** on Setscrew		1	
C.721	**Washer,** Shakeproof, under Nut		1	
C15665	**HORN SLIP RING ASSEMBLY ON INNER COLUMN (348252/B)**		1	
C.10317	**Slip Ring only** (348276)	38–36	1	
C.15666	**Cable,** complete with Contact (54944464)	38–37	1	
C.8761	**Contact,** for Cable (348266)	38–38	1	
C.8156	**Spring,** inside top half of Rotor, for Cable tension	38–39	1	
C.10319	**Rotor,** bottom half (348275)	38–40	1	**Supplied only in pairs to Part No. 8295 (54300638)**
C.10320	**Rotor,** top half (348274)	38–41	1	
C.12194	**Dowler Eyelet,** for connection of Cable from Slip Ring to Horn Push (188818)	38–42	1	
C.8157	**CONTACT ASSEMBLY FOR HORN SLIP RING (348262/A)**		1	
C.8158	**Contact Holder** (348269)	38–43	2	
C.8159	**Contact** (348268)	38–44	1	
C.8161	**Bolt,** securing Contact to Holders (136444)	38–45	1	
C.8160	**Nut,** on Bolt (348270)	38–46	1	
C.10650	**Rubber Sleeve,** insulating Contact from Steering Column Outer Tube (348272)	38–47	1	
C.13024	**Insulating Strip** (Fibre) protecting Contact from above (348486)	38–48	1	
C.20685	**DIRECTION INDICATOR/HEADLAMP FLASHER SWITCH (34563/A-85SA)**	38–49	1	Fitted from Chassis No. 1A.1001 to 1A.7000 1A.20001 to 1A.20359
C.25255	**DIRECTION INDICATOR/HEADLAMP FLASHER SWITCH (34563/H-85SA)**		1	Fitted from Chassis No. 1A.7001 to 1A.10326 1A.20360 to 1A.20479
C.25554	**DIRECTION INDICATOR/HEADLAMP FLASHER SWITCH (35530/B-131.SA)**		1	Fitted from Chassis No. 1A.10327 R.H.D. 1A.20480 L.H.D.
C.25555	**DIRECTION INDICATOR/HEADLAMP FLASHER SWITCH (35530A/131SA)**		1	**SUPERSEDES C.25554**
C.25668	**Insulating Strip,** protecting Direction Indicator Switch		1	Fitted to Chassis No. 1A.7815 and subs. 1A.20377 and subs.
BD.22686	**Stud,** securing Insulating Strip to Upper Steering Column		2	
C.22457	**STRIKER PLATE, RETURNING DIRECTION INDICATOR CONTROL TO NEUTRAL POSITION**	38–50	1	Fitted from Chassis No. 1A.1001 to 1A.10326 1A.20001 to 1A.20479

STEERING COLUMN

Part No.	Description	Plate No.	No. per Unit	Remarks
C.25538	**DRIVE CLIP, FOR STRIKER ON DIRECTION INDICATOR SWITCH**		1	Fitted to Chassis No. 1A.10327 and subs. 1A.20480 and subs.
C.14007/1	**Screw,** Set, securing Striker Plate or Drive Clip		2	
C.23008	**Washer,** Special, on Setscrews		2	
C.723/A	**Washer,** Shakeproof, on Setscrews		2	
C.27315	**UPPER SWITCH COVER ASSEMBLY AT CENTRE OF STEERING WHEEL**		1	**FOR CARS FITTED WITH STANDARD TRANSMISSION** Fitted to Chassis No. 1K.1001 and subs. 1K.30001 and subs.
C.16575	**LOWER SWITCH COVER ASSEMBLY AT CENTRE OF STEERING WHEEL**		1	
C.25927	**UPPER SWITCH COVER ASSEMBLY AT CENTRE OF STEERING WHEEL**		1	**FOR CARS FITTED WITH OVERDRIVE TRANSMISSION** Fitted to Chassis No. 1K.1001 and subs. 1K.30001 and subs.
C.16563	**LOWER SWITCH COVER ASSEMBLY AT CENTRE OF STEERING WHEEL**		1	
C.20779	**UPPER SWITCH COVER ASSEMBLY AT CENTRE OF STEERING WHEEL (WITH INDICATOR PLATE INSCRIBED "P.R.N.D.L.") (54361227)**	38–51	1	Fitted from Chassis No. 1A.1001 to 1A.3252 1A.1001 to 1A.20256 Except 1A.3096 to 1A.3098 1A.3232, 1A.3247 to 1A.3249
C.22538	**UPPER SWITCH COVER ASSEMBLY AT CENTRE OF STEERING WHEEL (WITH INDICATOR PLATE INSCRIBED "P.R.N.D_2.D_1.L.") (54361545)**		1	Fitted to Chassis No. 1A.3253 and subs. 1A.20257 and subs. Also to 1A.3096 to 1A.3098 1A.3232, 1A.3247 to 1A.3249
C.20788	**LOWER SWITCH COVER ASSEMBLY AT CENTRE OF STEERING WHEEL (54360620)**	38–52	1	
C.16460	**Screw,** Set, clamping Upper and Lower Switch Covers		2	
BD.13713/2	**Screw,** Set, securing Lower Switch Cover to Steering Column		3	
C.15830	**Washer,** Plain, on Setscrews		5	

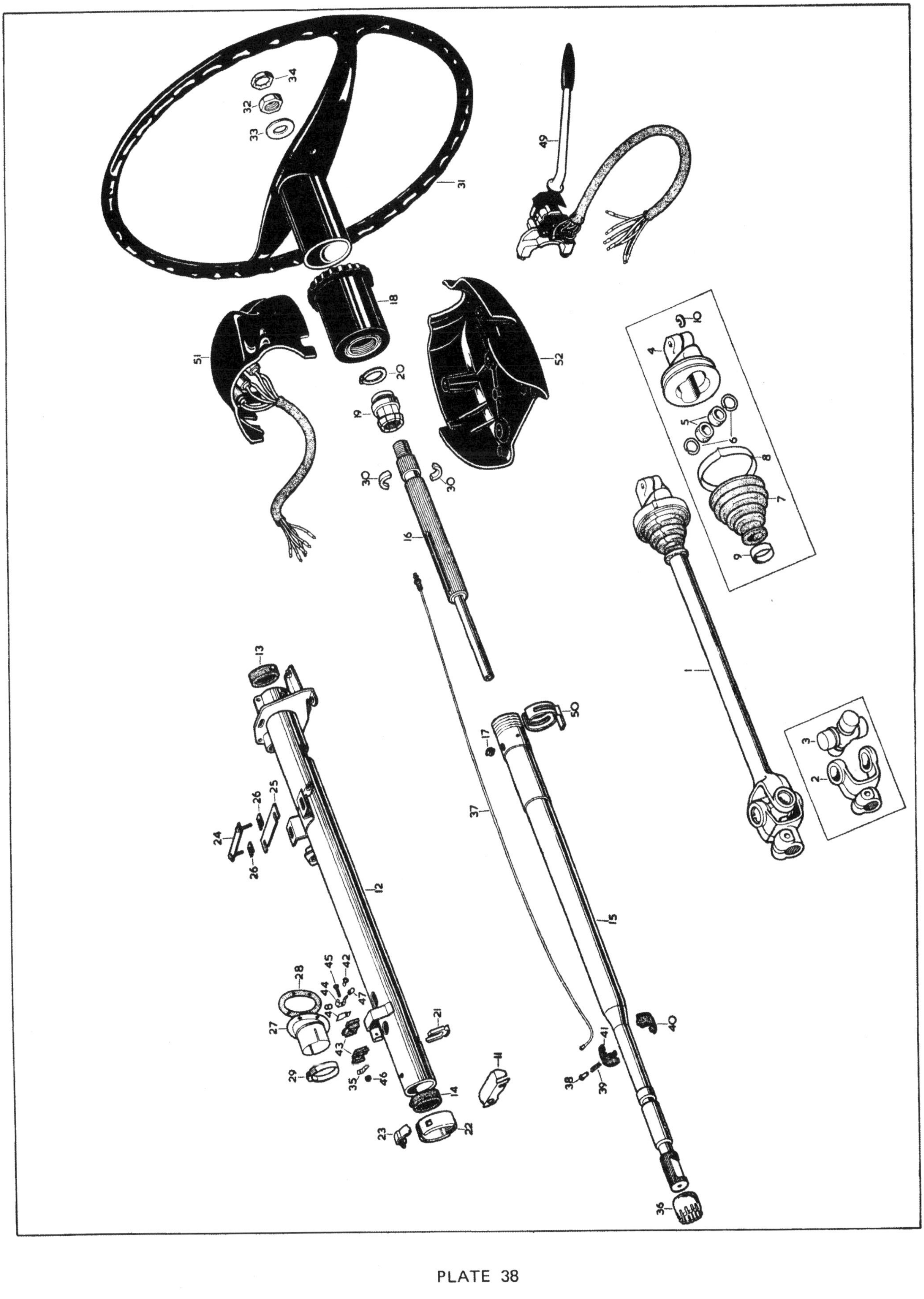

PLATE 38

AUTOMATIC TRANSMISSION CONTROL

Part No.	Description	Plate No.	No. per Unit	Remarks
C.20779	**UPPER SWITCH COVER ASSEMBLY AT CENTRE OF STEERING WHEEL (WITH INDICATOR PLATE INSCRIBED "P.R.N.D.L.") (54361227)**	39–1	1	Fitted from Chassis No. 1A.1001 to 1A.3252 1A.20001 to 1A.20256 Except 1A.3096 to 1A.3098 1A.3232, 1A.3247 to 1A.3249
C.22538	**UPPER SWITCH COVER ASSEMBLY AT CENTRE OF STEERING WHEEL (WITH INDICATOR PLATE INSCRIBED "P.R.N.D_2.D_1.L.") (54361545)**		1	Fitted to Chassis No. 1A.3253 and subs. 1A.20257 and subs. Also to 1A.3096 to 1A.3098 1A.3232, 1A.3247 to 1A.3249
C.20788	**LOWER SWITCH COVER ASSEMBLY AT CENTRE OF STEERING WHEEL (54360620)**	39–2	1	
C.16460	**Screw,** Set, clamping Upper and Lower Switch Covers		2	
C.15830	**Washer,** Plain, on Setscrews		2	
BD.13713/2	**Screw,** Set, securing Lower Switch Cover to Upper Steering Column		3	
C.15830	**Washer,** Plain, on Setscrews		3	
C.26229	**LEVER AND BALL PIN ASSEMBLY ON SELECTOR SHAFT**	39–3	1	
C.29015	**Locking Screw,** securing Lever and Ball Pin Assembly to Selector Housing Shaft		1	
C.21166	**HOUSING AND SHAFT ASSEMBLY FOR SELECTOR LEVER**	39–4	1	
C.15639	**Grommet,** in Selector Housing, for Indicator Arm Connecting Link	39–5	1	
C.15862	**Circlip,** securing Housing and Shaft to Bracket on Steering Column	39–6	1	
C.15896	**Washer,** behind Circlip	39–7	1	
C.20879	**GATE, POSITIONING SELECTOR LEVER**	39–8	1	Fitted from Chassis No. 1A.1001 to 1A.3252 1A.20001 to 1A.20256 Except 1A.3096 to 1A.3098 1A.3232, 1A.3247 to 1A.3249
C.26180	**GATE, POSITIONING SELECTOR LEVER**		1	Fitted to Chassis No. 1A.3253 and subs. 1A.20257 and subs. Also to 1A.3096 to 1A.3098 1A.3232, 1A.3247 to 1A.3249
UFS.119/7R	**Screw,** Set, securing Gate to Mounting Bracket on Steering Column Tube		2	
C.8667/17	**Nut,** on Setscrews		2	
C.16541	**Distance Piece,** at mounting of Gate	39–9	2	
C.22109	**LEVER ASSEMBLY, OPERATING SELECTOR (COMPLETE WITH KNOB)**	39–10	1	Fitted from Chassis No. 1A.1001 to 1A.3252 1A.20001 to 1A.20256 Except 1A.3096 to 1A.3098 1A.3232, 1A.3247 to 1A.3249

AUTOMATIC TRANSMISSION CONTROL

Part No.	Description	Plate No.	No. per Unit	Remarks
C.23038	**LEVER ASSEMBLY, OPERATING SELECTOR (COMPLETE WITH KNOB)**		1	Fitted to Chassis No. 1A.3253 and subs. 1A.20257 and subs. Also to 1A.3096 to 1A.3098 1A.3232, 1A.3247 to 1A.3249
C.20727/16	**Pivot Pin,** for Selector Lever	39–11	1	Fitted from Chassis No. 1A.1001 to 1A.10483 1A.20001 to 1A.20483
C.26745	**Clevis Pin,** pivoting Selector Lever		1	Fitted to Chassis No. 1A.10484 and subs. 1A.20484 and subs.
L.102/4U	**Split Pin,** retaining Clevis Pin		1	
C.13361	**Washer,** behind Split Pin		1	
C.900	**Spring,** in Selector Housing, tensioning Lever	39–12	1	
C.903	**Shim,** under Spring	39–13	As req'd	
C.16578	**BRACKET FOR INDICATOR ARM**	39–14	1	
UFS.119/4R	**Screw,** Set, securing Bracket		1	
UFN.119/L	**Nut,** on Setscrew		1	
AW.102/T	**Washer,** Plain, under Nut		1	
C.20734	**LINK, CONNECTING INDICATOR ARM AND SELECTOR LEVER**	39–15	1	Fitted from Chassis No. 1A.1001 to 1A.12647 1A.20001 to 1A.20583
C.27834	**LINK, CONNECTING INDICATOR ARM AND SELECTOR LEVER**		1	Fitted to Chassis No. 1A.12648 and subs 1A.20584 and subs
C.15637	**ARM ASSEMBLY FOR GEAR INDICATOR**	39–16	1	Fitted from Chassis No. 1A.1001 to 1A.12647 1A.20001 to 1A.20583
C.27836	**ARM ASSEMBLY FOR GEAR INDICATOR**		1	Fitted to Chassis No. 1A.12648 and subs 1A.20584 and subs
C.15639	**Grommet,** in Indicator Arm, for Connecting Link	39–17	1	
C.16725	**TRANSFER LEVER ASSEMBLY FOR OPERATION OF GEAR CONTROL CABLE**	39–18	1	R.H. DRIVE CARS ONLY
C.16727	**Bracket,** mounting Transfer Lever to Dash	39–19	1	
C.15887	**Nut Plate Assembly,** fixing Bracket		1	
UFS.131/6R	**Screw,** Set, securing Bracket to Dash		2	
FG.105/X	**Washer,** Spring, on Setscrews		2	
C.17381	**Stiffener,** below Transfer Lever Bracket		1	
UFB.131/11R	**Bolt,** through Bracket and Stiffener, pivoting Transfer Lever		1	
C.15814	**Distance Tube,** on Bolt		1	
UFN.131/L	**Nut,** on Bolt		1	
FG.105/X	**Washer,** Spring, under Nut		1	
C.16810	**LINK, FROM OPERATING SHAFT TO TRANSFER LEVER**	39–20	1	R.H. DRIVE CARS ONLY
C.16809	**Ball Joint** (Amal) connecting Link to Operating Shaft	39–21	1	
C.15684	**Ball Joint** (Amal) connecting Link to Transfer Lever	39–22	1	
UFN.125/L	**Nut,** locking Ball Joints to Link		2	
UFN.125/L	**Nut,** securing Ball Joint to Transfer Lever		1	
C.724	**Washer,** Shakeproof, under Nut		1	

AUTOMATIC TRANSMISSION CONTROL

Part No.	Description	Plate No.	No. per Unit	Remarks
C.20933	**TRANSFER LEVER AND BRACKET ASSEMBLY FOR OPERATION OF GEAR CONTROL CABLE**		1	**L.H. DRIVE CARS ONLY** Fitted from Chassis No. 1A.20001 to 1A.20267
UFS.131/4R	**Screw**, Set, securing Bracket to Body		2	
UFN.131/L	**Nut**, on Setscrews		2	
C.725	**Washer**, Shakeproof, under Nuts		2	
C.22210	**TRANSFER LEVER AND BRACKET ASSEMBLY FOR OPERATION OF GEAR CONTROL CABLE**		1	**L.H. DRIVE CARS ONLY** Fitted to Chassis No. 1A.20268 and subs
C.22282	**Stiffening Plate Assembly**, at fixings of Bracket		1	
BD.13565/2	**Bolt**, securing Bracket to Steering Column Support Bracket		3	
C.725	**Washer**, Shakeproof, on Bolts		3	
C.20947	**LINK, FROM OPERATING SHAFT TO TRANSFER LEVER**		1	**L.H. DRIVE CARS ONLY**
C.16809	**Ball Joint** (Amal) on Connecting Link		2	
UFN.125/L	**Nut**, locking Ball Joint to Link		2	
C.20929	**GEAR CONTROL CABLE FROM TRANSMISSION LEVER TO TRANSMISSION UNIT**	39–23	1	
C.16809	**Ball Joint** (Amal) fixing Cable to Transfer Lever	39–24	1	
UFN.125/L	**Nut**, locking Ball Joint to Cable		1	
C.5574	**Grommet**, in Dash, for Control Cable (740811)		1	
C.15599	**ABUTMENT BRACKET, ON DASH, FOR GEAR CONTROL CABLE**	39–25	1	
C.16814	**Nut Plate Assembly**, fixing Abutment Bracket		1	
UFS.131/6R	**Screw**, Set, securing Abutment Bracket		2	
FG.105/X	**Washer**, Spring, on Setscrews		2	
FW.105/T	**Washer**, Plain, under Spring Washers		2	
C.12690	**Nut**, securing Cable to Abutment Bracket		2	
C.16809	**BALL JOINT, SECURING GEAR CONTROL CABLE TO TRANSMISSION UNIT**	39–26	1	
UFN.125/L	**Nut**, securing Ball Joint to Cable		1	
C.21131	**ABUTMENT BRACKET AND CLAMP ASSEMBLY, ON HOUSING**	39–27	1	**R.H. DRIVE CARS ONLY**
C.21130	**ABUTMENT BRACKET AND CLAMP ASSEMBLY ON HOUSING**		1	**L.H. DRIVE CARS ONLY**
UFS.131/8R	**Screw**, Set, securing Cable in Clamp		1	
C.725	**Washer**, Shakeproof, on Setscrew		1	
UCS.131/6R	**Screw**, Set, securing Bracket Assembly to Converter Housing		2	
C.5574	**Grommet**, for Control Cable at Gearbox Tunnel	39–28	1	

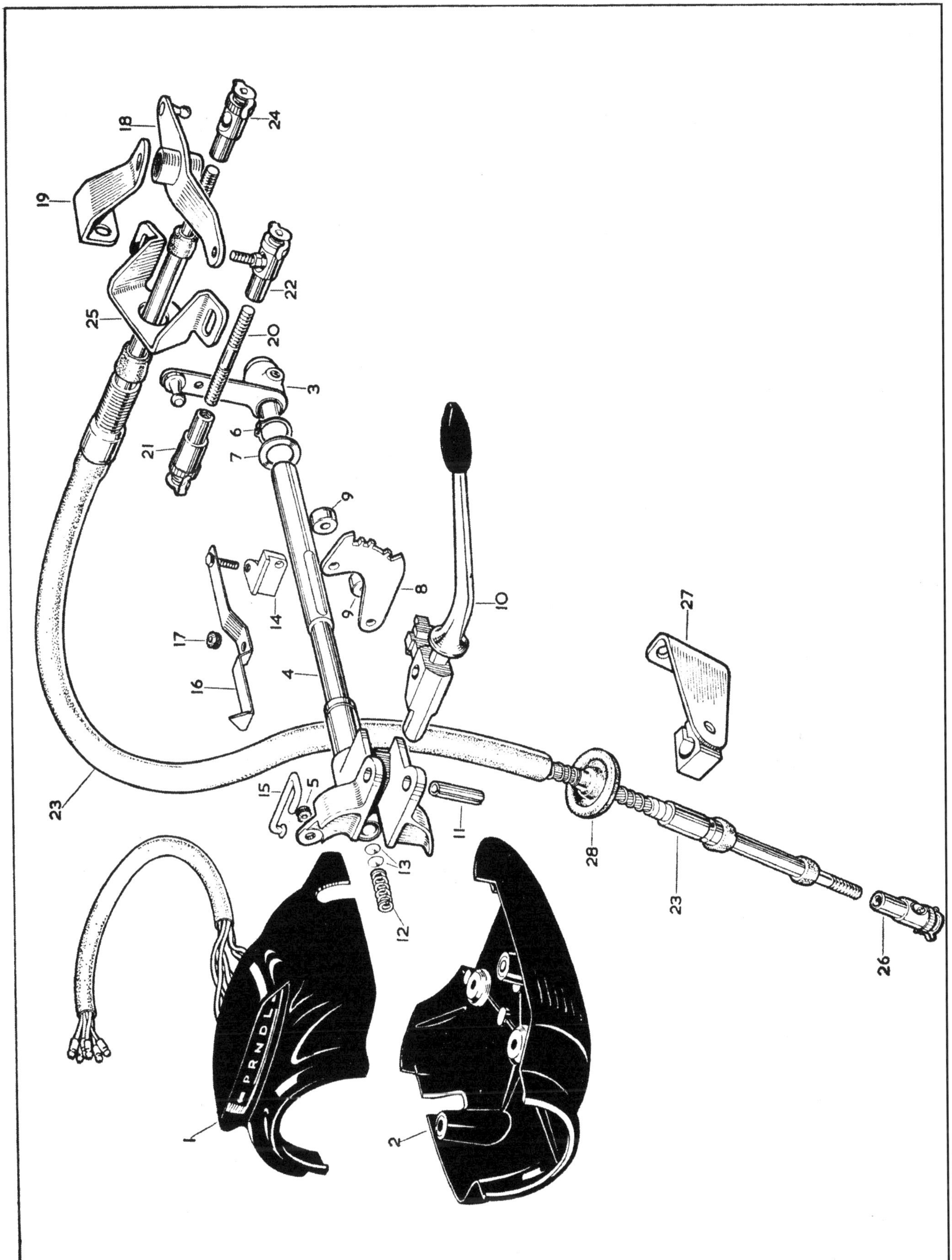

PLATE 39

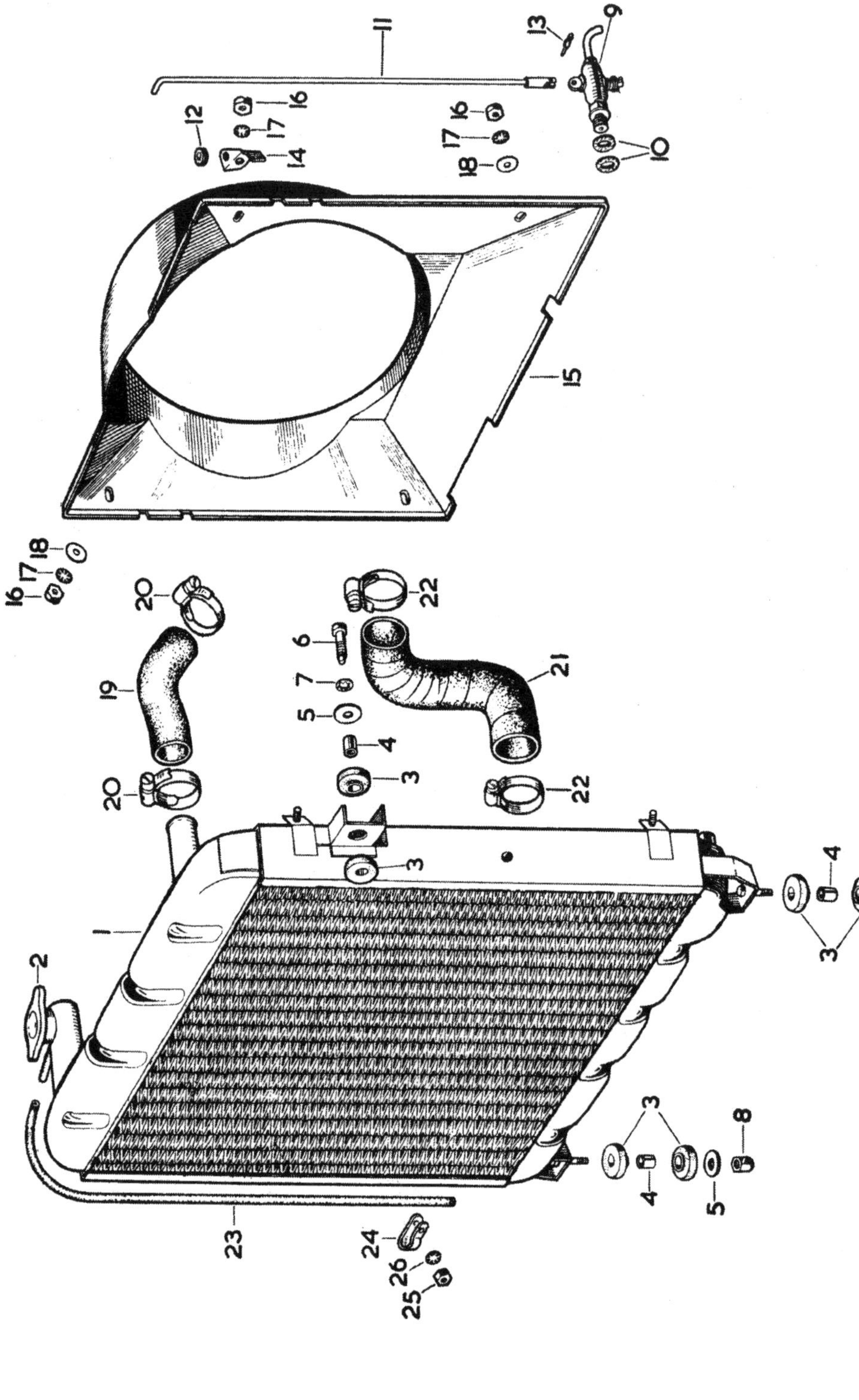

PLATE 40

Part No.	Description	Plate No.	No. per Unit	Remarks
C.27470	**RADIATOR BLOCK ASSEMBLY**		1	**FOR CARS FITTED WITH STANDARD AND OVERDRIVE TRANSMISSION**
C.24917	**RADIATOR BLOCK ASSEMBLY**	40–1	1	**THIS RADIATOR BLOCK INCORPORATES A COOLER FOR TRANSMISSION UNIT OIL** Fitted from Chassis No. 1A.1001 to 1A.12039 1A.20001 to 1A.20545
	NOTE: When fitting Radiator Block C.24917 to cars prior to Chassis No. 1A.6392 (R.H. Drive) or 1A.20336 (L.H. Drive), Cowl C.24964 will also be required.			
C.27181	**RADIATOR BLOCK ASSEMBLY**		1	**FOR CARS FITTED WITH AUTOMATIC TRANSMISSION** Fitted to Chassis No. 1A.12040 and subs 1A.20546 and subs
C.18460	**Filler Cap** (4 lbs. pressure)		1	
C.18485	**Filler Cap** (9 lbs. pressure)	40–2	1	**FOR EXPORT TO MEXICO**
C.8975	**Rubber Pad**, mounting Radiator Block	40–3	8	
C.8976	**Distance Tube**, through Rubber Pads	40–4	4	
C.410	**Special Washer**, retaining Mounting Pads	40–5	4	
C.12222	**Bolt**, securing Radiator at top	40–6	2	
C.726	**Washer**, Shakeproof, on Bolts	40–7	2	
C.410	**Special Washer**, packing Radiator at top		As req'd	
C.8667/3	**Nut**, Self-locking, securing Radiator at bottom	40–8	2	
C.4439/1	**DRAIN TAP FOR RADIATOR BLOCK**	40–9	1	
C.981	**Washer** (Fibre) on Drain Tap	40–10	As req'd	
C.8973	**CONTROL ROD ASSEMBLY, OPERATING DRAIN TAP**	40–11	1	
C.1042	**Grommet**, on Control Rod	40–12	2	
L.103/7U	**Split Pin**, securing Control Rod to Drain Tap	40–13	1	
C.11776	**BRACKET, ON FAN COWL, FOR DRAIN TAP CONTROL ROD**	40–14	1	
C.21028	**COWL, ON RADIATOR BLOCK, AROUND WATER PUMP FAN**	40–15	1	Fitted from Chassis No. 1A.1001 to 1A.6391 1A.20001 to 1A.20335
C.24964	**COWL, ON RADIATOR BLOCK, AROUND WATER PUMP FAN**		1	Fitted to Chassis No. 1A.6392 and subs 1A.20336 and subs
UFN.231/L	**Nut**, securing Cowl, Drain Tap Control Rod Bracket and Overflow Pipe Clip to Radiator	40–16	4	
C.725	**Washer**, Shakeproof, under Nuts	40–17	4	
BD.541/5	**Washer**, Plain, under Shakeproof Washers, at R.H. side of Cowl	40–18	2	

COOLING SYSTEM

Part No.	Description	Plate No.	No. per Unit	Remarks
C.27182	**HOSE FROM TOP OF RADIATOR BLOCK TO WATER OUTLET PIPE EXTENSION ON ENGINE**	40–19	1	
C.2905/5	**Clip,** securing Top Radiator Hose	40–20	2	
C.21389	**HOSE FROM BOTTOM OF RADIATOR BLOCK TO WATER PUMP**	40–21	1	
C.2905/5	**Clip,** securing Bottom Radiator Hose	40–22	2	
6705	**RUBBER OVERFLOW PIPE FROM TOP OF RADIATOR BLOCK**	40–23	1	
C.1040/2	**Clip,** securing Overflow Pipe at bottom of Cowl	40–24	1	

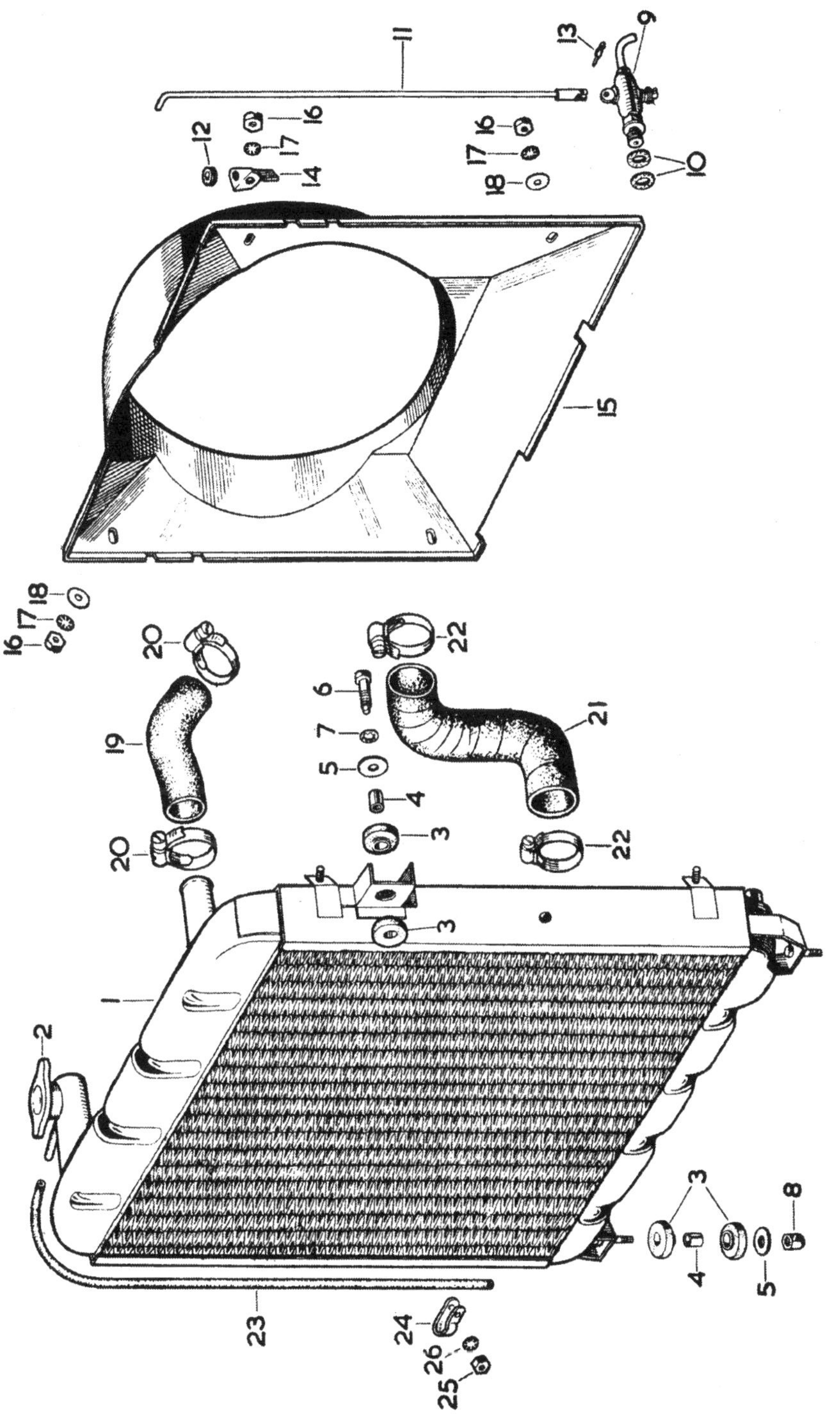

PLATE 40

EXHAUST SYSTEM

Part No.	Description	Plate No.	No. per Unit	Remarks
C.21822	**R.H. EXHAUST DOWN PIPE ASSEMBLY**	41–1	1	
C.1759	**Flexible Pipe,** on Down Pipe (12" long)	41–1A	1	
C.18401	**Flange,** for Down Pipe	41–1B	1	
C.21823	**L.H. EXHAUST DOWN PIPE ASSEMBLY**	41–2	1	
C.1759	**Flexible Pipe,** on Down Pipe (12" long)	41–2A	1	
C.18401	**Flange,** for Down Pipe	41–2B	1	
C.18405/1	**Sealing Ring,** between Down Pipes and Exhaust Manifolds	41–3	2	
C.17916	**Nut,** securing Down Pipes to Exhaust Manifolds		8	
FW.106/T	**Washer,** Plain, under Nuts		8	
C.21797	**BALANCE PIPE ASSEMBLY, LINKING DOWN PIPES TO SILENCERS**		1	Fitted from Chassis No. 1A.1001 to 1A.2041 1A.20001 to 1A.20126
C.21800	**Flexible Pipe,** on Balance Pipe (6" long)		1	
C.13063	**Clip,** securing Balance Pipe to Down Pipes and Silencers		4	
UFB.131/22R	**Bolt,** through Clips		4	
FW.105/T	**Washer,** Plain, on Bolts		8	
UFN.131/L	**Nut,** on Bolts		4	
C.22418	**FRONT EXHAUST PIPE ASSEMBLY, FROM DOWN PIPES TO SILENCERS**	41–4	2	Fitted to Chassis No. 1A.2042 and subs 1A.20127 and subs
C.13063	**Clip,** securing Front Exhaust Pipes to Down Pipes and Silencers	41–5	4	
UFB.131/22R	**Bolt,** through Clips		4	
FW.105/T	**Washer,** Plain, on Bolts		8	
UFN.131/L	**Nut,** on Bolts		4	
C.22420	**BALANCE PIPE ASSEMBLY, JOINING FRONT EXHAUST PIPES**	41–6	1	Fitted to Chassis No. 1A.2042 and subs 1A.20127 and subs
C.21800	**Flexible Pipe,** on Balance Pipe (6" long)	41–6A	1	
C.22422	**Clip,** securing Balance Pipe to Front Exhaust Pipes	41–7	2	
UFB.131/20R	**Bolt,** through Clips		2	
FW.105/T	**Washer,** Plain, on Bolts		4	
UFN.131/L	**Nut,** on Bolts		2	
C.21782	**EXHAUST SILENCER**	41–8	2	Fitted from Chassis No. 1A.1001 to 9933 1A.20001 to 1A.20470
C.8397	**Rubber Mounting,** fixing Silencer to Body	41–9	2	
UFS.131/6R	**Screw,** Set, securing Rubber Mountings to Body and Silencers		6	
FW.105/T	**Washer,** Plain, on Setscrews		6	
UFN.131/L	**Nut,** on Setscrews		6	
C.725	**Washer,** Shakeproof, under Nuts		6	
C.21803	**R.H. INTERMEDIATE PIPE**	41–10	1	
C.21804	**L.H. INTERMEDIATE PIPE**	41–11	1	
C.21783	**R.H. EXHAUST MUFFLER**	41–12	1	Fitted from Chassis No. 1A.1001 to 1A.9933 1A.20001 to 1A.20470
C.21784	**L.H. EXHAUST MUFFLER**	41–13	1	
C.17197	**Rubber Mounting,** fixing Exhaust Mufflers to rear Wheel Arches	41–14	2	
UFS.131/7R	**Screw,** Set, securing Rubber Mountings to Wheel Arches and Mufflers		8	
FW.105/T	**Washer,** Plain, on Setscrews		8	
UFN.131/L	**Nut,** on Setscrews		8	
C.11820/3	**Palnut,** locking Nuts		8	

Part No.	Description	Plate No.	No. per Unit	Remarks
C.22064	**R.H. TAIL PIPE**	41–15	1	Fitted from Chassis No. 1A.1001 to 1A.9933 1A.20001 to 1A.20470
C.22065	**L.H. TAIL PIPE**	41–16	1	
C.26167	**EXHAUST SILENCER**		2	Fitted to Chassis No. 1A.9934 and subs 1A.20471 and subs
C.8397	**Rubber Mounting,** fixing Silencers to Body		2	
UFS.131/6R	**Screw,** Set, securing Rubber Mounting to Silencers and Body		6	
UFN.131/L	**Nut,** on Setscrews		6	
C.725	**Washer,** Shakeproof, under Nuts		6	
FW.105/T	**Washer,** Plain, under Shakeproof Washers		6	
C.26260	**R.H. INTERMEDIATE PIPE ASSEMBLY**		1	
C.26261	**L.H. INTERMEDIATE PIPE ASSEMBLY**		1	
C.17197	**Rubber Mounting,** fixing Intermediate Pipes to Rear Wheel Arches		2	Fitted from Chassis No. 1A.9934 to 1A.10304 1A.20471 to 1A.20478
UFS.131/7R	**Screw,** Set, securing Rubber Mountings to Wheel Arches and Intermediate Pipes		8	
UFN.131/L	**Nut,** on Setscrews		8	
FW.105/T	**Washer,** Plain, under Nuts		8	
C.11820/3	**Palnut,** locking Nuts		8	
1363	**Rubber Mounting,** for Intermediate Pipes		2	Fitted to Chassis No. 1A.10305 and subs 1A.20479 and subs
C.26657	**Bracket,** fixing Rubber Mountings at Rear Wheel Arches		2	
UFS.131/6R	**Screw,** Set, securing Rubber Mountings to Brackets		2	
FW.105/T	**Washer,** Plain, on Setscrews		4	
UFN.131/L	**Nut,** on Setscrews		2	
C.725	**Washer,** Shakeproof, under Nuts		2	
UFS.131/7R	**Screw,** Set, securing Brackets to Body and Intermediate Pipes to Rubber Mountings		8	
UFN.131/L	**Nut,** on Setscrews		8	
FW.105/T	**Washer,** Plain, under Nuts		8	
C.11820/3	**Palnut,** locking Nuts		8	
C.26171	**TAIL PIPE**		2	Fitted to Chassis No. 1A.9934 and subs 1A.20471 and subs
DA.406/6S	**Screw,** Self-Tapping, securing Tail Pipes to Intermediate Pipes		4	
C.21811	**R.H. FRONT HEAT SHIELD FOR EXHAUST PIPE**	41–17	1	
C.21812	**L.H. FRONT HEAT SHIELD FOR EXHAUST PIPE**	41–18	1	
C.18325	**Bracket Assembly,** fixing Heat Shields to Body	41–19	8	
BD.1814/2	**Pop Rivet,** securing Heat Shield Brackets to Body		32	
UFS.125/4R	**Screw,** Set, securing Heat Shields to Brackets		8	
FW.104/T	**Washer,** Plain, on Setscrews		8	
C.724	**Washer,** Shakeproof, on Setscrews		8	
C.21813	**INTERMEDIATE HEAT SHIELD FOR EXHAUST PIPES**	41–20	2	
C.18325	**Bracket Assembly,** fixing Heat Shields to Body	41–21	4	
BD.1814/2	**Pop Rivet,** securing Heat Shield Brackets to Body		16	
UFS.125/4R	**Screw,** Set, securing Heat Shields to Brackets		4	
FW.104/T	**Washer,** Plain, on Setscrews		4	
C.724	**Washer,** Shakeproof, on Setscrews		4	

EXHAUST SYSTEM

Part No.	Description	Plate No.	No. per Unit	Remarks
C.21814	**REAR HEAT SHIELD FOR R.H. EXHAUST SILENCER**	41–22	1	
UFS.125/4R	**Screw,** Set, securing R.H. Rear Heat Shield to Body		3	
FW.104/T	**Washer,** Plain, on Setscrews		3	
C.724	**Washer,** Shakeproof, on Setscrews		3	
C.18322	**REAR HEAT SHIELD FOR L.H. EXHAUST SILENCER**	41–23	1	
UFS.125/4R	**Screw,** Set, securing L.H. Rear Heat Shield to Body		4	
FW.104/T	**Washer,** Plain, on Setscrews		4	
C.724	**Washer,** Shakeproof, on Setscrews		4	

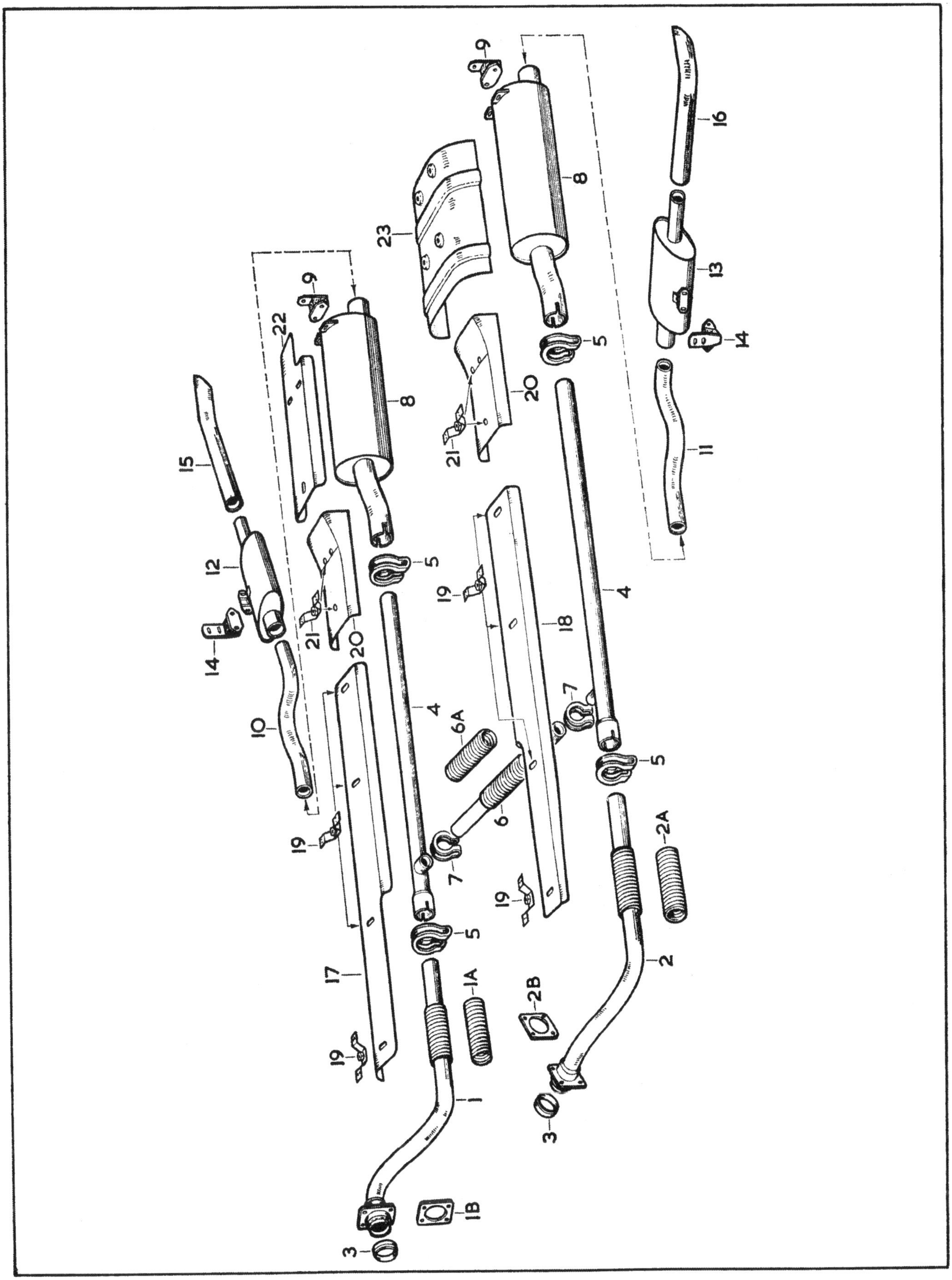

PLATE 41

FUEL SYSTEM

Part No.	Description	Plate No.	No. per Unit	Remarks
C.17565	PETROL TANK ASSEMBLY	42–1	1	
C.990/1	**Filter Assembly**, at bottom of Petrol Tank	42–2	1	
C.26310	**Gasket**, for Filter	42–3	1	
C.25089/2	**Air Vent Extension Tube**	42–4	1	
C.10269	**Clip**, securing Tube to Petrol Tank Flange	42–5	1	
BD.9898	**Rubber Grommet**, on Filler Neck	42–6	1	
	NOTE: For Details of Auxiliary Petrol Tank, please refer to Section for Optional Extras			
C.23601	PETROL FILLER CAP ASSEMBLY	42–7	1	
C.23603	**Sealing Ring**, in Filler Cap		1	
C.23604	**Spring Clip**, locking Filler Cap to Filler Neck		1	
UCS.819/3H	**Screw**, Set, securing Spring Clip to Cap		1	
AG.102/X	**Washer**, Spring, on Setscrew		1	
C.3053	RUBBER PAD, AT MOUNTING OF PETROL TANK	42–8	6	
C.6390	**Distance Tube**, through Mounting Pads	42–9	3	
C.8737/3	**Nut**, Self-locking, securing Petrol Tank		3	
C.410	**Special Washer**, under Nuts		3	
C.24418	**Pad** (Polyurethane) between Petrol Tank and Boot Floor		1	
C.15476	PETROL TANK ELEMENT UNIT (TF.1104/042)	42–10	1	
C.937	**Gasket**, under Element Unit	42–11	1	
NS.313/3H	**Screw**, Set, securing Element Unit to Tank		6	
C.2296/9	**Washer**, Copper, on Setscrews		12	
C.25076	PETROL PUMP (AUF.303)	42–12	1	
C.801	**Stud**, mounting Petrol Pump		2	
C.10126	**Grommet**, on Studs		2	
C.20323	**Distance Tube**, in Grommets		2	
C.8150/33	**Nut**, securing Petrol Pump		2	
FW.205/T	**Washer**, Plain, on Studs		4	
BD.541/5	**Special Washer**, on Studs		4	
C.16355	NYLON PIPE FROM PETROL TANK TO PUMP	42–13	1	
C.2708	**Grommet**, in Boot Floor, for Petrol Pipe	42–14	1	
C.21680	FLEXIBLE PIPE FROM PETROL PUMP TO REAR FEED PIPE	42–15	1	
C.11488	**Banjo Bolt**, securing Pipes to Petrol Pump (AUC.2698)		2	
C.11489	**Fibre Washer**, at each side of Banjos (AUC.2141)		4	
C.21682	**Nut**, securing Flexible Pipe to Boot Floor	42–16	1	
C.728	**Washer**, Shakeproof, under Nut	42–17	1	
C.21524	REAR PETROL FEED PIPE ASSEMBLY	42–18	1	Fitted from Chassis No. 1A.1001 to 1A.6366 1A.20001 to 1A.20335

Part No.	Description	Plate No.	No. per Unit	Remarks
C.24713	**REAR PETROL FEED PIPE ASSEMBLY**		1	Fitted to Chassis No. 1A.6367 and subs 1A.20336 and subs
C.21717	**Olive,** seating Rear Pipe to Pipe from Pump		1	
C.21718	**Nut,** securing Rear Pipe to Pipe from Pump		1	
C.21747	**UNION (DOUBLE-ENDED) CONNECTING REAR AND FRONT PIPES**	42–19	1	
C.21717	**Olive,** seating pipes to Double-Ended Union	42–20	2	
C.21718	**Nut,** securing Pipes to Double-Ended Union	42–21	2	
C.21523	**FRONT PETROL FEED PIPE ASSEMBLY**	42–22	1	Fitted from Chassis No. 1A.1001 to 1A.6366 1A.20001 to 1A.20335
C.24712	**FRONT PETROL FEED PIPE ASSEMBLY**		1	Fitted to Chassis No. 1A.6367 and subs 1A.20336 and subs
C.26110	**FRONT PETROL FEED PIPE ASSEMBLY**		1	Cars fitted with Power Assisted Steering C.26018. Fitted to Chassis No. 1A.13158 and subs (also 1A.13003)
C.28413	**FRONT PETROL FEED PIPE ASSEMBLY**		1	**Supersedes C.24712**
C.21717	**Olive,** seating Front Pipe in Petrol Filter	42–23	1	
C.21719	**Nut,** securing Front Pipe to Petrol Filter	42–24	1	
C.10268	**CLIP, SECURING PETROL PIPES**	42–25	6	
C.10269	**CLIP, SECURING PETROL PIPE AT REAR**	42–26	2	
C.3727	**CLIP, SECURING PETROL PIPES**	42–27	3	
C.1040/9	**CLIP, SECURING PIPE TO EXHAUST TAIL-PIPE BRACKET ON BODY**	42–28	1	
UFS.419/4H	**Screw,** Set, securing Clip to Bracket		1	
UFN.119/L	**Nut,** on Setscrew		1	
C.723/A	**Washer,** Shakeproof, on Setscrew		1	
C.1040/6	**CLIP, SECURING PIPE TO R.H. FRONT WING VALANCE**	42–29	1	
DZZ.408/6C	**Screw,** Self-Tapping, securing Clip		1	
C.13681	**PETROL FILTER ASSEMBLY (AC.7950001)**	42–30	1	
C.13705	**Banjo Bolt,** securing Petrol Pipe to Filter		1	
C.784	**Washer,** Fibre, on Banjo Bolt		2	

FUEL SYSTEM

Part No.	Description	Plate No.	No. per Unit	Remarks
C.27588	**PETROL FILTER ASSEMBLY (AC.7971243)**		1	Supersedes C.13681
11450	**Filter Casting** (7971137)		1	Supersedes C.13681
11951	**Sealing Washer,** between Filter Casting and Element (854471)		1	Supersedes C.13681
C.28080	**Filter Element** (Fibre) complete with Sealing Washers (854347)		1	Supersedes C.13681
11950	**Sealing Washer,** between Filter Casting and Glass Bowl (C.854361)		1	Supersedes C.13681
7300	**Glass Bowl,** (1523620)		1	Supersedes C.13681
7301	**Retaining Strap Assembly** (5592050)		1	Supersedes C.13681
C.14726	**BRACKET, MOUNTING PETROL FILTER TO R.H. FRONT WING VALANCE**	42–31	1	
UFS.131/6R	**Screw,** Set, securing Filter to Bracket and Bracket to Wing Valance		4	
UFN.131/L	**Nut,** on Setscrews		4	
FG.105/X	**Washer,** Spring, under Nuts		4	
C.21241	**CARBURETTER FEED PIPE ASSEMBLY (BETWEEN CARBURETTERS AND FILTER)**	42–32	1	

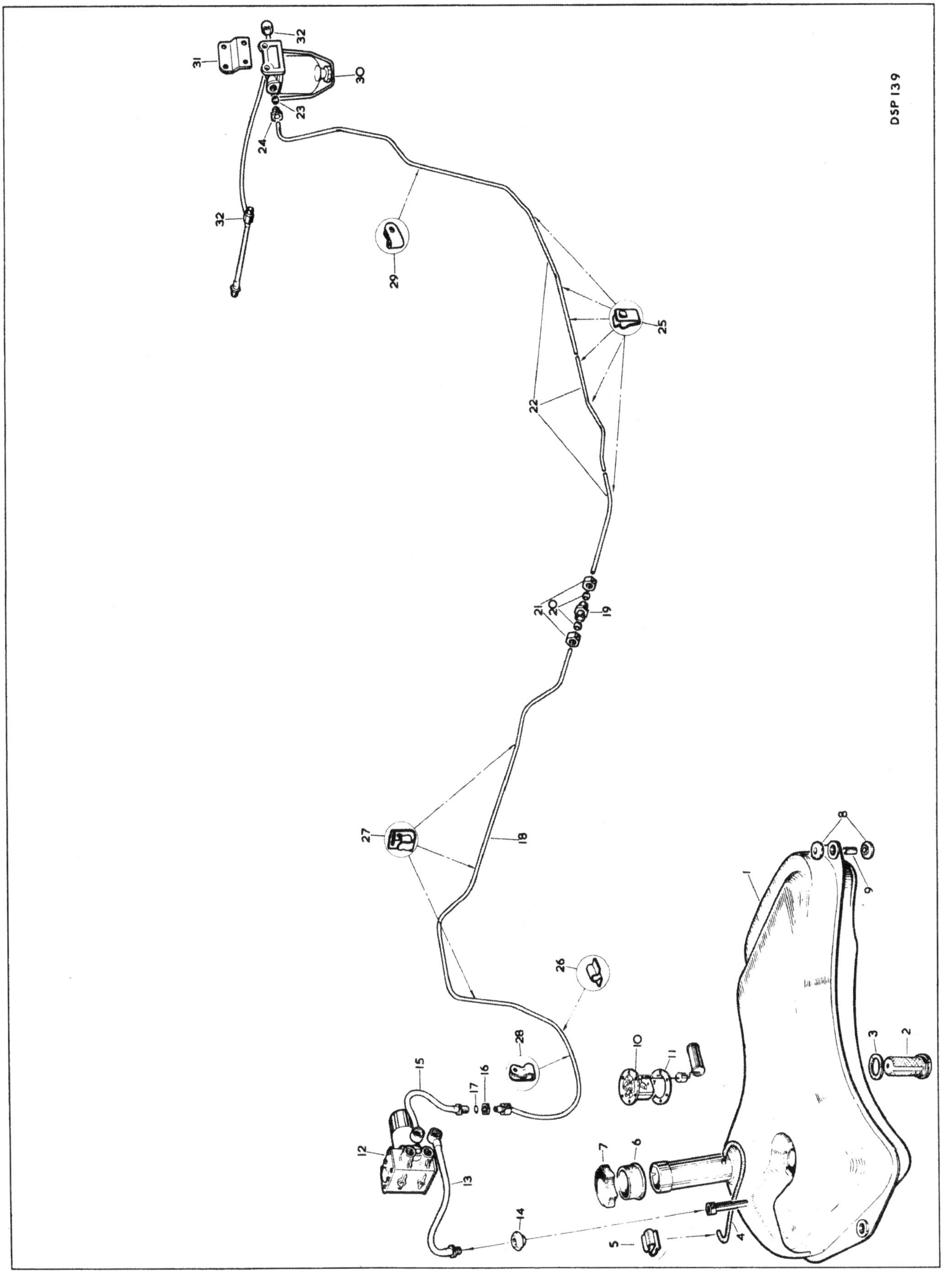
DSP 139
1
2
3
4
5
6
7
8
9
10
11
12
13
14
15
16
17
18
19
20
21
22
23
24
25
26
27
28
29
30
31
32

PLATE 42

BODY SHELL

Part No.	Description	Plate No.	No. per Unit	Remarks
BD.24281	**BODY SHELL ASSEMBLY (371/2950)**		1	
9963	**Underframe Assembly** (371/2024)		1	
8126	**Body Side Assembly**, R.H. (371/1010)		1	
8127	**Body Side Assembly**, L.H. (371/1011)		1	
8121	**Roof Panel Assembly** (371/1003)		1	
8007/A	**Front Wing Assembly**, R.H. (40710/1140)		1	**Superseded by 11141, 11142, and 11143**
8008/A	**Front Wing Assembly**, L.H. (40710/1141)		1	
11141	**Front Wing Assembly**, R.H. (40710/1144)		1	**Supersedes 8007/A and 8008/A**
11142	**Front Wing Assembly**, L.H. (40710/1145)		1	
11143	**Panel**, joining Front Wings at top (10710/1251)		1	
10529	**Baffle**, for R.H. Front Wing (10710/0186)		1	
10530	**Baffle**, for L.H. Front Wing (10710/0187)		1	
8141	**Petrol Filler Box Assembly** (371/221)		1	
6626	**Reinforcement Assembly**, Bonnet Lock (371/160)		1	
8027	**Tonneau Panel**, Side, R.H. (371/230)		1	
8028	**Tonneau Panel**, Side, L.H. (371/231)		1	
8142	**Reinforcement**, Back Light (371/1126)		1	
8145	**Reinforcement**, closing Cantrail Centre, R.H. (171/1128)		1	
8146	**Reinforcement**, closing Cantrail Centre, L.H. (171/1129)		1	
8147	**Panel Assembly**, Windscreen Header (371/1116)		1	
6627	**Closing Plate**, Valance to Wing, Joining Front Wings under Radiator Grille opening (171/758)		1	
6665	**Capping Strip** for Side of Front Wings (271/142)		2	
6666	**Capping Strip** for Front of Front Wings (271/143)		1	
6602	**Stay Bracket**, for R.H. Front Wing (171/238)		1	
6603	**Stay Bracket**, for L.H. Front Wing (171/239)		1	
6604	**Gusset**, reinforcing R.H. Stay Bracket (171/258)		1	
6605	**Gusset**, reinforcing L.H. Stay Bracket (171/259)		1	
8151	**Reinforcement Assembly**, for Facia Panel (371/1114)		1	
8152	**Lower Rail**, at R.H. side of Windscreen (371/1134)		1	
8153	**Lower Rail**, at L.H. side of Windscreen (371/1135)		1	
8437	**Closing Panel**, at R.H. side of Windscreen (171/1166)		1	
8438	**Closing Panel**, at L.H. side of Windscreen (171/1167)		1	
5100	**Bonnet Assembly** (371/014)		1	
8003	**Door Shell**, R.H. Front (371/1040)		1	
8004	**Door Shell**, L.H. Front (371/1041)		1	
8005	**Door Shell**, R.H. Rear (371/1046)		1	
8006	**Door Shell**, L.H. Rear (371/1047)		1	
BD.10404	**Panel Assembly**, Spare Wheel Cover (371/076)		1	
5112	**Trunk Lid Assembly** (371/085)		1	
5117	**Lid Assembly**, Scuttle Ventilator (371/100)		1	
BD.20652	**Gauze**, for Scuttle Ventilator Lid		1	
6635	**Operating Arm Assembly**, for Scuttle Ventilator Lid (271/200)		1	
7988	**Distance Piece**, for Scuttle Ventilator Lid (271/230)		2	
5118	**Door Assembly**, Petrol Filler (371/101)		1	
5125	**Striker Assembly**, Trunk Lid (371/130)		1	
5141	**Check Arm Assembly**, Rear Doors (271/100)		2	
8167	**Check Arm Assembly**, Front Doors (271/1100)		2	
7024	**Pin**, securing Check Arms (271/210)		4	
5144	**Hinge Assembly**, for R.H. side of Bonnet (271/116)		1	
5145	**Hinge Assembly**, for L.H. side of Bonnet (271/117)		1	
5147	**Hinge Assembly**, for Trunk Lid (271/144)		2	
5148	**Return Spring**, for Bonnet Catch (271/147)		1	
6743	**Peg Assembly**, for Bonnet Catch (271/148)		1	
6730	**Safety Hook Assembly**, for Bonnet Catch (271/153)		1	
6661	**Base Plate Assembly**, for Bonnet Catch (271/155)		1	
BD.12852	**Striker Plate**, for Bonnet Catch (271/158)		1	
BD.12853	**Catch Plate**, for Bonnet Catch (271/159)		1	
6664	**Distance Piece**, for Bonnet Catch (271/160)		2	
780	**Spring**, for Safety Hook of Bonnet Catch (253/820)		1	
782	**Pin**, for Safety Hook of Bonnet Catch (253/821)		1	
1504	**Return Spring**, for Petrol Filler Door (241/201)		1	
6716	**Attachment Bracket**, for Rear Wheel Covers (171/207)		2	
BD.17809	**Striker**, for R.H. Door Locks		2	
BD.17808	**Striker**, for L.H. Door Locks		2	
BD.18098	**Shakeproof Shim**, for Strikers (WB.3/10455)		4	
7358	**Cover Plate**, Dash Side (171/235)		2	
6721	**Cover Plate**, Valance to Grille opening (171/327)		1	
BD.17083	**Extension Panel**, for L.H. Exhaust Tail Pipe Cut-out		1	
BD.26186	**Extension Panel**, for R.H. Exhaust Tail Pipe Cut-out		1	

Part No.	Description	Plate No.	No. per Unit	Remarks
	BODY SHELL ASSEMBLY (371/2950) *(contd.)*			
BD.30083	**Bracket Assembly**, Inner, fixing Rear Bumper Mountings		2	Fitted to Chassis No. 1K.1001 and subs. 1K.30001 and subs.
BD.30088	**Bracket Assembly**, Outer, fixing Rear Bumper Mountings		2	
C.24119/2	**Nut**, Self-Locking, securing Brackets to Body		8	
BD.30068	**Extension Panel**, below Rear Bumper		1	
	SUPPLEMENTARY ITEMS NECESSARY TO COMPLETE BODY SHELL ASSEMBLY			
BD.15588/10	**Plastic Plug**, sealing hole in Rear Wheel Arches		2	
C.4928	**Plug**, for Jack Sockets, in Side Members		4	
C.13931	**Plug**, for redundant Accelerator fixing holes in Centre Dash		1	L.H. DRIVE CARS ONLY
BD.15588/5	**Plug**, for redundant Accelerator fixing holes in Centre Dash		2	R.H. DRIVE CARS ONLY
C.21811	**Heat Shield**, for Exhaust System, under R.H. Front Floor		1	
C.21812	**Heat Shield**, for Exhaust System, under L.H. Front Floor		1	
C.21813	**Heat Shield**, for Exhaust System, under Main Floor		2	
C.21814	**Heat Shield**, for Exhaust System, under R.H. Rear Seat Pan		1	
C.18322	**Heat Shield**, for Exhaust System, under L.H. Rear Seat Pan		1	
BD.24429	**Bracket**, on Dash, supporting Pedal Shaft Steady Bracket		1	
BD.29992	**COWL, ON OUTSIDE OF R.H. FRONT WING VALANCE, FOR MOUNTING OF BRAKE SERVO**		1	
UFS.125/4R	**Screw**, Set, securing Cowl to Wing Valance		4	Superseded by UFS.825/5H, C.724, DJZ.812/10C, BD.8536/3, and FW.104/T
UFS.125/5R	**Screw**, Set, securing Cowl to Wing Valance		4	
FW.104/T	**Washer**, Plain, on Setscrews		8	
C.724	**Washer**, Shakeproof, on Setscrews		8	
UFS.825/5H	**Screw**, Set securing Cowl to Wing Valance		2	Supersedes UFS.125/4R, UFS.125/5R FW.104/T, and C.724
C.724	**Washer**, Shakeproof, on Setscrews		2	
DJZ.812/10C	**Screw**, Self-Tapping, securing Cowl to Wing Valance		5	
BD.8536/3	**Spire Nut**, receiving Self-Tapping Screws		5	
FW.104/T	**Washer**, Plain, under Shakeproof Washers and on Self-Tapping Screws		7	
BD.24198	**GEARBOX COVER ASSEMBLY**		1	FITTED TO CARS WITH AUTOMATIC TRANSMISSION
BD.30277	**GEARBOX COVER**		1	REQUIRED FOR CARS FITTED WITH MANUAL OR OVERDRIVE TRANSMISSION. Fitted to Chassis No. 1A.12581 and subs. 1A.20565 and subs. also fitted to 1K.1001 and subs. 1K.30001 and subs.
DJZ.412/10C	**Screw**, Self-Tapping, securing Gearbox Cover to Floor		12	
BD.542/3	**Washer**, Plain, on Self-Tapping Screws		12	
BD.8536/3	**Spire Nut**, on Self-Tapping Screws		12	
BD.30986	**RUBBER GROMMET, ON GEARBOX COVER, FOR CHANGE-SPEED LEVER**		1	
BD.30985	**Retaining Ring**, holding Grommet to Gearbox Cover		1	
DAJ.406/10C	**Screw**, Self-Tapping, securing Retaining Ring to Gearbox Cover		6	
BD.11789	**SEAL ASSEMBLY FOR DIAPHRAGM AT REAR OF L.H. FRONT WING**		1	
BD.10229	**Rubber Seal**		1	
BD.10406	**Retainer**, for Seal (171/237)		1	
BD.25751	**Staple**, securing Seal to Retainer		4	

BODY SHELL

Part No.	Description	Plate No.	No. per Unit	Remarks
BD.11790	**SEAL ASSEMBLY FOR DIAPHRAGM AT REAR OF R.H. FRONT WING**		1	
BD.10229	**Rubber Seal**		1	
BD.10405	**Retainer,** for Seal (171/236)		1	
BD.25751	**Staple,** securing Seal to Retainer		4	
BD.1814/2	**Pop Rivet,** securing Retainers		12	
BD.17581	**DEFLECTOR PLATE SHIELDING FRONT WINDOW REGULATOR FROM WATER**		2	
BD.1814/2	**Pop Rivet,** securing Deflector Plates to Front Doors		4	
	SEALING PLUGS			
BD.15588/1	**Plastic Plug,** sealing Drain Holes (3/8" Diameter)		3	
BD.15588/10	**Plastic Plug,** sealing holes in Rear Wheel Arches		2	
BD.23386	**Steel Plug,** sealing Drain Holes (3/8" Diameter)		6	
BD.25150	**Steel Plug,** sealing Drain Holes (3/8" Diameter)		16	
BD.19721	**Steel Plug,** sealing Drain Holes (7/8" Diameter)		7	
BD.23387	**Steel Plug,** sealing Drain Holes (7/8" Diameter)		4	
BD.14771	**Rubber Plug,** in R.H. Side of Dash (1" Diameter)		1	

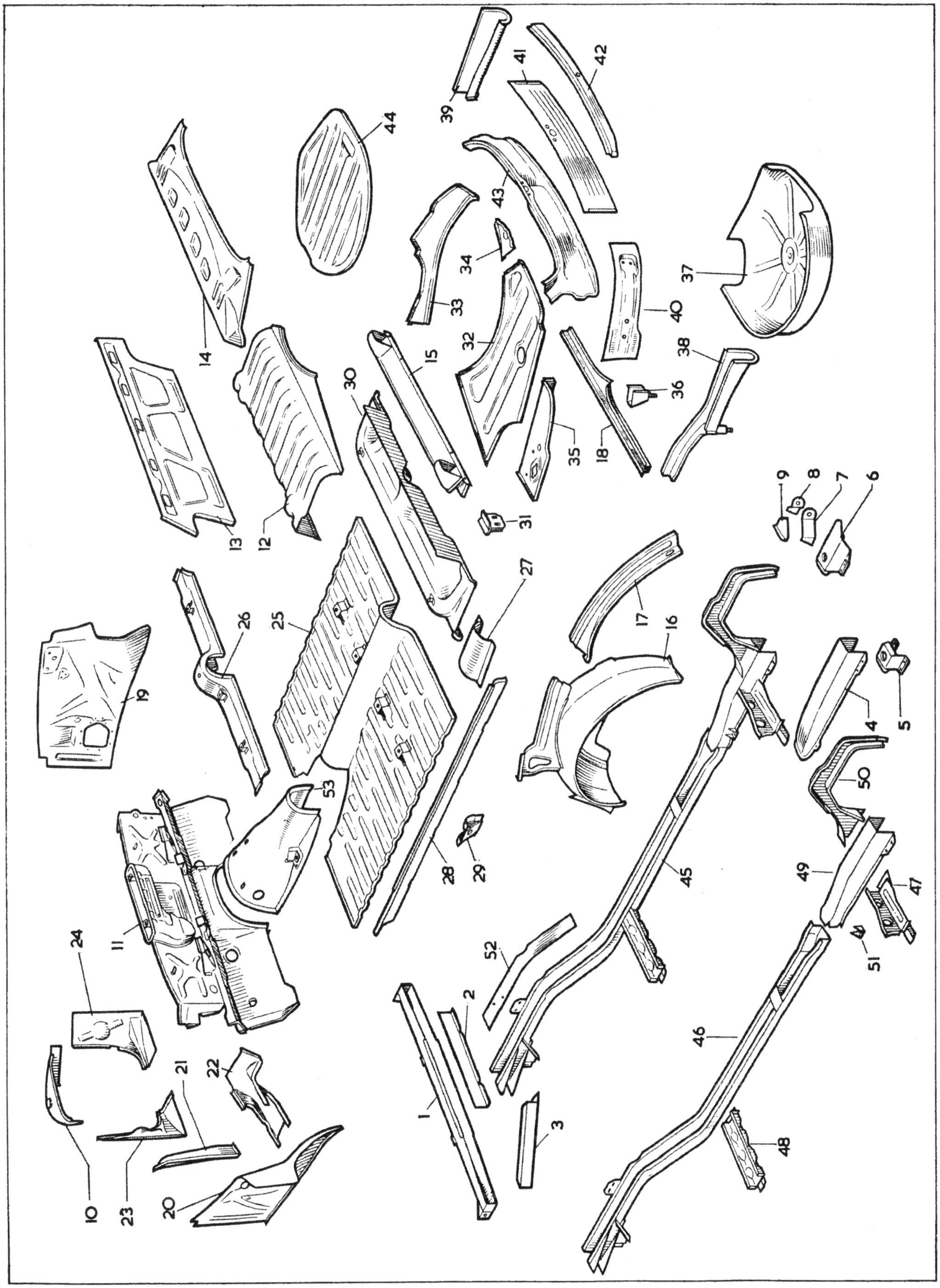
1
2
3
4
5
6
7
8
9
10
11
12
13
14
15
16
17
18
19
20
21
22
23
24
25
26
27
28
29
30
31
32
33
34
35
36
37
38
39
40
41
42
43
44
45
46
47
48
49
50
51
52
53

PLATE 43

BODY UNDERFRAME

Part No.	Description	Plate No.	No. per Unit	Remarks
9963	**UNDERFRAME ASSEMBLY (30710/2024)**		1	
5101	**Side Member Assembly,** R.H. (371/022)		1	
5102	**Side Member Assembly,** L.H. (371/023)		1	
6732	**Closing Plate,** for Underframe Side Member, R.H. (371/166)		1	
6733	**Closing Plate,** for Underframe Side Member, L.H. (371/167)		1	
5137	**Cross Member Assembly,** Front (371/716)	43–1	1	
4130	**Closing Plate,** for Front Cross Member (171/754)	43–2	1	
4128	**Brace,** at R.H. side of Front Cross Member (171/754)		1	
4129	**Brace,** at L.H. side of Front Cross Member (171/753)	43–3	1	
6582	**Reinforcing Channel,** for R.H. Rear Road Spring (171/710)		1	
6583	**Reinforcing Channel,** for L.H. Rear Road Spring (171/711)	43–4	1	
6741	**Upper Clamp Assembly,** for L.H. Rear Road Spring (371/745)	43–5	1	
6742	**Upper Clamp Assembly,** for R.H. Rear Road Spring (371/748)	43–6	1	
6727	**Inner Bracket,** for mounting Panhard Rod (171/828)	43–7	1	**SUPPLIED AS AN ASSEMBLY TO PART NO. 8933 (371/1174)**
6726	**Outer Bracket,** for mounting Panhard Rod (171/827)	43–8	1	
6728	**Top Plate,** for Panhard Rod Bracket (171/833)	43–9	1	
7081	**Valance Assembly,** under R.H. Front Wing (471/100)		1	
7082	**Valance Assembly,** under L.H. Front Wing (471/101)		1	
7359	**Filler Panel,** Valance to Radiator Grille opening, Upper (171/348)	43–10	1	
9722	**Dash Centre Assembly** (371/2027)	43–11	1	
10019	**Front Floor Assembly** (30710/2052)		1	
4121	**Cover Panel,** inside Trunk, over Rear Axle (171/715)	43–12	1	
4120	**Support Panel,** for Rear Seat Squab (171/218)	43–13	1	
171/303	**Bracket,** fixing Support Panel		2	
4116	**Panel,** Parcel Shelf (171/206)	43–14	1	
271/146	**Seal,** for Parcel Shelf Panel		2	
5134	**Rear Cross Member Assembly** (371/704)	43–15	1	
8113	**Rear Wheel Arch Assembly,** R.H. (371/216)		1	
8114	**Rear Wheel Arch Assembly,** L.H. (371/217)	43–16	1	
4114	**Reinforcement Panel,** Trunk Side, R.H. (171/172)		1	
4115	**Reinforcement Panel,** Trunk Side, L.H. (171/173)	43–17	1	
8115	**Trunk and Spare Wheel Floor Assembly** (371/220)		1	
4125	**Cross Member,** for Front of Trunk Floor (171/721)	43–18	1	
6835	**Spring Clip,** under Trunk Floor, fixing Spare Wheel Cover (271/701)		1	
6836	**Support Bracket Assembly,** for Trunk Lid Hinges (371/131)		2	
171/283	**Angle,** at L.H. side of Trunk, fixing Petrol Pump		1	
7311	**Clip,** securing Electrical Harness (171/298)		6	
7324	**Angle,** Jack Mounting (171/315)		1	
7325	**Bracket,** Jack Mounting (171/318)		1	
7378	**VALANCE ASSEMBLY, UNDER R.H. FRONT WING (371/194)**	43–19	1	
7081	**Valance Panel,** R.H. (471/100)		1	
8110	**VALANCE ASSEMBLY, UNDER L.H. FRONT WING (371/1157)**	43–20	1	
7082	**Valance Panel,** L.H. (471/101)		1	
7244	**Reinforcing Plate,** R.H. Side Member (171/848)		1	
7245	**Reinforcing Plate,** L.H. Side Member (171/849)		1	
6729	**Clip,** Air Intake Pipe (171/858)		1	
6694	**Bracket,** for attachment of Front Brake Hoses (171/864)		2	
7092	**Clip,** Cable Harness (153/855)		12	
7811	**Clip,** Cable Harness (171/298)		2	
7314	**Weld Nut** (UWN.163/HG)		11	
7313	**Weld Nut** (UWN.103/HG)		5	
7312	**Clinch Nut** (UNF.201/X)		6	
6749	**Support,** for Radiator Block at R.H. side (371/140)		1	
6750	**Support,** for Radiator Block at L.H. side (371/141)	43–21	1	

Part No.	Description	Plate No.	No. per Unit	Remarks
	UNDERFRAME ASSEMBLY (30710/2024) *(contd.)*			
6717	**Longitudinal Stiffener** to R.H. 'A' Post (171/226)		1	
6718	**Longitudinal Stiffener** to L.H. 'A' Post (171/227)	43–22	1	
7220	**Filler Panel**, from L.H. Wing Valance to Radiator Grille opening, Lower (171/347)	43–23	1	
7219	**Filler Panel**, from R.H. Wing Valance to Radiator Grille opening, Lower (171/346)	43–24	1	
6719	**Angle**, R.H., for attachment of Facia reinforcement (171/294)		1	
6720	**Angle**, L.H., for attachment of Facia reinforcement (171/295)		1	
6895	**Reinforcement**, Dash Front, R.H. (171/178)		1	
6896	**Reinforcement**, Dash Front, L.H. (171/179)		1	
10019	**FRONT FLOOR ASSEMBLY (30710/2052)**			
4103	**Panel only**, Front Floor (171/052)	43–25	1	NOT SERVICED
10021	**Intermediate Cross Member Assembly** (31680/0715)	43–26	1	
10020	**Rear Seat Pan Assembly** (30710/2066)		1	
7901	**Reinforcement**, Floor to Seat Pan (171/862)	43–27	1	
7003	**Inner Sill Panel**, R.H. (171/704)		1	
7032	**Inner Sill Panel**, L.H. (171/705)	43–28	1	
7004	**Diaphragm**, for Inner Sill Panels (171/766)	43–29	2	
6986	**Bracket**, R.H., supporting Handbrake Shaft (371/734)		1	
6987	**Bracket**, L.H., supporting Handbrake Shaft (371/735)		1	
10020	**REAR SEAT PAN ASSEMBLY (30710/2066)**	43–30	1	
BD.25142	**Bracket**, for Rear Brake Hose		1	
BD.25570	**Bracket Assembly**, mounting Exhaust Silencers	43–31	2	
8115	**TRUNK FLOOR AND SPARE WHEEL CONTAINER ASSEMBLY (371/220)**		1	
4124	**Inner Panel**, Trunk Floor, L.H. (171/719)	43–32	1	
8118	**Inner Panel**, Trunk Floor, R.H. (171/882)	43–33	1	
6735	**Rear Panel Assembly**, Trunk Floor (371/709)	43–34	1	
8117	**Outer Panel**, Trunk Floor, L.H. (171/881)	43–35	1	
6740	**Bracket Assembly**, mounting Petrol Tank, Front (371/725)	43–36	1	
4104	**Spare Wheel Container** (171/074)	43–37	1	
5138	**Longitudinal Member Assembly**, at L.H. side of Trunk Floor (371/717)	43–38	1	
4126	**Longitudinal Member Assembly**, at R.H. side of Trunk Floor (171/722)	43–39	1	
8119	**Tonneau Side Panel Assembly**, at R.H. side of Trunk Floor (371/218)		1	
8120	**Tonneau Side Panel Assembly**, at L.H. side of Trunk Floor (371/219)	43–40	1	
7689	**Reinforcement**, at R.H. side of Trunk Floor (171/724)		1	
4127	**Reinforcement**, for Lower Rear Body Panel (171/725)	43–41	1	
6038	**Rear Body Panel**, Lower (171/002)	43–42	1	
6734	**Panel Assembly**, at rear of Spare Wheel Floor (371/702)	43–43	1	
7360	**Bracket Assembly**, for attachment of Rear Bumper (371/706)		2	
6984	**Fixing**, for Spare Wheel (271/711)		1	
6608	**Spring Latch**, on Trunk Floor, fixing Spare Wheel Cover (271/181)		1	
TLP/D/BS419	**Pop Rivet**, securing Spring Latch		2	
BD.10404	**SPARE WHEEL COVER ASSEMBLY (371/076)**	43–44	1	
BD.10958	**Stud**, fixing Spare Wheel Cover (271/182)		1	
C.723/A	**Washer**, Shakeproof, on Stud		1	

BODY UNDERFRAME

Part No.	Description	Plate No.	No. per Unit	Remarks
5101	**R.H. SIDE MEMBER ASSEMBLY (371/022)**		1	
5102	**L.H. SIDE MEMBER ASSEMBLY (371/023)**	43–45	1	
7079	**Side Member only,** R.H. (471/102)		1	
7080	**Side Member only,** L.H. (471/103)	43–46	1	
6736	**Rear Jacking Tube Assembly,** for R.H. Side Member (371/712)		1	
6737	**Rear Jacking Tube Assembly,** for L.H. Side Member (371/713)	43–47	1	
6738	**Front Jacking Tube Assembly,** for R.H. Side Member (371/714)		1	
6739	**Front Jacking Tube Assembly,** for L.H. Side Member (371/715)	43–48	1	
6722	**Longitudinal Member,** at rear of R.H. Side Member (171/700)		1	
6723	**Longitudinal Member,** at rear of L.H. Side Member (171/701)	43–49	1	
6724	**Extension,** at rear of R.H. Side Member (171/702)		1	
6725	**Extension,** at rear of L.H. Side Member (171/703)	43–50	1	
C.11889	**Bracket,** Front Mounting Plate Rear Springs (171/852)	43–51	4	
6584	**Bracket,** fixing Radiator at R.H. side (371/726)		1	
6585	**Bracket,** fixing Radiator at L.H. side (371/727)		1	
6648	**Bracket,** mounting Front Suspension (371/754)		2	
6732	**Closing Plate,** for Underframe Side Member, R.H. (371/166)		1	
6733	**Closing Plate,** for Underframe Side Member, L.H. (371/167)	43–52	1	
BD.24283	**Reinforcing Bracket,** for Panhard Rod Mounting		1	
BD.24198	**GEARBOX COVER ASSEMBLY**	43–53	1	

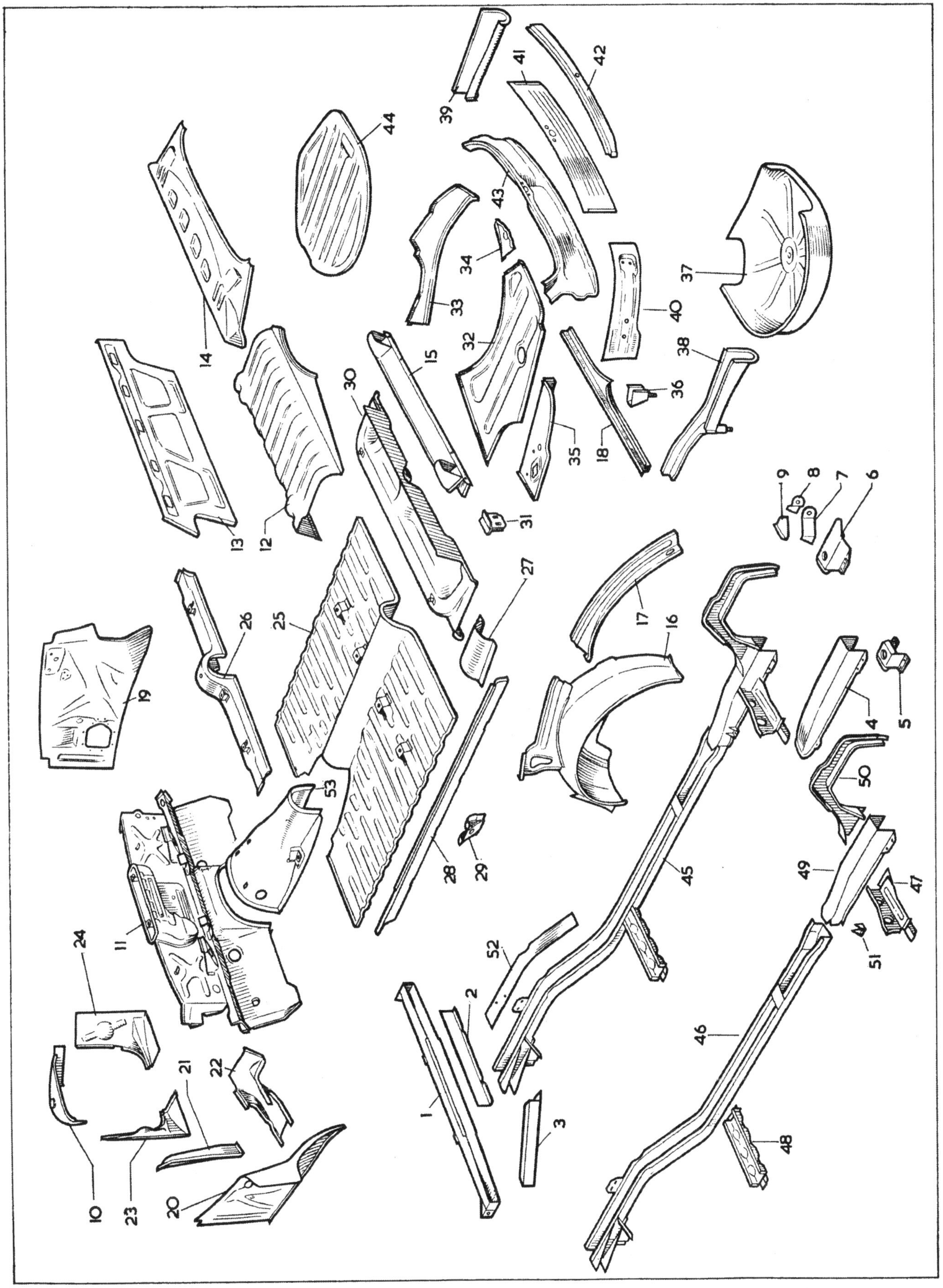
1
2
3
4
5
6
7
8
9
10
11
12
13
14
15
16
17
18
19
20
21
22
23
24
25
26
27
28
29
30
31
32
33
34
35
36
37
38
39
40
41
42
43
44
45
46
47
48
49
50
51
52
53

PLATE 43

BODY PANELS

Part No.	Description	Plate No.	No. per Unit	Remarks
8121	**ROOF PANEL ASSEMBLY (371/1003)**	44–1	1	
8122	**DRIP MOULDING AT FRONT R.H. SIDE OF ROOF (171/1152)**		1	
8123	**DRIP MOULDING AT FRONT L.H. SIDE OF ROOF (171/1153)**		1	
8124	**DRIP MOULDING AT REAR R.H. SIDE OF ROOF (171/1154)**		1	
8125	**DRIP MOULDING AT REAR L.H. SIDE OF ROOF (171/1155)**		1	
8126	**BODY SIDE ASSEMBLY, R.H. (371/1010)**		1	
8127	**BODY SIDE ASSEMBLY, L.H. (371/1011)**		1	
8128	**"A" Post Panel Assembly,** R.H. (371/1128)		1	
8129	**"A" Post Panel Assembly,** L.H. (371/1129)		1	
8130	**"B" and "C" Post Panel Assembly,** R.H. (371/1130)		1	
8131	**"B" and "C" Post Panel Assembly,** L.H. (371/1131)		1	
8132	**Closing Plate,** for R.H. "B" and "C" Post (171/1160)		1	
8133	**Closing Plate,** for L.H. "B" and "C" Post (171/1161)		1	
8134	**"D" Post Panel Assembly,** R.H. (371/1132)		1	
8135	**"D" Post Panel Assembly,** L.H. (371/1133)		1	
4109	**Door Sill Panel,** R.H. (171/112)		1	
4110	**Door Sill Panel,** L.H. (171/113)		1	
8136	**Cantrail Panel,** R.H. (171/1158)		1	
8137	**Cantrail Panel,** L.H. (171/1159)		1	
6786	**Dash Side Panel Assembly,** R.H. (371/028)		1	
6787	**Dash Side Panel Assembly,** L.H. (371/029)		1	
7712	**Retainer,** for R.H. Door Seals (171/166)		3	
7713	**Retainer,** for L.H. Door Seals (171/167)		3	
8122	**Drip Moulding,** Front Portion, R.H. (171/1152)		1	
8124	**Drip Moulding,** Rear Portion, R.H. (171/1154)		1	
8123	**Drip Moulding,** Front Portion, L.H. (171/1153)		1	
8125	**Drip Moulding,** Rear Portion, L.H. (171/1155)		1	
8128	**"A" POST PANEL ASSEMBLY, R.H. (371/1128)**		1	
8912	**"A" POST PANEL ASSEMBLY, L.H. (371/1129)**	44–2	1	
8130	**"B" AND "C" POST PANEL ASSEMBLY, R.H. (371/1130)**		1	
6981	**Bracket,** on "B" and "C" Post, fixing Door Check Strap (171/118)		1	
UWN.203/HG	**Weld Nut**		5	
171/1199	**Tapping Plate,** for Front Door Lock Striker		1	
171/1200	**Retainer,** for Tapping Plate on R.H. "B" and "C" Post		1	
8131	**"B" AND "C" POST PANEL ASSEMBLY, L.H. (371/1131)**	44–3	1	
6981	**Bracket,** on "B" and "C" Post, fixing Door Check Strap		1	
UWN.203/HG	**Weld Nut**		5	
171/1199	**Tapping Plate,** for Front Door Lock Striker		1	
171/1201	**Retainer,** for Tapping Plate on L.H. "B" and "C" Post		1	

Part No.	Description	Plate No.	No. per Unit	Remarks
8132	**CLOSING PLATE FOR R.H. "B" AND "C" POST (171/1160)**		1	
8133	**CLOSING PLATE FOR L.H. "B" AND "C" POST (171/1161)**	44–4	1	
8134	**"D" POST PANEL ASSEMBLY, R.H. (371/1132)**		1	
6857	**Spring,** for Dzus Fastener Securing R.H. Rear Wheel Cover (S.6/250)		2	
275	**Rivet,** securing Springs (RS.083/R.05)		4	
8138	**Tapping Plate,** for Rear Door Lock Striker (171/1199)		1	
8139	**Retainer,** for Tapping Plate (171/1200)		1	
8135	**"D" POST PANEL ASSEMBLY, L.H. (371/1133)**	44–5	1	
6857	**Spring,** for Dzus Fastener securing L.H. Rear Wheel Cover (S.6/250)		2	
275	**Rivet,** securing Springs (RS.083/R.05)		4	
8138	**Tapping Plate,** for Rear Door Lock Striker (171/1199)		1	
8140	**Retainer,** for Tapping Plate (171/1201)		1	
4109	**DOOR SILL PANEL, R.H. (171/112)**		1	
4110	**DOOR SILL PANEL, L.H. (171/113)**	44–6	1	
8136	**CANTRAIL PANEL, R.H. (371/1168)**		1	
8137	**CANTRAIL PANEL, L.H. (371/1169)**	44–7	1	
8437	**Closing Panel,** Windscreen side, R.H. (171/1166)		1	
8438	**Closing Panel,** Windscreen side, L.H. (171/1167)		1	
6786	**DASH SIDE PANEL ASSEMBLY, R.H. (371/028)**		1	
6787	**DASH SIDE PANEL ASSEMBLY, L.H. (371/029)**	44–8	1	
8007/A	**R.H. FRONT WING ASSEMBLY (40710/1140)**		1	SUPERSEDED BY 11141, 11142, AND 11143
7092	**Clip,** for Harness to Lamps (153/855)		1	SUPERSEDED BY 11141, 11142, AND 11143
8008/A	**L.H. FRONT WING ASSEMBLY (40710/1141)**	44–9	1	SUPERSEDED BY 11141, 11142, AND 11143
7092	**Clip,** for Harness to Lamps (153/855)		1	SUPERSEDED BY 11141, 11142, AND 11143
11141	**R.H. FRONT WING ASSEMBLY (40710/1144)**		1	SUPERSEDES 8007/A AND 8008/A
11142	**L.H. FRONT WING ASSEMBLY (40710/1145)**		1	SUPERSEDES 8007/A AND 8008/A
11143	**Panel,** joining Front Wings at top (10710/1251)		1	SUPERSEDES 8007/A AND 8008/A

BODY PANELS

Part No.	Description	Plate No.	No. per Unit	Remarks
	L.H. FRONT WING ASSEMBLY (40710/1145) *(contd.)*			
10529	**Baffle,** for R.H. Front Wing (10710/0186)		1	
10530	**Baffle,** for L.H. Front Wing (10710/0187)		1	
6721	**ACCESS PANEL FOR RADIATOR GRILLE (171/327)**	44–9A	1	
8141	**PETROL FILLER BOX ASSEMBLY (371/221)**	44–10	1	
5168	**Hinge Bracket and Arms Assembly** (371/137)		1	
UMF.161/H.08	**Screw,** Set, securing Hinge Bracket		2	
WP.161	**Washer,** Plain, on Setscrews		2	
WL.161	**Lock Washer,** on Setscrews		2	
5118	**PETROL FILLER DOOR ASSEMBLY ON L.H. REAR WING (371/101)**	44–11	1	
1504	**Return Spring,** for Petrol Filler Door (241/201)		1	
6626	**REINFORCEMENT PANEL FOR BONNET LOCK (371/160)**	44–12	1	
BD.12855	**RETURN SPRING FOR BONNET CATCH (271/147)**		1	
BD.12842	**PEG, CENTRING BONNET CATCH (271/148)**		1	
BD.12852	**STRIKER PLATE FOR BONNET CATCH (271/158)**		1	
BD.12853	**CATCH PLATE FOR BONNET CATCH (271/159)**		1	
BD.18303	**BASE PLATE FOR BONNET CATCH (271/155)**		1	
BD.12584	**SPACING BUSH FOR BONNET CATCH (271/160)**		2	
6730	**SAFETY HOOK FOR BONNET CATCH (271/153)**		1	
780	**TENSION SPRING FOR SAFETY HOOK ON BONNET CATCH (253/820)**		1	
BD.12860	**PIVOT PIN SECURING SAFETY HOOK TO BONNET (253/821)**		1	
8027	**R.H. TONNEAU SIDE PANEL ASSEMBLY (371/230)**		1	
8028	**L.H. TONNEAU SIDE PANEL ASSEMBLY (371/231)**	44–13	1	
BD.17083	**REAR EXTENSION PANEL FOR L.H. EXHAUST TAIL PIPE CUT-OUT**	44–14	1	

Part No.	Description	Plate No.	No. per Unit	Remarks
BD.26186	**EXTENSION PANEL ASSEMBLY, BELOW BUMPER, FOR R.H. EXHAUST TAIL PIPE CUT-OUT**	44–15	1	
8142	**REINFORCEMENT PANEL FOR TOP OF BACKLIGHT (371/1126)**	44–16	1	
8877	**REINFORCEMENT PANEL FOR BOTTOM OF BACKLIGHT (171/1125)**	44–16A	1	
8143	**REINFORCEMENT PANEL, CLOSING R.H. "D" POST (171/1148)**		1	
8144	**REINFORCEMENT PANEL, CLOSING L.H. "D" POST (171/1149)**	44–17	1	
8145	**REINFORCEMENT PANEL, CLOSING CANTRAIL AT R.H. CENTRE (171/1128)**		1	
8146	**REINFORCEMENT PANEL, CLOSING CANTRAIL AT L.H. CENTRE (171/1129)**	44–18	1	
8147	**WINDSCREEN HEADER PANEL ASSEMBLY (371/1116)**	44–19	1	
8148	**Tapping Plate**, fixing Interior Mirror (271/241)		1	
8149	**Tapping Plate**, fixing Sun Visors (171/180)		2	
8150	**Retainer**, for Sun Visor Tapping Plates (171/1116)		2	
6627	**CLOSING PLATE, FROM VALANCE, JOINING FRONT WINGS UNDER RADIATOR GRILLE OPENING (171/758)**	44–20	1	
6665	**CAPPING STRIP AT SIDE OF FRONT WINGS (271/142)**	44–21	2	
6666	**CAPPING STRIP AT FRONT OF FRONT WINGS (271/143)**	44–22	1	
6602	**STAY BRACKET, FOR R.H. FRONT WING (171/238)**		1	
6603	**STAY BRACKET, FOR L.H. FRONT WING (171/239)**	44–23	1	
6604	**GUSSET, REINFORCING R.H. STAY BRACKET (171/258)**		1	
6605	**GUSSET, REINFORCING L.H. STAY BRACKET (171/259)**	44–24	1	
8151	**REINFORCING PANEL ASSEMBLY FOR FACIA (371/1114)**	44–25	1	

BODY PANELS

Part No.	Description	Plate No.	No. per Unit	Remarks
8152	**RAIL AT BOTTOM R.H. SIDE OF WINDSCREEN (371/1134)**		1	
8153	**RAIL AT BOTTOM L.H. SIDE OF WINDSCREEN (371/1135)**	44–26	1	
BD.18734	**Bracket,** mounting Flasher Unit (271/1141)		1	
BD.25278	**BONNET, COMPLETE**	44–27	1	
7381	**Reinforcement Assembly,** at Front of Bonnet (371/103)		1	
5144	**HINGE ASSEMBLY AT R.H. SIDE OF BONNET (271/116)**		1	
5145	**HINGE ASSEMBLY AT L.H. SIDE OF BONNET (371/117)**		1	
BD.12751	**Return Spring,** for Bonnet Hinges		2	
10040	**Bolt,** securing Hinges to Bonnet (UCF.201/H.08)		4	
10041	**Bolt,** securing Hinges to Body (UMF.201/H.10)		6	
9520	**Washer,** Plain, on Bolts (WP.201)		10	
7935	**Washer,** Shakeproof, on Bolts (WS.201)		10	
8003/A	**R.H. FRONT DOOR SHELL ASSEMBLY (371/1040)**		1	
171/1112	**Bracket,** supporting Glass Channels at rear		1	
8158	**Felt,** in Bracket, for Glass Stop (271/1116)		1	
6573	**Guide Spring,** for Front Door Check Arm (269/116)		1	
7322	**Rivet,** semi-tubular, securing Guide Spring (RT.122/007)		2	
7021	**Tapped Plate,** fixing Lock on R.H. Front Door (271/132)		1	
7321	**Reinforcement Bracket,** for Check Arm (171/323)		1	
8004/A	**L.H. FRONT DOOR SHELL ASSEMBLY (371/1041)**	44–28	1	
7022	**Tapped Plate,** fixing Lock on L.H. Front Door (271/133)		1	
	All other items are as for R.H. Front Door Shell 8003/A			
5154	**Lower Hinge,** for R.H. Front Door (271/216)		1	
5153	**Upper Hinge,** for R.H. Front Door (271/217)		1	
5154	**Upper Hinge,** for L.H. Front Door (271/216)	44–29	1	
5153	**Lower Hinge,** for L.H. Front Door (271/217)	44–30	1	
7959	**Grease Nipple,** on Door Hinges (271/229)		4	
1490	**Bolt,** securing Hinges to Door (230/381)		12	
BD.11806	**Shroud,** for Water Drain Holes in Front Doors		8	
BD.1814/1	**Pop Rivet,** securing Shrouds		16	
BD.17581	**Deflector Plate,** preventing ingress of Water to Window Regulator mechanism		2	
BD.18189	**Sealing Strip** (Flintkote) for Deflector Plate		2	
BD.1814/2	**Pop Rivet,** securing Deflector Plate to Inner Door Panel		4	
6785	**Bolt,** Phillip's Head, securing Door Hinges to 'A' Posts (UCF.203H/QR.08)		8	
6784	**Bolt,** Hexagon Head, securing Door Hinges to 'A' Posts (UCF.201/H.10)		4	
FG.105/X	**Washer,** Spring, on Bolts (WL.201)		4	
BD.18351/1	**Shim,** for Hinge		As req'd	
BD.18351/2	**Shim,** for Hinge		As req'd	

Part No.	Description	Plate No.	No. per Unit	Remarks
8167	**CHECK ARM FOR FRONT DOORS (271/1100)**		2	
7024	**Pin,** securing Check Arm (271/210)		2	
BD.18296	**WINDOW REGULATOR ASSEMBLY IN R.H. FRONT DOOR (271/1102)**		1	
BD.18297	**WINDOW REGULATOR ASSEMBLY IN L.H. FRONT DOOR (271/1103)**	44–31	1	
UFS.425/2.5H	**Screw,** Set, securing Regulators (UMF.161/P.05)		16	
C.724	**Washer,** Shakeproof, on Setscrews		16	
BD.19581	**LOCK ASSEMBLY (LESS REMOTE CONTROL) ON R.H. FRONT DOOR**		1	
BD.19577	**Remote Control and Link Assembly,** for R.H. Door Lock		1	
BD.19580	**LOCK ASSEMBLY (LESS REMOTE CONTROL) ON L.H. FRONT DOOR**	44–32	1	
BD.19576	**Remote Control and Link Assembly,** for L.H. Door Lock	44–32A	1	
BD.19573	**Clip,** securing Link to Lock		2	
BD.19575	**Wavy Washer,** under Links		2	
BD.19574	**Flat Washer,** under Wavy Washers		2	
BD.29609	**Buffer,** on Door Panel, preventing vibration on Remote Control (299/728)		4	
BD.17820/4	**Screw,** Phillips Head, securing Locks		8	
7031	**Screw,** Binding Head, securing Remote Control (SZ.101/B.06)		6	
7206	**Spire Nut,** for Screws (NS.103/U.8)		6	
BD.17808	**STRIKER FOR L.H. FRONT DOOR LOCK**	44–33	1	
BD.17809	**STRIKER FOR R.H. FRONT DOOR LOCK**		1	
BD.18098	**Shakeproof Shim,** for Strikers (WB.3/10455)		2	
BD.17821/7	**Screw,** Set, securing Strikers to "B" and "C" Posts		6	
BD.17338	**Shim,** for Door Strikers		As req'd	
8005/A	**R.H. REAR DOOR SHELL ASSEMBLY (371/1046)**		1	
6573	**Guide Spring,** for Rear Door Check Arm (269/116)		1	
7322	**Rivet,** semi-tubular, securing Guide Spring (RT.122/007)		2	
7321	**Reinforcement Bracket,** for Check Arm (171/323)		1	
7021	**Tapped Plate,** fixing Lock on R.H. Rear Door (271/132)		1	
8006/A	**L.H. REAR DOOR SHELL ASSEMBLY (371/1047)**	44–34	1	
7022	**Tapped Plate,** fixing Lock on L.H. Rear Door (271/133)		1	
	All other items are as for R.H. Rear Door Shell 8005/A			

BODY PANELS

Part No.	Description	Plate No.	No. per Unit	Remarks
	L.H. REAR DOOR SHELL ASSEMBLY (371/1047) *(contd.)*			
5151	**Lower Hinge,** for R.H. Rear Door (271/212)		1	
8163	**Upper Hinge,** for R.H. Rear Door (271/1104)		1	
5152	**Lower Hinge,** for L.H. Rear Door (271/213)	44–35	1	
8164	**Upper Hinge,** for L.H. Rear Door (271/1105)	44–36	1	
7959	**Grease Nipple,** on Door Hinges (271/229)		4	
1490	**Bolt,** securing Lower Hinge to Doors (230/381)		6	
9725	**Screw,** securing Upper Hinges to Doors (UCF.203H/QR.16)		8	
8166	**Shim,** for Upper Hinges (171/1202)		As req'd	
6785	**Bolt,** Phillips Head, securing Rear Door Hinges to "B" and "C" Posts (UCF.203H/QR.08)		12	
BD.11806	**Shroud,** for Water Drain Holes in Rear Doors		8	
BD.1814/1	**Pop Rivet,** securing Shrouds		16	
BD.17990	**Filler Piece,** on Hinge Face, at Waist Level of R.H. Rear Door		1	
BD.17989	**Filler Piece,** on Hinge Face, at Waist Level of L.H. Rear Door		1	
BD.1814/1	**Pop Rivet,** securing Filler Pieces to Doors		4	
5141	**CHECK ARM FOR REAR DOORS (271/100)**		2	
7024	**Pin,** securing Check Arms (271/210)		2	
BD.18298	**WINDOW REGULATOR ASSEMBLY, IN R.H. REAR DOOR (271/1112)**		1	
BD.18299	**WINDOW REGULATOR ASSEMBLY, IN L.H. REAR DOOR (271/1113)**	44–37	1	
UFS.425/2.5H	**Screw,** Set, securing Regulators (UMF.161/P.05)		12	
C.724	**Washer,** Shakeproof, on Setscrews		12	
BD.19583	**LOCK ASSEMBLY (LESS REMOTE CONTROL) ON R.H. REAR DOOR**		1	NOT REQUIRED WITH CHILD'S SAFETY LOCK
BD.19579	**Remote Control and Link Assembly,** for R.H. Door Lock		1	
BD.19582	**LOCK ASSEMBLY (LESS REMOTE CONTROL) ON L.H. REAR DOOR**	44–38	1	NOT REQUIRED FOR CHILD'S SAFETY LOCK
BD.19578	**Remote Control and Link Assembly,** for L.H. Door Lock	44–38A	1	
BD.19573	**Clip,** securing Links to Locks		2	
BD.19575	**Wavy Washer,** under Links		2	
BD.19574	**Flat Washer,** under Wavy Washers		2	
BD.29609	**Buffer,** on Door Panels, preventing vibration on Remote Controls (299/278)		4	
BD.17820/4	**Screw,** Phillips Head, securing Locks		8	
7031	**Screw,** Binding Head, securing Remote Controls (SZ.101/B.06)		6	
7206	**Spire Nut,** for Screws (NS.103/U.8)		6	
BD.17808	**STRIKER FOR L.H. REAR DOOR LOCK**	44–39	1	
BD.17809	**STRIKER FOR R.H. REAR DOOR LOCK**		1	
BD.18098	**Shakeproof Shim,** for Strikers (WB.3/10455)		2	
BD.17821/7	**Screw,** Set, securing Strikers to "D" Posts		6	
BD.17338	**Shim,** for Door Strikers		As req'd	

Part No.	Description	Plate No.	No. per Unit	Remarks
BD.18103	**L.H. REAR DOOR LOCK (WB.3/10826) (COMPLETE WITH REMOTE CONTROL INCORPORATING CHILD'S SAFETY LOCK)**		1	SUPPLIED TO SPECIAL ORDER ONLY
BD.18104	**R.H. REAR DOOR LOCK (WB.3/10825) (COMPLETE WITH REMOTE CONTROL INCORPORATING CHILD'S SAFETY LOCK)**		1	SUPPLIED TO SPECIAL ORDER ONLY
BD.13348	**Distance Piece**, between Remote Control Plates and Inner Door Panels		6	SUPPLIED TO SPECIAL ORDER ONLY
AS.102/5D	**Screw**, Set, securing Remote Control Plates to Door Panels		6	SUPPLIED TO SPECIAL ORDER ONLY
AW.102/T	**Washer**, Plain, on Setscrews		6	SUPPLIED TO SPECIAL ORDER ONLY
C.723/A	**Washer**, Shakeproof, on Setscrews		6	SUPPLIED TO SPECIAL ORDER ONLY
BD.13372	**L.H. OUTER ESCUTCHEON, OPERATING CHILD'S SAFETY CATCH**		1	SUPPLIED TO SPECIAL ORDER ONLY
BD.13373	**R.H. OUTER ESCUTCHEON, OPERATING CHILD'S SAFETY CATCH**		1	SUPPLIED TO SPECIAL ORDER ONLY
BD.13350	**SPRING, BEHIND OUTER ESCUTCHEONS, FOR CHILD'S SAFETY LOCK**		2	SUPPLIED TO SPECIAL ORDER ONLY
5112	**TRUNK LID ASSEMBLY (371/085)**	44–40	1	
6838	**Outside Panel only** (171/083)		1	
7000	**Inside Panel Assembly** (371/084)		1	
5125	**STRIKER ASSEMBLY FOR TRUNK LID (371/130)**		1	
5147	**TRUNK LID HINGE ASSEMBLY (271/144)**		2	
6622	**Coil Spring**, for Hinges (WB.7/9464)		4	
5117	**VENTILATOR LID ASSEMBLY ON SCUTTLE (371/100)**	44–41	1	
BD.17011	**R.H. REAR WHEEL COVER ASSEMBLY (CUT-AWAY TYPE) (371/226)**		1	
6123	**Dzus Fastener**, on Rear Wheel Cover (AJ.6-40)		2	
6625	**Grommet**, securing Dzus Fastener to Rear Wheel Cover (GA.6-350)		2	
BD.17012	**L.H. REAR WHEEL COVER ASSEMBLY (CUT-AWAY TYPE) (371/227)**	44–42	1	
6123	**Dzus Fastener**, on Rear Wheel Cover (AJ.6-40)		2	
6625	**Grommet**, securing Dzus Fasteners to Rear Wheel Cover (GA.6-350)		2	
6572	**SUPPORT BRACKET FOR STEERING COLUMN MOUNTING (171/289)**		2	

BODY PANELS

Part No.	Description	Plate No.	No. per Unit	Remarks
BD.10407	**COVER PLATE OVER STEERING COLUMN CUT-OUT ON CENTRE DASH (171/231)**		1	
C.8971	**COVER PLATE OVER PEDAL MOUNTING CUT-OUT ON CENTRE DASH (171/265)**		1	
UFS.131/5R	**Screw,** Set, securing Cover Plate to Dash		4	
UFN.131/L	**Nut,** on Setscrews		4	
C.725	**Washer,** Shakeproof, under Nuts		4	
FW.105/T	**Washer,** Plain, under Shakeproof Washers		4	

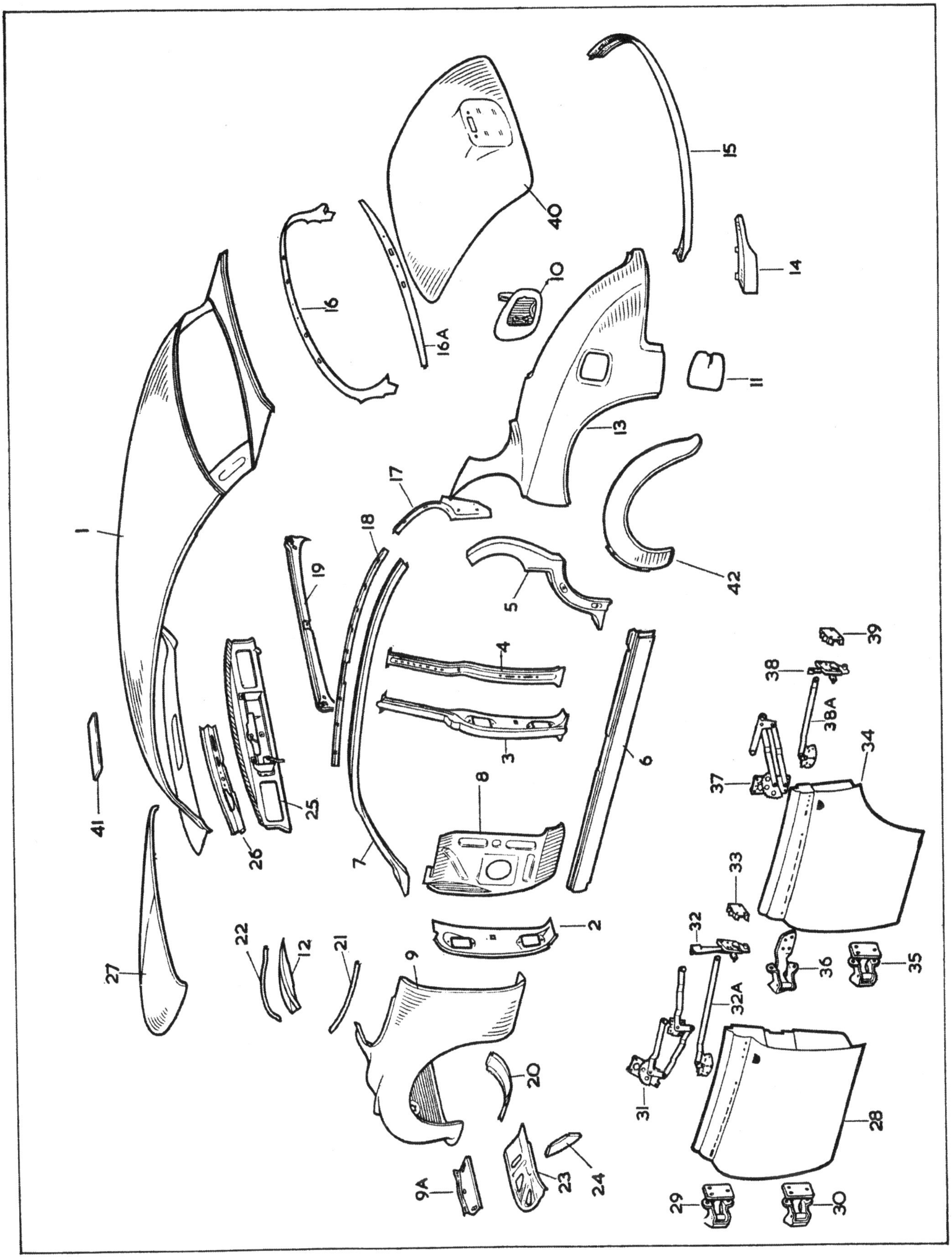
1
2
3
4
5
6
7
8
9
9A
10
11
12
13
14
15
16
16A
17
18
19
20
21
22
23
24
25
26
27
28
29
30
31
32
32A
33
34
35
36
37
38
38A
39
40
41
42

PLATE 44

BODY FITTINGS

Part No.	Description	Plate No.	No. per Unit	Remarks
BD.17170	**FRONT BUMPER, COMPLETE**		1	
BD.17172	**Bar only,** Front Bumper		1	
BD.10982	**Number Plate Carrier** (70/51156)		1	
BD.9869	**Bolt,** securing Number Plate Carrier to Front Bumper (70/51515)		2	
BD.9872	**Distance Piece,** between Front Bumper and Number Plate Carrier (70/51239)		2	
UFN.131/L	**Nut,** on Bolts		2	
FG.105/X	**Washer,** Spring, under Nuts		2	
FW.205/T	**Washer,** Plain, behind Spring Washers		2	
BD.10995	**Angle Bracket,** at each end of Bumper (70/51145)		2	
BD.9870	**Bolt,** securing Angle Brackets to Front Bumper (70/51516)		2	
UFN.137/L	**Nut,** on Bolts		2	
FG.106/X	**Washer,** Spring, under Nuts		2	
FW.206/T	**Washer,** Plain, behind Spring Washers		2	
BD.10996	**L.H. Outer Bracket,** fixing Bumper to Body (70/51144)		1	
BD.10997	**R.H. Outer Bracket,** fixing Bumper to Body (70/51143)		1	
UFB.143/12R	**Bolt,** securing Outer Brackets to Angle Brackets		2	
FW.107/T	**Washer,** Plain, on Bolts		4	
UFN.143/L	**Nut,** on Bolts		2	
C.727	**Washer,** Shakeproof, under Nuts		2	
BD.17176	**L.H. Inner Bracket,** fixing Bumper to Body		1	
BD.17177	**R.H. Inner Bracket,** fixing Bumper to Body		1	
UFS.137/7R	**Screw,** Set, securing Inner Brackets		2	
BD.11006	**Grip Nut,** receiving Setscrews for Inner Brackets		2	
C.726	**Washer,** Shakeproof, on Setscrews		2	
FW.106/T	**Washer,** Plain, under Shakeproof Washers		2	
BD.10985	**Over-Rider Assembly,** for Front Bumper (70/90393)		2	
BD.9868/1	**Beading,** between Over-Riders and Bumper (70/50881)		4	
BD.10990	**Bolt,** securing Over-Riders and Inner Brackets to Bumper (70/51517)		2	
C.8351	**Spire Nut,** securing Bolts to Over-Riders (SFW.0607)		2	
UFN.137/L	**Nut,** securing Over-Riders and Inner Brackets to Bumper		2	
FG.106/X	**Washer,** Spring, under Nuts		2	
FW.206/T	**Washer,** Plain, under Spring Washers		2	
UFS.143/8R	**Screw,** Set, securing Front Bumper to Frame		8	
FW.107/T	**Washer,** Plain, on Setscrews		8	
C.727	**Washer,** Shakeproof, on Setscrews		8	
BD.25273	**FRONT BUMPER ASSEMBLY, COMPLETE**		1	Fitted to Chassis No. 1K.1001 and subs. 1K.30001 and subs.
BD.23851	**Bumper Bar Assembly only**	45–1	1	
BD.23961	**Inner Bracket,** at R.H. side of Bumper Bar	45–2	1	
BD.23960	**Inner Bracket,** at L.H. side of Bumper Bar	45–3	1	
UFS.143/7R	**Screw,** Set, securing Inner Brackets to Bumper Bar		2	
UFN.143/1	**Nut,** on Setscrews		2	
C.727	**Washer,** Shakeproof, under Nuts		2	
FW.107/T	**Washer,** Plain, under Shakeproof Washer		2	
BD.23963	**Outer Bracket,** at R.H. side of Bumper Bar	45–4	1	
BD.23962	**Outer Bracket,** at L.H. side of Bumper Bar	45–5	1	
UFS.137/7R	**Screw,** Set, securing Outer Brackets to Bumper Bar		2	
UFN.137/L	**Nut,** on Setscrews		2	
C.726	**Washer,** Shakeproof, under Nuts		2	
FW.106/T	**Washer,** Plain, under Shakeproof Washer		2	
BD.26636	**R.H. Over-Rider Assembly**	45–6	1	
BD.26635	**L.H. Over-Rider Assembly**	45–7	1	
BD.25318	**Beading,** between Over-Riders and Bumper Bar	45–8	4	
BD.21172	**Grip Nut,** in Over-Riders, retaining Studs (SM/D/08/1/8)		2	
BD.21173	**Stud,** locating Over-Riders on Bumper Bar (70/51704)	45–9	2	
BD.21174	**Spacer,** on Studs (X.70/51682)	45–10	2	
UFN.137/1	**Nut,** on Studs, securing Over-Riders to Bumper Bar	45–11	2	

Part No.	Description	Plate No.	No. per Unit	Remarks
	FRONT BUMPER ASSEMBLY, COMPLETE *(contd.)*			
FG.106/X	**Washer**, Spring, under Nuts	45–12	2	
FW.206/T	**Washer**, Plain, under Spring Washers	45–13	2	
UFS.143/8R	**Screw**, Set, securing Front Bumper Assembly to Underframe	45–14	8	
C.727	**Washer**, Shakeproof, on Setscrews		8	
FW.107/7	**Washer**, Plain, under Shakeproof Washers		8	
C.25340	**FRONT NUMBER PLATE CARRIER ASSEMBLY**	45–15	1	Fitted to Chassis No. 1K.1001 and subs. 1K.30001 and subs. **NOT REQUIRED FOR U.S.A.**
BD.25319	**Carrier Bracket only**	45–16	2	
UFS.431/5H	**Screw**, Set, securing Carrier Brackets to Number Plate Carrier		4	
UFN.131/1	**Nut**, on Setscrews		4	
C.725X	**Washer**, Shakeproof, under Nuts		4	
FW.105/T	**Washer**, Plain, under Shakeproof Washer		4	
UFS.131/5R	**Screw**, Set, securing Corner Bracket to Bumper		2	
UFN.131/L	**Nut**, on Setscrews		2	
FG.105/X	**Washer**, Spring, under Nuts		2	
FW.105/T	**Washer**, Plain, under Spring Washers		2	
BD.25894	**FRONT NUMBER PLATE CARRIER**		1	Fitted to Chassis No. 1K.1001 and subs. 1K.30001 and subs. **FOR CARS EXPORTED TO U.S.A. ONLY**
BD.29239	**Carrier Bracket**		1	
UFS.431/5H	**Screw**, Set, securing Number Plate Carrier to Bracket		1	
UFS.431/8H	**Screw**, Set, securing Carrier Bracket to Front Bumper		1	
FW.105/T	**Washer**, Plain, on Setscrews		3	
C.19317	**Distance Piece**, on Setscrew		1	
UFN.131/L	**Nut**, on Setscrews		2	
C.725	**Washer**, Shakeproof, under Nuts		2	
BD.25969	**COVER PLATE ON UNDERFRAME FRONT CROSS MEMBER**		1	Fitted to Chassis No. 1K.1001 and subs. 1K.30001 and subs.
DAZ.410/6C	**Screw**, Self-Tapping, securing Cover Plate		6	
AW.102/T	**Washer**, Plain, on Self-Tapping Screws		6	
BD.9875	**HEADLAMP MOTIF ON FRONT WINGS**	45–17	2	
BD.17290	**Thread-Cutting Fix**, securing Motif (78/189)	45–18	4	
BD.17339	**GRILLE ASSEMBLY IN FOG LAMP RECESSES**		2	**NOT REQUIRED WHEN FOG LAMPS ARE FITTED**
BD.17340	**Distance Piece**, for Grilles		2	
BD.17341	**Rubber Buffer**, cushioning Grilles		8	
UFS.131/8R	**Screw**, Set, securing Grilles to Front Wings		2	
C.725	**Washer**, Shakeproof, on Setscrews		2	
FW.205/T	**Washer**, Bevelled, under Shakeproof Washers		2	
BD.24188	**RADIATOR GRILLE ASSEMBLY, COMPLETE**		1	
BD.24189	**Surround Assembly**, for Radiator Grille	45–19	1	
BD.24190	**L.H. Radiator Grille Assembly**	45–20	1	
BD.24191	**R.H. Radiator Grille Assembly**	45–21	1	
UFS.419/3H	**Screw**, Set, securing Grilles to Surround		2	
UFN.119/L	**Nut**, securing Grilles to Surround		4	
C.723/A	**Washer**, Shakeproof, on Setscrews and under Nuts		6	
AW.102/T	**Washer**, Plain, under Shakeproof Washers		6	
BD.14380/3	**Stud**, fixing Radiator Grille at top	45–22	1	
AW.102/T	**Washer**, Plain, on Stud		1	
BD.12370	**Cup Washer**, on Stud		1	
UFN.119/L	**Nut**, securing Radiator Grille		7	
C.723/A	**Washer**, Shakeproof, under Nuts		6	
BD.17150	**Fixing Strip**, for Radiator Grille, at sides and bottom	45–23	5	

BODY FITTINGS

Part No.	Description	Plate No.	No. per Unit	Remarks
BD.24190	**L.H. RADIATOR GRILLE ASSEMBLY**	45–20	1	
BD.24202	Vane No. 1		1	
BD.24204	Vane No. 2		1	
BD.24206	Vane No. 3		1	
BD.24208	Vane No. 4		1	
BD.24210	Vane No. 5		1	
BD.24212	Vane No. 6		1	
BD.24214	Vane No. 7		1	
BD.24216	Vane No. 8		1	
BD.24218	Vane No. 9		1	
BD.24220	Vane No. 10		1	
BD.24222	Vane No. 11		1	
BD.24191	**R.H. RADIATOR GRILLE ASSEMBLY**	45–21	1	
BD.24203	Vane No. 1		1	
BD.24205	Vane No. 2		1	
BD.24207	Vane No. 3		1	
BD.24209	Vane No. 4		1	
BD.24211	Vane No. 5		1	
BD.24213	Vane No. 6		1	
BD.24215	Vane No. 7		1	
BD.24217	Vane No. 8		1	
BD.24219	Vane No. 9		1	
BD.24221	Vane No. 10		1	
BD.24223	Vane No. 11		1	
BD.10224	**RUBBER BUFFER, AT FRONT CORNERS, FOR BONNET STOP**		2	
UFS.131/6R	**Screw**, Set, holding Rubber Buffers		2	
UFN.231/L	**Locknut**, on Setscrews		2	
BD.4920/3	**REST RUBBER AT REAR OF BONNET (22" LONG)**		1	
BD.4951	**RUBBER BUFFER AT SIDES OF BONNET APERTURE**		2	
BD.10221	**CONTROL WIRE, COMPLETE, OPERATING BONNET LOCK**	45–24	1	
C.4443	**Adjuster**, for Bonnet Lock Control	45–25	1	
C.3880	**Abutment**, for Cable Adjuster	45–26	1	
FW.104/T	**Washer**, Plain, at bottom of Abutment		1	
UFN.125/L	**Nut**, securing Abutment		1	
C.724	**Washer**, Shakeproof, under Nut		1	
BD.24164	**SUN VISOR, COMPLETE**	45–27	2	Superseded by BD.26702, BD.26703, and BD.25231
BD.24165	Friction Pivot Assembly	45–28	2	
BD.26703	**L.H. SUN VISOR ASSEMBLY COMPLETE**		1	R.H. DRIVE CARS ONLY Supersedes BD.24164
BD.26701	Sun Visor Assembly, with Vanity Mirror		1	R.H. DRIVE CARS ONLY Supersedes BD.24164
BD.24165	Friction Pivot Assembly		1	R.H. DRIVE CARS ONLY Supersedes BD.24164
BD.25231	**R.H. SUN VISOR ASSEMBLY COMPLETE**		1	R.H. DRIVE CARS ONLY Supersedes BD.24164
BD.25219	Sun Visor Assembly		1	R.H. DRIVE CARS ONLY Supersedes BD.24164
BD.24165	Friction Pivot Assembly		1	R.H. DRIVE CARS ONLY Supersedes BD.24164

Part No.	Description	Plate No.	No. per Unit	Remarks
BD.25231	**L.H. SUN VISOR ASSEMBLY COMPLETE**		1	**L.H. DRIVE CARS ONLY** Supersedes BD.24164
BD.25219	Sun Visor Assembly		1	
BD.24165	Friction Pivot Assembly		1	
BD.26702	**R.H. SUN VISOR ASSEMBLY COMPLETE**		1	
BD.26700	Sun Visor Assembly, with Vanity Mirror		1	
BD.24165	Friction Pivot Assembly		1	
BD.24165	**FRICTION PIVOT ASSEMBLY**	45–28	2	
BD.24166	Friction Arm Assembly		2	
BD.17713	Base, for Friction Arms		2	
BD.9635	Spring, on Friction Arms		2	
BD.541/32	Washer, Special, on Friction Arms		4	
C.8737/1	Nut, Self-locking, securing Friction Arms		2	
BD.13642/3	Screw, Set, securing Sun Visors		6	
C.2017/1	Washer, Packing, for Visor Base		As req'd	
BD.24593	End Plug		2	
BD.20716	**WINDSCREEN GLASS (TOUGHENED)**	45–29	1	**R.H. DRIVE CARS ONLY** Fitted from Chassis No.• 1A.1001 to 1A.9416
BD.20715	**WINDSCREEN GLASS (TOUGHENED)**	45–29	1	**L.H. DRIVE CARS ONLY** Fitted from Chassis No. 1A.20001 to 1A.20502
BD.28586	**WINDSCREEN GLASS (TOUGHENED, WITH ZEBRA ZONE)**		1	Fitted to Chassis No. 1A.9417 and subs. 1A.20503 and subs.
BD.29575	**WINDSCREEN GLASS (LAMINATED)**	45–29	1	Fitted to Chassis No. 1A.12233 and subs. 1A.20548 and subs.
BD.29575/1	**WINDSCREEN GLASS ("SUNDYM" GLASS, LAMINATED)**	45–29	1	Fitted to Chassis No. 1A.12366 and subs. 1A.20552 and subs.
BD.29575/2	**WINDSCREEN GLASS (SHADED TYPE LAMINATED)**	45–29	1	
BD.26888	**RUBBER SEAL FOR WINDSCREEN GLASS**	45–30	1	Supersedes BD.16908
BD.17291	**RUBBER INSERT, FOR SEAL (121" LONG)**	45–31	1	
BD.16909	**CHROMIUM PLATED FINISHER, L.H. FOR WINDSCREEN**	45–32	1	
BD.16910	**CHROMIUM PLATED FINISHER, R.H. FOR WINDSCREEN**	45–33	1	
BD.5366	**JOINT PIECE AT CENTRE OF CHROMIUM FINISHERS**	45–34	2	
BD.24427	**LICENCE HOLDER ON WINDSCREEN**	45–35	1	

BODY FITTINGS

Part No.	Description	Plate No.	No. per Unit	Remarks
C.20697	**INTERIOR MIRROR ASSEMBLY ON WINDSCREEN HEADER PANEL (62577)**	45–36	1	
C.20697/1	**Base only**		1	
C.16407/2	**Head only**		1	
BD.13642/3	**Screw**, Set, securing Mirror to Windscreen Header Panel		3	
BD.29607	**STAINLESS STEEL FINISHER ON L.H. 'A' POST**	45–37	1	
BD.29608	**STAINLESS STEEL FINISHER ON R.H. 'A' POST**	45–38	1	
BD.14508	**L.H. AIR DUCT HOSE FOR WINDSCREEN DEMISTER**	45–39	1	
BD.14607	**Rubber Elbow,** connecting L.H. Hose to Heater aperture in Scuttle	45–40	1	
BD.14507	**R.H. AIR DUCT HOSE FOR WINDSCREEN DEMISTER**	45–41	1	
BD.10343	**Rubber Elbow,** connecting R.H. Hose to Heater aperture in Scuttle	45–42	1	
BD.17163	**R.H. DEMISTER NOZZLE ASSEMBLY FOR CONNECTION OF AIR DUCT HOSES**	45–43	1	
BD.17164	**L.H. DEMISTER NOZZLE ASSEMBLY FOR CONNECTION OF AIR DUCT HOSES**	45–44	1	
DA.406/8C	**Screw,** Self-Tapping, securing Demister Nozzles		4	
BD.9877	**RUBBER SEAL, FOR RECIRCULATION DOOR ON SCUTTLE FACE PANEL**		1	
BD.17761	**KNOB FOR VENTILATOR OPERATING LEVER**		1	
BD.17764	**Locknut,** securing Knob to Operating Lever		1	
BD.9876	**RUBBER SEAL FOR VENTILATOR LID AT TOP OF SCUTTLE**	45–45	1	
BD.10370	**Drain Tube,** for Ventilator		1	
BD.12487	**SHROUD, IN SCUTTLE VENTILATOR PREVENTING ENTRY OF WATER INTO HEATER BOX**		1	
UCS.519/3H	**Screw,** Set, securing Shroud		3	
BD.5440/4	**Square Nut,** on Setscrews		3	
BD.20652	**GAUZE, FOR SCUTTLE VENTILATOR LID**	45–46	1	
UFS.819/4H	**Screw,** Set, securing Gauze		3	
C.723/A	**Washer,** Shakeproof, on Setscrews		3	
AW.102/T	**Washer,** Plain, under Shakeproof Washers		3	

Part No.	Description	Plate No.	No. per Unit	Remarks
BD.12010	**BRACKET ASSEMBLY, LOWER, MOUNTING UPPER STEERING COLUMN**		1	R.H. DRIVE CARS ONLY
BD.12011	**Gasket**, under Bracket		1	R.H. DRIVE CARS ONLY
BD,18587	**BRACKET ASSEMBLY, LOWER, MOUNTING UPPER STEERING COLUMN**		1	L.H. DRIVE CARS ONLY
BD.18588	**Gasket**, under Bracket		1	L.H. DRIVE CARS ONLY
UFN.125/L	**Nut**, on Screw in Lower Bracket		1	L.H. DRIVE CARS ONLY
C.724	**Washer**, Shakeproof, under Nut		1	L.H. DRIVE CARS ONLY
FW.104/T	**Washer**, Plain, under Shakeproof Washer		1	L.H. DRIVE CARS ONLY
UFS.125/4R	**Screw**, Set, securing Lower Bracket to Dash		3	
C.724	**Washer**, Shakeproof, on Setscrews		3	
C.2905/4	**Clip**, securing Upper Column in Bracket		1	
BD.14986	**TREAD PLATE ON FRONT DOOR SILLS**	45–47	2	
DAZ.504/8C	**Screw**, Self-Tapping, securing Tread Plates		10	
BD.4951	**RUBBER BUFFER FOR FRONT DOOR STOP**	45–48	2	
BD.17994	**SCUTTLE SIDE CASING ASSEMBLY, L.H.**	45–49	1	R.H. DRIVE CARS ONLY
BD.17996	**SCUTTLE SIDE CASING ASSEMBLY, R.H.**	45–50	1	R.H. DRIVE CARS ONLY
BD.17997	**SCUTTLE SIDE CASING ASSEMBLY, L.H.**		1	L.H. DRIVE CARS ONLY
BD.17995	**SCUTTLE SIDE CASING ASSEMBLY, R.H.**		1	L.H. DRIVE CARS ONLY
DAC.604/14C	**Screw**, Self-Tapping, securing Side Casings		6	
BD.532/1	**Cup Washer**, on Screws		6	
BD.17619	**SCUTTLE TOP CASING ASSEMBLY, L.H.**	45–51	1	R.H. DRIVE CARS ONLY
BD.17617	**SCUTTLE TOP CASING ASSEMBLY, R.H.**	45–52	1	R.H. DRIVE CARS ONLY. NOT REQUIRED WHEN STEERING COLUMN LOCK IS FITTED
BD.25093	**SCUTTLE TOP CASING ASSEMBLY, R.H.**		1	R.H. DRIVE CARS ONLY. FOR USE ONLY WHEN STEERING COLUMN LOCK IS FITTED
BD.25196	**SCUTTLE TOP CASING ASSEMBLY, L.H.**		1	L.H. DRIVE CARS ONLY Fitted from Chassis No. 1A.20001 to 1A.20267
BD.25092	**SCUTTLE TOP CASING ASSEMBLY, L.H.**		1	L.H. DRIVE CARS ONLY Fitted to Chassis No. 1A.20268 and subs.
BD.17618	**SCUTTLE TOP CASING ASSEMBLY, R.H.**		1	L.H. DRIVE CARS ONLY
DAC.606/8C	**Screw**, Self-Tapping, securing Top Casings (½" long)		7	
DAC.606/12C	**Screw**, Self-Tapping, securing Top Casings (¾" long)		2	
BD.532/2	**Cup Washer**, on Self-Tapping Screws		9	

BODY FITTINGS

Part No.	Description	Plate No.	No. per Unit	Remarks
BD.25698	L.H. FRONT RIGID SEAT ASSEMBLY, COMPLETE		1	Fitted from Chassis No. 1A.1001 to 1A.2622 1A.20001 to 1A.20233
BD.25699	R.H. FRONT RIGID SEAT ASSEMBLY, COMPLETE		1	
	NOTE: BD.25698 and BD.25699 covers the supply of a fully trimmed Seat complete with Cushion, Armrest and Seat Slides			
BD.24537	KIT OF PARTS, TO CONVERT L.H. RIGID SEAT TO L.H. RECLINING SEAT		1	For conversion of L.H. front Seat Assembly BD.25698
BD.24538	KIT OF PARTS, TO CONVERT R.H. RIGID SEAT TO R.H. RECLINING SEAT		1	For conversion of R.H. front Seat Assembly BD.25699
BD.26015	L.H. FRONT RIGID SEAT ASSEMBLY, COMPLETE		1	Fitted from Chassis No. 1A.2623 to 1A.6316 1A.20234 to 1A.20335
BD.26016	R.H. FRONT RIGID SEAT ASSEMBLY, COMPLETE		1	
	NOTE: BD.26015 and BD.26016 covers the supply of a fully trimmed Seat complete with Cushion, Armrest and Seat Slides			
BD.27392	L.H. FRONT RIGID SEAT ASSEMBLY, COMPLETE		1	Fitted to Chassis No. 1A.6317 and subs. 1A.20336 and subs.
BD.27393	R.H. FRONT RIGID SEAT ASSEMBLY, COMPLETE		1	
	NOTE: BD.27392 and BD.27393 covers the supply of a fully trimmed Seat complete with Cushion, Armrest and Seat Slides			
BD.25795	KIT OF PARTS TO CONVERT L.H. RIGID SEAT TO L.H. RECLINING SEAT		1	For conversion of L.H. Front Seat Assemblies BD.26015 or BD.27392
BD.25796	KIT OF PARTS TO CONVERT R.H. RIGID SEAT TO R.H. RECLINING SEAT		1	For conversion of R.H. Front Seat Assemblies BD.26016 or BD.27393
BD.25688	L.H. FRONT RECLINING SEAT ASSEMBLY, COMPLETE		1	Fitted from Chassis No. 1A.1001 to 1A.2622 1A.20001 to 1A.20233
BD.25689	R.H. FRONT RECLINING SEAT ASSEMBLY, COMPLETE	45–53	1	
	NOTE: BD.25688 and BD.25689 covers the supply of a fully trimmed Seat complete with Cushion, Armrest and Seat Slides			
BD.26005	L.H. FRONT RECLINING SEAT ASSEMBLY, COMPLETE		1	Fitted from Chassis No. 1A.2623 to 1A.6316 1A.20234 to 1A.20335
BD.26006	R.H. FRONT RECLINING SEAT ASSEMBLY, COMPLETE		1	
	NOTE: BD.26005 and BD.26006 covers the supply of a fully trimmed Seat complete withCushion, Armrest and Seat Slides			

Part No.	Description	Plate No.	No. per Unit	Remarks
BD.27386	**L.H. FRONT RECLINING SEAT ASSEMBLY, COMPLETE**		1	Fitted from Chassis No. 1A.6317 to 1A.12239; 1A.20336 to 1A.20547
BD.27387	**R.H. FRONT RECLINING SEAT ASSEMBLY, COMPLETE**		1	
	NOTE: BD.27386 and BD.27387 covers the supply of a fully trimmed Seat complete with Cushion, Armrest and Seat Slides			
BD.29962	**L.H. FRONT RECLINING SEAT ASSEMBLY, COMPLETE**		1	Fitted to Chassis No. 1A.12240 and subs. 1A.20548 and subs.
BD.29961	**R.H. FRONT RECLINING SEAT ASSEMBLY, COMPLETE**		1	
	NOTE: BD.29961 and BD.29962 covers the supply of a fully trimmed Seat complete with Cushion, Armrest and Seat Slides			
BD.33933	**L.H. FRONT RIGID SEAT ASSEMBLY COMPLETE**		1	**REQUIRED FOR CARS FITTED WITH AUTOMATIC TRANSMISSION** Fitted to Chassis No. 1K.1001 and subs. 1K.30001 and subs.
BD.33932	**R.H. FRONT RIGID SEAT ASSEMBLY COMPLETE**		1	
	NOTE: BD.33933 and BD.33932 covers the supply of a fully trimmed Seat complete with Cushion, and Seat Slides			
BD.33937	**L.H. FRONT RIGID SEAT ASSEMBLY COMPLETE**		1	**REQUIRED FOR CARS FITTED WITH MANUAL OR OVERDRIVE TRANSMISSION** Fitted to Chassis No. 1K.1001 and subs. 1K.30001 and subs.
BD.33936	**R.H. FRONT RIGID SEAT ASSEMBLY COMPLETE**		1	
	NOTE: BD.33937 and BD.33936 covers the supply of a fully trimmed Seat complete with Cushion, and Seat Slides			
BD.24081	**TIE BRACKET, FOR CONVERSION OF RECLINING SEATS TO RIGID SEATS**		4	
UFS.125/4R	**Screw,** Set, securing Tie Brackets to Seat Base and Squab Pivot		4	
C.724	**Washer,** Shakeproof, on Setscrews		4	
FW.104/T	**Washer,** Plain, under Shakeproof Washers		4	
BD.24904	**CUSHION ASSEMBLY FOR L.H. FRONT SEAT**		1	Fitted from Chassis No. 1A.1001 to 1A.12239; 1A.20001 to 1A.20547
BD.24905	**CUSHION ASSEMBLY FOR R.H. FRONT SEAT**		1	
BD.29966	**CUSHION ASSEMBLY FOR L.H. FRONT SEAT**		1	Fitted to Chassis No. 1A.12240 and subs. 1A.20548 and subs.
BD.29965	**CUSHION ASSEMBLY FOR R.H. FRONT SEAT**		1	
BD.24826	**ARMREST ASSEMBLY, COMPLETE, FOR FRONT SEATS**		2	Fitted from Chassis No. 1A.1001 to 1A.6316; 1A.20001 to 1A.20335

BODY FITTINGS

Part No.	Description	Plate No.	No. per Unit	Remarks
BD.27398	**ARMREST ASSEMBLY, COMPLETE, FOR FRONT SEATS**		2	Fitted from Chassis No. 1A.6317 to 1A.12239 1A.20336 to 1A.20547
BD.30531	**ARMREST ASSEMBLY, COMPLETE FOR FRONT SEATS**		2	Fitted to Chassis No. 1A.12240 and subs. 1A.20548 and subs.
BD.23848	**CONTROL LEVER ASSEMBLY FOR SEAT SLIDE**	45–54	2	Fitted from Chassis No. 1A.1001 to 1A.2622 1A.20001 to 1A.20233
BD.26606	**L.H. CONTROL LEVER FOR R.H. SEAT SLIDE**		1	Fitted from Chassis No. 1A.2623 to 1A.12239 1A.20234 to 1A.20547 (BD.26606 and BD.26607)
BD.26607	**R.H. CONTROL LEVER FOR L.H. SEAT SLIDE**		1	
BD.29421	**L.H. CONTROL LEVER FOR R.H. SEAT SLIDE**		1	Fitted to Chassis No. 1A.12240 and subs. 1A.20548 and subs. (BD.29421 and BD.29422)
BD.29422	**R.H. CONTROL LEVER FOR L.H. SEAT SLIDE**		1	
BD.26612	**KNOB ON CONTROL LEVERS**		2	Fitted to Chassis No. 1A.2623 and subs. 1A.20234 and subs.
BD.23834	**CROSS BAR, FOR SEAT SLIDES**	45–55	2	Fitted from Chassis No. 1A.1001 to 1A.2622 1A.20001 to 1A.20233
BD.26609	**CROSS BAR, FOR SEAT SLIDES**		2	Fitted from Chassis No. 1A.2623 to 1A.12239 1A.20234 to 1A.20547
BD.30147	**CROSS BAR, FOR SEAT SLIDES**		2	Fitted to Chassis No. 1A.12240 and subs. 1A.20548 and subs.
UFS.119/5R	**Screw**, Set, securing Cross Bar to Control Lever		2	
UFS.119/5R	**Screw**, Set, securing Cross Bar and Control Lever to Plungers		4	
AW.102/C	**Washer**, Brass, on Setscrews		6	
C.8667/17	**Locknut**, on Setscrews		6	
BD.23891	**PLUNGER, LOCKING SEAT IN SELECTED POSITION**	45–56	4	
BD.22685	**Spring**, tensioning Plungers	45–57	4	
FW.105/T	**Washer**, retaining Springs		4	
BD.25544	**Roll Pin**, behind Washers, retaining Plungers		4	
C.2155	**Spacing Washer**		2	
BD.11416	**SPRING, ASSISTING FORWARD MOVEMENT OF FRONT SEATS**	45–58	4	Fitted from Chassis No. 1A.1001 to 1A.12239 1A.20001 to 1A.20547
BD.29301	**SPRING, ASSISTING FORWARD MOVEMENT OF FRONT SEATS**		4	Fitted to Chassis No. 1A.12240 and subs. 1A.20548 and subs.
BD.24629	**Sleeve** (P.V.C.) on Springs		4	

Part No.	Description	Plate No.	No. per Unit	Remarks
BD.26608	**BRACKET STEADYING CONTROL LEVER**		2	
DZZ.410/8C	**Screw**, Self-Tapping, securing Bracket to Seat Frame		4	
BD.25646	**Grommet** (Polythene) inside Bracket		2	
BD.23863	**SLIDE RATCHET FOR FRONT SEATS**	45–59	4	Fitted from Chassis No. 1A.1001 to 1A.2622, 1A.20001 to 1A.20233
BD.24861	**Slide Assembly**, for Front Seats	45–60	4	
BD.23853	**Trunnion**, on Front Seat Slides	45–61	4	
BD.24093	**Clevis Pin**, securing Trunnion to Front Seats	45–62	4	
BD.541/17	**Washer**, Brass, on Clevis Pins		4	
L.102/6U	**Split Pin**, retaining Clevis Pins		4	
UFB.131/13R	**Bolt**, securing Slide Ratchet and Slide to Front Floor		8	
BD.24295	**Packing Piece**, between Ratchet and Mounting Bracket		As req'd	
C.18698	**Distance Piece**, between Slide Assembly and Ratchet		8	
C.21069	**Spacer**, between Ratchet and Mounting Bracket		16	
FW.105/T	**Washer**, Plain, on Bolts and under Nuts		12	
C.8667/2	**Locknut**, on rear Bolts		4	
BD.25956	**SLIDE RATCHET FOR FRONT SEATS**		4	Fitted to Chassis No. 1A.2623 and subs. 1A.20234 and subs.
BD.23854	**Slide**, for Front Seats		4	
BD.23853	**Trunnion**, on Front Seat Slides		4	
BD.26192	**Stop**, on forward end of Slides		4	
BD.24093	**Clevis Pin**, securing Trunnion to Front Seats		4	
BD.541/17	**Washer**, Brass, on Clevis Pin		4	
L.102/6U	**Split Pin**, retaining Clevis Pin		4	
UFB.131/13R	**Bolt**, securing Slide Ratchet to Slide		8	
BD.18698	**Distance Piece**, between Slide and Slide Ratchet		8	
FW.105/T	**Washer**, Plain, on Bolts and under Nuts		12	
C.8667/2	**Nut**, Self-Locking, on Bolts		8	
UFS.131/6R	**Screw**, Set, securing Slide Ratchet to Floor Brackets		8	
BD.24295	**Packing Piece**, between Slide Ratchet and Floor Brackets		As req'd	
C.740	**Washer**, Shakeproof, on Setscrews		8	
FW.105/T	**Washer**, Plain, under Shakeproof Washers		8	
BD.24626	**LEVER, FOR FRONT RECLINING SEAT SQUAB RAKE CONTROL**	45–63	2	Fitted from Chassis No. 1A.1001 to 1A.12239, 1A.20001 to 1A.20548
BD.31258	**LEVER, FOR FRONT RECLINING SEAT SQUAB RAKE CONTROL**		2	Fitted to Chassis No. 1A.12240 and subs. 1A.20549 and subs.
BD.26839	**Knob**, for R.H. Squab Rake Control Lever		1	Fitted from Chassis No. 1A.1001 to 1A.12438, 1A.20001 to 1A.20556
BD.26838	**Knob**, for L.H. Squab Rake Control Lever		1	
BD.31054	**Knob**, for R.H. Squab Rake Control Lever		1	Fitted to Chassis No. 1A.12439 and subs. 1A.20557 and subs.
BD.31055	**Knob**, for L.H. Squab Rake Control Lever		1	
DZC.006/6C	**Screw**, Self-Tapping, securing Knobs to Levers		2	
UFS.319/4H	**Screw**, Set, securing Levers to Cross Rods		2	
AG.102/X	**Washer**, Spring, on Setscrews		2	
BD.30488	**Collar**, at base of Levers		2	
BD.29380	**Washer**, behind Collars		2	
BD.29307	**COVER (PLASTIC) OVER SQUAB PIVOT AT R.H. SIDE OF FRONT SEATS**		2	
BD.29308	**COVER (PLASTIC) OVER SQUAB PIVOT AT L.H. SIDE OF FRONT SEATS**		2	
DAZ.410/8C	**Screw**, Self-Tapping, securing Covers		4	
BD.541/28	**Washer**, Plain, on Self-Tapping Screws		4	

BODY FITTINGS

Part No.	Description	Plate No.	No. per Unit	Remarks
BD.25815	**CROSS ROD ASSEMBLY FOR SQUAB RAKE CONTROL ON R.H. FRONT RECLINING SEAT**	45–65	1	Fitted from Chassis No. 1A.1001 to 1A.12239 1A.20001 to 1A.20547
BD.25816	**CROSS ROD ASSEMBLY FOR SQUAB RAKE CONTROL ON L.H. FRONT RECLINING SEAT**		1	
BD.24377	**Fork**, on Cross Rods, withdrawing Plungers from Rake Slides	45–66	4	
BD.22294	**Screw**, Allen Head, securing Forks to Cross Rods	45–67	4	
BD.23893	**Plunger**, locking Squab in selected position	45–68	4	
BD.24602	**Spring**, tensioning Plungers	45–69	4	
BD.541/31	**Washer**, Special, retaining Springs		4	
L.102/6U	**Split Pin**, behind Washers		4	
BD.24056	**PIVOT PLATE AT FRONT OF SEATS**		2	
BD.24029	**Bush** (Nylon) in Pivot Plates		16	
BD.26860	**Pivot Pin** (Upper) securing Pivot Plates to Seat Frames		4	
FW.104/T	**Washer**, Steel, on Pivot Pins		4	
L.102/6U	**Split Pin**, retaining Pivot Pins		4	
BD.25971	**SEAT ANCHOR BRACKET ASSEMBLY**		2	
BD.25989	**Pivot Pin** (Lower) securing Seat Anchor Bracket to Pivot Plate		4	
FW.104/C	**Washer**, Brass, on Pivot Pins		8	
L.102/6U	**Split Pin**, retaining Pivot Pins		4	
UFS.131/8R	**Screw**, Set, securing Seat Anchor Bracket to Floor		4	
C.740	**Washer**, Shakeproof, on Setscrews		4	
BD.542/7	**Washer**, Special, under Shakeproof Washers		4	
BD.25973	**Packing Piece**, under Seat Anchor Bracket		As req'd	
BD.27653	**EYE BOLT, IN FRONT FLOOR, FOR SAFETY HARNESS**		4	
BD.23149	**Distance Piece**, on Eye Bolts		4	
BD.28351	**Washer** (Canvas/Rubber) under Distance Pieces		4	
BD.11036	**TACKING STRIP FOR "A" POSTS**		2	
BD.17409	**TACKING STRIP, LOWER, FOR "B" AND "C" POSTS**		4	
BD.11038	**TACKING STRIP, LOWER, FOR "D" POSTS**		2	
BD.17410	**TACKING STRIP, UPPER, FOR "D" POSTS**		2	
BD.29570	**TACKING STRIP, UPPER, FOR REAR QUARTERS**		2	
BD.11975	**TACKING STRIP FOR SEAT PAN, SILLS AND "A" POSTS**		6	
BD.8093/3	**Screwnail**, securing Tacking Strips (¾" long)		72	
BD.8093/1	**Screwnail**, securing Tacking Strips (1" long)		50	

Part No.	Description	Plate No.	No. per Unit	Remarks
BD.19722	**L.H. FACIA BOARD ASSEMBLY (COMPLETE WITH CUBBY BOX LID)**	46–1	1	R.H. DRIVE CARS ONLY
BD.17194	**R.H. FACIA BOARD ASSEMBLY**	46–2	1	R.H. DRIVE CARS ONLY
BD.18290	**Lid,** for Cubby Box on L.H. Facia		1	
BD.12735	**Quadrant Assembly,** for Cubby Box Lid		1	
MC.103/4K	**Woodscrew,** securing Quadrant and Stop Plate		4	
BD.12358	**Stop Plate Assembly**		1	
BD.573	**Hinge,** for Cubby Box Lid		2	
MC.103/4K	**Woodscrew,** securing Hinges		8	
BD.656/1	**Ball Catch,** retaining Lid in closed position		2	
BD.656/2	**Plate,** for Ball Catch and Lock		3	
M.102/3K	**Woodscrew,** securing Plates		6	
BD.17191	**L.H. FACIA BOARD ASSEMBLY**		1	L.H. DRIVE CARS ONLY
BD.19723	**R.H. FACIA BOARD ASSEMBLY (COMPLETE WITH CUBBY BOX LID)**		1	L.H. DRIVE CARS ONLY
BD.18290	**Lid,** for Cubby Box on R.H. Facia		1	
BD.12735	**Quadrant Assembly,** for Cubby Box Lid		1	
MC.103/4K	**Woodscrew,** securing Quadrant and Stop Plate		4	
BD.12358	**Stop Plate Assembly**		1	
BD.573	**Hinge,** for Cubby Box Lid		2	
MC.103/4K	**Woodscrew,** securing Hinges		8	
BD.656/1	**Ball Catch,** retaining Lid in closed position		2	
BD.656/2	**Plate,** for Ball Catch and Lock		3	
M.102/3K	**Woodscrew,** securing Plates		6	
BD.5300	**LOCK FOR CUBBY BOX LID (WB.7/5631)**		1	
BD.8072	**Barrel,** for Lock (WB.7/5410)		1	
MC.204/4R	**Woodscrew,** securing Lock to Cubby Box Lid		3	
BD.12326	**HANDLE ON CUBBY BOX LID**		1	
MC.303/4R	**Woodscrew,** securing Handle		1	
BD.19089	**STUD PLATE ASSEMBLY AT L.H. SIDE OF FACIA BOARD**		1	For R.H. Facia on R.H. Drive Cars or L.H. Facia on L.H. Drive Cars
BD.19090	**STUD PLATE ASSEMBLY AT R.H. SIDE OF FACIA BOARD**		1	For R.H. Facia on R.H. Drive Cars or L.H. Facia on L.H. Drive Cars
M.104/3R	**Woodscrew,** securing Stud Plates to Facia Boards		4	
BD.17583	**Reinforcing Angle,** at bottom of Facia, fixing Scuttle Top Casing		1	
UFN.119/L	**Nut,** securing Stud Plate to Reinforcing Angle		2	
C.723/A	**Washer,** Shakeproof, under Nuts		2	
BD.17822	**Bracket,** fixing Scuttle Top Casings		1	For L.H. Facia on R.H. Drive cars
BD.17823	**Bracket,** fixing Scuttle Top Casings		1	For R.H. Facia on L.H. Drive cars
MJ.206/4R	**Woodscrew,** securing Bracket and Angle to Facia Board		8	
BD.17722	**Outer Bracket,** fixing Scuttle Top Casings		1	
MJ.208/5R	**Woodscrew,** securing Outer Bracket		2	
BD.17321	**INNER FIXING BRACKET ASSEMBLY FOR FACIA BOARDS**		2	
BD.17466	**L.H. OUTER FIXING BRACKET ASSEMBLY FOR FACIA BOARDS**		1	

BODY FITTINGS

Part No.	Description	Plate No.	No. per Unit	Remarks
BD.17467	**R.H. OUTER FIXING BRACKET ASSEMBLY FOR FACIA BOARDS**		1	
M.108/6R	**Woodscrew,** securing Inner Fixing Brackets to Facia Boards		4	
UFS.119/4R	**Screw,** Set, securing Inner Fixing Brackets to Dash		4	
C.723A	**Washer,** Shakeproof, on Setscrews		4	
AW.102/T	**Washer,** Plain, under Shakeproof Washers		4	
MJ.208/4R	**Woodscrew** securing Outer Fixing Brackets to Facia Boards		4	
UFN.125/L	**Nut,** securing Outer Fixing Brackets to Dash		2	
C.724	**Washer,** Shakeproof, under Nuts		2	
FW.104/T	**Washer,** Plain, under Shakeproof Washers		2	
BD.19220	**SUPPORT BRACKET FOR FACIA BOARDS**		2	
MJ.208/5R	**Woodscrew,** securing Support Brackets to Facia Boards		4	
BD.18002	**STEADY BRACKET FOR L.H. FACIA BOARD**		1	
BD.18003	**STEADY BRACKET FOR R.H. FACIA BOARD**		1	
MJ.204/4R	**Woodscrew,** securing Steady Brackets to Facias		2	
DA.404/8C	**Screw,** Self-Tapping, securing Steady Brackets to Body		2	
BD.19725	**L.H. CUBBY BOX**	46–3	1	R.H. DRIVE CARS ONLY
BD.19726	**R.H. CUBBY BOX**		1	L.H. DRIVE CARS ONLY
MJ.206/3R	**Woodscrew,** securing Cubby Box		7	
BD.17190	**Retainer,** for Cubby Box	46–4	1	
M102/4K	**Woodscrew,** securing Retainer		3	
BD.19288	**INSERT BOARD FOR L.H. CUBBY BOX**		1	R.H. DRIVE CARS ONLY
BD.19289	**INSERT BOARD FOR R.H. CUBBY BOX**		1	L.H. DRIVE CARS ONLY
C.18117	**LAMP, COMPLETE, INSIDE CUBBY BOX (56078/D.681)**	46–5	1	
8851	**Lens** (54572189)		1	
7880	**Bulb** (12V.6W. Festoon)		1	
C.18118	**SWITCH, OPERATING CUBBY BOX LAMP**	46–6	1	
C.17384	**CABLE SHIELD, UNDER DASH**		1	L.H. DRIVE CARS ONLY
DZZ.408/6C	**Screw,** Self-Tapping, securing Cable Shield		2	L.H. DRIVE CARS ONLY
AW.102/T	**Washer,** Plain, on Self-Tapping Screws		2	L.H. DRIVE CARS ONLY
BD.17579	**SCREEN RAIL CAPPING ASSEMBLY**	46–7	1	
BD.17569	**Fixing Plate Assembly,** for Capping		2	
M.106/3R	**Woodscrew,** securing Fixing Plates to Cappings		4	
UFN.125/L	**Nut,** securing Screen Rail Capping to Facia		2	
C.724	**Washer,** Shakeproof, under Nuts		2	
BD.541/20	**Special Washer,** under Shakeproof Washers		2	

Part No.	Description	Plate No.	No. per Unit	Remarks
BD.32872	**WINDSCREEN RAIL AND CRASH PAD ASSEMBLY**		1	Fitted to Chassis No. 1K.1001 and subs. 1K.30001 and subs.
UFN.125/L	**Nut,** securing Windscreen Rail and Crash Pad Assembly to Body		2	
C.724	**Washer,** Shakeproof, under Nuts		2	
BD.541/20	**Washer,** Plain, under Shakeproof Washers		2	
BD.17658	**CAPPING FOR L.H. 'A' POST**		1	
BD.17659	**CAPPING FOR R.H. 'A' POST**	46–8	1	
DAC.604/12C	**Screw,** Self-Tapping, at lower fixings of Cappings		2	
BD.532/1	**Cup Washer,** on Self-Tapping Screws		2	
DAZ.404/12C	**Screw,** Self-Tapping, at upper fixings of Cappings		2	
BD.17673	**CAPPING AT TOP OF L.H. SIDE OF WINDSCREEN**	46–9	1	
BD.17674	**CAPPING AT TOP R.H. SIDE OF WINDSCREEN**	46–10	1	
DAC.604/12C	**Screw,** Self-Tapping, securing Cappings		6	
BD.543/1	**Cup Washer,** on Screws		6	
BD.17313	**CAPPING ASSEMBLY AT FRONT OF L.H. CANTRAIL**		1	
BD.17314	**CAPPING ASSEMBLY AT FRONT OF R.H. CANTRAIL**	46–11	1	
BD.17776	**END FINISHER FOR FRONT CANTRAIL CAPPINGS**	46–12	2	
M.102/4R	**Woodscrew,** securing End Finishers		2	
BD.17319	**CAPPING ASSEMBLY AT REAR OF L.H. CANTRAIL**		1	
BD.17320	**CAPPING ASSEMBLY AT REAR OF R.H. CANTRAIL**	46–13	1	
BD.17988	**Fixing Plate,** at rear end of Cappings		2	
M.104/3R	**Woodscrew,** securing Fixing Plate to Cappings		4	
BD.17315	**CAPPING, BETWEEN CANTRAIL CAPPINGS, FOR MOUNTING INTERIOR LIGHTS ON 'B/C' POSTS**	46–14	2	Fitted from Chassis No. 1A.1001 to 1A.3299 1A.20001 to 1A.20256
DA.506/20C	**Screw,** Self-Tapping, securing Cappings		4	
BD.26159	**CAPPING, BETWEEN CANTRAIL CAPPINGS, FOR MOUNTING INTERIOR LIGHTS ON 'B/C' POSTS**		2	Fitted to Chassis No. 1A.3300 and subs. 1A.20257 and subs.
DA.506/20C	**Screw,** Self-Tapping, securing Cappings		4	
BD.22695	**UPPER FACIA FOR 'B/C' POSTS**	46–15	2	
BD.14130	**Spring Clip,** securing Facias to 'B/C' Posts		4	
MJ.206/3R	**Woodscrew,** securing Clips to Facias		8	

BODY FITTINGS

Part No.	Description	Plate No.	No. per Unit	Remarks
BD.17216	TRIM PANEL ASSEMBLY FOR L.H. 'B/C' POST		1	Superseded by BD.25479
BD.11309	**Clip,** securing Trim Panel to 'B/C' Post		5	
BD.2702	**Retainer,** for Clips		5	
BD.17317	**Waist Facia Assembly,** for 'B/C' Post		1	
M.104/4R	**Woodscrew,** securing Waist Facia to Trim Panel		2	
BD.17217	TRIM PANEL ASSEMBLY FOR R.H. 'B/C' POST	46–16	1	
BD.11309	**Clip,** securing Trim Panel to 'B/C' Post		5	
BD.2702	**Retainer,** for Clips		5	
BD.17317	**Waist Facia Assembly,** for 'B/C' Post	46–17	1	
M.104/4R	**Woodscrew,** securing Waist Facia to Trim Panel		2	
BD.25479	TRIM PANEL ASSEMBLY FOR 'B/C' POSTS		2	Supersedes BD.17216 and BD.17217
BD.24953	**Clip,** securing Trim Panels to 'B/C' Posts		12	
BD.17317	**Waist Facia Assembly,** for 'B/C' Posts		2	
M.104/4R	**Woodscrew,** securing Waist Facias to Trim Panels		4	
BD.17202	FACIA ASSEMBLY FOR L.H. FRONT DOOR		1	
BD.17203	FACIA ASSEMBLY FOR R.H. FRONT DOOR	46–18	1	
DA.506/20C	**Screw,** Self-Tapping, securing Facias to Inner Door Panels		8	
BD.18525	CAPPING FOR L.H. FRONT DOOR		1	Fitted to Chassis No. 1A.1001 and subs. 1A.20001 and subs.
BD.18526	CAPPING FOR R.H. FRONT DOOR	46–19	1	Fitted to Chassis No. 1A.1001 and subs. 1A.20001 and subs.
DAC.604/12C	**Screw,** Self-Tapping, securing Cappings		8	
BD.543/1	**Cup Washer,** on Screws		8	
BD.17282/2	WEATHER STRIP BETWEEN CAPPINGS AND DOOR GLASSES (18 5/8" LONG)		2	
BD.22312	**Clip,** securing Weather Strip		8	
BD.18525	CAPPING FOR L.H. FRONT DOOR		1	Fitted to Chassis No. 1K.1001 and subs. 1K.30001 and subs.
BD.18526	CAPPING FOR R.H. FRONT DOOR		1	
BD.17203/2	**Hardwood Fillet,** under L.H. Door Capping		1	
BD.17202/2	**Hardwood Fillet,** under R.H. Door Capping		1	
DA.506/20C	**Screw,** Self-Tapping, securing Fillet to Door		8	
DAC.604/12C	**Screw,** Self-Tapping, securing Door Cappings to Fillets		8	
BD.543/1	**Cup Washer,** on Self-Tapping Screws		8	
BD.17282/1	WEATHER STRIP BETWEEN CAPPINGS AND DOOR GLASSES (18 1/8" LONG)		2	
BD.22884	**Clip,** securing Weather Strip		8	
BD.30400	TRIMMED ROLL ASSEMBLY, BELOW L.H. FRONT DOOR CAPPING		1	
BD.30399	TRIMMED ROLL ASSEMBLY, BELOW R.H. FRONT DOOR CAPPING		1	

Part No.	Description	Plate No.	No. per Unit	Remarks
BD.17204	**FACIA ASSEMBLY FOR L.H. REAR DOOR**		1	
BD.17205	**FACIA ASSEMBLY FOR R.H. REAR DOOR**	46–20	1	
DA.506/20C	**Screw,** Self-Tapping, securing Facias to Inner Door Panels		8	
BD.18527	**CAPPING FOR L.H. REAR DOOR**		1	Fitted to Chassis No. 1A.1001 and subs. 1A.20001 and subs.
BD.18528	**CAPPING FOR R.H. REAR DOOR**	46–21	1	
DAC.604/12C	**Screw,** Self-Tapping, securing Cappings		8	
BD.543/1	**Cup Washer,** on Screws		8	
BD.18527	**CAPPING, FOR L.H. REAR DOOR**		1	Fitted to Chassis No. 1K.1001 and subs. 1K.30001 and subs.
BD.18528	**CAPPING, FOR R.H. REAR DOOR**		1	
BD.17205/2	**Hardwood Fillet** under L.H. Door Capping		1	
BD.17204/2	**Hardwood Fillet** under R.H. Door Capping		1	
DA.506/20C	**Screw,** Self-Tapping, securing Fillet to Door		8	
DAC.604/12C	**Screw,** Self-Tapping, securing Door Cappings to Fillets		8	
BD.543/1	**Cup Washer,** on Self-Tapping Screws		8	
BD.30402	**TRIMMED ROLL ASSEMBLY, BELOW L.H. REAR DOOR CAPPING**		1	Fitted to Chassis No. 1K.1001 and subs. 1K.30001 and subs.
BD.30401	**TRIMMED ROLL ASSEMBLY, BELOW R.H. REAR DOOR CAPPING**		1	
BD.17282/1	**WEATHER STRIP BETWEEN CAPPINGS AND DOOR GLASSES (18 1/8" LONG)**		2	
BD.22884	**Clip,** securing Weather Strip		8	
BD.17282/1	**WEATHER STRIP BETWEEN CAPPINGS AND DOOR GLASSES (18 1/8" LONG)**		2	
BD.22312	**Clip,** securing Weather Strip		8	
BD.17125	**NEWSPAPER TRAY AT BOTTOM OF INSTRUMENT PANEL**	46–22	1	
BD.18121	**CAPPING ASSEMBLY AT FRONT EDGE OF NEWSPAPER TRAY**	46–23	1	
BD.17642	**Chrome Finisher,** at L.H. end of Capping	46–24	1	
BD.17643	**Chrome Finisher,** at R.H. end of Capping	46–25	1	
M.102/4R	**Woodscrew,** securing Finishers to Cappings		2	
BD.19089	**Stud Plate Attachment,** fixing Capping to Attachment Bracket at L.H. side		2	
BD.19090	**Stud Plate Assembly** fixing Capping to Attachment Bracket at R.H. side		1	
M.104/4R	**Woodscrew,** securing Fixing Plates to Cappings		4	
BD.17681	**Attachment Bracket,** L.H., fixing Capping to Facia		1	
BD.17680	**Attachment Bracket,** R.H., fixing Capping to Facia		1	
MJ.204/4R	**Woodscrew,** securing Attachment Brackets to Facias		2	
UFN.119/L	**Nut,** securing Cappings to Attachment Brackets		2	
C.723/A	**Washer,** Shakeproof, under Nuts		2	

BODY FITTINGS

Part No.	Description	Plate No.	No. per Unit	Remarks
BD.18650	**AIR DISTRIBUTOR BOX UNDER DASH**	46–26	1	
BD.16983/2	**Felt Seal**, for Air Distributor Box		1	
UFS.125/4R	**Screw**, Set, securing Air Distributor Box to Dash		2	
UFS.125/7R	**Screw**, Set, securing Air Distributor Box to Dash		2	
C,724	**Washer**, Shakeproof, on Setscrews		4	
FW.104/T	**Washer**, Plain, on Setscrews		2	
BD.24352	**CONSOLE ASSEMBLY, COMPLETE**	46–27	1	**REQUIRED FOR CARS WITH AUTOMATIC TRANSMISSION**
BD.24353	**Radio Panel Assembly**	46–28	1	Fitted to Chassis No.
BD.24269	**Facia Assembly**, mounting Heater/Demister Controls and Radio Control Unit	46–29	1	1A.1001 and subs.
MJ.204/4R	**Woodscrew**, securing Facia to Radio Panel		6	1A.20001 and subs.
BD.17567	**Loudspeaker Grille Assembly**	46–30	1	1K.1001 and subs.
BD.21801	**Spring Clip**, securing Grille to Panel		4	1K.30001 and subs.
AW.104/T	**Washer**, Plain, under Shakeproof Washers		4	
UCS.416/4H	**Screw**, Set, securing Console to Newspaper Tray		3	
C.2501/2	**Washer**, on Setscrews		3	
BD.21072	**Knurled Knob**, securing Console to Transmission Cover	46–31	2	
BD.17590	**Bezel**, for Knobs	46–32	2	
BD.25314	**Blanking Piece**, behind Loudspeaker Grille		1	**NOT REQUIRED WHEN RADIO IS FITTED**
BD.25315	**Blanking Piece**, sealing Aperture on underside of Console		1	
BD.24363	**Radio Panel Finisher Assembly**, blanking Radio Aperture		1	
BD.24578	**Clamp**, for Radio Panel Finisher		2	
MJ.204/4R	**Woodscrew**, securing Clamps		2	
BD.31195	**CONSOLE ASSEMBLY, COMPLETE**		1	**REQUIRED FOR CARS WITH MANUAL OR OVERDRIVE TRANSMISSION**
BD.31196	**Radio Panel Assembly**		1	Fitted to Chassis No.
BD.31386	**Loudspeaker Mounting Panel Trimmed Assembly**		1	1A.12581 and subs.
BD.25350	**Loudspeaker Grille Assembly**		1	1A.20565 and subs.
BD.26854	**Gasket**, for Loudspeaker Grille		1	1K.1001 and subs. 1K.30001 and subs.
BD.31389/2	**Distance Piece** (long), for Loudspeaker Grille		4	Not Required When Loudspeaker Is Fitted
AW.102/T	**Washer**, Plain, under Nuts		8	
BD.31389/1	**Distance Piece** (short), for Loudspeaker Grille		4	Required When Loudspeaker Is Fitted
AW.102/T	**Washer**, Plain, under Nuts		4	
UFN.119/L	**Nut**, securing Grille to Panel		4	
C.723A	**Washer**, Shakeproof, under Nuts		4	
UCS.416/4H	**Screw**, Set, securing Console to Newspaper Tray		3	
C.2501/2	**Washer**, on Setscrews		3	
BD.21072	**Knurled Knob**, securing Console to Transmission Cover		2	
BD.17590	**Bezel**, for Knobs		2	
BD.24447	**ASH TRAY ASSEMBLY, IN CONSOLE**	46–33	1	
BD.24426	**Bracket**, mounting Ash Tray	46–34	1	
BD.24567	**CHROME FINISHER AROUND RADIO CONTROL UNIT APERTURE**	46–35	1	
BD.24363	**CAPPING ASSEMBLY IN RADIO CONTROL UNIT APERTURE**	46–36	1	**NOT REQUIRED WHEN RADIO IS FITTED**
BD.24578	**Clamp**, retaining Capping	46–37	2	
MJ.204/4R	**Woodscrew**, securing Clamps		2	

Part No.	Description	Plate No.	No. per Unit	Remarks
	CARPETS FOR R.H. DRIVE CARS (1201)			**REQUIRED FOR CARS FITTED WITH AUTOMATIC TRANSMISSION**
BD.18653	**Carpet,** for L.H. Toe Board	47–1	1	Fitted from Chassis No. 1A.1001 to 1A.4508
BD.19044	**Carpet,** for R.H. Toe Board	47–2	1	
BD.18655	**Carpet,** for Dipper Pad	47–3	1	
BD.24731	**Carpet,** for L.H. Front Floor	47–4	1	
BD.24732	**Carpet,** for R.H. Front Floor	47–5	1	
BD.18655	**Carpet,** for Dipper Pad		1	Fitted to Chassis No. 1A.4509 and subs.
BD.27314	**Carpet,** for L.H. Toe Board and Front Floor		1	
BD.27315	**Carpet,** for R.H. Toe Board and Front Floor		1	
BD.27985	**Carpet,** for Gearbox	47–6	1	Fitted to Chassis No. 1A.1001 and subs. Also 1K.1001 to 1K.3224
BD.18670	**Carpet,** for centre of Cross Member under Front Seats	47–7	1	
BD.18669	**Carpet,** for sides of Cross Member under Front Seats	47–8	2	
BD.18667	**Carpet,** for L.H. Door Sill	47–9	1	
BD.18668	**Carpet,** for R.H. Door Sill	47–10	1	
BD.18666	**Carpet,** for Propeller Shaft Tunnel	47–11	1	
BD.24733	**Carpet,** for L.H. Rear Floor	47–12	1	
BD.24734	**Carpet,** for R.H. Rear Floor	47–13	1	
BD.18671	**Carpet,** for L.H. Rear Seat Valance	47–14	1	
BD.18672	**Carpet,** for R.H. Rear Seat Valance	47–15	1	
	CARPETS FOR L.H. DRIVE CARS (1201)			**REQUIRED FOR CARS FITTED WITH AUTOMATIC TRANSMISSION**
BD.19043	**Carpet,** for L.H. Toe Board	47–16	1	Fitted from Chassis No. 1A.20001 to 1A.20282
BD.18659	**Carpet,** for R.H. Toe Board	47–17	1	
BD.24729	**Carpet,** for L.H. Front Floor	47–18	1	
BD.24730	**Carpet,** for R.H. Front Floor	47–19	1	
BD.27312	**Carpet,** for L.H. Toe Board and Front Floor		1	Fitted to Chassis No. 1A.20283 and subs.
BD.27313	**Carpet,** for R.H. Toe Board and Front Floor		1	
BD.27985	**Carpet,** for Gearbox	47–6	1	Fitted to Chassis No. 1A.20001 and subs. Also 1K.30001 to 1K.30077
BD.18670	**Carpet,** for centre of Cross Member under Front Seats	47–7	1	
BD.18669	**Carpet,** for sides of Cross Member under Front Seats	47–8	2	
BD.18667	**Carpet,** for L.H. Door Sill	47–9	1	
BD.18668	**Carpet,** for R.H. Door Sill	47–10	1	
BD.18666	**Carpet,** for Propeller Shaft Tunnel	47–11	1	
BD.24733	**Carpet,** for L.H. Rear Floor	47–12	1	
BD.24734	**Carpet,** for R.H. Rear Floor	47–13	1	
BD.18671	**Carpet,** for L.H. Rear Seat Valance	47–14	1	
BD.18672	**Carpet,** for R.H. Rear Seat Valance	47–15	1	
	CARPETS FOR R.H. DRIVE CARS (1201)			**REQUIRED FOR CARS WITH MANUAL OR OVERDRIVE TRANSMISSION**
BD..32001	**Carpet,** for L.H. Bell Housing		1	
BD.32000	**Carpet,** for R.H. Bell Housing		1	
BD.18655	**Carpet,** for Dipper Pad		1	Fitted to Chassis No. 1A.12581 and subs. Also 1K.1001 to 1K.3224
BD.26414	**Carpet,** for L.H. Front Floor		1	
BD.31264	**Carpet,** for R.H. Front Floor		1	
BD.32002	**Carpet,** for Gearbox		1	
BD.32002/1	**Carpet,** for Gearbox Top Cover		1	
BD.32002/2	**Carpet,** for Gearbox Side Cover R.H.		1	
BD.32002/3	**Carpet,** for Gearbox Side Cover L.H.		1	
BD.18670	**Carpet,** for centre of Cross Member under Front Seats		1	
BD.18669	**Carpet,** for sides of Cross Member under Front Seats		1	
BD.18667	**Carpet,** for L.H. Door Sill		1	
BD.18668	**Carpet,** for R.H. Door Sill		1	
BD.18666	**Carpet,** for Propeller Shaft Tunnel		1	
BD.26421	**Carpet,** for L.H. Rear Floor		1	
BD.26422	**Carpet,** for R.H. Rear Floor		1	
BD.18671	**Carpet,** for L.H. Rear Seat Valance		1	
BD.18672	**Carpet,** for R.H. Rear Seat Valance		1	
	CARPETS FOR L.H. DRIVE CARS (1201)			**REQUIRED FOR CARS WITH MANUAL OR OVERDRIVE TRANSMISSION**
BD.32001	**Carpet,** for L.H. Bell Housing		1	
BD.32000	**Carpet,** for R.H. Bell Housing		1	
BD.32002/1	**Carpet,** for Gearbox Top Cover		1	Fitted to Chassis No. 1A.20565 and subs. Also 1K.30001 to 1K.30077
BD.32002/2	**Carpet,** for Gearbox Side Cover R.H.		1	
BD.32002/3	**Carpet,** for Gearbox Side Cover L.H.		1	
BD.31263	**Carpet,** for L.H. Front Floor		1	
BD.26413	**Carpet,** for R.H. Front Floor		1	
BD.32002	**Carpet,** for Gearbox		1	
BD.18670	**Carpet,** for centre of Cross Member under Front Seats		1	

BODY FITTINGS

Part No.	Description	Plate No.	No. per Unit	Remarks
	CARPETS FOR L.H. DRIVE CARS (1201) *(contd.)*			
BD.18669	**Carpet**, for sides of Cross Member under Front Seats		1	
BD.18667	**Carpet**, for L.H. Door Sill		1	
BD.18668	**Carpet**, for R.H. Door Sill		1	
BD.18666	**Carpet**, for Propeller Shaft Tunnel		1	
BD.26421	**Carpet**, for L.H. Rear Floor		1	
BD.26422	**Carpet**, for R.H. Rear Floor		1	
BD.18671	**Carpet**, for L.H. Rear Seat Valance		1	
BD.18672	**Carpet**, for R.H. Rear Seat Valance		1	
	CARPETS FOR R.H. DRIVE CARS (NYLON)			**REQUIRED FOR CARS WITH AUTOMATIC TRANSMISSION**
BD.34886	**Carpet**, for L.H. Front Floor		1	Fitted to Chassis No. 1K.3225 and subs.
BD.34885	**Carpet**, for R.H. Front Floor		1	
BD.34906	**Carpet**, for Gearbox		1	
BD.34877	**Carpet**, for centre of Cross Member under Front Seats		1	
BD.34891	**Carpet**, for L.H. side of Cross Member under Front Seats		1	
BD.34890	**Carpet**, for R.H. side of Cross Member under Front Seats		1	
BD.34879	**Carpet**, for L.H. Door Sill		1	
BD.34880	**Carpet**, for R.H. Door Sill		1	
BD.34878	**Carpet**, for Propeller Shaft Tunnel		1	
BD.34888	**Carpet**, for L.H. Rear Floor		1	
BD.34887	**Carpet**, for R.H. Rear Floor		1	
BD.34881	**Carpet**, for L.H. Rear Seat Valance		1	
BD.34882	**Carpet**, for R.H. Rear Seat Valance		1	
	CARPETS FOR L.H. DRIVE CARS (NYLON)			**REQUIRED FOR CARS WITH AUTOMATIC TRANSMISSION**
BD.34884	**Carpet** for L.H. Front Floor		1	Fitted to Chassis No. 1K.30078 and subs.
BD.34883	**Carpet**, for R.H. Front Floor		1	
BD.34889	**Carpet**, for Gearbox		1	
BD.34877	**Carpet**, for centre of Cross Member under Front Seats		1	
BD.34891	**Carpet**, for L.H. side of Cross Member under Front Seats		1	
BD.34890	**Carpet**, for R.H. side of Cross Member under Front Seats		1	
BD.34879	**Carpet**, for L.H. Door Sill		1	
BD.34880	**Carpet**, for R.H. Door Sill		1	
BD.34878	**Carpet**, for Propeller Shaft Tunnel		1	
BD.34871	**Carpet**, for L.H. Rear Floor		1	
BD.34870	**Carpet**, for R.H. Rear Floor		1	
BD.34881	**Carpet**, for L.H. Rear Seat Valance		1	
BD.34882	**Carpet**, for R.H. Rear Seat Valance		1	
	CARPETS FOR R.H. DRIVE CARS (NYLON)			**REQUIRED FOR CARS WITH MANUAL OR OVERDRIVE TRANSMISSION**
BD.34873	**Carpet**, for L.H. Bell Housing		1	Fitted to Chassis No. 1K.3225 and subs.
BD.34872	**Carpet**, for R.H. Bell Housing		1	
BD.34869	**Carpet**, for L.H. Front Floor		1	
BD.34859	**Carpet**, for R.H. Front Floor		1	
BD.34874	**Carpet**, for Gearbox		1	
BD.34877	**Carpet**, for centre of Cross Member under Front Seats		1	
BD.34891	**Carpet**, for L.H. side of Cross Member under Front Seats		1	
BD.34890	**Carpet**, for R.H. side of Cross Member under Front Seats		1	
BD.34880	**Carpet**, for L.H. Door Sill		1	
BD.34879	**Carpet**, for R.H. Door Sill		1	
BD.34878	**Carpet**, for Propeller Shaft Tunnel		1	
BD.34888	**Carpet**, for L.H. Rear Floor		1	
BD.34887	**Carpet**, for R.H. Rear Floor		1	
BD.34882	**Carpet**, for L.H. Rear Seat Valance		1	
BD.34881	**Carpet**, for R.H. Rear Seat Valance		1	
	CARPETS FOR L.H. DRIVE CARS (NYLON)			**REQUIRED FOR CARS WITH MANUAL OR OVERDRIVE TRANSMISSION**
BD.34873	**Carpet**, for L.H. Bell Housing		1	Fitted to Chassis No. 1K.30078 and subs.
BD.34872	**Carpet**, for R.H. Bell Housing		1	
BD.34858	**Carpet**, for L.H. Front Floor		1	
BD.34866	**Carpet**, for R.H. Front Floor		1	
BD.34874	**Carpet**, for Gearbox		1	
BD.34877	**Carpet**, for centre of Cross Member under Front Seats		1	
BD.34891	**Carpet**, for L.H. side of Cross Member under Front Seats		1	

Part No.	Description	Plate No.	No. per Unit	Remarks
	CARPETS FOR L.H. DRIVE CARS (Nylon) *(contd.)*			
BD.34890	**Carpet**, for R.H. side of Cross Member under Front Seats		1	REQUIRED FOR CARS WITH MANUAL OR OVERDRIVE TRANSMISSION
BD.34880	**Carpet**, for L.H. Door Sill		1	
BD.34879	**Carpet**, for R.H. Door Sill		1	
BD.34878	**Carpet**, for Propeller Shaft Tunnel		1	Fitted to Chassis No. 1K. 30078 and subs.
BD.34871	**Carpet**, for L.H. Rear Floor		1	
BD.34870	**Carpet**, for R.H. Rear Floor		1	
BD.34882	**Carpet**, for L.H. Rear Seat Valance		1	
BD.34881	**Carpet**, for R.H. Rear Seat Valance		1	
BD.13102	**BOW TYPE CARPET FASTENER**	47–20	8	Fitted from Chassis No. 1A.1001 to 1A.4508 1A.20001 to 1A.20282
BD.11769	**Seal**, in Floor, for Bow Type Carpet Fasteners	47–21	8	
BD.28927	**STUD FASTENER FOR CARPETS**		20	Fitted to Chassis No. 1A.4509 and subs. 1A.20283 and subs.
BD.22665	**Socket**, in floor receiving Stud Fastener		20	
	FELT, UNDER CARPETS			REQUIRED FOR CARS FITTED WITH AUTOMATIC TRANSMISSION
BD.24740	**Felt**, under R.H. Front Floor Carpet		1	
BD.24741	**Felt**, under L.H. Front Floor Carpet		1	
BD.24746	**Felt**, under Gearbox Carpet		1	
BD.24745	**Felt**, over R.H. Cross Brace on Floor		1	
BD.24744	**Felt**, over L.H. Cross Brace on Floor		1	
BD.12809	**Felt**, under Front Seat Support Packing		4	
BD.17828	**Felt**, over R.H. Front Door Sill		1	
BD.17827	**Felt**, over L.H. Front Door Sill		1	
BD.17830	**Felt**, over R.H. Rear Door Sill		1	
BD.17829	**Felt**, over L.H. Rear Door Sill		1	
BD.25026	**Felt**, for Propeller Shaft Tunnel		1	
BD.24742	**Felt**, under R.H. Rear Floor Carpet		1	
BD.24743	**Felt**, under L.H. Rear Floor Carpet		1	
BD.12822	**Felt**, for Front of Rear Seat		1	
BD.12827	**Felt**, for Rear Seat Pan		2	
	FELT, UNDER CARPETS			REQUIRED FOR CARS WITH MANUAL OR OVERDRIVE TRANSMISSION
BD.26373	**Felt**, under R.H. Front Floor Carpet		1	
BD.26374	**Felt**, under L.H. Front Floor Carpet		1	
BD.24746	**Felt**, under Gearbox Carpet		1	
BD.24745	**Felt**, over R.H. Cross Member		1	
BD.24744	**Felt**, over L.H. Cross Member		1	
BD.17828	**Felt**, over R.H. Front Door Sill		1	
BD.17827	**Felt**, over L.H. Front Door Sill		1	
BD.17830	**Felt**, over R.H. Rear Door Sill		1	
BD.17829	**Felt**, over L.H. Rear Door Sill		1	
BD.26375	**Felt**, under R.H. Rear Floor Carpet		1	
BD.26376	**Felt**, under L.H. Rear Floor Carpet		1	
BD.12822	**Felt**, for Front of Rear Seat		1	
BD.12827	**Felt**, for Rear Seat Pan		2	
BD.26582	**Felt**, at Scuttle Side		2	
	INSULATING FELTS			REQUIRED FOR CARS FITTED WITH AUTOMATIC TRANSMISSION
BD.17835	**Hardura Felt**, for L.H. Toe Board (22½" x 16½")		1	L.H. DRIVE CARS ONLY
BD.17836	**Hardura Felt**, for R.H. Toe Board (21" x 13½")		1	
BD.13210	**Hardura Felt**, for L.H. Toe Board (22½" x 16½")		1	R.H. DRIVE CARS ONLY
BD.13211	**Hardura Felt**, for R.H. Toe Board (21" x 13½")		1	
BD.24749	**Hardura Felt**, for Gearbox Top Collar (29½" x 17")		1	
BD.13213	**Hardura Felt**, for R.H. side of Gearbox (12" x 6½")		1	
BD.24747	**Hardura Felt**, for Gearbox, rear		1	
BD.24748	**Hardura Felt**, for Gearbox, front		1	
BD.24750	**Hardura Felt**, for Rear Tunnel		1	
BD.17841	**Hardura Felt**, for L.H. Scuttle Top Panel		1	L.H. DRIVE CARS ONLY
BD.17842	**Hardura Felt**, for R.H. Scuttle Top Panel		1	
BD.17839	**Hardura Felt**, for L.H. Scuttle Top Panel		1	R.H. DRIVE CARS ONLY
BD.17840	**Hardura Felt**, for R.H. Scuttle Top Panel		1	
BD.17881	**Hardura Felt**, for Heater Box (centre upper)		1	
BD.17868	**Hardura Felt**, for Heater Box (centre)		1	
BD.17879	**Hardura Felt**, for Heater Box (side)		1	

BODY FITTINGS

Part No.	Description	Plate No.	No. per Unit	Remarks
	INSULATING FELTS *(contd.)*			
BD.17880	**Hardura Felt,** for Heater Box (face)		1	
BD.17882	**Hardura Felt,** for centre of Scuttle Top Panel		1	
BD.17883	**Hardura Felt,** for L.H. side of Screen Rail		1	
BD.17884	**Hardura Felt,** for R.H. side of Screen Rail		1	
BD.17875	**Hardura Felt,** for face of L.H. side of Cubby Box		1	
BD.17876	**Hardura Felt,** for face of R.H. side of Cubby Box		1	
BD.17877	**Hardura Felt,** for centre of Instrument Board, L.H.		1	
BD.17878	**Hardura Felt,** for centre of Instrument Board, R.H.		1	
BD.17874	**Hardura Felt,** for Top of R.H. Dash Panel		1	
BD.17873	**Hardura Felt,** for Top of L.H. Dash Panel		1	
BD.20537	**Hardura Felt,** for Steering Column Bracket (4" x 6")		1	
BD.17833	**Rubberised Felt,** for L.H. Toe Board (21½" x 16")		1	L.H. DRIVE CARS ONLY
BD.17834	**Rubberised Felt,** for R.H. Toe Board (18" x 13¼")		1	L.H. DRIVE CARS ONLY
BD.13229	**Rubberised Felt,** for L.H. Toe Board (21½" x 16")		1	R.H. DRIVE CARS ONLY
BD.13230	**Rubberised Felt,** for R.H. Toe Board (18" x 13¼")		1	R.H. DRIVE CARS ONLY
BD.13231	**Rubberised Felt,** for R.H. side of Gearbox Collar (15" x 8")		1	
BD.13232	**Rubberised Felt,** for R.H. side of Gearbox (5½" x 9¼")		1	
BD.13233	**Rubberised Felt,** for L.H. side of Gearbox Collar (16" x 8½")		1	
BD.17866	**Rubberised Felt,** for Heater Box		1	
BD.17867	**Rubberised Felt,** for Heater Box		1	
BD.17860	**Rubberised Felt,** for outer face of L.H. side of Cubby Box		1	
BD.17862	**Rubberised Felt,** for outer face of R.H. side of Cubby Box		1	
BD.17861	**Rubberised Felt,** for inner face of L.H. side of Cubby Box		1	
BD.17863	**Rubberised Felt,** for inner face of R.H. side of Cubby Box		1	
BD.17858	**Rubberised Felt,** for top R.H. Dash Panel		1	
BD.17859	**Rubberised Felt,** for top L.H. Dash Panel		1	
BD.17864	**Rubberised Felt,** for upper face of Centre Instrument Board		1	
BD.17865	**Rubberised Felt,** for upper face of Centre Instrument Board		1	
BD.26582	**Rubberised Felt,** for Scuttle Sides (10¾" x 10½")		2	
BD.13244	**Rubberised Felt,** under Scuttle Top Panel (7½" x 5½")		4	Superseded by BD.28847 (2 off)
BD.13245	**Rubberised Felt,** under Scuttle Top Panel (7½" x 4")		2	Superseded by BD.28847 (2 off)
BD.13246	**Rubberised Felt,** under Scuttle Top Panel (7½" x 3")		2	Superseded by BD.28847 (2 off)
BD.13247	**Rubberised Felt,** under Scuttle Top Panel (5¾" x 5½")		2	Superseded by BD.28848
BD.13248	**Rubberised Felt,** under Scuttle Top Panel (5½" x 4½")		1	Superseded by BD.28848
BD.13249	**Rubberised Felt,** under Scuttle Top Panel (5½" x 3½")		1	Superseded by BD.28848
BD.13251	**Rubberised Felt,** R.H., under Scuttle Top Panel (6½" x 6")		2	Superseded by BD.28849
BD.13252	**Rubberised Felt,** R.H., under Scuttle Top Panel (5½" x 5")		1	Superseded by BD.28849
BD.13255	**Rubberised Felt,** R.H., under Scuttle Top Panel (5½" x 3½")		1	Superseded by BD.28849
BD.13250	**Rubberised Felt,** L.H., under Scuttle Top Panel (6½" x 6")		2	Superseded by BD.28850
BD.13253	**Rubberised Felt,** L.H., under Scuttle Top Panel (5" x 4½")		1	Superseded by BD.28850
BD.13254	**Rubberised Felt,** L.H., under Scuttle Top Panel (5½" x 3½")		1	Superseded by BD.28850
	INSULATING FELTS			REQUIRED FOR CARS WITH STANDARD MANUAL OR OVERDRIVE TRANSMISSION
BD.13210	**Hardura Felt,** for L.H. Toe Board (22½" x 16½")		1	R.H. DRIVE CARS ONLY
BD.13211	**Hardura Felt,** for R.H. Toe Board (21" x 13½")		1	R.H. DRIVE CARS ONLY
BD.17835	**Hardura Felt,** for L.H. Toe Board (22½" x 16½")		1	L.H. DRIVE CARS ONLY
BD.17836	**Hardura Felt,** for R.H. Toe Board (21" x 13½")		1	L.H. DRIVE CARS ONLY
BD.32003	**Hardura Felt,** for R.H. side of Bell Housing		1	
BD.32004	**Hardura Felt,** for L.H. side of Bell Housing		1	
BD.32005	**Hardura Felt,** for R.H. Gearbox Collar		1	
BD.32006	**Hardura Felt,** for L.H. Gearbox Collar		1	
BD.32007	**Hardura Felt,** for R.H. Gearbox side		1	
BD.32008	**Hardura Felt,** for L.H. Gearbox side		1	
BD.32009	**Hardura Felt,** for Gearbox Top		1	
BD.13213	**Hardura Felt,** for Gearbox Side (12" x 6½")		1	
BD.24750	**Hardura Felt,** for Rear Tunnel		1	
BD.17839	**Hardura Felt,** for L.H. Scuttle Top Panel		1	R.H. DRIVE CARS ONLY
BD.17840	**Hardura Felt,** for R.H. Scuttle Top Panel		1	R.H. DRIVE CARS ONLY
BD.17841	**Hardura Felt,** for L.H. Scuttle Top Panel		1	L.H. DRIVE CARS ONLY
BD.17842	**Hardura Felt,** for R.H. Scuttle Top Panel		1	L.H. DRIVE CARS ONLY
BD.17881	**Hardura Felt,** for Heater Box (centre upper)		1	
BD.17879	**Hardura Felt,** for Heater Box (side)		1	
BD.17880	**Hardura Felt,** for Heater Box (face)		1	
BD.17882	**Hardura Felt,** for centre of Scuttle Top Panel		1	
BD.17883	**Hardura Felt,** for L.H. side of Screen Rail		1	
BD.17884	**Hardura Felt,** for R.H. side of Screen Rail		1	

Part No.	Description	Plate No.	No. per Unit	Remarks
	INSULATING FELTS *(contd.)*			
BD.17833	**Rubberised Felt,** for L.H. Toe Board (21½" x 16")		1	L.H. DRIVE CARS ONLY
BD.17834	**Rubberised Felt,** for R.H. Toe Board (20 3/16" x 14")		1	
BD.13229	**Rubberised Felt,** for L.H. Toe Board (21½" x 16")		1	R.H. DRIVE CARS ONLY
BD.13230	**Rubberised Felt,** for R.H. Toe Board (18" x 13¼")		1	
BD.32011	**Rubberised Felt,** for R.H. side of Bell Housing		1	
BD.32012	**Rubberised Felt,** for L.H. side of Bell Housing		1	
BD.32013	**Rubberised Felt,** for R.H. Gearbox Collar		1	
BD.32014	**Rubberised Felt,** for L.H. Gearbox Collar		1	
BD.32015	**Rubberised Felt,** for R.H. Gearbox side		1	
BD.32016	**Rubberised Felt,** for Gearbox Top		1	
BD.13232	**Rubberised Felt,** for R.H. side of Gearbox		1	
	ANTI-DRUM MATERIAL			REQUIRED FOR CARS FITTED WITH AUTOMATIC TRANSMISSION
BD.28847	**Pad** (Foam Rubber) under Scuttle Top Panel (6½" x 5½")		2	Supersedes BD.13244, BD.13245, & BD.13246
BD.28848	**Pad** (Foam Rubber) under Scuttle Top Panel (5½" x 5½")		1	Supersedes BD.13247, BD.13248, & BD.13249
BD.28849	**Pad,** R.H. (Foam Rubber) under Scuttle Top Panel (5" x 3½")		1	Supersedes BD.13251, BD.13252, & BD.13255
BD.28850	**Pad,** L.H. (Foam Rubber) under Scuttle Top Panel (5" x 3½")		1	Supersedes BD.13250, BD.13253, & BD.13254
BD.24751	**Flintkote,** for R.H. Front Floor		1	
BD.24752	**Flintkote,** for L.H. Front Floor		1	
BD.24753	**Flintkote,** for R.H. Rear Floor		1	
BD.24754	**Flintkote,** for L.H. Rear Floor		1	
BD.24793	**Flintkote,** for Rear Tunnel		1	
	ANTI-DRUM MATERIAL			REQUIRED FOR CARS WITH STANDARD MANUAL OR OVERDRIVE TRANSMISSION
BD.11665	**Flintkote,** for Rear Squab Back		1	
BD.18147	**Flintkote,** for sides of Rear Seat Squab		2	
BD.12805	**Flintkote,** for lower front of Gearbox Cover		1	
BD.12806	**Flintkote,** for lower rear of Gearbox Cover		1	
BD.12808	**Flintkote,** for top front of Gearbox		1	
BD.27282	**Dedshete,** for R.H. Front Floor		1	
BD.27283	**Dedshete,** for L.H. Front Floor		1	
	TRIM DETAILS			
BD.18040	**Vynide,** Rear Seat Pan, Front (8" x 74")		1	
BD.24736	**Vynide,** Front Seat Channel Support (4" x 9½")		4	
BD.18041	**Vynide,** Rear Seat Side (29½" x 5½")		1	
BD.18042	**Vynide,** Rear Seat Side (29½" x 5½")		1	
BD.18111	**Vynide,** Front Door, Rear Top (5½" x 8")		2	
BD.18109	**Vynide,** Front Door, Centre Top (3" x 5½")		2	
BD.18110	**Vynide,** Front Door, Centre Top (6½" x 17")		2	
BD.18112	**Vynide,** Rear Door, Centre (5½" x 8½")		2	
BD.18077	**Vynide,** Heater Duct facing (3" x 13")		2	
BD.24737	**Vynide,** Crossmember, Centre (12" x 19")		2	
BD.23944	**TRIM PANEL ASSEMBLY ON WINDSCREEN HEADER RAIL**	48–1	1	
DA.404/6C	**Screw,** Self-Tapping, securing Trim Panel		4	
BD.23946	**HEAD LINING**		1	
BD.23928	**TRIM ROLL ASSEMBLY ABOVE R.H. CANTRAIL**	48–2	1	

BODY FITTINGS

Part No.	Description	Plate No.	No. per Unit	Remarks
BD.23929	**TRIM ROLL ASSEMBLY ABOVE L.H. CANTRAIL**		1	
DA.404/8C	**Screw,** Self-Tapping, securing Trim Rolls		10	
BD.26174	**Chrome Finisher,** at rear of R.H. Trim Roll	48–3	1	
BD.26173	**Chrome Finisher,** at rear of L.H. Trim Roll		1	
DA.404/6C	**Screw,** Self-Tapping, securing Finishers		2	
BD.23932	**TRIM PANEL ASSEMBLY, BETWEEN TRIM ROLL AND R.H. CANTRAIL CAPPING**	48–4	1	
BD.23933	**TRIM PANEL ASSEMBLY, BETWEEN TRIM ROLL AND L.H. CANTRAIL CAPPING**		1	
DA.404/8C	**Screw,** Self-Tapping, securing Trim Panels		10	
C.15441	**INTERIOR LIGHT AT TOP OF CENTRE PILLARS (5911)**	48–5	2	Fitted from Chassis No. 1A.1001 to 1A.3299 1A.20001 to 1A.20256
9592	**Bridge Moulding** (01877)		2	
8245	**Plastic Lens,** complete with Chrome Shield (12434)		2	
989	**Bulb** (12 volt, 6 watt) (989)		2	
C.22304	**INTERIOR LIGHT AT TOP OF CENTRE PILLARS (6352)**		2	Fitted to Chassis No. 1A.3300 and subs. 1A.20257 and subs.
10184	**Lens Shield, and Plinth Assembly** (03737)		2	
10185	**Base Assembly** (03738)		2	
7880	**Bulb** (12 volt, 6 watt Festoon) (254)		2	
DZC.404/6C	**Screw,** Self-Tapping, securing Lamps		4	
BD.23940	**R.H. TRIM ROLL ASSEMBLY AT REAR OF ROOF**	48–6	1	
BD.23941	**L.H. TRIM ROLL ASSEMBLY AT REAR OF ROOF**		1	
BD.18068	**Clip,** retaining Trim Rolls		10	Fitted from Chassis No. 1A.1001 to 1A.2078 1A.20001 to 1A.20134
BD.25974	**Clip,** retaining Trim Rolls		10	Fitted to Chassis No. 1A.2079 and subs. 1A.20135 and subs.
BD.23936	**Bracket,** fixing R.H. Rear Trim Roll	48–7	1	
BD.23937	**Bracket,** fixing L.H. Rear Trim Roll		1	
DA.404/6C	**Screw,** Self-Tapping, securing Brackets		8	
C.19731	**INTERIOR LIGHT BETWEEN REAR TRIM ROLLS (6182)**	48–8	1	Fitted from Chassis No. 1A.1001 to 1A.3584 1A.20001 to 1A.20264 (applies to C.19731 through BD.25204)
9593	**Backplate Assembly** (01893)		1	
8245	**Plastic Lens,** complete with Chrome Shield (12434)		1	
7880	**Bulb** (12 volt, 6 watt Festoon)		1	
BD.25212	**Blind Fix,** for Interior Lamp	48–9	2	
BD.25204	**Fixing Pin,** securing Interior Lamp	48–10	2	

Part No.	Description	Plate No.	No. per Unit	Remarks
C.22304	**INTERIOR LIGHT BETWEEN REAR TRIM ROLLS (6352)**		1	Fitted to Chassis No. 1A.3585 and subs. 1A.20265 and subs. 1K.1001 and subs. 1K.30001 and subs.
10184	**Lens, Shield, and Plith Assembly** (03737)		1	
10185	**Base Assembly** (03738)		1	
7880	**Bulb** (12 volt, 6 watt Festoon) (254)		1	
DJJ.504/8C	**Screw,** Self-Tapping, securing Interior Light		2	
BD.8633/9	**Spire Nut,** receiving Self-Tapping Screws		2	
BD.22522	**TRIM BLOCK ASSEMBLY, MOUNTING INTERIOR LAMP BETWEEN REAR TRIM ROLLS**	48–11	1	REQUIRED ONLY WITH INTERIOR LIGHT C.19731
BD.23938	**TRIM PANEL ASSEMBLY, ABOVE AND AT SIDES OF BACKLIGHT**	48–12	1	
BD.18068	**Clip,** retaining Trim Panel		5	
DA.404/8C	**Screw,** Self-Tapping securing Trim Panel		10	
AW.104/T	**Washer,** Plain, on Self-Tapping Screws		10	
BD.16991	**BACKLIGHT GLASS**	48–13	1	
BD.28329	**BACKLIGHT GLASS ("SUNDYM" TYPE) (ALTERNATIVE TO BD.16991)**		1	
BD.21942	**BACKLIGHT GLASS (CLEAR)**		1	ELECTRICALLY HEATED
BD.22742	**BACKLIGHT GLASS ("SUNDYM" TYPE) (ALTERNATIVE TO BD.21942)**		1	
BD.16992	**RUBBER SEAL AROUND BACKLIGHT GLASS**	48–14	1	NOT REQUIRED WHEN ELECTRICALLY HEATED BACKLIGHT IS FITTED
BD.27127	**RUBBER SEAL AROUNG BACKLIGHT GLASS**		1	FOR USE ONLY WHEN ELECTRICALLY HEATED BACKLIGHT IS FITTED
BD.16993	**CHROME FINISHER AT L.H. SIDE OF SEAL**	48–15	1	
BD.16994	**CHROME FINISHER AT R.H. SIDE OF SEAL**	48–16	1	
BD.5366	**JOINT PIECE FOR CHROME FINISHERS**	48–17	2	
BD.24816	**TRIM PANEL ASSEMBLY, BELOW BACKLIGHT**	48–18	1	
DAJ.404/14C	**Screw,** Self-Tapping, securing Trim Panel to Parcel Shelf Trim		6	
BD.25097	**TRIM PANEL ASSEMBLY, ON PARCEL SHELF**	48–19	1	NOT REQUIRED WHEN RADIO SPEAKER IS FITTED TO PARCEL SHELF
BD.14130	**Clip,** retaining Trim Panel		3	
BD.524/3	**Bifurcated Rivet,** securing Clips		6	

BODY FITTINGS

Part No.	Description	Plate No.	No. per Unit	Remarks
BD.25098	**TRIM PANEL ASSEMBLY, ON PARCEL SHELF**		1	**FOR USE ONLY WHEN RADIO SPEAKER IS FITTED TO PARCEL SHELF**
BD.14130	**Clip**, retaining Trim Panel		3	
BD.524/3	**Bifurcated Rivet**, securing Clips		6	
DA.504/8C	**Screw**, Self-Tapping, securing Trim Panel		2	
BD.17422	**CUSHION ASSEMBLY FOR REAR SEAT**	48–20	1	Fitted to Chassis No. 1A.1001 and subs. 1A.20001 and subs.
BD.17468	**REAR SEAT SQUAB ASSEMBLY**	48–21	1	
BD.32573	**REAR SEAT SQUAB ASSEMBLY**		1	Fitted to Chassis No. 1K.1001 and subs. 1K.30001 and subs.
BD.32570	**CUSHION ASSEMBLY FOR REAR SEAT**		1	
BD.17474	**CENTRE ARMREST ASSEMBLY FOR REAR SEAT**	48–22	1	Fitted to Chassis No. 1A.1001 and subs. 1A.20001 and subs.
BD.32578	**CENTRE ARMREST ASSEMBLY FOR REAR SEAT**		1	Fitted to Chassis No. 1K.1001 and subs. 1K.30001 and subs.
BD.10202	**Hinge Assembly**, at L.H. side of Armrest		1	
BD.10203	**Hinge Assembly**, at R.H. side of Armrest		1	
BD.10565/3	**Screw**, Set, securing Armrest to Squab		6	
FW.104/T	**Washer**, Plain, on Setscrews		6	
UFN.125/L	**Nut**, on Setscrews		6	
C.724	**Washer**, Shakeproof, on Setscrews		6	
BD.17693	**TRIM PANEL ASSEMBLY FOR L.H. REAR QUARTER**	48–23	1	
BD.17694	**TRIM PANEL ASSEMBLY FOR R.H. REAR QUARTER**	48–24	1	
BD.11309	**Clip**, securing Lower Trim Panels		4	
BD.2702	**Retainer**, for Clips		4	
DAZ.404/10C	**Screw**, Self-Tapping, securing Lower Trim Panels		4	
BD.18174	**UPPER TRIM PANEL ASSEMBLY FOR L.H. REAR WHEEL ARCH**	48–25	1	
BD.18175	**UPPER TRIM PANEL ASSEMBLY FOR R.H. REAR WHEEL ARCH**	48–26	1	
BD.14988	**TREAD PLATE ON REAR DOOR SILLS**	48–27	2	
DAZ.504/8C	**Screw**, Self-Tapping, securing Tread Plates		6	
BD.4951	**RUBBER BUFFER FOR REAR DOOR STOP**	48–28	2	
BD.18039	**MAT ASSEMBLY ON TRUNK FLOOR**	48–29	1	
BD.1286	**DURABLE DOT FASTENER SECURING MAT TO TRUNK FLOOR**		6	
BD.1286/1	**Button only**		6	
BD.1286/2	**Socket only**		6	
BD.1286/3	**Stud only**		6	
BD.1814/2	**Rivet**, securing Fasteners to Floor		6	

Part No.	Description	Plate No.	No. per Unit	Remarks
BD.9878	**COVER, ON TRUNK FLOOR, GIVING ACCESS TO PETROL TANK GAUGE UNIT**		1	
BD.9888	**LOCK, COMPLETE, FOR TRUNK LID (WB.3/6853)**	48–30	1	
UFS.319/4H	**Screw**, Set, securing Trunk Lid Lock		4	
C.723/A	**Washer**, Shakeproof, on Setscrews		4	
BD.11572	**RUBBER SEAL FOR TRUNK LID LOCK**	48–31	1	
BD.11472	**ESCUTCHEON FOR TRUNK LID LOCK**	48–32	1	
DAC.606/6C	**Screw**, Self-Tapping, securing Escutcheon		2	
BD.11321	**SHIM FOR TRUNK LID LOCK STRIKER**		As req'd	
BD.14100/1	**SEALING RUBBER AROUND TRUNK LID APERTURE (73" LONG)**	48–33	2	
BD.6581	**RUBBER PLUG IN SLOTS OF REAR NUMBER PLATE PANEL**	48–34	4	
BD.27005	**CHROMIUM PLATED BEADING SURROUNDING REAR NUMBER PLATE PANEL**	48–35	1	
BD.20887	**Nylon Bolt**, securing Beading	48–36	6	
BD.21040	**Special Nut**, on Nylon Bolts		6	
BD.22444	**Sealing Ring** (Polythene) under Nuts		6	
BD.10809	**Staple**, securing Beading at bottom	48–37	4	
C.20630	**REAR NUMBER PLATE ILLUMINATION LAMP**	48–38	1	NOT REQUIRED FOR CARS EXPORTED TO FRANCE
C.20631	**REAR NUMBER PLATE ILLUMINATION LAMP**		1	FOR CARS EXPORTED TO FRANCE
BD.9868/3	**Seal**, between Number Plate Lamp and Trunk Lid	48–39	1	
BD.24521	**'V8' EMBLEM ON TRUNK LID**	48–40	1	
BD.25659	**Friction Bush**, securing Emblem		2	
BD.32126	**'250' EMBLEM ON TRUNK LID**	48–41	1	Fitted only from Chassis No. 1K.1001 and subs. 1K.30001 and subs.
BD.25659	**Friction Bush**, securing Emblem		2	
BD.24520	**'DAIMLER' EMBLEM ON TRUNK LID**	48–42	1	
BD.17725	**'AUTOMATIC' EMBLEM ON TRUNK LID**	48–43	1	Fitted only from Chassis No. 1A.1001 to 1A.5410 1A.20001 to 1A.20310
BD.10735	**Spire Fixing**, securing Emblems	48–44	9	
BD.15293	**Rubber Seal**, under Spire Fixings	48–45	9	
BD.12488	**CASING ASSEMBLY AT L.H. SIDE OF LUGGAGE COMPARTMENT**	48–46	1	

BODY FITTINGS

Part No.	Description	Plate No.	No. per Unit	Remarks
BD.17851	**CASING ASSEMBLY AT R.H. SIDE OF LUGGAGE COMPARTMENT**	48–47	1	
DAC.604/8C	**Screw,** Self-Tapping, securing Casings		18	
BD.532/1	**Cup Washer,** on Screws		18	
BD.11007	**SPARE WHEEL CLAMP ASSEMBLY**	48–48	1	
BD.10637	**Handle,** operating Spare Wheel Clamp		1	
BD.11008	**Screw**		1	
BD.10639	**Plate,** on Screw, holding down Spare Wheel		1	
C.704/1	**Mills Pin,** securing Handle to Screw		1	
BD.10993	**TOOL CONTAINER ASSEMBLY**	48–49	1	
BD.4321	**RUBBER STOP FOR PETROL FILLER DOOR AND REAR WHEEL COVER**		8	
BD.17179	**REAR BUMPER ASSEMBLY, COMPLETE**		1	Fitted to Chassis No. 1A.1001 and subs. 1A.20001 and subs.
BD.17180	**Bar only,** Rear Bumper		1	
BD.11006	**Grip Nut,** in Outer Brackets of Rear Bumper		4	
BD.17185	**Over-Rider Assembly,** L.H. for Rear Bumper		1	
BD.17186	**Over-Rider Assembly,** R.H. for Rear Bumper		1	
BD.9868/1	**Beading,** between Over-Riders and Rear Bumper (70/50881)		4	
BD.10990	**Bolt,** securing Over-Riders (70/51517)		2	
C.8351	**Spire Nut,** securing Bolts to Over-Riders (SFW.0607)		2	
NN.137/L	**Nut,** on Bolts		2	
FG.106/X	**Washer,** Spring, under Nuts		2	
FW.206/T	**Washer,** Plain, under Spring Washers		2	
BD.24549	**'D' EMBLEM AT CENTRE OF REAR BUMPER**		1	Fitted to Chassis No. 1A.1001 and subs. 1A.20001 and subs.
BD.4984	**Washer,** Rubber, between Emblem and Bumper		2	
UFN.119/L	**Nut,** securing Emblem to Rear Bumper		1	
AG.102/X	**Washer,** Spring, under Nut		1	
AW.102/T	**Washer,** Plain, under Spring Washer		1	
BD.9949/1	**SEALING RUBBER BETWEEN REAR BUMPER AND BODY (6' LONG)**		1	Fitted to Chassis No. 1A.1001 and subs. 1A.20001 and subs.
BD.9950	**Retainer,** securing Sealing Rubber to Rear Bumper		16	
BD.18080	**Chromium Plated Finisher,** at L.H. End of Sealing Rubber		1	
BD.18081	**Chromium Plated Finisher,** at R.H. End of Sealing Rubber		1	
BD.9927	**REAR BUMPER MOUNTING (RUBBER)**		4	Fitted to Chassis No. 1A.1001 and subs. 1A.20001 and subs.
UFN.131/L	**Nut,** securing Rubber Mountings to Studs on Body		8	
C.725	**Washer,** Shakeproof, under Nuts		8	
BD.9925	**BRACKET, FOR OUTER MOUNTING OF REAR BUMPER**		2	
BD.9603/1	**Bolt,** securing Brackets and Bumper to Rubber Mountings		4	
C.728	**Washer,** Shakeproof, on Bolts		4	
FW.108/T	**Washer,** Plain, on Inner Body only		2	
BD.9604/1	**Bolt,** securing Rear Bumper to Mounting Brackets		4	
FW.106/T	**Washer,** Plain, on Bolts		4	
C,726	**Washer,** Shakeproof, on Bolts		4	
FW.108/T	**Washer,** Packing, between Mounting Rubbers and Mounting Brackets		As req'd	

Part No.	Description	Plate No.	No. per Unit	Remarks
	FIXINGS FOR EXTERIOR TOOLS			
BD.11136	**Clip**, securing Wheel Brace		1	
DAJ.208/8C	**Screw**, Self-Tapping, securing Clip		1	
BD.11074	**Clip**, securing Wheel Brace		2	
BD.11137	**Plate**, reinforcing Clips		2	
BD.1814/4	**Pop Rivet**, securing Clips		4	
BD.11055	**Clip**, securing Jack		1	
DAJ.208/8C	**Screw**, Self-Tapping, securing Jack Clip		1	
BD.12443	**Felt** (1/8" x 9½" x 2½") for Jack Housing		1	
BD.11785	**RUBBER BUFFER AT EACH SIDE OF TRUNK LID**		2	
DAC.504/8C	**Screw**, Self-Tapping, securing Rubber Buffers		2	
BD.532/1	**Cup Washer**, on Screws		2	
BD.30081	**REAR BUMPER ASSEMBLY COMPLETE**		1	Fitted to Chassis No. 1K.1001 and subs. 1K.30001 and subs.
BD.30144	**Bumper Bar Assembly only**	48–50	1	
BD.30145	**R.H. Over-Rider Assembly**	48–51	1	
BD.30146	**L.H. Over-Rider Assembly**	48–52	1	
BD.30394	**Beading** (long) between Over-Riders and Bumper Bar	48–53	2	
BD.30395	**Beading** (short) between Over-Rider and Bumper Bar		2	
BD.30393	**Distance Piece**, inside Bumper Bar, receiving Bolt securing Over-Riders	48–54	2	
UFB.137/16R	**Bolt**, securing Over-Riders to Bumper Bar	48–55	2	
FG.106/X	**Washer**, Spring on Bolt	48–56	2	
FW.106/T	**Washer**, Plain, under Spring Washer	48–57	2	
BD.30088	**BRACKET ASSEMBLY, OUTER, MOUNTING REAR BUMPER TO BODY**	48–58	2	Fitted to Chassis No. 1K.1001 and subs. 1K.30001 and subs.
BD.30083	**BRACKET ASSEMBLY, INNER, MOUNTING REAR BUMPER TO BODY**	48–59	2	
C.24119/2	**Nut**, Self-Locking, securing Fixing Brackets to Body		8	
BD.30091	**BUFFER ASSEMBLY, ON FIXING BRACKETS**	48–60	4	Fitted to Chassis No. 1K.1001 and subs. 1K.30001 and subs.
BD.30090	**BUFFER BRACKET, OUTER, ON REAR BUMPER**	48–61	2	
BD.9604/2	**Screw**, Set, securing Outer Bracket to Rear Bumper	48–62	4	
FG.106/X	**Washer**, Spring, on Setscrews		4	
BD.542/6	**Washer**, Special, under Spring Washer		4	
UFN.150/L	**Nut**, securing Bumper to Body at Inner and Outer Mountings	48–63	4	
FG.108/X	**Washer**, Spring, under Nuts	48–64	4	
BD.33118	**Washer Plate**, under Spring Washers on Inner Mountings	48–65	2	
FW.108/T	**Washer**, Plain, under Spring Washers on Outer Mountings	48–66	2	
BD.9949/1	**SEALING RUBBER BETWEEN REAR BUMPER AND BODY (72" LONG)**	48–67	1	Fitted to Chassis No. 1K.1001 and subs. 1K.30001 and subs.
BD.9950	**Retainer**, securing Sealing Rubber to Bumper	48–68	16	
BD.33876	**Chrome Finisher**, at R.H. end of Sealing Rubber	48–69	1	
BD.33877	**Chrome Finisher**, at L.H. end of Sealing Rubber	48–70	1	
UFN.119/L	**Nut**, securing Finishers to Bumper		2	
C.723/A	**Washer**, Shakeproof, under Nuts		2	
BD.32692	**Washer**, Special, under Shakeproof Washers		2	

BODY FITTINGS

Part No.	Description	Plate No.	No. per Unit	Remarks
	SEALING RUBBERS			
BD.18267	**Sealing Rubber**, for L.H. "A" Post		1	
BD.18268	**Sealing Rubber**, for R.H. "A" Post		1	
BD.19155	**Sealing Rubber**, for L.H. "B" Post		1	
BD.19156	**Sealing Rubber**, for R.H. "B" Post		1	
BD.18266	**Sealing Rubber**, for L.H. "C" Post		1	
BD.18265	**Sealing Rubber**, for R.H. "C" Post		1	
BD.18269	**Sealing Rubber**, for L.H. "D" Post		1	
BD.18270	**Sealing Rubber**, for R.H. "D" Post		1	
BD.18271	**Sealing Rubber**, for Front Door Sills (37" long)		2	
BD.18272	**Sealing Rubber**, for Rear Door Sills (24" long)		2	
BD.18273/1	**Sealing Rubber**, for Wheel Arches (18" long)		2	
BD.17847/1	**Sealing Rubber**, for Rear Wheel Cover (58" long)		2	
BD.26835	**Rubber Plug**, between Window Frame and Shut Pillar of Front Doors		2	Fitted to Chassis No. 1A.3859 and subs. 1A.20269 and subs.
BD.26836	**Rubber Pad**, sealing Window Frame to Hinge Pillar of Rear Doors		2	(as above)
	RETAINERS FOR SEALING RUBBERS			
BD.10663	**Retainer**, for Seal on L.H. "C" Post		1	
BD.10664	**Retainer**, for Seal on R.H. "C" Post		1	
BD.10665	**Retainer**, for Seal on L.H. "A" Post		1	
BD.10666	**Retainer**, for Seal on R.H. "A" Post		1	
BD.1814/2	**Pop Rivet**, securing Seal Retainers		22	
BD.17126	CHROME FINISHER AT REAR OF L.H. DRIP BEAD	49–1	1	
BD.17127	CHROME FINISHER AT REAR OF R.H. DRIP BEAD	49–2	1	
BD.17128	CHROME FINISHER AT FRONT OF L.H. DRIP BEAD	49–3	1	
BD.17129	CHROME FINISHER AT FRONT OF R.H. DRIP BEAD	49–4	1	
BD.17139	JOINT PIECE AT CENTRE OF DRIP BEAD FINISHERS	49–5	2	
DAC.504/6C	**Screw**, Self-Tapping, securing Joint Pieces		2	
BD.23927	'D' MOTIF ON BONNET	49–6	1	NOT REQUIRED ON CARS EXPORTED TO FRANCE
UCS.419/3H	**Screw**, Set, securing Motif to Bonnet		2	Fitted from Chassis No. 1A.1001 to 1A.10233 1A.20001 to 1A.20475
C.723/A	**Washer**, Shakeproof, on Setscrews		2	(as above)
BD.12370	**Cup Washer**, on Setscrew (front)		1	(as above)
1744	**Cup Washer**, on Setscrew (rear)		1	(as above)
BD.28623/1	**Stud**, mounting Motif to Bonnet at front		1	Fitted to Chassis No. 1A.10234 and subs. 1A.20476 and subs. 1K.1001 and subs. 1K.30001 and subs.
UCN.119/L	**Nut**, on Stud		1	(as above)
C.723/A	**Washer**, Shakeproof, under Nut		1	(as above)
AW.102/T	**Washer**, Plain, under Shakeproof Washer		1	(as above)
UCS.419/3H	**Screw**, Set, securing Motif to Bonnet at rear		1	(as above)
C.723/A	**Washer**, Shakeproof, on Setscrew		1	(as above)
1744	**Cup Washer**, under Shakeproof Washer		1	(as above)
BD.23955	CHROME MOULDING AT CENTRE OF BONNET	49–7	1	FOR USE ONLY WHEN 'D' MOTIF IS FITTED
BD.4919/13	**Special Setscrew**, securing Moulding	49–8	11	(as above)
UCN.116/L	**Nut**, on Setscrews		11	(as above)
C.722	**Washer**, Shakeproof, under Nuts		11	(as above)
AW.102/T	**Washer**, Plain, under Shakeproof Washers		11	(as above)

Part No.	Description	Plate No.	No. per Unit	Remarks
BD.26085	**CHROME MOULDING AT CENTRE OF BONNET**		1	NOT REQUIRED WHEN 'D' MOTIF IS FITTED
BD.4919/13	**Special Setscrew,** securing Moulding		12	
UCN.116/L	**Nut,** on Setscrews		12	
C.722	**Washer,** Shakeproof, under Nuts		12	
AW.102/T	**Washer,** Plain, under Shakeproof Washers		12	
BD.9718	**CHROME BEAD AT L.H. SIDE OF BONNET**	49–9	1	
BD.9719	**CHROME BEAD AT R.H. SIDE OF BONNET**	49–10	1	
BD.11546	**Staple,** fixing Bonnet Side Beads	49–11	24	
BD.26567	**INSULATING PAD (GLASS FIBRE) UNDER L.H. SIDE OF BONNET**		1	Fitted to Chassis No. 1A.2866 and subs. 1A.20249 and subs.
BD.26568	**INSULATING PAD (GLASS FIBRE) UNDER R.H. SIDE OF BONNET**		1	
BD.26566	**Retainer Strip,** for Pads at rear of Bonnet		2	
BD.1814/2	**Pop Rivet,** securing Retainer Strips		6	
	NOTE: The Insulating Pads may be fitted to cars prior to the stated Chassis Nos. by using the items detailed above			
BD.9720	**CHROME BEAD FOR L.H. SCUTTLE PANEL**	49–12	1	
BD.9721	**CHROME BEAD FOR R.H. SCUTTLE PANEL**	49–13	1	
BD.11719	**Clip,** securing Beads	49–14	6	Fitted from Chassis No. 1A.1001 to 1A.9958 1A.20001 to 1A.20470
DA.404/8C	**Screw,** Self-Tapping, securing Clips		6	
BD.28982	**Clip,** securing Beads		4	Fitted to Chassis No. 1A.9959 and subs. 1A.20471 and subs. 1K.1001 and subs. 1K.30001 and subs.
DAZ.404/8C	**Screw,** Self-Tapping, securing Clips		4	
BD.17087	**CHROME BEAD FOR L.H. FRONT DOOR**	49–15	1	
BD.17088	**CHROME BEAD FOR R.H. FRONT DOOR**	49–16	1	
BD.26706/2	**Retainer and Rivet,** securing Beads to Front Doors	49–17	10	
BD.9848	**Stud,** securing Beads at front end	49–18	2	
UCN.116/L	**Nut,** on Studs		2	
C.722	**Washer,** Shakeproof, under Nuts		2	
BD.16720	**OUTSIDE HANDLE FOR L.H. FRONT DOOR (WB.1/8042)**	49–19	1	
8186	**Locking Barrel** (complete with Keys) for L.H. Front Door Lock (WB.1/8017)		1	
8192	**Circlip,** securing Locking Barrels in Handles (WB.RS.84/3)		1	
8184	**Push Button** (WB.1/8015)		1	
8188	**Operating Lever Assembly** (WB.1/10132)		1	

BODY FITTINGS

Part No.	Description	Plate No.	No. per Unit	Remarks
BD.16721	**OUTSIDE HANDLE FOR R.H. FRONT DOOR (WB.1/8043)**	49–20	1	
8187	**Locking Barrel** (complete with Keys) for R.H. Front Door Lock (WB.1/8018)		1	
8192	**Circlip**, securing Locking Barrels in Handles (WB.RS.84/3)		1	
8184	**Push Button** (WB.1/8015)		1	
8189	**Operating Lever Assembly**, for locking R.H. Front Door Handle (WB.1/10131)		1	
BD.18126	**Seating Washers**, for Outside Door Handles		2	
UFN.119/L	**Nut**, securing Outside Door Handles		4	
C.723/A	**Washer**, Shakeproof, under Nuts		4	
AW.102/T	**Washer**, Plain, under Shakeproof Washers		4	(USE ONLY AS REQUIRED)
BD.9846	**CHROME BEAD (LOWER) ON "B/C" POSTS**	49–21	2	Fitted from Chassis No. 1A.1001 to 1A.13140 1A.20001 to 1A.20617
UFN.119/L	**Nut**, securing Beads to "B/C" Posts		2	
AW.102/T	**Washer**, Plain, under Nuts		2	
C.723/A	**Washer**, Shakeproof, under Nuts		2	
BD.31273	**CHROME BEAD (LOWER) ON "B/C" POSTS**		2	Fitted to Chassis No. 1A.13141 and subs. 1A.20618 and subs. (Also to certain cars prior to the above Chassis Nos.) 1K.1001 and subs. 1K.30001 and subs.
BD.26989	**Friction Bush**, securing Chrome Bead to "B/C" Posts		4	
BD.17086	**CHROME BEAD FOR REAR DOORS**	49–22	2	
BD.26706/2	**Retainer and Rivet**, securing Beads to Rear Doors	49–23	10	
BD.17497	**Clip**, securing Beads at front end	49–24	2	
BD.1814/1	**Pop Rivet**, securing Clips		2	
BD.16722	**OUTSIDE HANDLE FOR L.H. REAR DOOR**	49–25	1	
8190	**Push Button**, for Handle (1/8027)		1	
BD.18126	**Seating Washer**, for Handle		1	
BD.16723	**OUTSIDE HANDLE FOR R.H. REAR DOOR**	49–26	1	
8191	**Push Button**, for Handle (1/8028)		1	
BD.18126	**Seating Washer**, for Handle		1	
UFN.119/L	**Nut**, securing Outside Door Handles		4	
C.723/A	**Washer**, Shakeproof, under Nuts		4	
AW.102/T	**Washer**, Plain, under Shakeproof Washers		4	(USE ONLY AS REQUIRED)
BD.9725	**CHROME BEAD FOR L.H. SIDE OF TONNEAU**	49–27	1	

Part No.	Description	Plate No.	No. per Unit	Remarks
BD.9726	**CHROME BEAD FOR R.H. SIDE OF TONNEAU**	49–28	1	
BD.11719	**Clip,** securing Beads to Tonneau	49–29	26	Fitted from Chassis No. 1A.1001 to 1A.9958 1A.20001 to 1A.20470
BD.11712/1	**Pop Rivet,** securing Clips		2	
BD.11712/2	**Pop Rivet,** securing Clips		24	
BD.26706/1	**Retainer and Rivet,** securing Beads to Tonneau		2	Fitted to Chassis No. 1A.9959 and subs. 1A.20471 and subs. 1K.1001 and subs. 1K.30001 and subs.
BD.26706/2	**Retainer and Rivet,** securing Beads to Tonneau		18	
BD.17093	**CHROME BEAD AT TOP OF L.H. FRONT DOOR**	49–30	1	Fitted from Chassis No. 1A.1001 to 1A.11618 1A.20001 to 1A.20534
BD.25962	**CHROME BEAD AT TOP OF L.H. FRONT DOOR**		1	Fitted to Chassis No. 1A.11619 and subs. 1A.20535 and subs. 1K.1001 and subs. 1K.30001 and subs.
BD.17094	**CHROME BEAD AT TOP OF R.H. FRONT DOOR**	49–31	1	Fitted from Chassis No. 1A.1001 to 1A.11618 1A.20001 to 1A.20534
BD.17497	**Clip,** securing Top Beads to Front Doors	49–32	14	
BD.11712/2	**Pop Rivet,** securing Clips		14	
BD.25963	**CHROME BEAD AT TOP OF R.H. FRONT DOOR**		1	Fitted to Chassis No. 1A.11619 and subs. 1A.20535 and subs. 1K.1001 and subs. 1K.30001 and subs.
DAJ.204/6C	**Screw,** Self-Tapping, securing Chrome Beads		8	
2179	**Rubber Buffer,** on Self-Tapping Screws		2	
DAZ.406/6C	**Screw,** Self-Tapping, securing Chrome Beads		2	
BD.17090	**CHROME BEAD (UPPER) ON "B/C" POSTS**	49–33	2	Fitted from Chassis No. 1A.1001 to 1A.11618 1A.20001 to 1A.20534
BD.17497	**Clip,** securing Beads to "B/C" Posts	49–34	4	
BD.11712/1	**Pop Rivet,** securing Clips		4	
BD.16752	**FINISHER FOR "B/C" POSTS**	49–35	2	
DAC.504/6C	**Screw,** Self-Tapping, securing Finishers		8	
BD.29321	**FINISHER FOR "B/C" POSTS**		2	Fitted to Chassis No. 1A.11619 and subs. 1A.20535 and subs. 1K.1001 and subs. 1K.30001 and subs.
DAC.504/6C	**Screw,** Self-Tapping, securing Finishers		8	
BD.17091	**CHROME BEAD AT TOP OF L.H. REAR DOOR**	49–36	1	Fitted from Chassis No. 1A.1001 to 1A.11618 1A.20001 to 1A.20534
BD.17092	**CHROME BEAD AT TOP OF R.H. REAR DOOR**	49–37	1	
BD.17497	**Clip,** securing Top Beads to Rear Doors	49–38	14	
BD.11712/1	**Pop Rivet,** securing Clips		14	

BODY FITTINGS

Part No.	Description	Plate No.	No. per Unit	Remarks
BD.25964	CHROME BEAD AT TOP OF L.H. REAR DOOR		1	Fitted to Chassis No. 1A.11619 and subs. 1A.20535 and subs. 1K.1001 and subs. 1K.30001 and subs.
BD.25965	CHROME BEAD AT TOP OF R.H. REAR DOOR		1	
BD.26706/4	**Retainer and Rivet,** securing Chrome Beads		8	
DAZ.406/6C	**Screw,** Self-Tapping, securing Chrome Beads		2	
BD.18284	L.H. FRONT DOOR (COMPLETE WITH ALL FITTINGS)		1	Fitted from Chassis No. 1A.1001 to 1A.1160 1A.20001 to 1A.20026
BD.11596	**Mutocel,** on Front Doors (20½" x 16½")		1	
BD.11597	**Mutocel,** on Front Doors (10" x 9½")		1	
BD.11599	**Felt,** on Front Doors (12" x 3")		1	
BD.18285	R.H. FRONT DOOR (COMPLETE WITH ALL FITTINGS)		1	
BD.11596	**Mutocel,** on Front Door (20½" x 16½")		1	
BD.11597	**Mutocel,** on Front Door (10" x 9½")		1	
BD.11599	**Felt,** on Front Door (12" x 3")		1	
BD.22608	L.H. FRONT DOOR (COMPLETE WITH ALL FITTINGS)		1	Fitted to Chassis No. 1A.1161 and subs. 1A.20027 and subs.
BD.11596	**Mutocel,** on Front Door (20½" x 16½")		1	
BD.11597	**Mutocel,** on Front Door (10" x 9½")		1	
BD.11599	**Felt,** on Front Door (12" x 3")		1	
BD.22609	R.H. FRONT DOOR (COMPLETE WITH ALL FITTINGS)		1	
BD.11596	**Mutocel,** on Front Door (20½" x 16½")		1	
BD.11597	**Mutocel,** on Front Door (10" x 9½")		1	
BD.11599	**Felt,** on Front Door (12" x 3")		1	
BD.33478	L.H. FRONT DOOR (COMPLETE WITH ALL FITTINGS)		1	Fitted to Chassis No. 1K.1001 and subs. 1K.30001 and subs.
BD.11596	**Mutocel,** on Front Door (20½" x 16½")		1	
BD.11597	**Mutocel,** on Front Door (10" x 9½")		1	
BD.11599	**Felt,** on Front Door (12" x 3")		1	
BD.33479	R.H. FRONT DOOR (COMPLETE WITH ALL FITTINGS)		1	
BD.11596	**Mutocel,** on Front Door (20½" x 16½")		1	
BD.11597	**Mutocel,** on Front Door (10" x 9½")		1	
BD.11599	**Felt,** on Front Door (12" x 3")		1	
BD.19580	LOCK IN L.H. FRONT DOOR	50–1	1	
BD.19576	**Remote Control and Link Assembly,** for L.H. Front Door Lock	50–2	1	
BD.19581	LOCK IN R.H. FRONT DOOR		1	
BD.19577	**Remote Control and Link Assembly,** for R.H. Front Door Lock		1	
DJZ.410/6C	**Screw,** Self-Tapping, securing Remote Controls		6	
BD.8633/2	**Spire Nut,** on Self-Tapping Screws		6	
BD.19573	**Clip,** securing Links to Locks		2	
BD.19575	**Wavy Washer,** under Links		2	
BD.19574	**Flat Washer,** under Wavy Washers		2	
BD.18341	**Grommet,** on Door Panel, preventing Vibration of Remote Controls		4	

Part No.	Description	Plate No.	No. per Unit	Remarks
BD.17808	**STRIKER FOR L.H. FRONT DOOR LOCK**	50–3	1	
BD.17809	**STRIKER FOR R.H. FRONT DOOR LOCK**		1	
BD.17821/7	**Screw,** Set, securing Strikers to "B/C" Posts		6	
BD.18098	**Shim,** Shakeproof, for Strikers		2	
BD.17338	**Shim,** for Strikers		As req'd	
BD.10728	**INSIDE HANDLE FOR FRONT DOORS (WB.1/3859)**	50–4	2	
BD.16721	**OUTSIDE HANDLE (LOCKING) FOR R.H. FRONT DOOR (WB.1/8043)**		1	
8187	**Locking Barrel** (complete with keys) for R.H. Front Door Lock (WB.1/8018)		1	
8192	**Circlip,** securing Locking Barrels in Handles (WB.RS.84/3)		1	
8185	**Push Button** (WB.1/8016)		1	
11240	**Return Spring,** for Push Button (RS.41-432)		1	
8189	**Operating Lever Assembly,** for locking R.H. Front Door Handle (WB.1/10131)		1	
UFS.119/7R	**Screw,** Set, securing Base Plate to Handles		2	
C.723/A	**Washer,** Shakeproof, on Setscrews		2	
BD.16720	**OUTSIDE HANDLE (LOCKING) FOR L.H. FRONT DOOR (WB.1/8042)**	50–5	1	
8186	**Locking Barrel** (complete with keys) for L.H. Front Door Lock (WB.1/8017)		1	
8192	**Circlip,** securing Locking Barrels in Handles (WB.RS.84/3)		1	
8184	**Push Button** (WB.1/8015)		1	
11240	**Return Spring,** for Push Button (RS.41-432)		1	
8188	**Operating Lever Assembly,** for Locking L.H. Front Door Handle (WB.1/10132)		1	
UFS.119/7R	**Screw,** Set, securing Base Plate to Handles		2	
C.723/A	**Washer,** Shakeproof on Setscrews		2	
BD.18126	**Seating Washer,** at rear end of Front Outside Door Handles	50–6	2	
UFN.119/L	**Nut,** securing Outside Door Handles		4	
C.723/A	**Washer,** Shakeproof, under Nuts		4	
BD.11657	**Rubberised Felt** (3" x 3") packing Door Handles		4	
BD.17381	**LINK, CONNECTING FRONT DOOR HANDLES TO LOCKS (WB.1/11516)**	50–7	2	
BD.23473	**Clip,** retaining Links to Door Handles and Locks (WB.1/11517)	50–8	4	
BD.10729	**HANDLE, OPERATING FRONT DOOR WINDOW REGULATOR (WB.1/6671)**	50–9	2	
BD.10730	**OUTER ESCUTCHEON FOR REGULATOR AND INSIDE FRONT DOOR HANDLES (W.128)**	50–10	4	
BD.10731	**INNER ESCUTCHEON FOR REGULATOR AND INSIDE FRONT DOOR HANDLES (W.129)**	50–11	4	

BODY FITTINGS

Part No.	Description	Plate No.	No. per Unit	Remarks
BD.10732	**SPRING BEHIND ESCUTCHEONS (41/379)**	50–12	4	
BD.10734	**Rubber Plug,** behind Handles		4	
BD.10733	**Peg,** securing Handles to Locks and Regulators		4	
BD.18521	**FRAME FOR WINDOW IN L.H. FRONT DOOR, COMPLETE WITH NO DRAUGHT VENTILATOR**		1	Fitted from Chassis No. 1A.1001 to 1A.1160 1A.20001 to 1A.20026
BD.18522	**FRAME, FOR WINDOW IN R.H. FRONT DOOR, COMPLETE WITH NO DRAUGHT VENTILATOR**		1	
BD.18516	**FRAME ONLY, FOR WINDOW IN L.H. FRONT DOOR**		1	
BD.18517	**FRAME ONLY, FOR WINDOW IN R.H. FRONT DOOR**		1	
BD.22604	**FRAME, FOR WINDOW IN L.H. FRONT DOOR, COMPLETE WITH NO DRAUGHT VENTILATOR**		1	Fitted to Chassis No. 1A.1161 and subs. 1A.20027 and subs. 1K.1001 and subs. 1K.30001 and subs.
BD.22605	**FRAME, FOR WINDOW IN R.H. FRONT DOOR, COMPLETE WITH NO DRAUGHT VENTILATOR**		1	
BD.20331	**FRAME ONLY, FOR WINDOW IN L.H. FRONT DOOR**	50–13	1	Fitted to Chassis No. 1A.1161 and subs. 1A.20027 and subs. 1K.1001 and subs. 1K.30001 and subs.
BD.20332	**FRAME ONLY, FOR WINDOW IN R.H. FRONT DOOR**		1	
BD.14399/3	**Screw,** Set, securing Window Frames at Waist Rail		8	
C.723/A	**Washer,** Shakeproof, on Setscrews		8	
AW.102/T	**Washer,** Plain, under Shakeproof Washers		8	
BD.20091/1	**Packing,** between Window Frame and Waist Rail (.048" thick)		As req'd	SELECTIVE SIZES
BD.20091/2	**Packing,** between Window Frame and Waist Rail (.128" thick)		As req'd	
BD.8291/7	**Screw,** Set, securing Window Frames to Doors at bottom		4	
UFN.125/L	**Nut,** on Setscrews		2	
C.724	**Washer,** Shakeproof, on Setscrews and under Nuts		4	
BD.541/20	**Washer,** Special, on Setscrews and under Shakeproof Washers		6	
BD.17055	**GLASS FOR WINDOWS IN FRONT DOORS**		2	Fitted from Chassis No. 1A.1001 to 1A.1160 1A.20001 to 1A.20026
BD.22580	**GLASS FOR WINDOWS IN FRONT DOORS**	50–14	2	Fitted to Chassis No. 1A.1161 and subs. 1A.20027 and subs. 1K.1001 and subs. 1K.30001 and subs.

Part No.	Description	Plate No.	No. per Unit	Remarks
BD.29556	REGULATOR CHANNEL FOR R.H. FRONT DOOR GLASS	50–15	1	
BD.29557	REGULATOR CHANNEL FOR L.H. FRONT DOOR GLASS		1	
BD.4714	**Everseal Strip**, in Regulator Channels		2	
BD.5195/2	FELT, IN TOP RAIL OF WINDOW FRAMES (19" LONG)	50–16	2	
BD.5195/1	FELT IN FORWARD RAIL OF WINDOW FRAMES (30" LONG)		2	Fitted from Chassis No. 1A.1001 to 1A.1160 1A.20001 to 1A.20026
BD.17134/2	FELT, IN REAR RAIL OF WINDOW FRAMES (30¾" LONG)		2	
BD.21070/1	SEAL, IN FORWARD RAIL OF WINDOW FRAMES (30" LONG)	50–17	2	Fitted to Chassis No. 1A.1161 and subs. 1A.20027 and subs. 1K.1001 and subs. 1K.30001 and subs.
BD.21070/2	SEAL, IN REAR RAIL OF WINDOW FRAMES (31" LONG)	50–18	2	
BD.16792	SEAL (PRESTIK) BETWEEN NO DRAUGHT VENTILATOR RAIL AND DOOR	50–19	2	
BD.20010/1	WEATHER STRIP BETWEEN TOP OF FRONT DOOR AND GLASS	50–20	2	Fitted from Chassis No. 1A.1001 to 1A.11618 1A.20001 to 1A.20534
BD.22312	**Clip**, securing Weather Strips	50–21	8	
BD.21073/5	OUTER SEAL FOR GLASS IN FRONT DOORS		2	Fitted to Chassis No. 1A.11619 and subs. 1A.20535 and subs. 1K.1001 and subs. 1K.30001 and subs.
BD.11605/1	DRAUGHT PIPING FOR FRONT DOORS (65½" LONG)		2	
BD.11605/2	DRAUGHT PIPING AT FRONT OF "B/C" POSTS (42½" LONG)		2	
BD.18267	SEALING RUBBER FOR L.H. FRONT DOOR HINGE PILLAR	50–22	1	
BD.18268	SEALING RUBBER FOR R.H. FRONT DOOR HINGE PILLAR		1	
BD.21361	SEALING RUBBER, ON HINGE FACE, AT WAIST LEVEL OF R.H. FRONT DOOR		1	

BODY FITTINGS

Part No.	Description	Plate No.	No. per Unit	Remarks
BD.21362	**SEALING RUBBER, ON HINGE FACE, AT WAIST LEVEL OF L.H. FRONT DOOR**	50–23	1	
BD.19155	**SEALING RUBBER FOR L.H. FRONT DOOR SHUT PILLAR**	50–24	1	
BD.19156	**SEALING RUBBER FOR R.H. FRONT DOOR SHUT PILLAR**		1	
BD.17355	**FILLER PIECE, ON SHUT FACE PANEL, AT WAIST LEVEL OF L.H. FRONT DOOR**	50–25	1	
BD.17356	**FILLER PIECE, ON SHUT FACE PANEL, AT WAIST LEVEL OF R.H. FRONT DOOR**		1	
BD.17357	**Plywood Packing,** under Filler Pieces		2	
DA.604/10C	**Screw,** Self-Tapping, securing Filler Pieces		4	
BD.26835	**Rubber Plug,** between Window Frame and Shut Pillar of Front Doors		2	Fitted to Chassis No. 1A.3859 and subs. 1A.20269 and subs. 1K.1001 and subs. 1K.30001 and subs.
BD.28005	**SEALING RUBBER FOR FRONT DOOR SILLS (37" LONG)**	50–26	2	**Supersedes BD.18271**
BD.17121	**NO DRAUGHT VENTILATOR, COMPLETE, FOR L.H. FRONT DOOR**	50–27	1	
BD.17122	**NO DRAUGHT VENTILATOR, COMPLETE, FOR R.H. FRONT DOOR**		1	
BD.17068	**Glass only,** for Front No Draught Ventilators		2	
BD.1300	**Everseal,** between Glass and Frame (27" long)		2	
BD.17112	**Locking Handle,** for R.H. Front No Draught Ventilator		1	
BD.17111	**Locking Handle,** for L.H. Front No Draught Ventilator	50–28	1	
BD.3897	**Rubber Washer,** behind Locking Handles	50–29	2	
BD.18073	**Peg,** securing Locking Handles to Spindle on Inner Frames	50–30	2	
BD.17114	**Bolt,** securing Locking Handles to Inner Frames	50–31	2	
BD.18171	**Fibre Washer,** on Bolts	50–32	2	
BD.17211	**Nut,** on Bolts	50–33	2	
BD.17249	**UPPER PIVOT ARM FOR NO DRAUGHT VENTILATOR IN L.H. FRONT DOOR**	50–34	1	
BD.17250	**UPPER PIVOT ARM FOR NO DRAUGHT VENTILATOR IN R.H. FRONT DOOR**		1	
BD.17253	**Pin,** for upper Pivots	50–35	2	
BD.17276/1	**Screw,** Set, securing Upper Pivot Arms to Window Frames		4	

Part No.	Description	Plate No.	No. per Unit	Remarks
BD.9647	**INDEX SLEEVE, FOR LOWER PIVOT SPINDLE, CONTROLLING OPENING ANGLE OF NO DRAUGHT VENTILATORS**	50–36	2	
BD.9657	**Bush,** on Lower Pivot Spindles, engaging with Index Sleeve	50–37	2	Fitted from Chassis No. 1A.1001 to 1A.4070 1A.20001 to 1A.20271 May be replaced by BD.26840 or BD.26841
BD.26840	**Bush,** on Lower Pivot Spindle of L.H. No Draught Ventilator, engaging with Index Sleeve		1	Fitted to Chassis No. 1A.4071 and subs. 1A.20272 and subs. 1K.1001 and subs. 1K.30001 and subs. May be used as replacement for BD.9657
BD.26841	**Bush,** on Lower Pivot Spindle of R.H. No Draught Ventilator, engaging with Index Sleeve		1	
BD.9651	**Spring,** on Lower Pivot Spindles, below Index Sleeves	50–38	2	
BD.18067	**Pin,** tensioning Spring		2	
UFN.119/L	**Nut,** locking Springs and Index Sleeves to Lower Pivot Spindles		4	
AW.102/T	**Washer,** Plain, under Nuts		2	
BD.17048	**SEALING RUBBER FOR NO DRAUGHT VENTILATOR IN L.H. FRONT DOOR**	50–39	1	
BD.17049	**SEALING RUBBER FOR NO DRAUGHT VENTILATOR IN R.H. FRONT DOOR**		1	
BD.24701	**CASING ASSEMBLY FOR L.H. FRONT DOOR**		1	
BD.24688	**Casing only** (complete with Flap but less Arm Rest)	50–40	1	
BD.18068	**Clip,** securing Casing to L.H. Front Door		22	
BD.24683	**Arm Rest,** L.H.	50–41	1	
DAZ.406/12C	**Screw,** Self-Tapping, securing Arm Rest		4	
BD.541/13	**Special Washer,** on Screws		4	
BD.24702	**CASING ASSEMBLY FOR R.H. FRONT DOOR**		1	
BD.24689	**Casing only** (complete with Flap but less Arm Rest)		1	
BD.18068	**Clip,** securing Casing to R.H. Front Door		22	
BD.24684	**Arm Rest,** R.H.		1	
DAZ.406/12C	**Screw,** Self-Tapping, securing Arm Rest		4	
BD.541/13	**Special Washer,** on Screw		4	
BD.24952/2	**Rubber Thrust Pad,** behind Casings, at Door Handle and Window Regulator Handle apertures		4	
DA.504/8C	**Screw,** Self-Tapping, securing Casings to top of Front Doors		8	
BD.18286	**L.H. REAR DOOR (COMPLETE WITH ALL FITTINGS)**		1	Fitted from Chassis No. 1A.1001 to 1A.1160 1A.20001 to 1A.20026
BD.11598	**Mutocel,** on Rear Door (16½" x 16")		1	
BD.18287	**R.H. REAR DOOR (COMPLETE WITH ALL FITTINGS)**		1	
BD.11598	**Mutocel,** on Rear Door (16½" x 16")		1	

BODY FITTINGS

Part No.	Description	Plate No.	No. per Unit	Remarks
BD.22614	**L.H. REAR DOOR (COMPLETE WITH ALL FITTINGS)**		1	Fitted to Chassis No. 1A.1161 and subs. 1A.10027 and subs.
BD.11598	**Mutocel,** on Rear Door (16½" x 16")		1	
BD.22615	**R.H. REAR DOOR (COMPLETE WITH ALL FITTINGS)**		1	
BD.11598	**Mutocel,** on Rear Door (16½" x 16")		1	
BD.33480	**L.H. REAR DOOR (COMPLETE WITH ALL FITTINGS)**		1	Fitted to Chassis No. 1K.1001 and subs. 1K.30001 and subs.
BD.11598	**Mutocel,** on Rear Door (16½" x 16")		1	
BD.33481	**R.H. REAR DOOR (COMPLETE WITH ALL FITTINGS)**		1	
BD.11598	**Mutocel,** on Rear Door (16½" x 16")		1	
BD.10728	**INSIDE HANDLE FOR REAR DOORS (WB.1/3859)**	51–1	2	
BD.19582	**LOCK IN L.H. REAR DOOR**	51–2	1	
BD.19578	**Remote Control and Link Assembly,** for L.H. Rear Door Lock	51–3	1	
BD.19583	**LOCK IN R.H. REAR DOOR**		1	
BD.19579	**Remote Control and Link Assembly,** for R.H. Rear Door Lock		1	
DJZ.410/6C	**Screw,** Self-Tapping, securing Remote Controls		6	
BD.8633/2	**Spire Nut,** on Self-Tapping Screws		6	
BD.19573	**Clip,** securing Links to Locks		2	
BD.19575	**Wavy Washer,** under Links		2	
BD.19574	**Flat Washer,** under Wavy Washers		2	
BD.18341	**Grommet,** on Door Panel, preventing vibration of Remote Control		4	
BD.17808	**STRIKER FOR L.H. REAR DOOR LOCK**	51–4	1	
BD.17809	**STRIKER FOR R.H. REAR DOOR LOCK**		1	
BD.17821/7	**Screw,** Set, securing Strikers to "D" Posts		6	
BD.18098	**Shim,** Shakeproof, for Strikers (271/1128)		2	
BD.17338	**Shim,** for Strikers		As req'd	
BD.16722	**OUTSIDE HANDLE FOR L.H. REAR DOOR (WB.1/8044)**	51–5	1	
8190	**Push Button,** for Rear Door Handles (WB.1/8027)		1	
11240	**Return Spring,** for Push Buttons (RS.41-432)		1	
8192	**Circlip,** retaining Push Buttons (RS.84-3)		1	
11241	**Washer,** under Circlips (RS.14-380)		1	
UFS.119/7R	**Screw,** Set, securing Base Plate to Handles		2	
C.723/A	**Washer,** Shakeproof on Setscrews		2	
BD.16723	**OUTSIDE HANDLE FOR R.H. REAR DOOR (WB.1/8045)**		1	
	Individual items are as for Handle BD.16722			
BD.18126	**Seating Washer,** at rear end of Rear Outside Door Handles	51–6	2	

Part No.	Description	Plate No.	No. per Unit	Remarks
	OUTSIDE HANDLE FOR R.H. REAR DOOR (WB.1/8045) *(contd.)*			
UFN.119/L	**Nut,** securing Outside Door Handles		4	
C.723/A	**Washer,** Shakeproof, under Nuts		4	
AW.102/T	**Washer,** Plain, under Shakeproof Washers		4	
BD.11657	**Felt** (3" x 3") packing Door Handles		4	
BD.17382	**LINK, CONNECTING L.H. REAR DOOR HANDLE TO LOCK (WB.1/8049)**	51–7	1	
BD.17383	**LINK, CONNECTING R.H. REAR DOOR HANDLE TO LOCK (WB.1/8048)**		1	
BD.23473	**Clip,** retaining Links to Door Handles and Locks (WB.1/11517)	51–8	4	
BD.10729	**HANDLE OPERATING REAR DOOR WINDOW REGULATOR (WB.1/6671)**	51–9	2	
BD.10730	**OUTER ESCUTCHEON FOR REGULATOR AND INSIDE REAR DOOR HANDLES (W.128)**	51–10	4	2 ONLY REQUIRED (FOR REGULATOR HANDLES) WHEN CHILD'S SAFETY LOCK IS FITTED
BD.13372	**OUTER ESCUTCHEON FOR L.H. INSIDE REAR DOOR HANDLE (OPERATING CHILD'S SAFETY LOCK)**		1	FOR USE ONLY WHEN CHILD'S SAFETY LOCK IS FITTED
BD.13373	**OUTER ESCUTCHEON FOR R.H. INSIDE REAR DOOR HANDLE (OPERATING CHILD'S SAFETY LOCK)**		1	
BD.10731	**INNER ESCUTCHEON FOR REGULATOR AND INSIDE REAR DOOR HANDLES (W.129)**	51–11	4	
BD.10732	**SPRING BEHIND ESCUTCHEON (41/379)**	51–12	4	2 ONLY REQUIRED (FOR REGULATOR HANDLES) WHEN CHILD'S SAFETY LOCK IS FITTED
BD.13350	**SPRING, BEHIND ESCUTCHEONS FOR INSIDE REAR DOOR HANDLES**		2	FOR USE ONLY WHEN CHILD'S SAFETY LOCK IS FITTED
BD.10734	**Rubber Plug,** behind Handles		4	
BD.10733	**Peg,** securing Handles to Locks and Regulators		4	
BD.18523	**FRAME, FOR WINDOW IN L.H. REAR DOOR COMPLETE WITH NO DRAUGHT VENTILATOR**		1	Fitted from Chassis No. 1A.1001 to 1A.1160 1A.20001 to 1A.20026
BD.18524	**FRAME, FOR WINDOW IN R.H. REAR DOOR, COMPLETE WITH NO DRAUGHT VENTILATOR**		1	
BD.18518	**FRAME ONLY, FOR WINDOW IN L.H. REAR DOOR**		1	Fitted from Chassis No. 1A.1001 to 1A.1160 1A.20001 to 1A.20026
BD.18519	**FRAME ONLY, FOR WINDOW IN R.H. REAR DOOR**		1	

BODY FITTINGS

Part No.	Description	Plate No.	No. per Unit	Remarks
BD.22610	**FRAME, FOR WINDOW IN L.H. REAR DOOR, COMPLETE WITH NO DRAUGHT VENTILATOR**		1	Fitted to Chassis No. 1A.1161 and subs. 1A.20027 and subs. 1K.1001 and subs. 1K.30001 and subs.
BD.22611	**FRAME, FOR WINDOW IN R.H. REAR DOOR, COMPLETE WITH NO DRAUGHT VENTILATOR**		1	
BD.20333	**FRAME ONLY, FOR WINDOW IN L.H. REAR DOOR**	51–13	1	Fitted to Chassis No. 1A.1161 and subs. 1A.20027 and subs. 1K.1001 and subs. 1K.30001 and subs.
BD.20334	**FRAME ONLY, FOR WINDOW IN R.H. REAR DOOR**		1	
BD.14399/3	**Screw,** Set, Special, securing Window Frames at Waist Rail		8	
C.723/A	**Washer,** Shakeproof, on Setscrews		8	
AW.102/T	**Washer,** Plain, for Packing, on Setscrews		8	
BD.20091/1	**Packing,** between Window Frame and Waist Rail (.048" thick)		As req'd	SELECTIVE SIZES
BD.20091/2	**Packing,** between Window Frame and Waist Rail (.128" thick)		As req'd	
UFS.125/7R	**Screw,** Set, securing Window Frames to Doors at bottom front fixing		2	
BD.27756/2	**Screw,** Set, securing Window Frames to Doors at bottom rear fixing		2	
UFN.125/L	**Nut,** on Setscrews		4	
C.724	**Washer,** Shakeproof, under Nuts		4	
BD.541/20	**Washer,** Special, under Shakeproof Washers		4	
BD.19691/1	**Packing Washer** (Plastic) at rear fixing (3/16" thick)		As req'd	SELECTIVE SIZES Fitted to Chassis No. 1A.6384 and subs. 1A.20336 and subs. 1K.1001 and subs. 1K.30001 and subs.
BD.19691/2	**Packing washer** (Plastic) at rear fixing (7/16" thick)		As req'd	
BD.19691/3	**Packing Washer** (Plastic) at rear fixing (1/16" thick)		As req'd	
BD.26836	**Rubber Pad,** sealing Window Frame to Hinge Pillar of Rear Doors at Waist		2	Fitted to Chassis No. 1A.3859 and subs. 1A.20269 and subs. 1K.1001 and subs. 1K.30001 and subs.
BD.5195/2	**FELT, IN TOP RAIL OF WINDOW FRAMES (19" LONG)**	51–14	2	
BD.17134/1	**FELT, IN FORWARD RAIL OF WINDOW FRAMES (28¼" LONG)**		2	Fitted from Chassis No. 1A.1001 to 1A.1160 1A.20001 to 1A.20026
BD.5195/1	**FELT, IN REAR RAIL OF WINDOW FRAMES (30" LONG)**		2	
BD.21070/4	**SEAL, IN FORWARD AND REAR RAILS OF WINDOW FRAMES (27½" LONG)**	51–15	4	Fitted to Chassis No. 1A.1161 and subs. 1A.20027 and subs. 1K.1001 and subs. 1K.30001 and subs.

Part No.	Description	Plate No.	No. per Unit	Remarks
BD.17056	GLASS FOR WINDOWS IN REAR DOORS		2	Fitted from Chassis No. 1A.1001 to 1A.1160 1A.20001 to 1A.20026
BD.22581	GLASS FOR WINDOWS IN REAR DOORS	51–16	2	Fitted to Chassis No. 1A.1161 and subs. 1A.20027 and subs. 1K.1001 and subs. 1K.30001 and subs.
BD.17060	REGULATOR CHANNEL FOR L.H. REAR DOOR GLASS	51–17	1	
BD.17061	REGULATOR CHANNEL FOR R.H. REAR DOOR GLASS		1	
BD.4714	**Everseal Strip**, in Regulator Channels		2	
BD.16792	SEAL (PRESTIK) BETWEEN NO DRAUGHT VENTILATOR RAIL AND DOOR	51–18	2	
BD.20010/2	WEATHER STRIP BETWEEN TOP OF REAR DOOR AND GLASS	51–19	2	Fitted from Chassis No. 1A.1001 to 1A.11618 1A.20001 to 1A.20534
BD.22312	**Clip**, securing Weather Strips		8	
BD.21073/3	OUTER SEAL FOR GLASS IN REAR DOORS		2	Fitted to Chassis No. 1A.11619 and subs. 1A.20535 and subs. 1K.1001 and subs. 1K.30001 and subs.
BD.11605/2	DRAUGHT PIPING AT REAR OF "B/C" POSTS (42½" LONG)		2	
BD.11605/3	DRAUGHT PIPING FOR REAR DOORS (70½" LONG)		2	
BD.18266	SEALING RUBBER FOR L.H. REAR DOOR HINGE PILLAR	51–20	1	
BD.18265	SEALING RUBBER FOR R.H. REAR DOOR HINGE PILLAR		1	
BD.18269	SEALING RUBBER FOR L.H. REAR DOOR SHUT PILLAR	51–21	1	

BODY FITTINGS

Part No.	Description	Plate No.	No. per Unit	Remarks
BD.18270	**SEALING RUBBER FOR R.H. REAR DOOR SHUT PILLAR**		1	
BD.17362	**FILLER PIECE, ON SHUT FACE PANEL, AT WAIST LEVEL OF L.H. REAR DOOR**	51–22	1	
BD.17363	**FILLER PIECE, ON SHUT FACE PANEL AT WAIST LEVEL OF R.H. REAR DOOR**		1	
BD.17364	**Hardwood Packing**, under Filler Piece		2	
DA.604/12C	**Screw**, Self-Tapping, securing Filler Pieces		4	
BD.28006	**SEALING RUBBER FOR REAR DOOR SILLS (24" LONG)**	51–23	2	**Supersedes BD.18272**
BD.18273/1	**SEALING RUBBER FOR REAR DOOR WHEEL ARCHES (18" LONG)**	51–24	2	
BD.17132	**NO DRAUGHT VENTILATOR, COMPLETE, FOR L.H. REAR DOOR**	51–25	1	
BD.17133	**NO DRAUGHT VENTILATOR, COMPLETE, FOR R.H. REAR DOOR**		1	
BD.17089	**Glass only**, for Rear No Draught Ventilators		2	
BD.1300	**Everseal**, between Glass and Frame (32" long)		2	
BD.4899/1	**Screw**, Set, securing N.D. Ventilators to Door Light Frames		10	
BD.17054	**HINGE PLATE FOR NO DRAUGHT VENTILATOR IN REAR DOORS**		2	
BD.17301	**ANGLE BRACKET, ON L.H. REAR DOOR LIGHT FRAME, FOR CATCH OPERATING NO DRAUGHT VENTILATOR**	51–26	1	
BD.17302	**ANGLE BRACKET, ON R.H. REAR DOOR LIGHT FRAME, FOR CATCH OPERATING NO DRAUGHT VENTILATOR**		1	
UCS.316/4H	**Screw**, Set, securing Angle Brackets to Frames		4	
C.722	**Washer**, Shakeproof, on Setscrews		4	
AW.102/T	**Washer**, Plain, on Setscrews		4	

Part No.	Description	Plate No.	No. per Unit	Remarks
BD.20086	**CATCH OPERATING NO DRAUGHT VENTILATOR IN L.H. REAR DOOR**	51–27	1	
7758	**Operating Arm**		1	
10447	**Shouldered Bolt,** securing Operating Arm and Catch Lever to Catch Body (61/682/312/8)		1	
10449	**Washer,** Double Coil Spring on Shouldered Bolt (61/682/312/11)		1	
10448	**Washer,** Plain, under Spring Washer (61/682/312/12)		1	
841	**Shouldered Bolt,** securing Operating Arm on Catch to Bracket on No Draught Ventilator (BSK.318)	51–28	1	
C.3563/1	**Nut,** locking Bolt	51–29	1	
BD.20087	**CATCH OPERATING NO DRAUGHT VENTILATOR IN R.H. REAR DOOR**		1	
7759	**Operating Arm**		1	
10447	**Shouldered Bolt,** securing Operating Arm and Catch Lever to Catch Body (61/682/312/8)		1	
10449	**Washer,** Double Coil Spring, on Shouldered Bolt (61/682/312/11)		1	
10448	**Washer,** Plain, under Spring Washer (61/682/312/12)		1	
841	**Shouldered Bolt,** securing Operating Arm on Catch to Bracket on No Draught Ventilator (BSK.318)	51–28	1	
C.3563/1	**Nut,** locking Bolt	51–29	1	
BD.17305	**Base,** for Catch on L.H. Door	51–30	1	
BD.17306	**Base,** for Catch on R.H. Door		1	
BD.5454/3	**Screw,** Set, securing Catch and Base to Angle Bracket		4	
C.3563	**Fibre Washer,** under Head of Bolts	51–31	2	
C.3563	**Fibre Washer,** between Lever and Bracket on No Draught Ventilator Frame	51–31	2	
BD.17072	**SEALING RUBBER FOR L.H. REAR NO DRAUGHT VENTILATOR**	51–32	1	
BD.17073	**SEALING RUBBER FOR R.H. REAR NO DRAUGHT VENTILATOR**		1	
BD.24703	**CASING ASSEMBLY FOR L.H. REAR DOOR**		1	
BD.17212	**Casing only** (less Arm Rest and Pocket)	51–33	1	
BD.18068	**Clip,** securing Casing to L.H. Rear Door		18	
BD.24717	**Arm Rest Assembly,** L.H.	51–34	1	
DAZ.406/12C	**Screw,** Self-Tapping, securing Arm Rest		4	
BD.541/13	**Washer,** Special, on Self-Tapping Screws		4	
BD.25280	**Pocket Assembly,** L.H.	51–35	1	
M.108/5R	**Woodscrew,** securing Pocket (long)		6	
M.108/4R	**Woodscrew,** securing Pocket (short)		5	
BD.17641	**Ash Tray,** in Arm Rest	51–36	1	
DAN.506/8C	**Screw,** Self-Tapping, securing Ash Tray		2	

BODY FITTINGS

Part No.	Description	Plate No.	No. per Unit	Remarks
BD.24704	**CASING ASSEMBLY FOR R.H. REAR DOOR**		1	
BD.17213	**Casing only** (less Arm Rest and Pocket)		1	
BD.18068	**Clip,** securing Casing to R.H. Rear Door		18	
BD.24718	**Arm Rest Assembly,** R.H.		1	
DAZ.406/12C	**Screw,** Self-Tapping, securing Arm Rest		4	
BD.541/13	**Washer,** Special, on Self-Tapping Screws		4	
BD.25279	**Pocket Assembly,** R.H.		1	
M.108/5R	**Woodscrew,** securing Pocket (long)		6	
M.108/4R	**Woodscrew,** securing Pocket (short)		5	
BD.17641	**Ash Tray,** in Arm Rest		1	
DAN.506/8C	**Screw,** Self-Tapping, securing Ash Tray		2	
BD.24952/2	**Rubber Thrust Pad,** behind Casing, at Door Handle and Window Regulator Handle apertures		2	
DA.504/8C	**Screw,** Self-Tapping, securing Casing to top of Rear Doors		8	

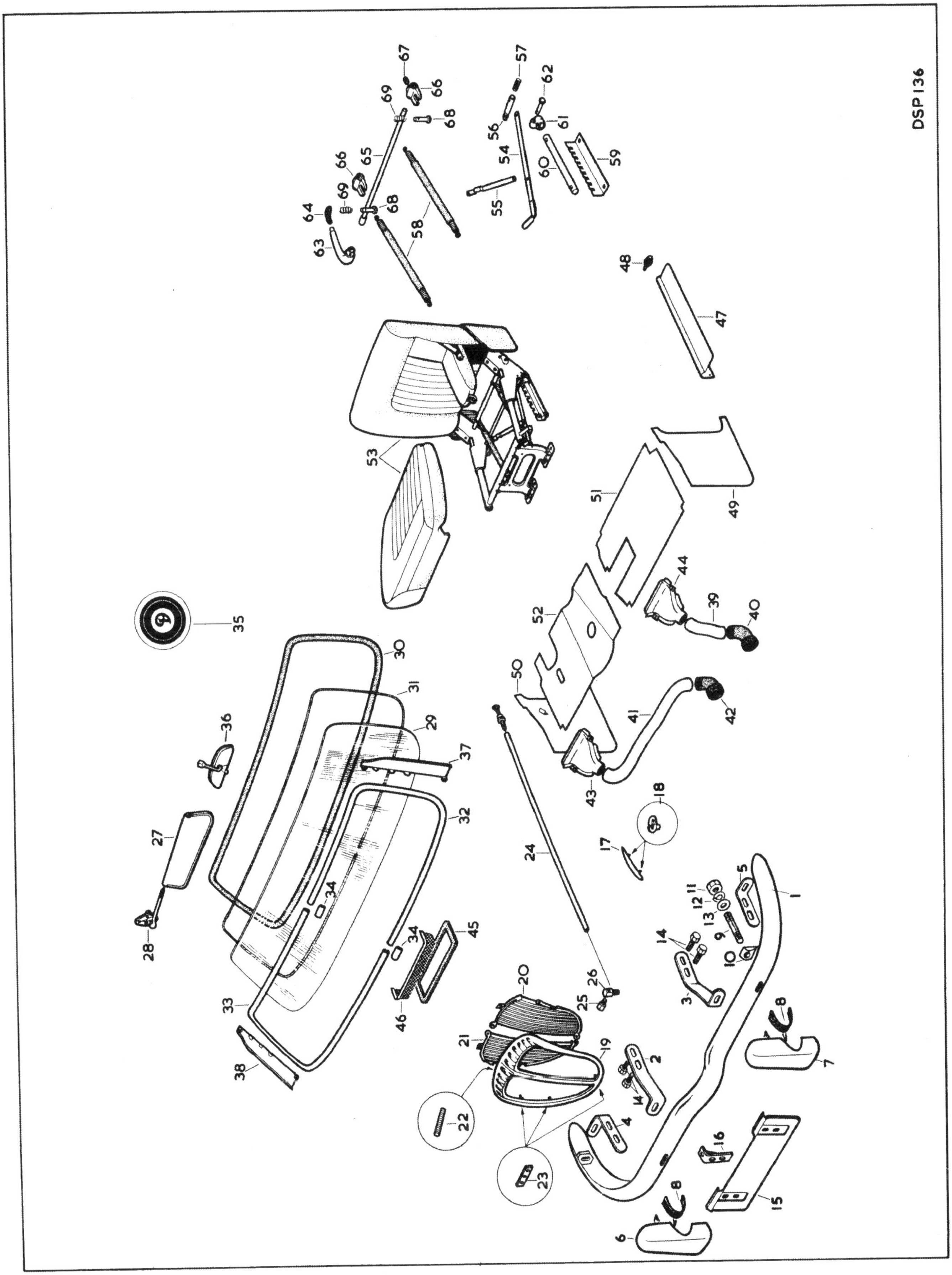

PLATE 45

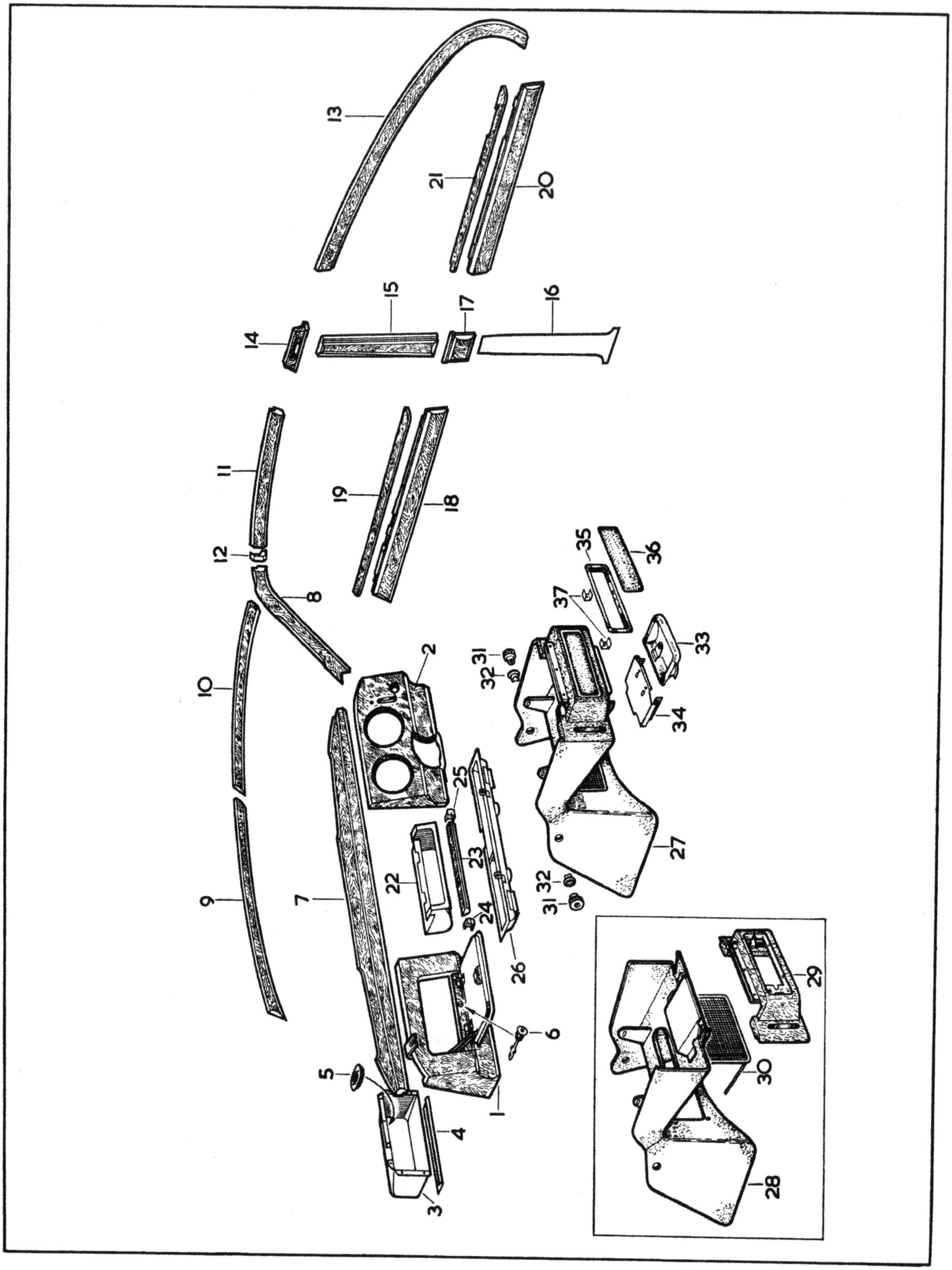
1
2
3
4
5
6
7
8
9
10
11
12
13
14
15
16
17
18
19
20
21
22
23
24
25
26
27
28
29
30
31
32
33
34
35
36
37

PLATE 46

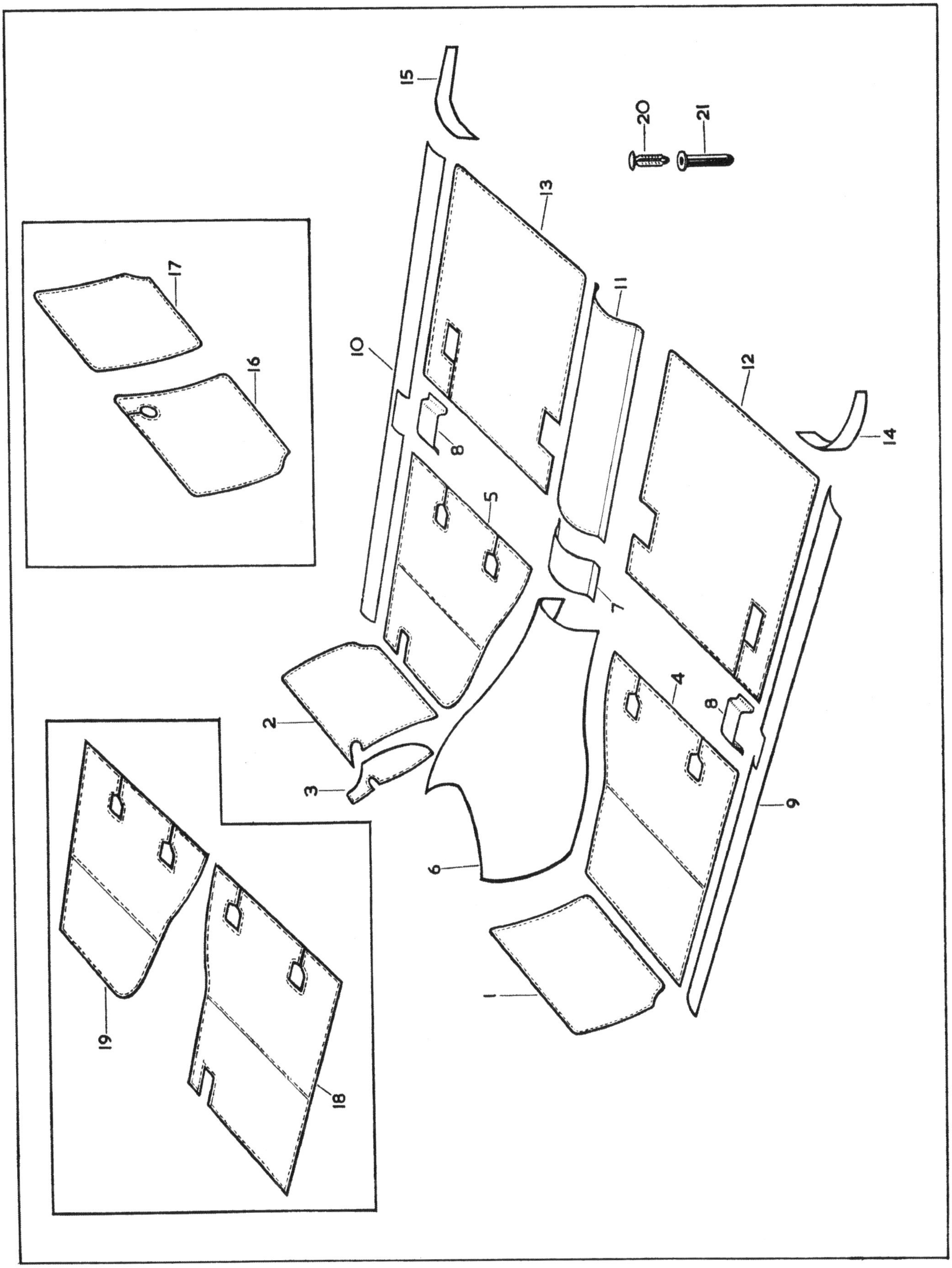

PLATE 47

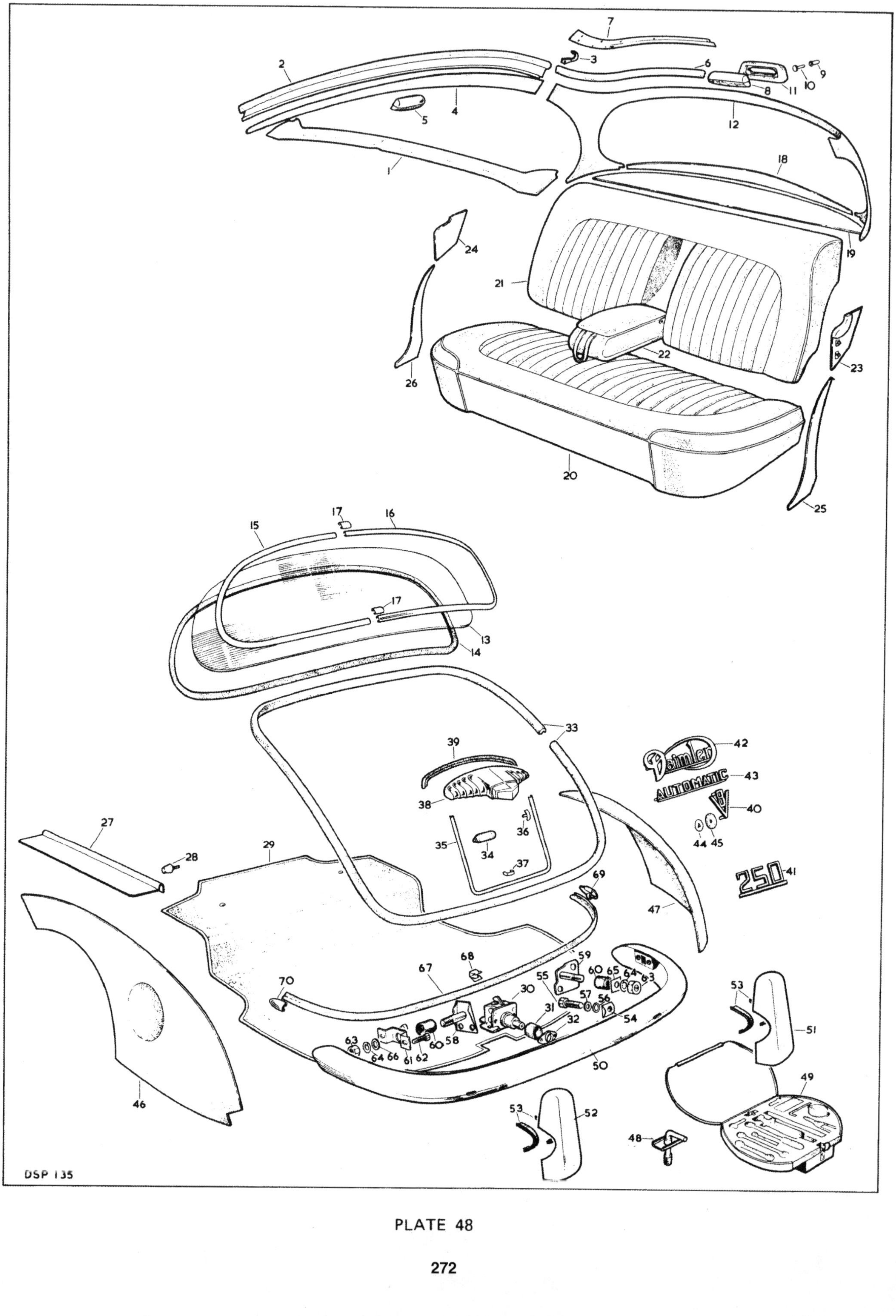
1
2
3
4
5
6
7
8
9
10
11
12
13
14
15
16
17
18
19
20
21
22
23
24
25
26
27
28
29
30
31
32
33
34
35
36
37
38
39
40
41
42
43
44
45
46
47
48
49
50
51
52
53
54
55
56
57
58
59
60
61
62
63
64
65
66
67
68
69
70
Daimler
AUTOMATIC
250
DSP 135

PLATE 48

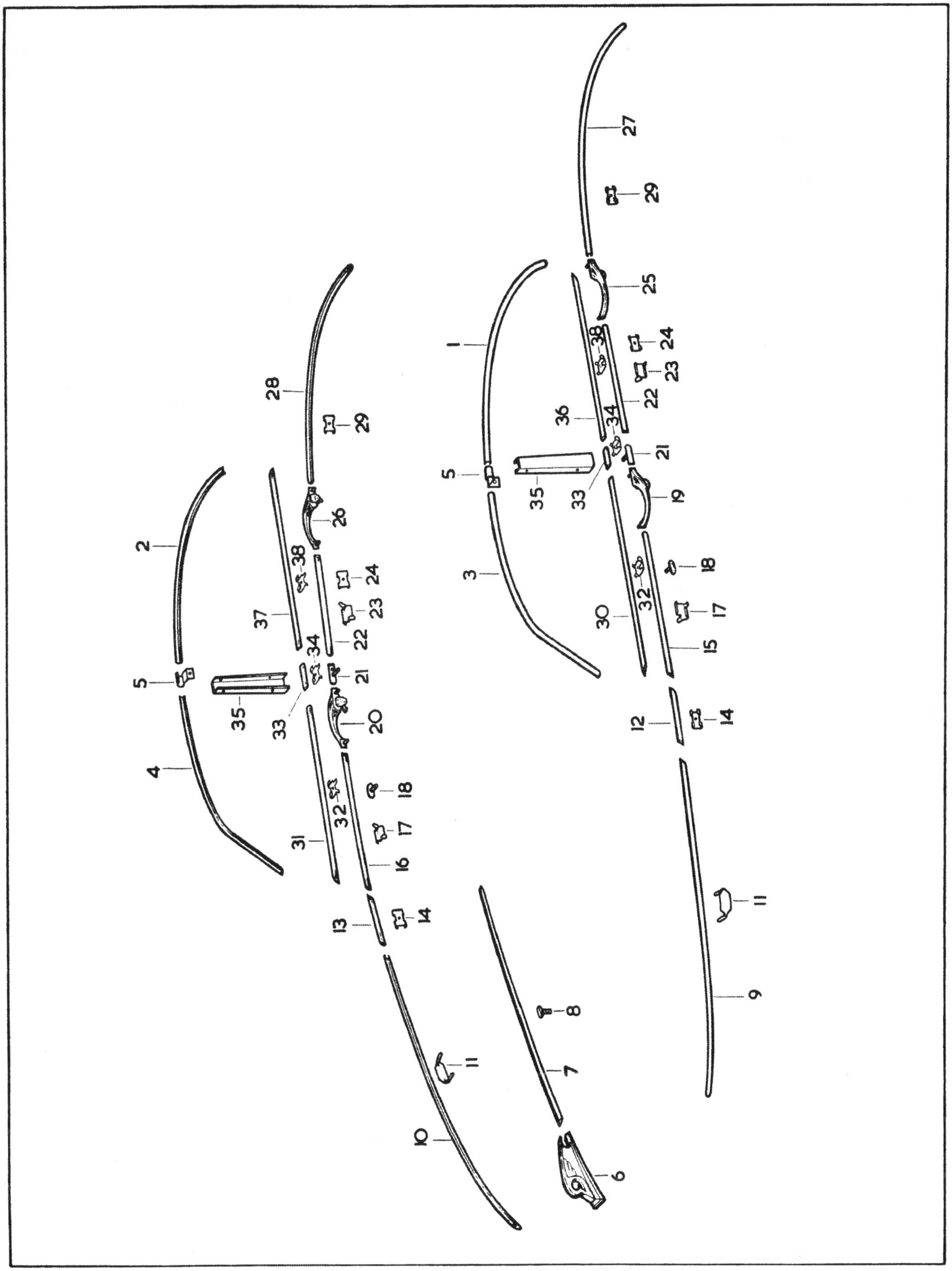
1
2
3
4
5
6
7
8
9
10
11
12
13
14
15
16
17
18
19
20
21
22
23
24
25
26
27
28
29
30
31
32
33
34
35
36
37
38

PLATE 49

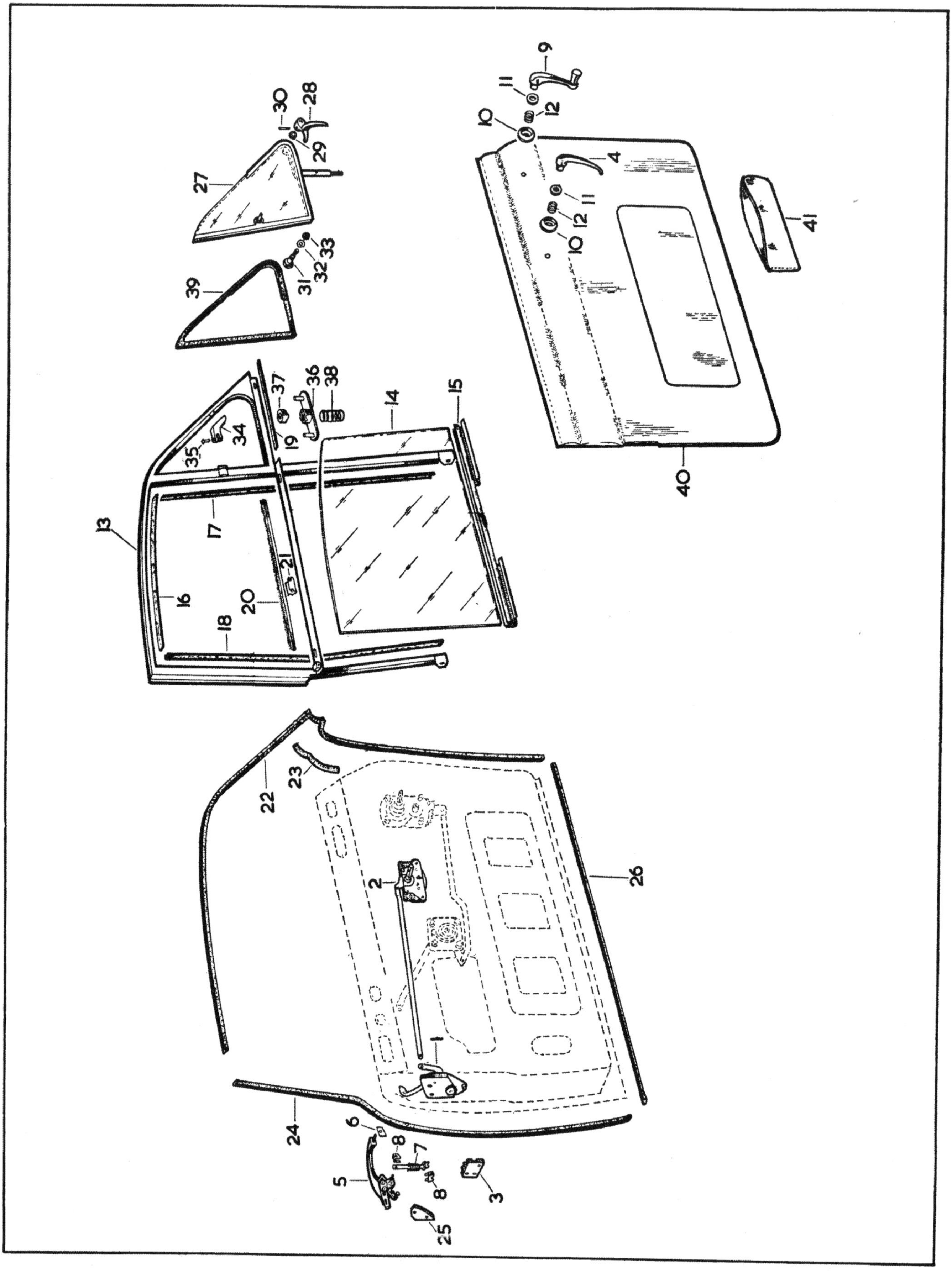

1
2
3
4
5
6
7
8
9
10
11
12
13
14
15
16
17
18
19
20
21
22
23
24
25
26
27
28
29
30
31
32
33
34
35
36
37
38
39
40
41

PLATE 50

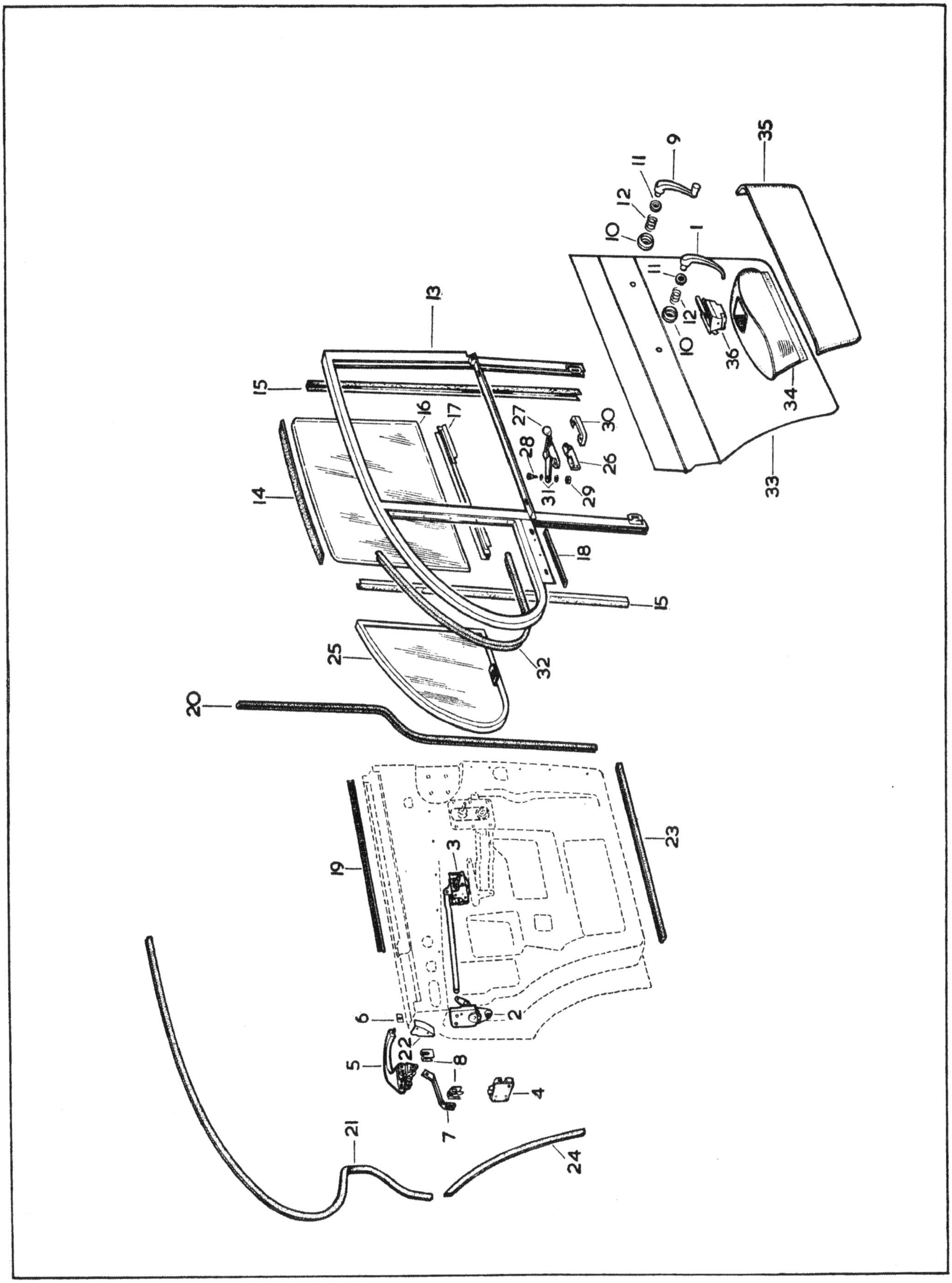

1
2
3
4
5
6
7
8
9
10
11
12
13
14
15
16
17
18
19
20
21
22
23
24
25
26
27
28
29
30
31
32
33
34
35
36

PLATE 51

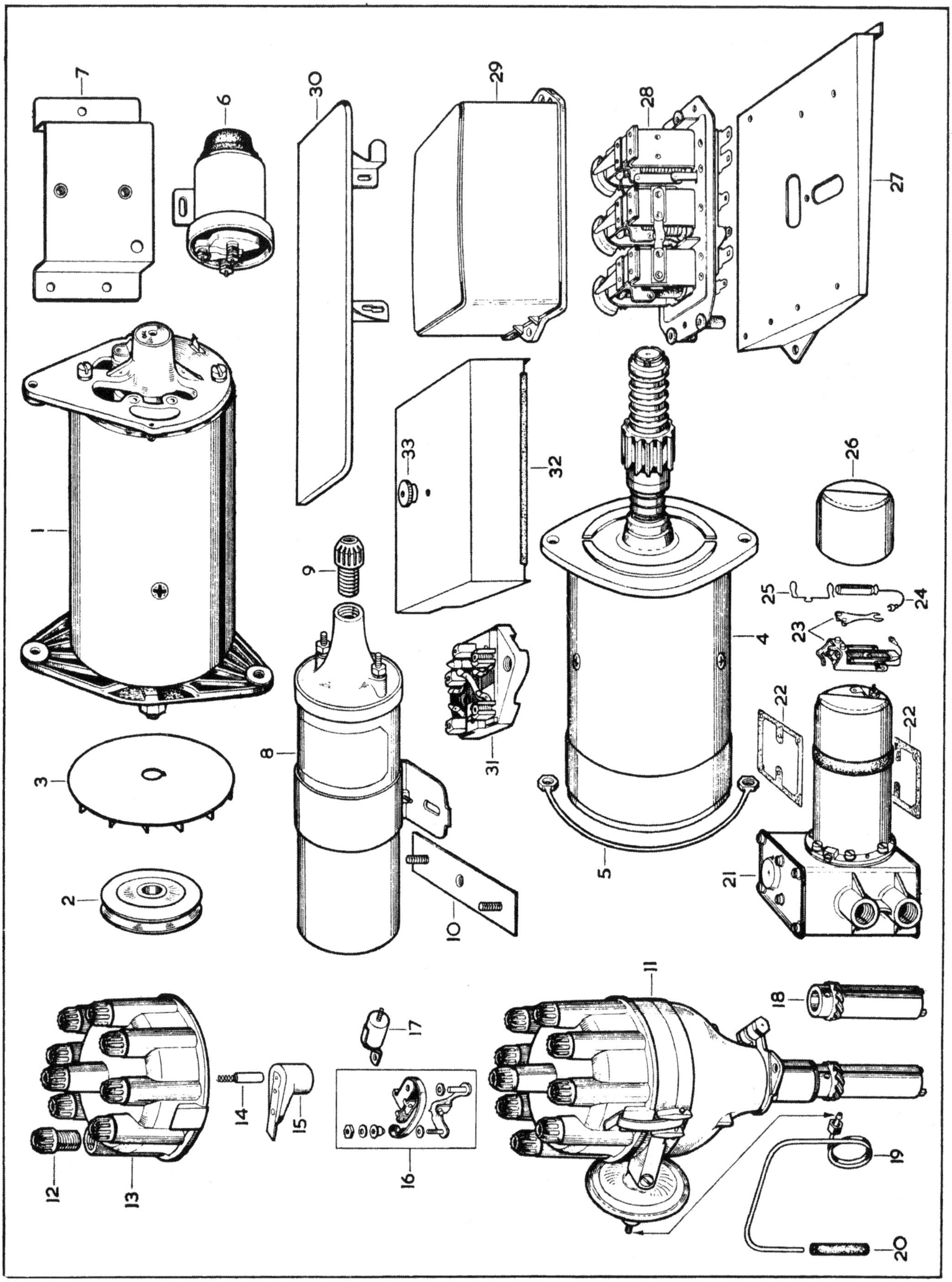
1
2
3
4
5
6
7
8
9
10
11
12
13
14
15
16
17
18
19
20
21
22
23
24
25
26
27
28
29
30
31
32
33

PLATE 52

Part No.	Description	Plate No.	No. per Unit	Remarks
C.20626	**DYNAMO (22716/H-C.40L)**	52–1	1	Fitted to Engine No. 7A.13231 and subs.
9698	**Brush** (54213117)		1 set	
9017	**Spring,** for Brush tension (227542)		1 set	
9699	**Bracket,** Commutator end (54212628)		1	
9704	**Bearing,** for C.E. Bracket (227818)		1	
9705	**Bracket,** Driving End (54211095)		1	
2756	**Bearing,** for D.E. Bracket (189307)		1	
9706	**Armature** (54213321)		1	
9707	**Field Coil** (54211803)		1 set	
UFB.131/8R	**Bolt,** securing Dynamo to Mounting Bracket		2	
FW.105/T	**Plain Washer,** at Dynamo Mounting		2	
UFN.131/L	**Nut,** on Bolts		2	
FG.105/X	**Washer,** Spring, under Nuts		2	
C.20912	**DYNAMO PULLEY**	52–2	1	Fitted to Engine No. 7A.13231 and subs.
C.21605	**FAN FOR DYNAMO PULLEY**	52–3	1	Fitted to Engine No. 7A.13231 and subs.
UCS.131/7R	**Screw,** Set, for adjustment of Dynamo Belt Tension		1	
401883	**Serrated Washer,** on Setscrew		1	
FG.105/X	**Washer,** Spring, on Setscrew		1	
C.25073	**ALTERNATOR (54021122 11.A.C.)**		1	Fitted to Engine No. 7K.1001 and subs.
10768	**Bolt,** through fixing (54112646)		3	
10769	**Nut,** for fixing Bolts (54131736)		3	
8586	**Nut,** on Shaft (198800)		1	
9365	**Washer,** under Shaft Nut (185085)		1	
2756	**Bearing,** (189307)		1	
10770	**Bracket Frame,** at driving end (54215209)		1	
10771	**Rotor** (54214557)		1	
12324	**Bracket,** at slip ring end (54217200)		1	
12325	**Heatsink,** with Cathode Base Diodes (54217551)		1	
12326	**Heatsink,** with Anode Base Diodes (54217552)		1	
10775	**Brush, Spring and Lucar** (54215076)		1 set	
10776	**Brushbox** (Bakelite) (54216364)		1	
8638	**Connector,** Lucar (Cut-out) (54190096)		1	
10777	**Stator** (54214441)		1	
C.22249	**Cooling Fan** (54214882)		1	
C.27447	**ALTERNATOR PULLEY**		1	Fitted to Engine No. 7K.1001 and subs.
C.27354	**BELT, DRIVING ALTERNATOR**		1	
C.27375	**ALTERNATOR MOUNTING BRACKET**		1	Fitted to Engine No. 7K.1001 and subs.
C.27522	**Mounting Plate,** at front of Engine for Alternator Mounting Bracket		1	
C.8511	**Stud,** in Engine Tappet Block for Alternator Mounting Plate		4	
UFN.131/L	**Nut,** on Studs, securing Alternator to Mounting Plate		4	
FG.105/X	**Washer,** Spring, under Nuts		4	
UFB.137/27R	**Bolt,** securing Alternator to Mounting Bracket		1	
C.8667/3	**Nut,** Self-locking, on Bolt		1	
C.27448	**Adjusting Link,** for Alternator		1	
UCS.131/10R	**Screw,** Set, securing Alternator to Adjusting Bracket		1	
FG.105/X	**Washer,** Spring, on Setscrew		1	
401883	**Washer,** Serrated, on Setscrew		1	
C.23028	**Locknut,** on Setscrew		1	
C.18203	**Stud,** in Alternator Mounting Bracket for Adjusting Link		1	
C.31207	**Steady Bracket,** for Alternator Shield		1	
FG.105/X	**Washer,** Spring, on Stud		1	
UFN.131/L	**Nut,** on Stud		1	
FW.105/T	**Washer,** Plain, on Stud		1	

ELECTRICAL EQUIPMENT

Part No.	Description	Plate No.	No. per Unit	Remarks
C.28439	**ALTERNATOR SHIELD ASSEMBLY**		1	Fitted to Engine No. 7K.1001 and subs.
C.8786	**STARTER MOTOR (25612A-M.418/G)**	52–4	1	Fitted to Engine No. 7A.1001 and subs. 7K.1001 and subs.
2804	**Brush** (255659)		1 set	
6546	**Bracket**, Commutator End (256748)		1	
2801	**Bushing**, for Commutator End Bracket (255491)		1	
6547	**Bracket**, Driving End (256495)		1	
2812	**Bushing**, for Driving End Bracket (270038)		1	
11045	**Spring**, for Brush tension (54257221)		1 set	
2803	**Pinion and Sleeve** (255649)		1	
2806	**Spring**, retaining Pinion (255728)		1	
12328	**Armature** (54241959)		1	
2810	**Field Coil** (256578)		1 set	
UFB.137/15R	**Bolt**, securing Starter Motor to Gearbox Support Plate		2	
FW.106/T	**Washer**, Plain, on Bolts		2	
FG.106/X	**Washer**, Spring, on Bolts		2	
C.22085	**Strap Assembly**, receiving Bolts	52–5	1	
C.20367	**SOLENOID FOR STARTER MOTOR (76445/H.2ST)**	52–6	1	Fitted to Chassis No. 1A.1001 and subs. 1A.20001 and subs. 1K.1001 to 1K.3672 1K.30001 to 1K.30083
C.30287	**SOLENOID FOR STARTER MOTOR (76772)**		1	Fitted to Chassis No. 1K.3673 and subs. 1K.30084 and subs.
UFS.119/4R	**Screw**, Set, securing Solenoid to Adaptor		2	
C.723/A	**Washer**, Shakeproof, on Setscrews		2	
C.12700	**ADAPTOR PLATE ASSEMBLY FOR SOLENOID**	52–7	1	
UFS.125/4R	**Screw**, Set, securing Adaptor		3	
C.724	**Washer**, Shakeproof, on Setscrews		3	
C.8749	**IGNITION COIL (45067/F-HA.12)**	52–8	1	**REQUIRED FOR CARS FITTED WITH DYNAMO C.20626**
C.5451	**Nut**, in centre of Coil, for attachment of Ignition Cable (408120)	52–9	1	
8638	**Lucar Blade** (54190096)		2	
C.23684	**IGNITION COIL (45169B/HA.12) (NEGATIVE EARTH)**		1	**REQUIRED FOR CARS FITTED WITH AN ALTERNATOR AT CHASSIS No.** 1K.1001 and subs. 1K.30001 and subs.
C.21506	**BRACKET ASSEMBLY, MOUNTING IGNITION COIL**	52–10	1	
UFS.119/3H	**Screw**, Set, securing Bracket		1	
UFN.125/L	**Nut**, securing Coil		2	
FW.104/T	**Washer**, Plain, under Nuts		2	
C.724	**Washer**, Shakeproof, under Nuts		2	
136104	**DISTRIBUTOR (40676K-20.D8)**	52–11	1	**8:1 COMPRESSION RATIO**
C.11906	**Nut**, for connection of High Tension Cables (410600)	52–12	9	
9708	**Cap** (54411740)	52–13	1	
6479	**Brush and Spring**, inside Cap (418856)	52–14	1	
9709	**Rotor** (425620)	52–15	1	
9710	**Contact Set** (54411772)	52–16	2	

Part No.	Description	Plate No.	No. per Unit	Remarks
	DISTRIBUTOR (40676K-20.D8) *(contd.)*			
9711	**Condenser** (54411935)	52–17	1	
9712	**Springs**, Auto Advance (54417470)		1 set	
9713	**Vacuum Unit** (54411675)		1	
2755	**Bearing** (189272)		1	
9714	**Bush** (423571)		1	
413819	**Washer** (Celeron) under Driving Collar		1	
C.20979	**DRIVING COLLAR AND GEAR**	52–18	1	Fitted from Engine No. 7A.1001 to 7A.5666
422524/32	**Mills Pin**, securing Driving Collar		1	
C.23504	**DRIVING COLLAR AND GEAR**		1	Fitted to Engine No. 7A.5667 and subs. 7K.1001 and subs.
422524/32	**Mills Pin**, securing Driving Collar and Gear to Distributor Shaft		1	
C.23505	**EXTENSION FOR DRIVING COLLAR AND GEAR**		1	
C.23968	**Circlip**, retaining Extension to Driving Collar and Gear		1	
307991	**SUCTION PIPE FOR DISTRIBUTOR VACUUM UNIT**	52–19	1	
C.14212	**Neoprene Pipe**, connecting Suction Pipe to L.H. Carburetter	52–20	1	
C.20576	**PETROL PUMP (AUB.152)**	52–21	1	Fitted from Chassis No. 1A.1001 to 1A.4692 1A.20001 to 1A.20286
1833	**Gasket**, under Lid (AUA.4646)	52–22	2	
9098	**Diaphragm Assembly** (AUB.6035)		1	**SUPPLIED WITH TWO GASKETS PART No. 9099**
9099	**Gasket**, at each side of Diaphragm (AUB.593)		2	
9320	**Spring**, on Diaphragm, for Armature (AUB.521)		1	
5846	**Filter** (AUA.4647)		1	
9100	**Rocker and Blade Assembly** (AUB.6022)	52–23	1	
1145	**Terminal Screw** (AUA.1468)		1	
7664	**Condenser** (AUA.5060)	52–24	1	
7665	**Clip**, holding Condenser (AUA.5059)	52–25	1	
9321	**End Cover** (AUB.626)	52–26	1	
C.22458	**PETROL PUMP (AUF.301)**		1	Fitted from Chassis No. 1A.4693 to 1A.6207 1A.20287 to 1A.20335
10202	**Coil Housing** (AUB.6080)		1	
9100	**Rocker and Blade Assembly** (AUB.6022)		1	
1145	**Terminal Screw** (AUA.1468)		1	
7664	**Condenser** (AUA.5060)		1	
7665	**Spring Clip**, holding Condenser (AUA.5059)		1	
10203	**End Cover** (AUB.707)		1	
10204	**Breather Elbow**, on End Cover (AUB.702)		1	
10205	**Cap**, for Breather (AUB.703)		1	
10206	**Ball**, under Cap (AUB.704)		1	
C.18916	**Lucar Blade**, for Feed and Earth Cable Connections to Pump (199279)		2	
10199	**Diaphragm Assembly** (AUB.6071)		1	
10201	**Roller**, for Diaphragm (AUA.1433)		11	
9320	**Spring**, for Diaphragm (AUB.521)		1	
10200	**Gasket**, under Diaphragm (AUB.706)		1	
AS.302/4D	**Screw**, Set, securing Coil Housing and Diaphragm to Valve Housing		6	
10186	**Valve Assembly** (AUB.6031)		2	
10188	**Filter**, for Inlet Valve (AUB.617)		1	
10767	**Sealing Washer**, for Valves and Filter (AUB.676)		3	
10189	**Cover**, for Valves (AUB.652)		2	
10190	**Plate**, retaining Valves (AUB.653)		1	

Part No.	Description	Plate No.	No. per Unit	Remarks
	PETROL PUMP (AUF.301) *(contd.)*			
10191	**Screw,** Self-Tapping, securing Plate (AUB.597)		2	
10193	**Cover,** for end of Valve Housing (AUA.574)		1	
10192	**Cork Gasket,** under Cover (AUA.573)		1	
5803	**Screw,** securing Cover (AUC.2694)		1	
5648	**Washer,** Spring, on Screw (AUC.2246)		1	
10194	**Dished Washer,** under Spring Washer (AUA.575)		1	
10195	**Anti-Surge Diaphragm** (Nylon) (AUB.696)		1	
10198	**Sealing Washer,** under Diaphragm Plate (AUB.675)		1	
10196	**Backing** (Rubber) for Anti-Surge Diaphragm (AUB.693)		1	
10197	**'O' Ring,** under Diaphragm Cover (AUB.711)		1	
AS.304/3D	**Screw,** Set, securing Diaphragm Cover		4	
C.25076	PETROL PUMP (AUF.303)		1	Fitted to Chassis No. 1A.6208 and subs. 1A.20336 and subs. 1K.1001 and subs. 1K.30001 and subs.
10966	**Spring,** for Diaphragm (AUB.758)		1	
10967	**Valve Assembly** (AUB.6061 or AUB.6062)		1	
	All other items are as for Petrol Pump C.22458			
C.801	**Stud,** mounting Petrol Pump		2	
C.8150/33	**Nut,** on Studs		2	
FW.205/T	**Washer,** Plain, under Nuts		4	
BD.541/5	**Special Washer,** under Nuts		4	
C.10126	**Grommet,** for Petrol Pump Mounting		2	
C.20323	**Distance Tube,** in Grommets		2	
C.2895	**Earth Cable,** for Petrol Pump (859499)		1	
UFS.119/4R	**Screw,** Set, securing Earth Cable		1	
UFN.119/L	**Nut,** on Setscrew		1	
C.723/A	**Washer,** Shakeproof, under Nut		1	
AW.102/T	**Washer,** Plain, under Shakeproof Washer		1	
C.21264	BRACKET ASSEMBLY, MOUNTING FUSE AND CONTROL BOXES AND HORN RELAY	52–27	1	
C.11218	**Rod Assembly,** fixing Fuse Box Cover		1	
C.12365/2	**Insulating Tube,** on Rod		1	
C.4548	**Grommet**		2	
UFS.119/4R	**Screw,** Set, securing Bracket		2	
UFS.119/7R	**Screw,** Set, securing Bracket		1	
BD.5815	**Distance Piece**		1	
UFN.119/L	**Nut,** on Setscrews		3	
C.723/A	**Washer,** Shakeproof, under Nuts		3	
AW.102/T	**Washer,** Plain, on Setscrew		1	
C.20627	CONTROL BOX (37342/D-RB.340)	52–28	1	
8558	**Cover** (54381342)	52–29	1	
9943	**Screw,** securing Cover		2	
8559	**Resistance** (Swamp) (54380438)		1	
9715	**Resistance** (Points) (54380084)		1	
UFS.119/9R	**Screw,** Set, securing Control Box to Bracket		3	
C.22253	CONTROL BOX (4TR) (37423D)		1	**REQUIRED FOR USE WITH ALTERNATOR C.25073**
C.11472/1	SHIELD ASSEMBLY, PROTECTING CONTROL BOX	52–30	1	

Part No.	Description	Plate No.	No. per Unit	Remarks
C.16510	**FUSE BOX (54038009-SF.6)**	52–31	1	
C.5638	**Fuse**, 35 amps (188218)		2	
C.13932	**Fuse**, 50 amps (188219)		2	
6471	**Terminal Screw** (121610)		5	
UCS.319/4H	**Screw**, Set, securing Fuse Box to Mounting Bracket		2	
C.21267	**COVER OVER FUSE BOX**	52–32	1	
C.4877/4	**Rubber Insulator**, under Cover		1	
C.5549	**Knob**, securing Cover	52–33	1	
C.15445	**INSTRUMENT PANEL ASSEMBLY**	53–1	1	
C.15785	**Fixing Screw**, for Instrument Panel	53–2	2	
BD.5453	**Washer** (Black Fibre), on Fixing Screws	53–3	2	
UFS.125/4R	**Screw**, Set, securing Hinges to Centre Dash		4	
UFS.419/2H	**Screw**, Set, securing Cable Clips to Pillars on Back Face of Instrument Panel		2	
BD.17747	**RUBBER SEAL, PREVENTING ENTRY OF LIGHT INTO CAR FROM INDICATOR STRIP**	53–4	2	
C.15289	**LIGHT FILTER FOR INDICATOR STRIP**	53–5	1	
C.15916	**CHROME FINISHER AT TOP OF INDICATOR STRIP**	53–6	1	
C.15485	**VOLTAGE CONTROL UNIT AT BACK OF INSTRUMENT PANEL (BR.1300/01)**	53–7	1	
BD.17710	**DIFFUSER SCREEN FOR INSTRUMENT PANEL**	53–8	1	
UFS.419/4H	**Screw**, Set, securing Diffuser Screen		2	
AW.102/T	**Washer**, Plain, on Setscrews		4	
UFN.119/L	**Nut**, on Setscrews		2	
C.723/A	**Washer**, Shakeproof, under Nuts		2	
C.15789	**CONNECTOR BLADE FOR EARTHING INSTRUMENT PANEL (54190038)**	53–9	3	
C.15790	**CONNECTOR CABLE FOR INSTRUMENT PANEL (54944612)**		1	
C.15609	**INDICATOR STRIP, AT BOTTOM OF INSTRUMENT PANEL, DETAILING SWITCH POSITIONS (54850103)**	53–10	1	R.H. DRIVE CARS ONLY
C.15610	**INDICATOR STRIP, AT BOTTOM OF INSTRUMENT PANELS, DETAILING SWITCH POSITIONS (54850104)**		1	L.H. DRIVE CARS ONLY
BD.540/19	**Screw**, Self-Tapping, securing Indicator Strip		4	
C.15787	**Bulb** (2 watt Lilliput) illuminating Indicator Strip (281)		3	

ELECTRICAL EQUIPMENT

Part No.	Description	Plate No.	No. per Unit	Remarks
C.15454	**IGNITION SWITCH (31962/B-S.45)**	53–11	1	
C.31114	**IGNITION SWITCH (39174-149S.A.)**		1	**REQUIRED FOR CARS FITTED WITH AN ALTERNATOR AT CHASSIS No.** 1K.1001 and subs. 1K.30001 and subs.
BD.10610	**Barrel**, for Ignition Switch (WB.1/6545)	53–12	1	
C.15667	**Adaptor Plate**, fixing Ignition Switch (54331297)	53–13	1	
C.18189	**ROAD LIGHTS SWITCH (34383/B-PRS.7)**	53–14	1	**NOT REQUIRED FOR U.S.A.**
C.15447	**Escutcheon**, for Road Lights Switch (54850095)	53–15	1	**NOT REQUIRED FOR U.S.A.**
C.18188	**ROAD LIGHTS SWITCH (34382/B-PRS.7)**		1	**FOR U.S.A. ONLY**
C.15449	**Escutcheon**, for Road Lights Switch (54850094)		1	**FOR U.S.A. ONLY**
C.15452	**Hollow Bolt**, securing Escutcheon (54332269)	53–16	1	
C.15451	**Nut**, on Bolt (156574)	53–17	1	
C.15928	**Washer**, under Nut (154604)	53–18	1	
C.27121	**RESISTANCE FOR ROAD LIGHTS SWITCH (54334484)**		1	
C.15450	**LEVER OPERATING ROAD LIGHTS SWITCH (54332270)**	53–19	1	
C.17513/1	**Sleeve** (PVC) on Lever		1	
C.15453	**STARTER SWITCH (31967/B-SS.5)**	53–20	1	
C.15455	**PANEL LIGHTS SWITCH (31969/A/D-57SA)**	53–21	1	
C.15589	**HEATER FAN SWITCH (31960/D-57SA)**	53–22	1	
C.15456	**WINDSCREEN WIPER SWITCH (31966/F-79SA)**	53–23	1	
C.15759	**WINDSCREEN WASHER SWITCH (31984/D-65SA)**	53–24	1	
C.15760	**MAP LIGHT SWITCH (31963/E-65SA)**	53–25	1	
C.15760	**INTERIOR LIGHT SWITCH (31963/E-65SA)**	53–26	1	
C.25678	**AMMETER (36321/B-CZU.60)**	53–27	1	**REQUIRED FOR CARS FITTED WITH DYNAMO C.20626**
9716	**Connector** (Lucar "M" Large) (54190217)		1	
8058	**Connector** (Lucar "M" Small) (199275)		2	
9717	**Connector** (Lucar "M" double) (Large) (54190310)		1	
987	**Bulb**, illuminating Ammeter (987)		1	

Part No.	Description	Plate No.	No. per Unit	Remarks
C.27309	**AMMETER (36358D)**		1	REQUIRED FOR CARS FITTED WITH AN ALTERNATOR AT CHASSIS No. 1K.1001 and subs. 1K.30001 and subs.
9716	**Connector** (Lucar "M" Large) (54190217)		1	
8058	**Connector** (Lucar "M" Small) (199275)		2	
9717	**Connector** (Lucar "M" double) (Large)		1	
987	**Bulb**, illuminating Ammeter (987)		1	
C.11724	**CHOKE CONTROL WARNING LIGHT ASSEMBLY (38147/A-WL.13)**	53–28	1	
987	**Bulb** (2.2 watt) (987)		1	
C.12833	**SWITCH OPERATING CHOKE WARNING LIGHT (31504/D-12.SA)**	53–29	1	See also Handbrake Controls
UFN.231/L	**Nut**, securing Switch to Mounting Bracket		1	
C.16179	**Earth Cable**, for Warning Light Switch (54945031)		1	
C.16197	**CABLE FOR CHOKE WARNING LIGHT (54945232)**		1	
C.15592	**ESCUTCHEON FOR CHOKE CONTROL LEVER**	53–30	1	
BD.11802	**PLATE ("COLD" "HOT" "RUN") FOR CHOKE CONTROL LEVER**	53–31	1	
UFN.119/L	**Nut**, securing Escutcheon and Plate to Facia		2	
C.723/A	**Washer**, Shakeproof, under Nuts		2	
C.16556	**MOUNTING BRACKET ASSEMBLY FOR CHOKE CONTROL LEVER**	53–32	1	
BD.1502/2	**Woodscrew**, securing Mounting Bracket to Facia Board		3	
C.15977	**LEVER OPERATING CHOKE CONTROL**	53–33	1	
BD.6235/2	**Screw**, Set, securing Lever to Mounting Bracket		1	
BD.541/9	**Washer**, Brass, on Setscrew		1	
C.8737/1	**Nut**, on Setscrew		1	
BD.11489	**Ball Bearing**, between Operating Lever and Mounting Bracket	53–34	1	
C.15989	**Knob**, on Control Lever	53–35	1	
UCS.311/2H	**Screw**, Set, securing Knob to Lever		1	
C.21375	**INNER CABLE OPERATING CHOKE LEVERS ON CARBURETTERS**	53–36	1	
C.23903	**Bolt**, securing Inner Cable to Operating Lever	53–37	1	
C.8737/9	**Nut**, Self-Locking, on Bolt		1	
C.21373	**OUTER CASING FOR OPERATING CABLE**	53–38	1	
C.21377	**Sleeve**, insulating Outer Casing from Warning Light	53–39	1	
BD.11524	**Grommet**, in Scuttle, for Outer Casing	53–40	1	
C.15998	**PIVOT FOR OUTER CABLE**	53–41	1	
UFS.119/4R	**Screw**, Set, securing Pivot		1	
C.5203/1	**Washer**, Special, on Setscrew		1	
BD.1708/2	**Washer**, Double Coil Spring, on Setscrew		1	

Part No.	Description	Plate No.	No. per Unit	Remarks
C.18641	**OIL PRESSURE GAUGE (ELECTRIC) (PE.2300/02)**	53–42	1	
C.15474	**Electric Element,** for Oil Pressure Gauge (PT.1801/06)	53–43	1	
C.15288	**Lucar Blade,** for Cable Connection to Oil Pressure Gauge (54190107)	53–44	1	
987	**Bulb** (2.2 watt) for illumination of Oil Pressure Gauge (987)		1	
C.16178	**WARNING LIGHT INDICATING HANDBRAKE "ON" POSITION AND LEVEL IN BRAKE FLUID CONTAINER (38091/A-WL.3)**	53–45	1	
987	**Bulb** (2.2 watt) (987)		1	
C.2607	**Rubber Sleeve,** protecting Warning Light Terminals		1	Fitted to Chassis No. 1A.10629 and subs. 1A.20495 and subs. (May be fitted to earlier cars) 1K.1001 and subs. 1K.30001 and subs.
C.19637	**ESCUTCHEON FOR HANDBRAKE AND BRAKE FLUID WARNING LIGHT**	53–46	1	
C.16184	**BRACKET MOUNTING WARNING LIGHT**	53–47	1	
BD.1502/1	**Woodscrew,** securing Bracket		2	
C.12833	**SWITCH OPERATING HANDBRAKE AND BRAKE FLUID WARNING LIGHT (31504/D-12.SA)**	53–48	1	
C.16179	**Earth Cable,** for Switch in Handbrake Mechanism		1	
C.15475	**PETROL GAUGE (BF.2200)**	53–49	1	
987	**Bulb** (2.2 watt) for illumination of Petrol Gauge (987)		1	
C.18640	**WATER TEMPERATURE GAUGE (BT.2200/01)**	53–50	1	Fitted from Chassis No. 1A.1001 and subs. 1A.20001 and subs. also from Chassis No. 1K.1001 to 1K.1980 1K.30001 to 1K.30045
C.29876	**WATER TEMPERATURE GAUGE (BT.2200/06)**		1	Fitted to Chassis No. 1K.1981 and subs. 1K.30046 and subs.
C.15472	**Electric Element,** for Water Temperature Gauge (TT.4801/00)	53–51	1	
987	**Bulb** (2.2 watt) for illumination of Water Temperature Gauge (987)		1	
C.15600	**SPEEDOMETER (SN.6317) (CALIBRATED IN MILES)**	53–52	1	**SUITABLE FOR CARS FITTED WITH 4.55 : 1 REAR AXLE**

Part No.	Description	Plate No.	No. per Unit	Remarks
15907	SPEEDOMETER (SN.6317/12) (CALIBRATED IN KILOS) NOTE: To ensure that the correct Speedometer is supplied by Daimler Spares Dept., IT IS IMPORTANT WHEN ORDERING, to quote the number printed on the dial of the displaced instrument. On some specially calibrated instruments, the number is printed on the back of the Speedometer Casing.		1	SUITABLE FOR CARS FITTED WITH 4.27 : 1 REAR AXLE
C.15468	REMOTE CONTROL OPERATING TRIP MILEAGE RECORDER ON SPEEDOMETER (41-741-152-01/9)	53–53	1	
8409	Screwed Escutcheon, securing Remote Control at winding end		1	
987	BULB (2.2 WATT) FOR ILLUMINATION OF SPEEDOMETER AND FOR IGNITION, LOW FUEL CONTENTS AND HEADLAMP MAIN BEAM WARNING LIGHTS (987)		5	
C.15929	FLEXIBLE SPEEDOMETER CABLE COMPLETE (DF.1104/19/Z.66)	53–54	1	R.H. DRIVE CARS ONLY
8241	Inner Cable only (DI.1110/00/Z.66)		1	
8242	Outer Cable only (DO.1104/18/66)		1	
C.25249	FLEXIBLE SPEEDOMETER CABLE COMPLETE (DF.1104/192/90)		1	L.H. DRIVE CARS ONLY
11074	Inner Cable only (DI.1110/00Z/90)		1	
11075	Outer Cable only (DO.1104/18/90)		1	
9720	REV. COUNTER ONLY (RV.7413/07)	53–55	1	Fitted to Chassis No. 1A.1001 and subs. 1A.20001 and subs.
8260	ELECTRIC CLOCK ONLY (CE.1111/00)	53–56	1	
987	Bulb (2.2 watt) for illumination of Rev. Counter (987)		2	
C.14996	GENERATOR FOR ELECTRIC REV. COUNTER (TV.1100/00)	53–57	1	
C.15287	REMOTE CONTROL FOR CLOCK ON REV. COUNTER (41-741-127-03/6)	53–58	1	
8409	Screwed Escutcheon, securing Remote Control at winding end		1	
11579	REV. COUNTER ONLY (RV.7413/14) IMPULSE TYPE WITH NEGATIVE EARTH		1	Fitted to Chassis No. 1K.1001 and subs. 1K.30001 and subs.
10488	ELECTRIC CLOCK ONLY (CE.1114/00) NEGATIVE EARTH		1	
987	Bulb (2.2 watt) for illumination of Rev. Counter (987)		2	
C.15287	REMOTE CONTROL FOR CLOCK ON REV. COUNTER (41-741-127-03/6)		1	
8409	Screwed Escutcheon, securing Remote Control at winding end		1	

ELECTRICAL EQUIPMENT

Part No.	Description	Plate No.	No. per Unit	Remarks
C.18638	**THERMOSTATIC CIGAR LIGHTER COMPLETE (L.15094.12V)**	53–59	1	
C.18638/1	**Body only**		1	
C.18638/2	**Element**		1	
C.18638/3	**Clamping Shell**		1	
C.15476	**PETROL TANK ELEMENT UNIT (TF.1104/040)**	53–60	1	
C.937	**Gasket,** for Element Unit	53–61	1	
C.9653	**SWITCH, BEHIND DOOR HINGES, OPERATING INTERIOR LIGHTS**	53–62	4	
UFN.131/L	**Nut,** securing Switches		4	
C.725	**Washer,** Shakeproof, under Nuts		4	
C.13633	**Washer,** Special, for Switches		4	
C.23877	**HEADLAMP, COMPLETE (59310/A-F.700)**		2	**FOR R.H. DRIVE CARS EXCEPT THOSE EXPORTED TO SWEDEN, GIBRALTAR, OR PORTUGUESE EAST AFRICA**
6512	**Outer Rim** (554440)	54–1	2	
6513	**Screw,** securing Outer Rim (198898)	54–2	2	
6524	**Washer,** retaining Screw (161385)	54–3	2	
8076	**Spire Nut,** on Screws (187117)	54–4	2	
6516	**Bracket,** fixing Outer Rim (555704)		2	
9266	**Rim,** retaining Sealed Beam Unit (54520549)	54–5	2	
5221	**Screw,** fixing Unit Retaining Rim (186137)	54–6	6	
9271	**Rim,** Unit Seating (54520446)	54–7	2	
9404	**Spring,** fixing Unit Seating Rim (54520514)	54–8	2	
10763	**Sealed Beam Unit** (54522480)	54–9	2	
7764	**Adaptor** (553738)	54–10	2	
9405	**Trimmer Screw** (54520451)	54–11	4	
11453	**Retainer,** for Trimmer Screws (54522798)		4	
10219	**Nut,** on Trimmer Screws (54520551)		4	
10218	**Sleeve,** on Trimmer Screws (54520550)		4	
9403	**Gasket,** Headlamp Body (54521487)	54–12	2	
3446	**Dust Excluder,** behind Headlamp Outer Rim (552906)	54–13	2	
C.20333	**HEADLAMP COMPLETE (58843/A-F.700)**		2	**FOR L.H. DRIVE CARS ONLY** Fitted from Chassis No. 1A.20001 to 1A.20292 Refer to list on page 287 for countries affected
9964	**Sealed Beam Unit** (54520481)		2	
	All other items as for Headlamp C.23877			
3446	**Dust Excluder,** behind Headlamp Outer Rim (552906)		2	
23640	**HEADLAMP COMPLETE (59193/A-F.700)**		2	**FOR L.H. DRIVE CARS ONLY** Fitted to Chassis No. 1A.20293 and subs. Refer to list on page 287 for countries affected
10465	**Sealed Beam Unit** (54522231)		2	
	All other items are as for Headlamp C.23877			
3446	**Dust Excluder,** behind Headlamp Outer Rim (552906)		2	
C.27010	**HEADLAMP, COMPLETE (59428A)**		2	**FOR L.H. DRIVE CARS ONLY** See applicability list on page 287
7762	**Light Unit** (54522683)		2	
7763	**Bulb** (410)		2	
7765	**Spring,** retaining Bulb (556442)		2	
	All other items are as for Headlamp C.23877			

Part No.	Description	Plate No.	No. per Unit	Remarks
C.16779	**HEADLAMP COMPLETE (58271/A-F.700EF)**		2	FOR L.H. DRIVE CARS EXPORTED TO ALGERIA, MOROCCO, TANGIERS, OR TUNISIA
8612	**Bulb**, 45/40 watt (Yellow) (411)		2	
	All other items as for Headlamp C.14441)			
C.16780	**HEADLAMP COMPLETE (58450/A-F.700E)**		2	FOR R.H. DRIVE CARS EXPORTED TO SWEDEN, GIBRALTAR, OR PORTUGUESE EAST AFRICA
8680	**Light Unit** (54520148)		2	
	All other items as for Headlamp C.14441			
C.16781	**HEADLAMP COMPLETE (58634/A-F.700)**		2	FOR L.H. DRIVE CARS EXPORTED TO AUSTRIA
8613	**Light Unit** (54520883)		2	
	All other items as for Headlamp C.14441			
UFS.319/4H	**Screw**, Set, securing Headlamps		8	
AW.102/T	**Washer**, Plain, on Setscrews		8	
C.723/A	**Washer**, Shakeproof, on Setscrews		8	
C.27011	**HEADLAMP, COMPLETE (59429A)**		2	FOR CARS EXPORTED TO FRANCE
7762	**Light Unit** (54522683)		2	
8612	**Bulb** (411)		2	
7765	**Spring**, retaining Bulb (556442)		2	
10219	**Nut**, for Trimmer Screws (54520551)		4	
10218	**Housing**, for Trimmer Nuts (54520550)		4	
	All other items are as for Headlamp C.23877			
C.29976	**HEADLAMP, COMPLETE (59408A/B)**		2	FOR CARS EXPORTED TO GERMANY
11909	**Light Unit** (54522909)		2	
7763	**Bulb**, Main (410)		2	
N.S.S.	Bulb, Pilot		2	
10220	**Bulb Holder**, Pilot (54573590)		2	
11886	**Interior**, for Bulb Holder (55378010)		2	
10219	**Nut**, for Trimmer Screws (54520551)		2	
10218	**Housing**, for Trimmer Nuts (54520550)		2	
	All other items are as for Headlamp C.23877			

HEADLAMP APPLICABILITY – L.H. DRIVE CARS ONLY

PART NO.	FOR USE IN
C.20333 or C.23640	Argentine, Bahrain, Bolivia, Brazil, Canada, China, Chile, Colombia, Cuba, Egypt, Guatemala, Hawaii, Iran, Iraq, Israel, Kuwait, Lebanon, Jordan, Mexico, Nicaragua, Panama, Peru, Philippines, Puerto Rico, Saudi Arabia, Sudan, Turkey, Uruguay, U.S.A., Venezuela.
C.27010	Belgium, Belgian Congo, Canary Isles, Czechoslovakia, Denmark, Dutch West Indies, Finland, Greece, Holland, Italy, Libya, Maderia, Norway, Portugal, Rumania, Spain, Sweden, Switzerland.

Part No.	Description	Plate No.	No. per Unit	Remarks
C.2639	**HEADLAMP DIPPER SWITCH (31284/D-FS.22/1)**	54–14	1	Fitted from Chassis No. 1A.1001 to 1A.2394 1A.20001 to 1A.20197
C17724	**Rubber Cap,** for Dipper Switch (54334682)	54–15	1	
AS.102/15/R	**Screw,** Set, mounting Dipper Switch		2	R.H. DRIVE CARS ONLY Fitted from Chassis No. 1A.1001 to 1A.2394
C.723	**Washer,** Shakeproof, on Setscrews		2	
C.5429	**Distance Piece,** on Setscrews		2	
C.4579/1	**Plate,** under Dipper Switch		1	
C.2707	**Grommet,** in Plate		1	
C.4578	**Gasket,** under Plate		1	
UFS.119/4R	**Screw,** Set, sealing redundant holes		2	
UFN.119/L	**Nut,** on Setscrews		2	
C.723/A	**Washer,** Shakeproof, under Nuts		2	
AS.102/15R	**Screw,** Set, mounting Dipper Switch		2	L.H. DRIVE CARS ONLY Fitted from Chassis No. 1A.20001 to 1A,20197
C.723	**Washer,** Shakeproof, on Setscrews		2	
C.5429	**Distance Piece,** on Setscrews		2	
C.4580	**Blanking Plate,** sealing redundant hole		1	
C.4578	**Gasket,** under Blanking Plate		1	
UFS.119/4R	**Screw,** Set, securing Blanking Plate		2	
UFN.119/L	**Nut,** on Setscrews		2	
C.723/A	**Washer,** Shakeproof, under Nuts		2	
C.22044	**HEADLAMP DIPPER SWITCH (34536/B-103.SA)**		1	Fitted to Chassis No. 1A.2395 and subs. 1A.20198 and subs. 1K.1001 and subs. 1K.30001 and subs.
C.22276	**Cap** (Rubber) for Dipper Switch (54339495)		1	
C.22042	**Bracket,** mounting Dipper Switch		1	R.H. DRIVE CARS ONLY Fitted to Chassis No. 1A.2395 and subs. 1K.1001 and subs.
C.4579/1	**Plate,** under Bracket		1	
C.2707	**Grommet,** in Plate		1	
C.4578	**Gasket,** under Plate		1	
UFS.319/4H	**Screw,** Set, securing Dipper Switch to Bracket		2	
UFS.119/8R	**Screw,** Set, securing Bracket and Plate to Body		2	
C.723/A	**Washer,** Shakeproof, on Setscrews		4	
C.5648	**Plug** (Rubber) sealing redundant Hole on L.H. Side of Body		1	
C.22043	**Bracket,** mounting Dipper Switch		1	L.H. DRIVE CARS ONLY Fitted to Chassis No. 1A.20198 and subs. 1K.30001 and subs.
UFS.319/4H	**Screw,** Set, securing Dipper Switch to Bracket		2	
UFS.119/4R	**Screw,** Set, securing Bracket to Body		2	
C.723/A	**Washer,** Shakeproof, on Setscrews		4	
C.4580	**Blanking Plate,** sealing redundant Hole on R.H. Side of Body		1	
C.4578	**Gasket,** under Blanking Plate		1	
UFS.119/4R	**Screw,** Set, securing Blanking Plate to Body		2	
UFN.119/L	**Nut,** on Setscrews		2	
C.723/A	**Washer,** Shakeproof, under Nuts		2	
C.14929	**SIDE LAMP, COMPLETE (52474/B-L.490)**	54–16	2	
3455	**Bulb Holder** (553780)		2	
989	**Bulb,** 6 watt (989)		2	
8064	**Terminal Sleeve** (555910)		2	
C.4549/2	**Screw,** Set, securing Side Lamps		2	
C.4691	**INDICATOR LENS FOR SIDE LAMPS (573449/A)**	54–17	2	
C.4098/2	**Spring Clip,** securing Indicator Lens (571888)	54–18	2	
BD.15307	**L.H. STOP/TAIL/FLASHER LAMP (RED/AMBER) (53726/E-L.627)**		1	NOT FOR U.S.A.
8080	**Lens,** Flasher (54570227)	54–19	1	
8081	**Lens,** Stop/Tail (54570229)	54–20	1	
8286	**Shield,** dividing Flasher Lens from Stop/Tail Lens (54570231)	54–21	1	
8287	**Plate,** mounting Flasher and Stop/Tail Lens (54570296)	54–22	1	
8174	**Chrome Clip,** between Flasher and Stop/Tail Lens (54570230)	54–23	1	
8082	**Screw,** Lens fixing (54110161)	54–24	1	

Part No.	Description	Plate No.	No. per Unit	Remarks
	L.H. STOP/TAIL/FLASHER LAMP (RED/AMBER) (53726/E-L.627) *(contd.)*			
8083	**Gasket,** Lens Seating (54570233)	54–25	1	
6538	**Interior Bulb Holder,** Flasher (573832)		1	
C.9126	**Bulb,** Flasher (382)	54–26	1	
5311	**Interior Bulb Holder,** Stop/Tail (573828)		1	
C.9125	**Bulb,** Stop/Tail (380)	54–27	1	
8084	**Reflex Reflector,** complete (574914)	54–28	1	
8085	**Plinth** (54570291)	54–29	1	
9721	**Gasket,** Lamp seating (54572361)	54–30	1	
BD.15308	**R.H. STOP/TAIL/FLASHER LAMP (RED/AMBER) (53727/E-L.627)**		1	NOT FOR U.S.A.
8087	**Plinth** (54570292)		1	
	All other items as for Lamp BD.15307			
BD.15307/1	**L.H. STOP/TAIL/FLASHER LAMP (RED/RED) (53757/E-L.627)**		1	U.S.A. ONLY
8089	**Lens,** Flasher (54570228)		1	
	For other items see BD.15307			
BD.15308/1	**R.H. STOP/TAIL/FLASHER LAMP (RED/RED) (53758/E-L.627)**		1	U.S.A. ONLY
8089	**Lens,** Flasher (54570228)		1	
8087	**Plinth** (54570292)		1	
	For other items see BD.15307			
UFN.125/L	**Nut,** securing Stop/Tail Lamps		2	
C.16062	**SWITCH OPERATING STOP LIGHTS (31802/D-HL.2)**	54–31	1	
C.15430	**FRONT FLASHER LAMP (52479/A-L.563)**		2	NOT FOR HOLLAND, U.S.A., OR SWITZERLAND
5322	**Rim,** complete with Lens (575289)	54–32	2	
5218	**Screw,** fixing Rim (113511)		2	
6524	**Washer,** retaining Screw (161385)		2	
5324	**Lens** (Amber) (575292)	54–33	2	
5285	**Wire,** retaining Lens (522869)	54–34	10	
5323	**Gasket,** seating Lens (575290)	54–35	2	
8261	**Seating Plate,** complete with Bulb Holder (575287)	54–36	2	
8064	**Sleeve,** Terminal (555910)		2	
C.9126	**Bulb,** 21 watt (382)		2	
8059	**Body** (Rubber) (54571149)	54–37	2	
C.15431	**FRONT FLASHER LAMP (52480/A-L.563)**		2	FOR U.S.A. ONLY
7667	**Rim,** complete with Lens (575298)		2	
7253	**Lens** (Clear) (575285)		2	
	All other details as for Flasher Lamp C.15430			
C.16340	**SIDE/FLASHER LAMP (52498/A-L.563)**		2	FOR HOLLAND AND SWITZERLAND ONLY
7667	**Rim,** complete with Lens (575298)		2	
5218	**Screw,** fixing Rim (113511)		2	
6524	**Washer,** retaining Screw (161385)		2	
7253	**Lens** (Clear) (575285)		2	
5285	**Wire,** retaining Lens (522869)		10	
5323	**Gasket,** seating Lens (575290)		2	
8262	**Seating Plate,** complete with Bulb holder (54571527)		2	

Part No.	Description	Plate No.	No. per Unit	Remarks
	SIDE/FLASHER LAMP (52498/A-L.563) *(contd.)*			
5311	**Interior Bulb Holder** (573828)		2	
8064	**Sleeve**, Terminal (555910)		2	
C.9125	**Bulb**, 6/24 watt (380)		2	
8059	**Body** (Rubber) (54571149)		2	
UCS.313/5H	**Screw**, Set, securing Front Flasher Lamps		6	Superseded by DAZ.406/8C and BD.8633/9
UCN.113/L	**Nut**, on Setscrews		6	
C.721	**Washer**, Shakeproof, under Nuts		6	
DAZ.406/8C	**Screw**, Self-Tapping, securing Front Flasher Lamps		6	Supersedes UCS.313/5H, UCN.113/L, and C.721
BD.8633/9	**Spire Nut**, receiving Self-Tapping Screws		6	
C.16729	**FLASHER UNIT (35020/A-FL.5)**	54–38	1	
UFN.125/L	**Nut**, securing Flasher Unit		1	
C.724	**Washer**, Shakeproof, under Nut		1	
C.26344	**CONTROL PANEL ASSEMBLY FOR HAZARD WARNING DEVICE (30270)**		1	FITTED AS STANDARD EQUIPMENT ON CARS EXPORTED TO U.S.A. at Chassis No. 1A.20472 and subs. 1K.30001 and subs. ALSO SUPPLIED, TO SPECIAL ORDER, FOR OTHER CARS IRRESPECTIVE OF COUNTRY
11471	**Flasher Unit** (54362147)		1	
11215	**Switch** (35646-45.SA)		1	
11216	**Warning Lamp Assembly** (38726-WL.13)		1	
11217	**Bulb Holder** (863511)		1	
987	**Bulb** (987)		1	
DAC.410/8C	**Screw**, Self-Tapping, securing Control Panel to Scuttle Casing Bracket		2	
C.723/A	**Washer**, Shakeproof, on Self-Tapping Screws		2	
BD.8633/6	**Spire Nut**, receiving Self-Tapping Screws		2	
C.20630	**NUMBER PLATE ILLUMINATION AND REVERSE LAMP (54145/B-L.720)**		1	NOT REQUIRED FOR CARS EXPORTED TO FRANCE
10008	**Housing** (Chromium Plated) (54574347)	54–39	1	
6529	**Screw**, for Bulb Holder Bracket (126474)		1	
6531	**Screw**, for Bulb Holder fixing (126544)		2	
6530	**Washer**, for Screws (185038)		3	
6538	**Interior**, Reverse Bulb Holder (573832)		1	
3455	**Interior**, Number Plate Bulb Holder (553780)		3	
3489	**Gasket**, Bulb Shield (573521)	54–40	1	
5310	**Shroud Assembly**, for Bulbs (573527)	54–41	1	
C.12194	**Terminal Sleeve** (188818)		5	
9719	**Lens** (54574394)	54–42	1	
3488	**Gasket**, Lens seating (573519)	54–43	1	
C.9126	**Bulb**, for Reverse Light, 21 watt (382)		1	
989	**Bulb**, illuminating Number Plate and Luggage Compartment, 6 watt (989)		3	
5309	**Bulb Holder** (573522)	54–44	1	
C.20631	**NUMBER PLATE ILLUMINATION AND REVERSE LAMP (54146/B-L.720)**		1	FOR CARS EXPORTED TO FRANCE
10007	**Lens** (54574402)		1	
	All other items as for Number Plate Lamp C.20630			
UCS.125/4R	**Screw**, Set, securing Number Plate Lamp		2	
UCS.125/6R	**Screw**, Set, securing Number Plate Lamp		2	
FW.104/T	**Washer**, Plain, on Setscrews		4	
C.724	**Washer**, Shakeproof, on Setscrews		4	
C.21008	**SWITCH, OPERATING REVERSE LAMP AND STARTER CUT-OUT (35EA-A.286B)**	54–45	1	

Part No.	Description	Plate No.	No. per Unit	Remarks
C.18117	**CUBBY BOX LAMP (56078E)**	54–46	1	
7880	**Bulb** (12v, 6w) Festoon		1	
8851	**Lens only** (54572189)		1	
C.18118	**Switch,** operating Cubby Box Lamp (34379/B)	54–47	1	
C.19008	**Earthing Strip,** for Switch		1	
C.19009	**Sleeve,** on Earthing Strip		1	
BD.1501/1	**Woodscrew,** securing Earthing Strip		1	
C.15288	**Lucar Blade,** for Cubby Box Lamp Connector (54190107)		1	
C.18120	**Insulator,** for Cubby Box Lamp		1	
C.19080	**L.H. HORN (LOW NOTE) (69127/A-WT.618)**	54–48	1	Fitted from Chassis No. 1A.1001 to 1A.9814 1A.20001 to 1A.20466
8483	**Bracket** (707095)		1	
8484	**Contact Set** (702687)		1	
8485	**Diaphragm** (702836)		1	
C.19081	**R.H. HORN (HIGH NOTE) (69128/A-WT.618)**	54–49	1	
8486	**Diaphragm** (702851)		1	
	All other items for R.H. Horn as for C.19080			
C.22255	**L.H. HORN (LOW NOTE) (54068078-9H)**		1	Fitted to Chassis No. 1A.9815 and subs. 1A.20467 and subs. 1K.1001 and subs. 1K.30001 and subs.
10209	**Bracket,** on Horn (54680686)		1	
C.22254	**R.H. HORN (HIGH NOTE) (54068079-9H)**		1	
10209	**Bracket,** on Horn (54680686)		1	
C.17489	**Mounting Bracket,** for L.H. Horn	54–50	1	
C.17490	**Mounting Bracket,** for R.H. Horn	54–51	1	
UFS.137/7R	**Screw,** Set, securing Brackets to Body		2	
UFN.137/L	**Nut,** on Setscrews		2	
C.726	**Washer,** Shakeproof, under Nuts		2	
FW.106/T	**Washer,** Plain, under Shakeproof Washers		2	
UFS.125/6R	**Screw,** Set, securing Horns		6	
UFN.125/L	**Nut,** on Setscrews		6	
C.724	**Washer,** Shakeproof, under Nuts		6	
C.26654	**HORN MUTE (RUBBER) IN HORN TRUMPET (54682523)**		2	**FITTED ONLY TO CARS EXPORTED TO HOLLAND**
C.17338	**HORN RELAY (33209/B-6RA)**	54–52	1	
C.17511	**Distance Piece,** at Relay Mounting		2	
UFS.419/9H	**Screw,** Set, securing Horn Relay		2	
C.20776	**STEERING COLUMN HORN RING ASSEMBLY (33612/A)**		1	Fitted from Chassis No. 1A.1001 to 1A.2529 1A.20001 to 1A.20219
5916	**Nut,** fixing Plate Stud (169290)		3	
9021	**Spring,** Horn Ring (54300943)		3	
C.18177	**Ring,** with Upper Contact (54300725)		1	
10062	**Clip,** in Horn Ring, fixing Cover (54150219)		2	
C.21273	**COVER (COMPLETE WITH MOTIF) OVER HORN CONTACTS (54301195)**		1	

ELECTRICAL EQUIPMENT

Part No.	Description	Plate No.	No. per Unit	Remarks
C.18979	**STEERING COLUMN HORN RING ASSEMBLY (33585/B-CC.2)**	54–53	1	Fitted to Chassis No. 1A.2530 and subs. 1A.20220 and subs. 1K.1001 and subs. 1K.30001 and subs.
9021	**Spring,** for Horn Ring (54300725)		3	
C.18177	**Ring,** with Upper Contact		1	
C.20530	**Screw,** fixing Horn Ring		3	
C.20531	**Washer,** on Screws		3	
C.23357	**HORN RING COVER**		1	Fitted from Chassis No. 1A.2530 to 1A.3010 1A.20220 to 1A.20253
C.20777	**HORN RING COVER ASSEMBLY**		1	Fitted to Chassis No. 1A.3011 and subs. 1A.20254 and subs. 1K.1001 and subs. 1K.30001 and subs.
C.22271	**Cover only** (54300845/A)	54–54	1	
C.22389	**Horn Push** (54301148)	54–55	1	
C.22273	**Return Spring,** for Horn Push (54301259)	54–56	1	
C.22274	**Contact Plate,** in Horn Push (54301256)	54–57	1	
UCS.416/9H	**Screw,** Set, securing Horn Ring Cover to Steering Wheel		4	Fitted to Chassis No. 1A.2530 and subs. 1A.20220 and subs.
C.722	**Washer,** Shakeproof, on Setscrews		4	
C.16176	**CARRIER ASSEMBLY FOR CABLE TO STEERING COLUMN HORN RING**		1	
MJ.206/3R	**Woodscrew,** securing Carrier to Back of Facia Board		2	
C.25255	**DIRECTION INDICATOR/HEADLAMP FLASHER SWITCH (34563/H-85.SA)**	54–58	1	Fitted from Chassis No. 1A.1001 to 1A.10326 1A.20001 to 1A.20479
10484	**Nylon Spring,** operating Self-Cancelling Device (54320721)		1	
10485	**Pivot Pin,** for Operating Arm (54336527)		1	
C.25555	**DIRECTION INDICATOR/HEADLAMP FLASHER SWITCH (35532)**		1	Fitted to Chassis No. 1A.10327 and subs. 1A.20480 and subs. 1K.1001 and subs. 1K.30001 and subs.
11316	**Striker Ring** (Nylon) (54322277)		1	
C.25668	**INSULATING STRIP, PROTECTING SWITCH LEADS**		1	Fitted to Chassis No. 1A.7815 and subs. 1A.20377 and subs. 1K.1001 and subs. 1K.30001 and subs.
BD.22686	**Stud,** securing Insulating Strip to Outer Steering Column		2	
C.27315	**UPPER SWITCH COVER ASSEMBLY AT CENTRE OF STEERING WHEEL (54361227)**	54–59	1	Fitted from Chassis No. 1A.1001 to 1A.3252 1A.20001 to 1A.20256 Except 1A.3096 to 1A.3098 1A.3232, 1A.3247 to 1A.3249 **CARS FITTED WITH AUTOMATIC TRANSMISSION**
C.15792	**Indicator Light Unit** (54360413)		1	
C.27167	**'D' Motif Plate**		1	
C.15787	**Bulb** (2 watt. Lilliput) (281)		3	
8411	**Bulb Holder,** complete with Cables (54945043)		1	
8092	**Rubber Pad** (54360531)		1	
C.24287	**SHIELD, PROTECTING INDICATOR LIGHTING CABLES**		1	**R.H. DRIVE CARS ONLY** Fitted to Chassis No. 1A.6662 and subs.
DZZ.408/6C	**Screw,** Self-Tapping, securing Shield to Dash		2	
AW.102/T	**Washer,** Plain, on Self-Tapping Screws		2	
C.16575	**LOWER SWITCH COVER AT CENTRE OF STEERING WHEEL (54360624)**	54–60	1	

Part No.	Description	Plate No.	No. per Unit	Remarks
C.20735	**FOG LAMP (55262/A-5WFT)**		2	NOT REQUIRED FOR CARS EXPORTED TO ALGERIA, FRANCE, MOROCCO, GERMANY, SWITZERLAND, OR U.S.A.
8060	**Screw,** fixing Fog Lamp Rim (127511)	54–61	2	
8062	**Light Unit,** "Fogranger" (Clear Lens) (54520142)	54–62	2	
9012	**Rim** (54521012)	54–63	2	
8061	**Screw,** fixing Rim to Light Unit (198903)	54–64	6	
323	**Bulb** (323) (12 v. 48 w.)		2	
8064	**Sleeve Terminal** (555910)		4	
8065	**Grommet,** for Cable Entry (572363)	54–65	2	
10216	**Nut,** fixing Lamp (181261)	54–66	2	
8067	**Washer,** Shakeproof, under Nuts (188350)	54–67	2	
8068	**Plate,** fixing Lamp (534732)	54–68	2	
8069	**Gasket,** seating Lamp (534722)	54–69	2	
C.20144	**FOG LAMP (55282/A)**		2	REQUIRED ONLY FOR CARS EXPORTED TO GERMANY
9011	**Light Unit** (54521826)		2	
9015	**Bulb Holder** (54521829)		2	
9016	**Bulb** (443) (12 v. 35 w.)		2	
	All other items as for Fog Lamp C.20735			
C.11904	**FOG LAMP (55174/E.SFT.576)**		2	REQUIRED ONLY FOR CARS EXPORTED TO SWITZERLAND
5296	**Light Unit** (554813)		2	
5295	**Adaptor,** for Bulb (554667)		2	
8064	**Terminal Sleeve** (555910)		4	
5288	**Clamp,** for Light Unit (535009)		2	
6659	**Cushion,** for Clamp (534383)		2	
6658	**Screw,** for Clamp (144921)		2	
5287	**Base** (534386)		2	
5220	**Fixing Nut** (180377)		2	
5223	**Lockwasher,** under Nut (188349)		2	
323	**Bulb** (12 v. 48 w.) (323)		2	
C.20736	**FOG LAMP (55263/A.5WFT)**		2	REQUIRED FOR CARS EXPORTED TO ALGERIA, FRANCE OR MOROCCO
8074	**Light Unit** (Yellow Lens) (54520625)		2	
	For other items see C.20735			
C.17439	**Plain Washer,** at Foglamp Mounting		2	NOT REQUIRED FOR CARS EXPORTED TO U.S.A. OR SWITZERLAND
8401	**LIGHT UNIT "RANGER" FOR FOG LAMP (54520077)**		2	REQUIRED TO CONVERT FOG LAMPS TO LONG-RANGE SPOT LAMPS
C.15892	**EMBELLISHER FOR FOG LAMPS (54520672)**	54–70	2	NOT REQUIRED FOR CARS EXPORTED TO U.S.A. OR SWITZERLAND
C.15486	**MAP LIGHT (52477/A)**	54–71	1	
989	**Bulb** (12v. 6w.) (989)		1	
BD.1501/1	**Woodscrew,** securing Map Light		2	
C.13503	**WINDSCREEN WIPER MOTOR ASSEMBLY (073144-DR.3)**		1	R.H. DRIVE CARS ONLY
7185	**Motor only** (75310/D/F)	54–72	1	
7187	**Brush Gear** (743171)		2	
3509	**Brushes** (729367)		1 set	
6551	**Spring,** for Brushes (735631)		2	
7188	**Cover,** End Terminals (743924)		1	
7189	**Armature** (743929)		1	
7190	**Coil** (744907)		1	
7191	**Shaft and Gear** (744913)		1	
6555	**Connecting Rod** (740563)		1	
7192	**Limit Switch** (743846)		1	
7193	**Rack and Cross Head** (744257)	54–73	1	

ELECTRICAL EQUIPMENT

Part No.	Description	Plate No.	No. per Unit	Remarks
C.13504	**WINDSCREEN WIPER MOTOR ASSEMBLY (073145-DR.3)**		1	**L.H. DRIVE CARS ONLY**
7186	**Motor only** (75311/D/F)		1	
	All other items as for Windscreen Wiper Motor C.13503			
C.14405	**WHEEL BOX ASSEMBLY FOR WINDSCREEN WIPER (72714/D)**	54–74	2	
6557	**Spindle and Gear** (740995)		2	
7767	**Front Bush** (Chromium Plated) (743889)		2	
7769	**Washer,** (Rubber) under Front Bush (740925)		2	
7768	**Rear Bush** (743890)		2	
C.5522/2	**Rubber Tube,** for Spindles (740711)		2	
5219	**Nut,** fixing Wheel Box (153511)		2	
C.8799	**OUTER CASING, FROM SCREEN WIPER MOTOR TO WHEEL BOX (743818)**	54–75	1	
C.8800	**OUTER CASING, FROM WHEEL BOX TO WHEEL BOX (743819)**	54–76	1	
C.5450	**OUTER CASING FOR WHEEL BOX EXTENSION (740746)**	54–77	1	
8697	**ARM FOR WINDSCREEN WIPER (54711436)**	54–78	2	**R.H. DRIVE CARS ONLY**
8698	**ARM FOR WINDSCREEN WIPER (54711437)**		2	**L.H. DRIVE CARS ONLY**
8183	**BLADE FOR WINDSCREEN WIPER ARMS (54711282)**	54–79	2	
BD.21942	**ELECTRICALLY HEATED BACKLIGHT (CLEAR GLASS)**		1	Fitted as Standard Equipment at Chassis No. 1K.1001 and subs. 1K.30001 and subs.
BD.22742	**ELECTRICALLY HEATED BACKLIGHT ("SUNDYM" GLASS)**		1	**SUPPLIED TO SPECIAL ORDER ONLY**
9653	**Fuse** (15 amp) (188220)		1	
C.19724	**Backlight Fuse Connector** (54930724)		1	
C.27878	**Grommet** (340168)		2	
BD.22655	**Clip,** for Fuse Holder		1	
DAZ.404/4C	**Screw,** Self-Tapping, securing Clip		1	
C.5204	**Earth Terminal**		1	
DZZ.408/6C	**Screw,** securing Earth Terminal		1	
C.2744	**SWITCH CONTROLLING ELECTRICALLY HEATED BACKLIGHT (031216)**		1	Fitted as Standard Equipment at Chassis No. 1K.1001 and subs. 1K.30001 and subs.
2836	**Knob,** for Switch (314437)		1	

Part No.	Description	Plate No.	No. per Unit	Remarks
C.26467	WARNING LAMP (AMBER) FOR HEATED BACKLIGHT CIRCUIT (38728/A-WL-13)		1	Fitted as Standard Equipment at Chassis No. 1K.1001 and subs. 1K.30001 and subs.
11333	**Bezel and Window Assembly** (54361476)		1	
987	**Bulb** (987)		1	
C.25883	**Escutcheon** (Marked "Rear Window") for Switch and Warning Lamp		1	
C.26651	RELAY, DIMMING WARNING LAMP (33285/A-6.R.A.)		1	Fitted as Standard Equipment at Chassis No. 1K.1001 and subs. 1K.30001 and subs.
DAZ.408/8C	**Screw**, Self-Tapping, securing Relay		2	
C.26473	**Connector Cable**, for Backlight Switch		1	R.H. DRIVE CARS ONLY
C.26474	**Connector Cable**, for Backlight Switch		1	L.H. DRIVE CARS ONLY
C.8792	BATTERY (BT.9A)	54–80	1	REQUIRED FOR CARS FITTED WITH A DYNAMO (D.C. GENERATOR)
C.27003	BATTERY (WITH REVERSED TERMINALS) (54028461.59/8)		1	REQUIRED FOR CARS FITTED WITH AN ALTERNATOR
C.10192	**Felt**, 5" x 11", under Battery		1	
C.10077	BATTERY TRAY	54–81	1	
C.14346	COVER ASSEMBLY OVER TOP OF BATTERY	54–82	1	
7760	REPLACEMENT KIT FOR BATTERY COVER FIXINGS (Comprising 2 each Clips, Pivot Pins and Circlips)		1	
C.10640/1	DRAIN TUBE FOR BATTERY TRAY (27" LONG)	54–83	1	
C.1040/17	**Clip**, securing Drain Tube		2	
DZZ.408/6C	**Screw**, Self-Tapping, securing Clip		2	
C.10186	STRAP ACROSS FRONT OF BATTERY	54–84	1	
C.10188/1	**Rubber Pad**, between Strap and Battery (21" long)	54–85	1	
C.10187	ROD, THROUGH STRAP, SECURING BATTERY	54–86	2	
UFN.125/K	**Nut**, on Fixing Rods		2	
FG.204/X	**Washer**, Spring, under Nuts		2	
C.2778	BATTERY CABLE (POSITIVE TO EARTH) (810709)		1	
C.19621	BATTERY CABLE (NEGATIVE) (792914)		1	Fitted to Chassis No. 1A.1001 and subs. 1A.20001 and subs.

ELECTRICAL EQUIPMENT

Part No.	Description	Plate No.	No. per Unit	Remarks
C.23688	**BATTERY CABLE (POSITIVE) (54936179)**		1	REQUIRED FOR CARS FITTED WITH AN ALTERNATOR AT CHASSIS No. 1K.1001 and subs. 1K.30001 and subs.
C.23685	**BATTERY CABLE (NEGATIVE) (54936004)**		1	
C.2609	**Screw,** securing Cables to Battery Posts (186111)		2	
6826	**Screw** (Oversize) securing Cables to Battery Posts (186114)		2	REQUIRED ONLY WHEN HOLE IN BATTERY POST IS ENLARGED
C.28363	**FORWARD HARNESS (54953548)**		1	
C.11099	**Grommet Plate,** for forward Harness		1	
C.2796	**Bracket,** holding Cable connectors to R.H. Front Wing		1	
C.2797	**Bracket,** holding Cable connectors to L.H. Front Wing		1	
UCS.313/4H	**Screw,** Set, securing Brackets		4	
UCN.113/L	**Nut,** on Setscrews		4	
C.721	**Washer,** Shakeproof, under Nuts		4	
C.20782	**R.H. BODY HARNESS (54933211)**		1	
C.26346	**ADAPTOR HARNESS FOR HAZARD WARNING DEVICE**		1	FITTED TO CARS EXPORTED TO U.S.A. at Chassis No. 1A.20472 and subs. 1K.30001 and subs. ALSO REQUIRED WHEN HAZARD WARNING DEVICE IS FITTED TO ANY OTHER CAR
C.2706	**Grommet,** protecting Harness at Dash Panel (859059)		1	
3571	**Snap connector** (850641)		1	
C.20783	**L.H. BODY HARNESS (54933212)**		1	
C.15668	**NUMBER PLATE HARNESS (54944614)**		1	
C.27599	**COVER, AROUND HAZARD WARNING CONTROL PANEL**		1	Fitted to Chassis No. 1A.20552 and subs. 1K.30001 and subs. **May be fitted to earlier cars equipped with a Hazard Warning Device**
	ELECTRICAL CABLES			
C.15790	**Cable,** Panel Connector (54944612)		1	Fitted from Chassis No. 1A.1001 to 1A.8060 20001 to 1A.20393
C.25609	**Cable,** Panel Connector (54938097)		1	Fitted to Chassis No. 1A.8061 and subs. 1A.20394 and subs.
C.27312	**Cable,** Panel Connector		1	Fitted to Chassis No. 1K.1001 and subs. 1K.30001 and subs.
C.15462	**Cable,** Petrol Tank Element Unit Connector (54944615)		1	
C.16004	**Cable,** Front Door Switch Connector (54944716)		1	
C.26097	**Cable,** Headlamp Flasher Cable		1	REQUIRED FOR CARS EXPORTED TO ITALY ONLY
C.15463	**Cable,** Fog Lamp Connector (54944611)		2	NOT REQUIRED FOR CARS EXPORTED TO U.S.A.
C.10092	**Cable,** Reverse Light Switch connector (828209)		1	NOT REQUIRED WHEN OVERDRIVE UNIT OR AUTOMATIC TRANSMISSION IS FITTED
C.16197	**Cable,** Choke Warning Light Switch (54945232)		1	
C.15465	**Cable,** Instrument Light Connector (54944613)		1	
C.21185/04	**Cable,** Coil to Distributor (54949245)		1	
C.21235	**Cable,** Solenoid to Starter (54933213)		1	
C.19621	**Cable,** Battery Negative (792914)		1	
C.16115	**Cable,** Brake Fluid Level Connector (54945032)		1	
C.21274	**Cable,** Brake W/L Connector (54933225)		1	
C.17770	**Cable,** Cubby Box Light Connector (54948226)		1	

Part No.	Description	Plate No.	No. per Unit	Remarks
	ELECTRICAL CABLES *(contd.)*			
C.15652	**Cable,** Pillar Interior Light Connector (54944609)		2	
C.16241	**Cable,** Brake Fluid Level W/L Extension		1	
C.21959	**Cable,** Starter Cut-Out Switch Connector		1	
C.15464	**Cable,** Horn Switch Connector (54944610)		1	
C.20240	**Cable,** Steering Column Lock Connector		1	REQUIRED ONLY WHEN STEERING COLUMN LOCK IS FITTED
	EARTH CABLES			
C.21334	**Earth Cable,** from Engine to Body (999709)		1	
C.2778	**Earth Cable,** Battery Positive (810709)		1	
C.16179	**Earth Cable,** for Handbrake Warning Light Switch (54945031)		1	
C.2895	**Earth Cable,** for Petrol Pump (859449)		1	
C.18699	**Earth Cable,** for Cigar Lighter (54949514)		1	
C.16179	**Earth Cable,** for Choke W/L Switch (54945031)		1	
	CABLE SLEEVES			
C.2783	**Sleeve,** for Headlamp Cables (859300)		2	
C.15555	**Sleeve,** for Flasher Lamp Cables (54944385)		2	
C.21516	**Sleeve,** for Forward Harness		1	
C.2607	**Sleeve,** for Leads to Handbrake W/L Switch		1	
	RUBBER GROMMETS			
C.2706	**Grommet,** for Screen Washer Tubes, Speedometer Cable, Bonnet Lock Control, Screen Wiper Cable, Dipper Switch Cable, Petrol Tank Cable, Boot Lid Cable, Door Switches, Brake Pipes, Steering Column Lock Cable and Side Lamps		See Remarks	13 required for R.H. Drive Cars 14 required for L.H. Drive Cars
C.17432	**Grommet,** for Screen Wiper Harness		1	
C.17432	**Grommet,** for Speedometer Cable		1	For L.H. Drive Cars
C.2707	**Grommet,** for Body Harness and Dip Switch on R.H.D. cars		See Remarks	3 required for L.H. Drive Cars 4 required for R.H. Drive Cars
C.12425	**Grommet,** for Gearbox Tunnel		1	
C.2708	**Grommet,** for Lamp Cables		2	
C.5574	**Grommet,** for Screen Wiper Motor		1	
C.977	**Grommet,** at rear of Body, for Body Harness		2	
C.974	**Grommet,** in Centre Pillars		2	
C.975	**Grommet,** in Centre Pillars		2	
C.1426	**Grommet,** for Forward Harness		1	
C.2850	**Grommet,** for Heater Controls, through Scuttle		2	
C.15749	**Grommet,** for Body Harness through Sills		4	
C.2708	**Grommet,** for Petrol Pipe through Boot Floor		1	
	RUBBER PLUGS AND BOOTS			
C.4650	**Rubber Boot,** on Starter Motor and Solenoid Terminals		3	
C.5648	**Rubber Plug,** sealing hole for L.H. Dipper Switch Cable		1	R.H. DRIVE CARS ONLY
C.5648	**Rubber Plug,** sealing redundant Speedo Cable Hole in Bulkhead		1	
C.5648	**Rubber Plug,** sealing hole for Screen Washer Tubing		1	
BD.14621	**Rubber Plug,** sealing Choke Control Cable Hole in Bulkhead		1	
C.5648	**Rubber Plug,** sealing redundant Cable Entry Hole in Bulkhead		1	
BD.14771	**Rubber Plug,** in Door Sills		2	
	PLASTIC PLUGS			
BD.15588/4	**Plastic Plug,** sealing Cable Entry Hole in Side Lamps		2	REQUIRED ONLY IN CARS EXPORTED TO SWITZERLAND

ELECTRICAL EQUIPMENT

Part No.	Description	Plate No.	No. per Unit	Remarks
	CABLE CLIPS			
C.1040/2	**Clip,** securing Radiator Overflow Pipe (Lower R.H. Fixing) and Starter Solenoid Cable		2	
C.1040/9	**Clip,** securing Heater Cable, Dip Switch Cable, Speedo Cable, Petrol Pipe and Brake Pipe		7	
C.1040/8	**Clip,** securing Dynamo Cable, Forward Harness and Speedo Cable		3	
C.1040/12	**Clip,** securing Handbrake Cable and Petrol Pipe		2	
C.1040/21	**Clip,** securing Hose to Engine Bracket		1	POWER ASSISTED STEERING ONLY
C.1040/25	**Clip,** securing Hose to Engine Bracket		1	POWER ASSISTED STEERING ONLY
C.1071/3	**Clip,** securing Rear Boot Harness and Front Lamp Cables		6	
C.5204	**Earth Terminal**		4	
C.1040/17	**Clip,** securing Forward Wing Harness to Dash		1	
C.1040/5	**Clip,** securing Forward Wing Harness to Dash		3	
C.1040/4	**Clip,** securing Chassis Harness and Washer Harness		3	
C.1040/16	**Clip,** securing Forward Wiring Harness to Lower Dash		2	
C.1040/25	**Clip,** securing Forward Wiring Harness to Upper Dash		1	
C.1040/31	**Clip,** securing Forward Wiring Harness to Upper Dash		1	
C.1040/3	**Clip,** securing Panel Harness		2	
C.1543/7	**Clip,** securing Speedo Cable		2	
C.1543/6	**Clip,** securing Boot Lid Harness		1	
C.16394	**Clip,** securing Interior Light Cables to Centre Pillars		8	
C.15970	**Clip,** securing Panel Harness		1	
C.15971	**Clip,** securing Panel Harness		1	
C.1543/2	**Clip,** securing Panel Harness		3	
C.1543/4	**Clip,** securing Panel Harness		1	
C.1040/1	**Clip,** securing Wiper Cable, Starter Solenoid Cable and Petrol Pipe		5	
	INSULATORS AND SLEEVES			
C.4877/2	**Insulator** (Rubber) (8" long)		1	R.H. DRIVE CARS ONLY
C.4877/4	**Insulator** (Rubber) (6" long)		2	
C.16212	**Sleeve** (P.V.C.) for Wiring Tags (1¾" long)		23	
C.16864	**Sleeve** (P.V.C.) for Wiring Tags (7/8" long)		12	
C.17077	**Sleeve** (P.V.C.) for Wiring Tags (1" long)		10	

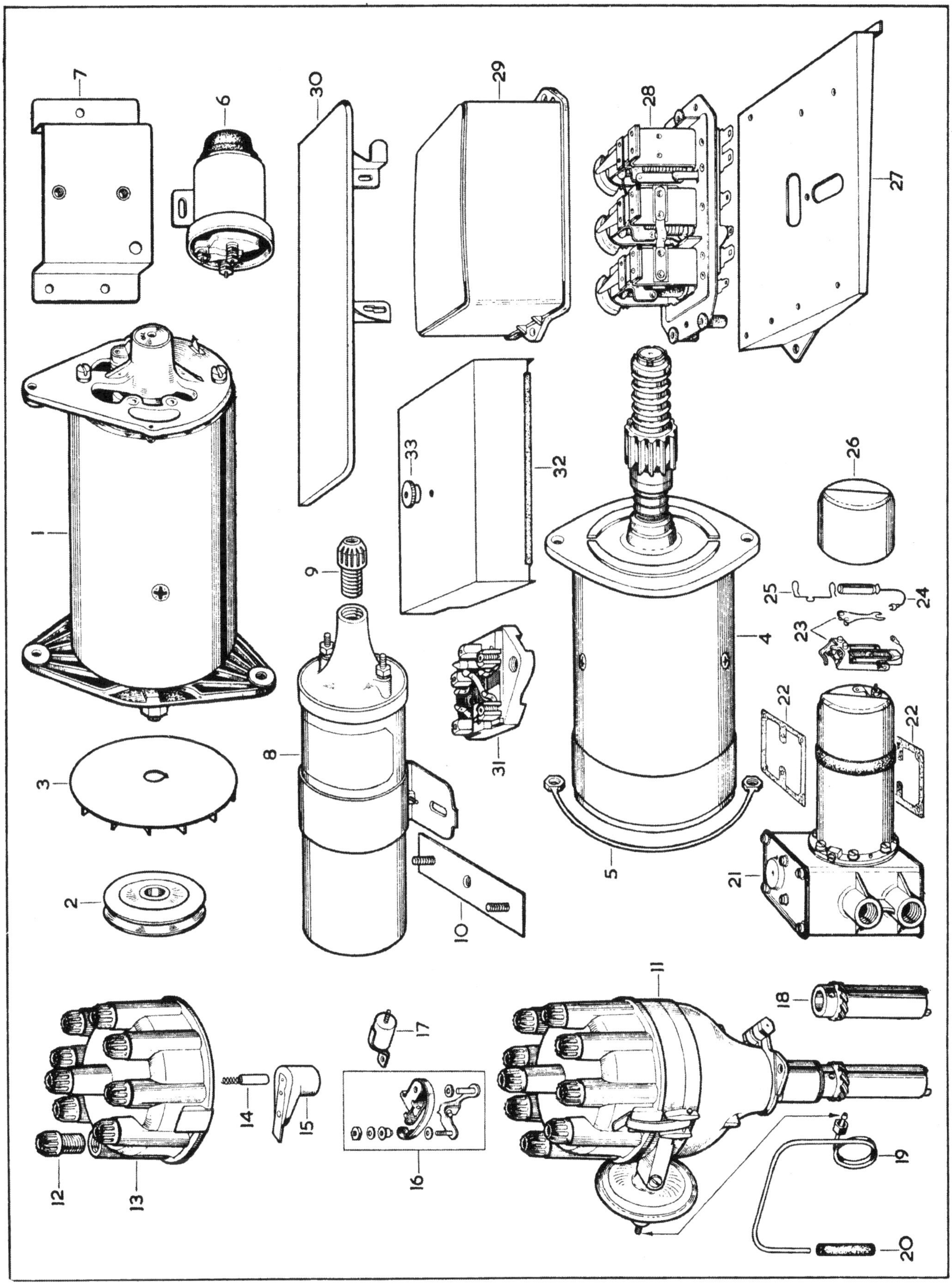
1
2
3
4
5
6
7
8
9
10
11
12
13
14
15
16
17
18
19
20
21
22
23
24
25
26
27
28
29
30
31
32
33

PLATE 52

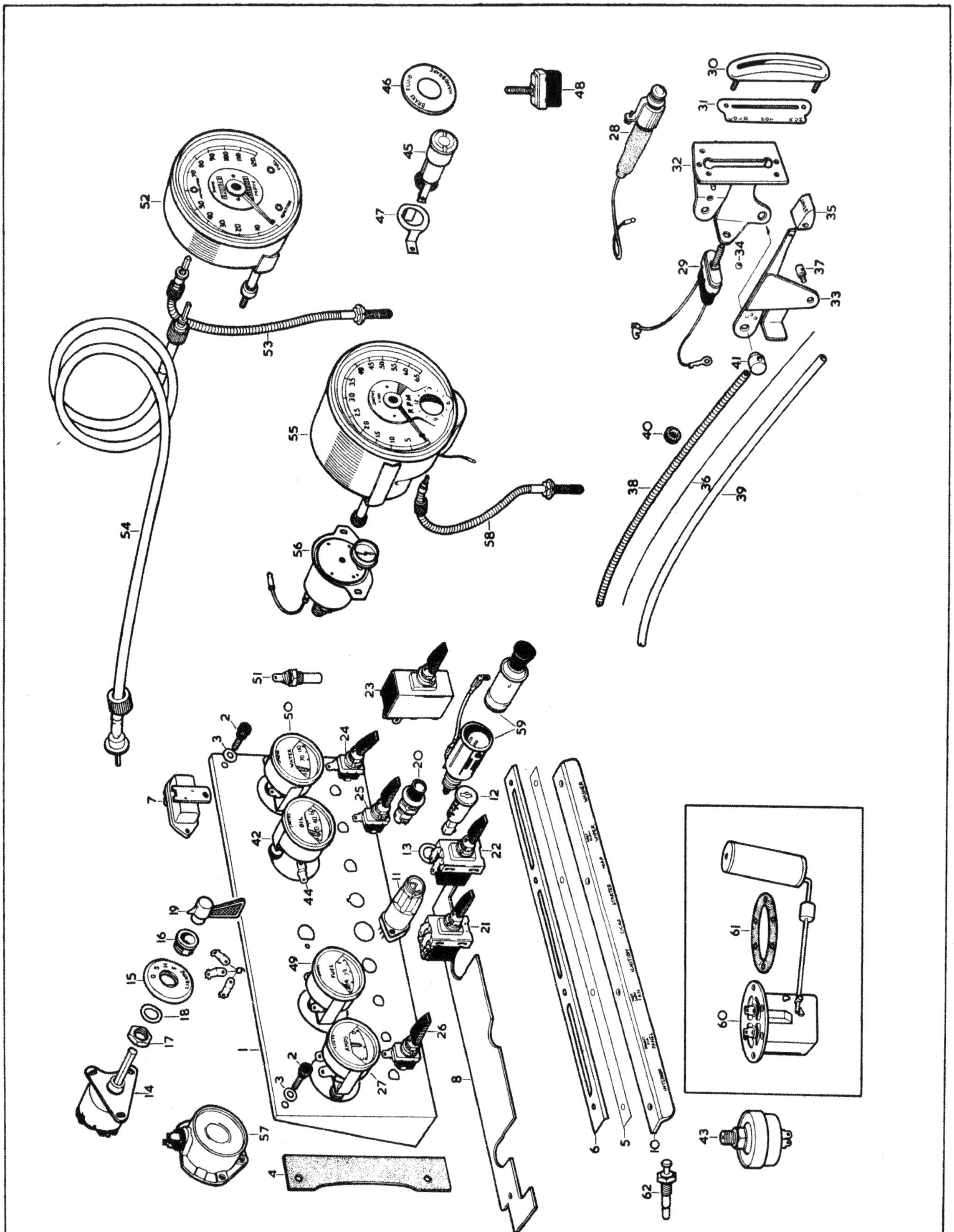

PLATE 53

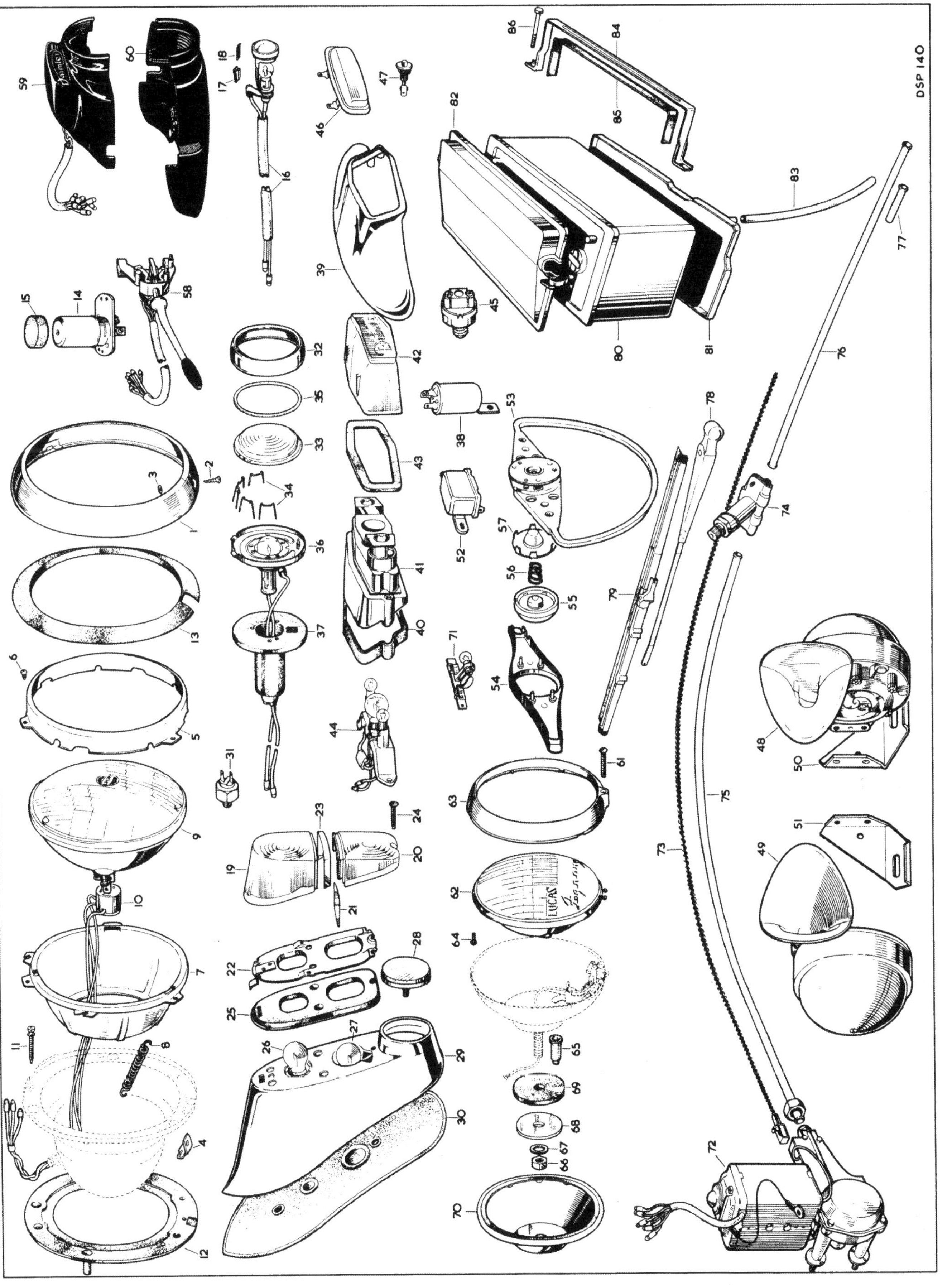

PLATE 54

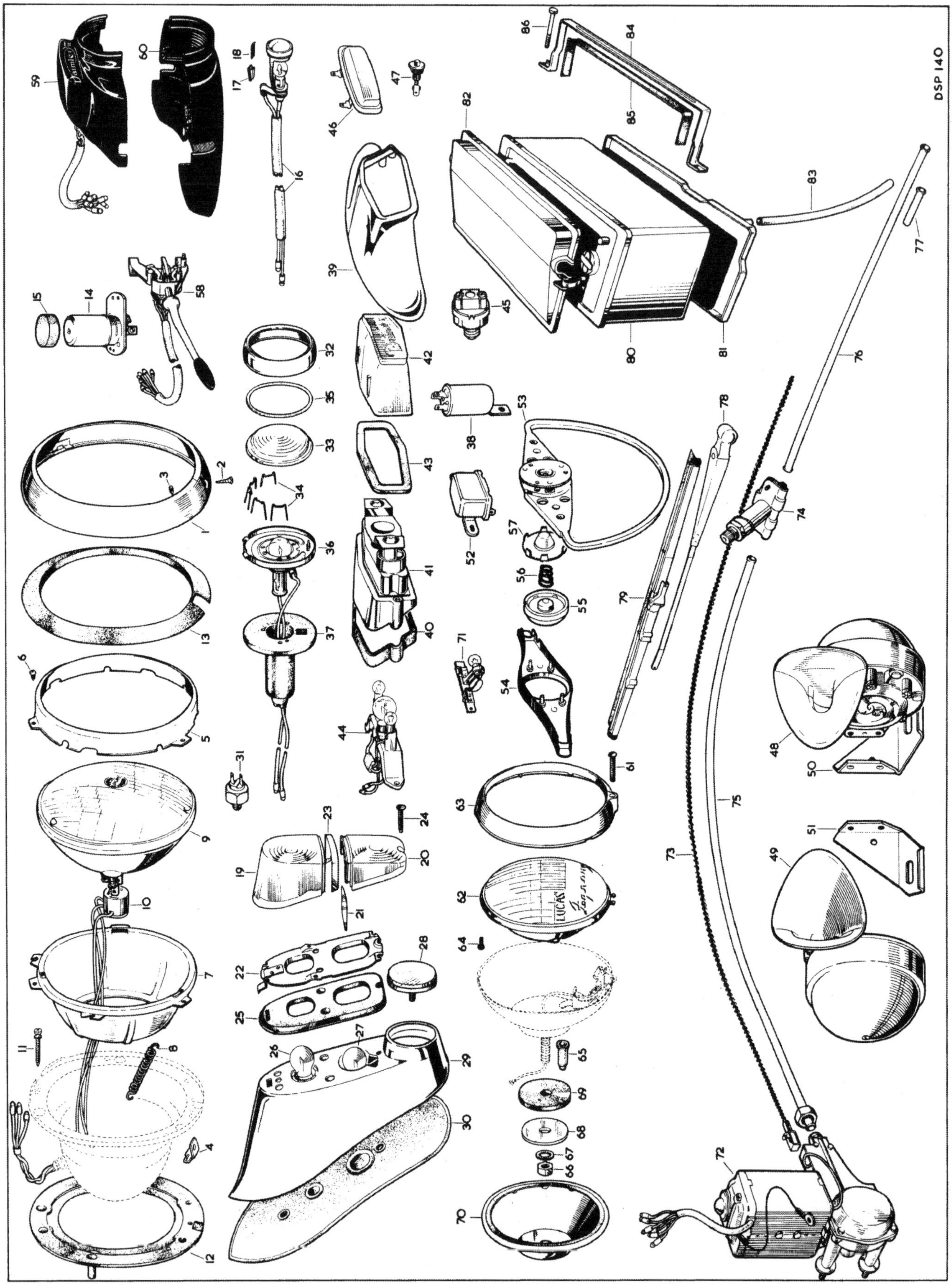

PLATE 54

Part No.	Description	Plate No.	No. per Unit	Remarks
C.1083	**SWITCH ON GEARBOX TOP COVER, OPERATING REVERSING LIGHT/OVERDRIVE CIRCUIT (31542-SS.10/1)**		2	
C.4531	**Gasket**, for Switches (163682)		As req'd	
C.26377	**Fibre Washer**, for Switches		As req'd	
C.23391	**OVERDRIVE OPERATING SWITCH (38404)**		1	
C.16600	**Bracket**, mounting Operating Switch to Outer Steering Column Tube		1	
UFS.119/4R	**Screw**, Set, securing Bracket		2	
UFN.119/L	**Nut**, on Setscrews		2	
C.723/A	**Washer**, Shakeproof, under Nuts		2	
10002	**SOLENOID ON OVERDRIVE UNIT (76522/H-11S) (65210)**		1	
9240	**Gasket**, for Solenoid (60121)		1	
C.15604	**SPEEDOMETER (SN.6317/04) (CALIBRATED IN MILES)**		1	FOR 4.55 : 1 REAR AXLE RATIO
C.15909	**SPEEDOMETER (SN.6317/14) (CALIBRATED IN KILOS)**		1	FOR 4.55 : 1 REAR AXLE RATIO
C.16332	**FLEXIBLE SPEEDOMETER CABLE (DF.1104/19Z/78)**		1	FOR R.H. DRIVE CARS
6898	**Inner Cable only** (DI.1110/00/78)		1	
8253	**Outer Cable only** (DO.1104/18/78)		1	
C.26432	**FLEXIBLE SPEEDOMETER CABLE (DF.1104/19Z/106)**		1	FOR L.H. DRIVE CARS
11350	**Inner Cable only** (DI.1110/00/78)		1	
11351	**Outer Cable only** (DO.1104/18/106)		1	
C.17432	**Grommet**, for Speedometer Cable (863123)		1	
C.28363	**FORWARD HARNESS (54953549)**		1	
C.11099	**Grommet Plate**, for Forward Harness		1	
C.2796	**Bracket**, holding Cable connectors to R.H. Front Wing		1	
C.2797	**Bracket**, holding Cable Connectors to L.H. Front Wing		1	
UCS.313/4H	**Screw**, Set, securing Brackets		4	
UNC.113/L	**Nut**, on Setscrews		4	
C.721	**Washer**, Shakeproof, under Nuts		4	
C.27102	**ELECTRICAL HARNESS FOR GEARBOX**		1	
C.27121	**RESISTANCE (LOCATED BEHIND INSTRUMENT PANEL) FOR OVERDRIVE INDICATOR (54334484)**		1	

Part No.	Description	Plate No.	No. per Unit	Remarks
C.25927	UPPER SWITCH COVER ASSEMBLY, AT CENTRE OF STEERING WHEEL		1	
C.15792	**Indicator Light Unit** (54360413)		1	
C.25924	**Unit Indicator** ("Overdrive") (54361967)		1	
C.15787	**Bulb** (12V 2W) (281)		3	
8411	**Bulb Holder**, complete with Cables (54945043)		1	
8092	**Rubber Pad** (54360531)		1	
	CABLE CLIPS			
C.1543/7	**Clip**, securing Speedometer Cable		2	
C.1543/4	**Clip**, securing Harness (Panel Connector)		1	
BD.22655	**Clip**, securing Line Fuse		1	
DAZ.404/4C	**Woodscrew**, for Line Fuse Clip		1	
C.25555	DIRECTION INDICATOR/HEADLAMP FLASHER SWITCH (35530A/1315A)		1	

Part No.	Description	Plate No.	No. per Unit	Remarks
C.20626	**DYNAMO (22716/H-C40L)**		1	Fitted to Engine No. 7A.1001 and subs.
9698	**Brush** (54213117)		1 set	
9017	**Spring**, tensioning Brushes (227542)		1 set	
9699	**Bracket**, Commutator end (54212628)		1	
9704	**Bearing**, for C.E. Bracket (227818)		1	
9705	**Bracket**, Driving end (54211095)		1	
2756	**Bearing**, for D.E. Bracket (189307)		1	
9706	**Armature** (54213321)		1	
9707	**Field Coil** (54211803)		1 set	
UFB.131/8R	**Bolt**, securing Dynamo to Mounting Bracket		2	
FW.105/T	**Washer**, Plain, at Dynamo Mounting		2	
UFN.131/L	**Nut**, on Bolts		2	
FG.105/X	**Washer**, Spring, under Nuts		2	
C.20912	**DYNAMO PULLEY**		1	
C.21605	**FAN FOR DYNAMO PULLEY**		1	
UCS.131/7R	**Screw**, Set, for adjustment of Dynamo Belt Tension		1	
401883	**Washer**, Serrated, on Setscrew		1	
FG.105/X	**Washer**, Spring, on Setscrew		1	
C.21264	**BRACKET ASSEMBLY, MOUNTING FUSE AND CONTROL BOXES AND HORN RELAY**		1	REQUIRED FOR CARS FITTED WITH DYNAMO C.20626 (C.21264 to AW.102/T)
C.11218	**Rod Assembly**, fixing Fuse Box Cover		1	
C.12365/2	**Insulating Tube**, on Rod		1	
C.4548	**Grommet**, for Wiring through Bracket		2	
UFS.119/4R	**Screw**, Set, securing Bracket		2	
UFS.119/7R	**Screw**, Set, securing Bracket		1	
BD.5815	**Distance Piece**		1	
UFN.119/L	**Nut**, on Setscrews		3	
C.723/A	**Washer**, Shakeproof, under Nuts		3	
AW.102/T	**Washer**, Plain, on Setscrew		1	
C.21267	**COVER, FOR BRACKET ASSEMBLY**		1	REQUIRED FOR BRACKET ASSEMBLY C.21264 (C.21267 to C.5549)
C.4877/4	**Rubber Insulator**, under Cover		1	
C.5549	**Knob**, securing Cover		1	
C.20627	**CONTROL BOX (37342/D-RB.340)**		1	REQUIRED FOR USE WITH DYNAMO C.20626
8558	**Cover** (54381342)		1	
9943	**Screw**, securing Cover		2	
8559	**Resistance** (Swamp) (54380438)		1	
9715	**Resistance** (Points) (54380084)		1	
UFS.119/9R	**Screw**, Set, securing Control Box to Bracket		3	
C.11472/1	**SHIELD ASSEMBLY, PROTECTING CONTROL BOX**		1	
C.16510	**FUSE BOX (54038009-SF.6)**		1	Fitted to Chassis No. 1A.1001 and subs. 1A.20001 and subs. 1K.1001 and subs. 1K.30001 and subs.
C.5638	**Fuse**, 35 Amp (188218)		2	
C.13932	**Fuse**, 50 Amp (188219)		2	
6471	**Terminal Screw** (121610)		5	
UCS.319/4H	**Screw**, Set, securing Fuse Box to Mounting Bracket		2	

ELECTRICAL EQUIPMENT
AUTOMATIC TRANSMISSION

Part No.	Description	Plate No.	No. per Unit	Remarks
C.25073	**ALTERNATOR (54021122 11 A.C.)**		1	Fitted to Engine No. 7K.1001 and subs.
10768	**Bolt**, through fixing (54112646)		3	
10769	**Nut**, for fixing Bolts (54131736)		3	
8586	**Nut**, on Shaft (198800)		1	
9365	**Washer**, under Shaft Nut (185085)		1	
2756	**Bearing** (189307)		1	
10770	**Bracket**, Frame, at Driving end (54215209)		1	
10771	**Rotor** (54214557)		1	
12324	**Bracket**, at slip ring end (54217200)		1	
12325	**Heatsink**, with Cathode Base Diodes (54217551)		1	
12326	**Heatsink**, with Anode Base Diodes (54217552)		1	
10775	**Brush**, Spring and Lucar (54215076)		1 set	
10776	**Brushbox** (Bakelite) (54216364)		1	
8638	**Connector**, 'Lucar' (Cut-out) (54190096)		1	
10777	**Stator** (54214441)		1	
C.22249	**Cooling Fan** (54214882)		1	
C.27447	**ALTERNATOR PULLEY**		1	
C.27354	**BELT, DRIVING ALTERNATOR**		1	
C.27375	**ALTERNATOR MOUNTING BRACKET**		1	
C.27522	**Mounting Plate**, at front of engine for Alternator Mounting Bracket		1	
C.8511	**Stud**, in Engine Tappet Block for Alternator Mounting Plate		4	
UFN.131/L	**Nut**, on Studs, securing Alternator Mounting Plate		4	
FG.105/X	**Washer**, Spring, under Nuts		4	
UFB.137/27R	**Bolt**, securing Alternator to Mounting Bracket		1	
C.8667/3	**Nut**, Self-locking, on Bolt		1	
C.27448	**Adjusting Link**, for Alternator		1	
UCS.131/10R	**Screw**, Set, securing Alternator to Adjusting Bracket		1	
FG.105/X	**Washer**, Spring, on Setscrew		1	
401883	**Washer**, Serrated, on Setscrew		1	
C.20328	**Locknut**, on Setscrew		1	
C.18203	**Stud**, in Alternator Mounting Bracket for Adjusting Link		1	
C.31207	**Steady Bracket**, for Alternator Shield		1	
FG.105/X	**Washer**, Spring, on Stud		1	
UFN.131/L	**Nut**, on Stud		1	
FW.105/T	**Washer**, Plain, on Stud		1	
C.28439	**ALTERNATOR SHIELD ASSEMBLY**		1	
C.28602	**BRACKET ASSEMBLY, MOUNTING FUSE AND CONTROL BOXES AND HORN RELAY**		1	REQUIRED FOR CARS FITTED WITH AN ALTERNATOR
C.27664	**Mounting Post**, for Cover, on Bracket		1	
UFS.119/2R	**Screw**, Set, securing Mounting Post to Bracket		1	
C.273A	**Washer**, Shakeproof, on Setscrew		1	
C.12365/3	**Insulation Sleeve**, on Mounting Post		1	
C.4548	**Grommet**, for Wiring through Bracket		2	
UFS.119/4R	**Screw**, Set, securing Bracket to Wing Valance		3	
UFN.119/L	**Nut**, on Setscrew		3	
C.723/A	**Washer**, Shakeproof, under Nuts		3	
AW.102/T	**Washer**, Plain, on Setscrew		3	
BD.32206/3	**Screw**, Self-Tapping, securing Horn Relay to Bracket Assembly		2	
BD.32206/3	**Screw**, Self-Tapping, securing Flasher Unit to Bracket Assembly		1	
BD.32206/5	**Screw**, Self-Tapping, securing Control Box to Bracket Assembly		2	
C.23376	**Distance Piece**, on Control Box Screws		2	

Part No.	Description	Plate No.	No. per Unit	Remarks
C.28604	**COVER FOR BRACKET ASSEMBLY**		1	**REQUIRED FOR BRACKET ASSEMBLY C.28602**
C.4877/4	**Rubber Insulator**, under Cover		1	
C.5549	**Knob**, fixing Cover		1	
C.22253	**CONTROL BOX (4TR) (37423A)**		1	**REQUIRED FOR USE WITH ALTERNATOR C.25073**
C.20781	**FORWARD HARNESS (54933209)**		1	Fitted from Chassis No. 1A.1001 to 1A.2394 1A.20001 to 1A.20197
C.22357	**FORWARD HARNESS (54934304)**		1	Fitted from Chassis No. 1A.2395 to 1A.8060 1A.20198 to 1A.20393
C.25610	**FORWARD HARNESS (54938110)**		1	Fitted to Chassis No. 1A.8061 and subs. 1A.20394 and subs.
C.28362	**FORWARD HARNESS (54953548)**		1	Fitted to Chassis No. 1K.1001 and subs. 1K.30001 and subs.
C.11099	**Grommet Plate**, for Forward Harness		1	
C.2796	**Bracket**, holding Cable connectors to R.H. Front Wing		1	
C.2797	**Bracket**, holding Cable connectors to L.H. Front Wing		1	
UCS.313/4H	**Screw**, Set, securing Brackets		4	
UCN.113/L	**Nut**, on Setscrews		4	
C.721	**Washer**, Shakeproof, under Nuts		4	
C.21959	**Wiring Assembly** (Cable) for Starter Cut-out Switch		1	
C.15602	**SPEEDOMETER (SN.6317/02)** **(CALIBRATED IN MILES)**		1	**SUITABLE FOR CARS FITTED WITH 4.55 : 1 RATIO REAR AXLE**
C.15604	**SPEEDOMETER (SN.6317/04)** **(CALIBRATED IN MILES)**		1	**SUITABLE FOR CARS FITTED WITH 4.27 : 1 RATIO REAR AXLE**
C.15908	**SPEEDOMETER (SN.6317/13)** **(CALIBRATED IN KILOS)**		1	**SUITABLE FOR CARS FITTED WITH 4.55 : 1 RATIO REAR AXLE** Fitted from Chassis No. 1A.1001 to 1A.8639 1A.20001 to 1A.20403
C.15909	**SPEEDOMETER (SN.6317/14)** **(CALIBRATED IN KILOS)**		1	**SUITABLE FOR CARS FITTED WITH 4.27 : 1 RATIO REAR AXLE** Fitted to Chassis No. 1A.8640 and subs. 1A.20404 and subs. 1K.1001 and subs. 1K.30001 and subs.
	NOTE: **To ensure that the correct Speedometer is supplied by Daimler Spares Dept., IT IS IMPORTANT WHEN ORDERING, to quote the number printed on the dial of the displaced instrument.**			
C.15468	**REMOTE CONTROL OPERATING TRIP MILEAGE RECORDER ON SPEEDOMETER (41-741-152-01/9)**		1	
8409	**Screwed Escutcheon**, securing Remote Control at Winding End		1	

Part No.	Description	Plate No.	No. per Unit	Remarks
987	BULB (2.2 WATT) FOR ILLUMINATION OF SPEEDOMETER AND FOR IGNITION, LOW FUEL CONTENTS AND HEADLAMP MAIN BEAM WARNING LIGHTS (987)		5	
C.15929	FLEXIBLE SPEEDOMETER CABLE COMPLETE (DF.1104/19/Z66)		1	R.H. DRIVE CARS ONLY
8241	Inner Cable only (DI.1110/00/66)		1	
8242	Outer Cable only (DO.1104/18/66)		1	
C.15930	FLEXIBLE SPEEDOMETER CABLE COMPLETE (DF.1104/19/Z.84)		1	L.H. DRIVE CARS ONLY
8243	Inner Cable only (DI.1110/00/84)		1	
8244	Outer Cable only (DO.1104/18/84)		1	
C.8792	BATTERY (BT.GA)		1	REQUIRED FOR CARS FITTED WITH A D.C. GENERATOR (DYNAMO)
C.27003	BATTERY (WITH REVERSED TERMINALS) (54028461.S9/8)		1	REQUIRED FOR CARS FITTED WITH AN ALTERNATOR
C.10192	Felt, 5" x 11", under Battery		1	
9720	REVOLUTION COUNTER (RV.7413/07)		1	
C.20779	UPPER SWITCH COVER ASSEMBLY AT CENTRE OF STEERING WHEEL (54361227)		1	Fitted from Chassis No. 1A.1001 to 1A.3252 1A.20001 to 1A.20256 (Except 1A.3096 to 1A.3098, 1A.3232, 1A.3247 to 1A.3249)
C.15792	Indicator Light Unit (54360413)		1	
C.20780	Gear Indicator (P.R.N.D.L.) (54361228)			
C.15787	Bulb (2 watt, Lilliput) (281)		3	
8411	Bulb Holder, complete with Cables (54945043)		1	
8092	Rubber Pad (54360531)		1	
C.22538	UPPER SWITCH COVER ASSEMBLY AT CENTRE OF STEERING WHEEL (54361545)		1	Fitted to Chassis No. 1A.3253 and subs. 1A.20257 and subs. (Also to 1A.3096 to 1A.3098, 1A.3232, 1A.3247 to 1A.3249) 1K.1001 and subs. 1K.30001 and subs.
C.22540	Gear Indicator (P.R.N.D_2.D_1.L.) (54361538)		1	
C.22539	Indicator Light Unit		1	
	All other items are as for Upper Switch Cover C.20779			

Part No.	Description	Plate No.	No. per Unit	Remarks
C.16778	**CASE ASSEMBLY FOR HEATER**	55–1	1	Fitted from Chassis No. 1A.1001 to 1A.2348 1A.20001 to 1A.20194
C.16776	**Lid,** for Heater Case	55–2	1	
C.21991	**CASE ASSEMBLY FOR HEATER**		1	Fitted to Chassis No. 1A.2349 and subs. 1A.20195 and subs. 1K.1001 and subs. 1K.30001 and subs.
C.16776	**Lid,** for Heater Case		1	
DZZ.406/4C	**Screw,** Self-Tapping, securing Lid to Case		11	
C.15587	**Spring,** steadying Air Control Flaps	55–3	2	
BD.8291/9	**Screw,** Set, securing Heater Case to Dash		1	
BD.8291/6	**Screw,** Set, securing Heater Case to Dash		1	
FG.104/X	**Washer,** Spring, on Setscrews		2	
C.9627	**Mounting Rubber,** for Heater Volute to Dash	55–4	3	
UFN.125/L	**Nut,** securing Heater to Dash at Volute		3	
FG.104/X	**Washer,** Spring, under Nuts		3	
C.18571	**Abutment Clamp,** holding Control Outer Cables to Bracket on Heater Case	55–5	1	
UCN.119/L	**Nut,** securing Abutment Clamp		1	
AW.102/T	**Washer,** Plain, under Nut		1	
AG.102/X	**Washer,** Spring, under Nut		1	
C.16774	**WATER RADIATOR FOR HEATER**	55–6	1	
C.15968	**SEAL (FELT) FOR HEATER RADIATOR**	55–7	1	
C.15742	**PACKING STRIP FOR HEATER RADIATOR (FELT 5/8" x 5/16" x 3" LONG)**	55–8	2	
C.15736	**PACKING STRIP FOR HEATER RADIATOR (FELT 5/8" x 5/16" x 2 1/8")**	55–9	2	
C.15755	**SEAL (POLYURETHANE) ON PIPES OF HEATER RADIATOR**	55–10	2	
C.15743	**SEAL (POLYURETHANE) FOR UPPER AIR CONTROL FLAP**	55–11	2	
C.15966	**SEAL, LONG (POLYURETHANE) FOR LOWER AIR CONTROL FLAP**	55–12	1	
C.15967	**SEAL, SHORT (POLYURETHANE) FOR LOWER AIR CONTROL FLAP**	55–13	2	
C.21976	**SEAL (POLYURETHANE) FOR LOWER AIR CONTROL FLAP**	55–14	2	
BD.13603	**SEAL FOR DEMISTER OUTLET AT REAR OF HEATER CASE**	55–15	1	
C.15741	**SEAL FOR HEATER SLOT IN SCUTTLE**	55–16	1	**Superseded by C.21805**

HEATER INSTALLATION

Part No.	Description	Plate No.	No. per Unit	Remarks
C.21805	SEAL FOR HEATER SLOT IN SCUTTLE		1	**Supersedes C.15741**
C.9624	SEAL BETWEEN HEATER AND DASH AT VOLUTE	55–17	1	
C.15711	HEATER MOTOR	55–18	1	
C.17428	**Resistance** (FHE.3103/01)		1	
C.15591	**Fan**, for Heater Motor	55–19	1	
C.10102	**Felt Washer**, between Heater Motor and Volute	55–20	1	
UFS.419/8H	**Screw**, Set, securing Heater Motor to Volute		3	
C.10103	**Rubber Washer**, on Setscrews, at each side of Motor Flange	55–21	6	
UFN.119/L	**Nut**, on Setscrews		3	
AW.102/T	**Washer**, Plain, under Nuts		3	
C.21191	FRONT FEED PIPE FROM TEE PIECE TO REAR FEED PIPE	55–22	1	
C.21557	**Clip**, securing Pipe	55–23	1	
C.14999/4	HOSE, CONNECTING FRONT AND REAR FEED PIPES	55–24	1	Fitted to Chassis No. 1A.1001 and subs. 1A.20001 and subs. Also fitted to Chassis No. 1K.1001 to 1K.1400 1K.30001 to 1K.30022
C.2905/2	**Clip**, securing Hose	55–25	2	
C.21186	REAR FEED PIPE FROM HOSE TO WATER VALVE	55–26	1	Fitted to Chassis No. 1A.1001 and subs. 1A.20001 and subs. Also fitted to Chassis No. 1K.1001 to 1K.1401 1K.30001 to 1K.30023
C.25703	HOSE, CONNECTING FRONT AND REAR FEED PIPES		1	**REQUIRED FOR CARS FITTED WITH MANUAL TRANSMISSION FROM CHASSIS No.** 1K.1402 and subs. 1K.30023 and subs. **REQUIRED FOR CARS FITTED WITH AUTO-TRANSMISSION FROM CHASSIS No.** 1K.1925 and subs. 1K.30044 and subs.
C.2905/2	**Clip**, securing Hose		2	
C.22695	REAR FEED PIPE FROM HOSE TO WATER VALVE		1	
C.16655	ELBOW, CONNECTING REAR FEED PIPE TO WATER VALVE	55–27	1	
C.2905/2	**Clip**, securing Elbow to Feed Pipe	55–28	1	
C.2905/1	**Clip**, securing Elbow to Valve	55–29	1	
C.16559	WATER VALVE ASSEMBLY (FHW.1901/2)	55–30	1	
C.9928	**Sealing Ring**, for Water Valve (CHS.4653)	55–31	1	
C.9931	**Screw**, securing Water Valve to Heater Case		2	
C.9932	**Washer**, Spring, on Screws		2	
C.16655	ELBOW, CONNECTING HEATER OUTLET TO REAR RETURN PIPE	55–32	1	
C.2905/1	**Clip**, securing Elbow to Heater	55–33	1	
C.2905/2	**Clip**, securing Elbow to Rear Return Pipe	55–34	1	

HEATER INSTALLATION

Part No.	Description	Plate No.	No. per Unit	Remarks
C.21187	**REAR RETURN PIPE, ON SCUTTLE**	55–35	1	
C.14999/2	**HOSE, CONNECTING REAR AND FRONT RETURN PIPES**	55–36	1	
C.2905/2	**Clip,** securing Hose	55–37	2	
C.21192	**FRONT RETURN PIPE FROM HOSE TO TEE PIECE**	55–38	1	
C.21196	**Clip,** securing Pipe	55–39	1	
C.21469	**LEVER ASSEMBLY FOR "HOT" AND "COLD" CONTROL**		1	
C.21464	**Lever only**	55–40	1	
C.21467	**Bracket only,** for Lever	55–41	1	
UFS.125/7R	**Screw,** Set, pivoting Control Lever		1	
C.8737/1	**Nut,** Self-locking, on Setscrews		1	
FN.104/T	**Washer,** Plain, under Nut		1	
C.15712	**Spacer,** on Setscrew		1	
BD.11721	**Trunnion,** for Inner Cable		1	
UCS.113/2R	**Screw,** Set, securing Inner Cable in Trunnion		1	Fitted from Chassis No. 1A.1001 to 1A.1097 1A.20001 to 1A.20018
C.8667/20	**Nut,** Self-locking, securing Trunnion to Lever		1	
AW.104/T	**Washer,** Plain, under Nut		1	
C.21604	**Clamp,** for Control Cable	55–42	1	
UCS.119/5R	**Screw,** Set, securing Clamp		1	
UCN.119/L	**Nut,** on Setscrew		1	
AG.102/X	**Washer,** Spring, under Nut		1	
C.21468	**LEVER ASSEMBLY FOR HEATER/ DEMISTER CONTROL**		1	
C.21466	**Bracket only,** for Lever	55–43	1	
	All other items as for C.21469			
C.21872	**LEVER ASSEMBLY FOR "HOT" AND "COLD" CONTROL**		1	
C.21873	**Lever only**		1	
C.21875	**Bracket only,** for Lever		1	
	All other items as for C.21469			Fitted to Chassis No. 1A.1098 and subs. 1A.20019 and subs. 1K.1001 and subs. 1K.30001 and subs.
C.21871	**LEVER ASSEMBLY FOR HEATER/ DEMISTER CONTROL**		1	
C.21873	**Lever only**		1	
C.21874	**Bracket only,** for Lever		1	
	All other items as for C.21469			
C.15719	**KNOB, ON "HOT/COLD" CONTROL LEVER**	55–44	1	
UCS.311/2H	**Screw,** Set, securing Knob to Lever		1	
C.21600	**KNOB, ON "HEATER/DEMISTER" CONTROL LEVER**	55–45	1	
UCS.311/3H	**Screw,** Set, securing Knob to Lever		1	
C.15714/4	**OUTER CABLE FOR "HOT/COLD" CONTROL**	55–46	1	

HEATER INSTALLATION

Part No.	Description	Plate No.	No. per Unit	Remarks
C.21877	**INNER CABLE FOR "HOT/COLD" CONTROL**	55–47	1	
C.10351	**Trunnion**, connecting Cable to Valve (CHS.4097)		1	
C.10352	**Screw**, securing Cable in Trunnion		1	
C.16614	**WIRE CABLE LINKING WATER VALVE TO FLAP ON HEATER CASE**		1	Fitted from Chassis No. 1A.1001 to 1A.2348 1A.20001 to 1A.20194
C.21992	**WIRE CABLE LINKING WATER VALVE TO FLAP ON HEATER CASE**		1	Fitted to Chassis No. 1A.2349 and subs. 1A.20195 and subs. 1K.1001 and subs. 1K.30001 and subs.
C.17409	**Trunnion**, for Joint of Link Cable to Valve and Flap (CH.36209/05)		2	
C.17443	**Screw**, securing Link Wire in Trunnions		2	
C.15715	**OUTER CABLE FOR HEATER/ DEMISTER CONTROL**	55–48	1	
C.15974	**INNER CABLE FOR HEATER/ DEMISTER CONTROL**	55–49	1	
C.13824	**Clip**, retaining Control Cable to Heater Flap	55–50	1	
C.2850	**Grommet**, in Dash, for Control Cables	55–51	2	
C.15592	**ESCUTCHEON FOR HEATER CONTROLS**	55–52	2	
C.21463	**Plate**, inscribed "Hot"/"Cold", behind L.H. Escutcheon	55–53	1	
C.21462	**Plate**, inscribed "Car"/"Screen", behind R.H. Escutcheon	55–54	1	
UFN.119/L	**Nut**, securing Escutcheons		4	
C.723A	**Washer**, Shakeproof, under Nuts		4	

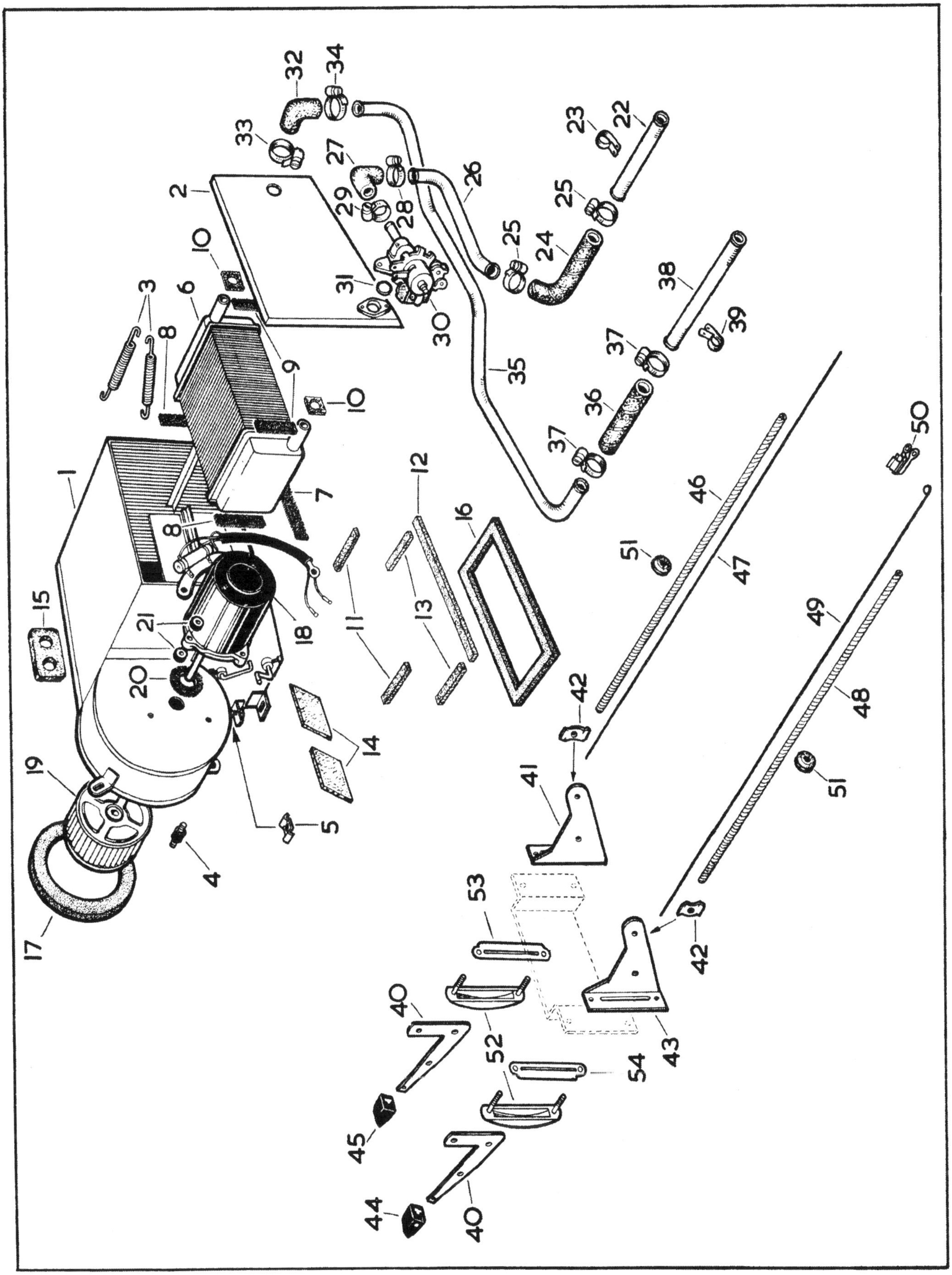

PLATE 55

WINDSCREEN WASHER

FITTED FROM CHASSIS No.1A.1001 to 1A.8060
1A.20001 to 1A.20393

Part No.	Description	Plate No.	No. per Unit	Remarks
C.17004	**WINDSCREEN WASHER UNIT ASSEMBLY (54071196) (2SJ)**		1	
8408	**Motor**, complete with Coupling (54711928)	56–1	1	
8035	**Pump Assembly** (295194)	56–2	1	
8036	**Moulded Cover**, for Reservoir (295138)	56–3	1	
8037	**Rubber Gasket**, between Cover and Reservoir (295042)	56–4	1	
8038	**Rubber Filler Cap**, in Cover (295043)	56–5	1	
8039	**Union**, securing Motor to Pump (54715154)	56–6	1	
8040	**Glass Reservoir** (295058)	56–7	1	
8041	**Bracket**, holding Reservoir (295061)	56–8	1	
8408	**MOTOR COMPLETE WITH COUPLING (54711928)**	56–1	1	
8042	**Contact Assembly**, complete with Adjusting Screw (295251)		1	
8043	**Brush Set** (295198)		1	
8044	**Armature** (295023)		1	
8990	**Lucar Connector** (54190172)		3	
C.15643	**TUBE (108" LONG) (54711658)**	56–9	1	
C.15644	**"T" PIECE AT JUNCTION OF FLUID TUBES (295091)**	56–10	1	
C.15645	**JET ASSEMBLY (54711657)**	56–11	2	
C.24782	**BRACKET ASSEMBLY, MOUNTING WINDSCREEN WASHER BRACKET**		1	
UFN.119/L	**Nut**, securing Bracket to Washer Bracket and to R.H. Front Wing Valance		4	
C.723/A	**Washer**, Shakeproof, under Nuts		4	
C.24751	**Stud**, fixing Washer Bracket at bottom		1	
UFN.119/L	**Nut**, securing Stud to Valance and Washer Bracket on Stud		2	
C.723/A	**Washer**, Shakeproof, under Nuts		2	

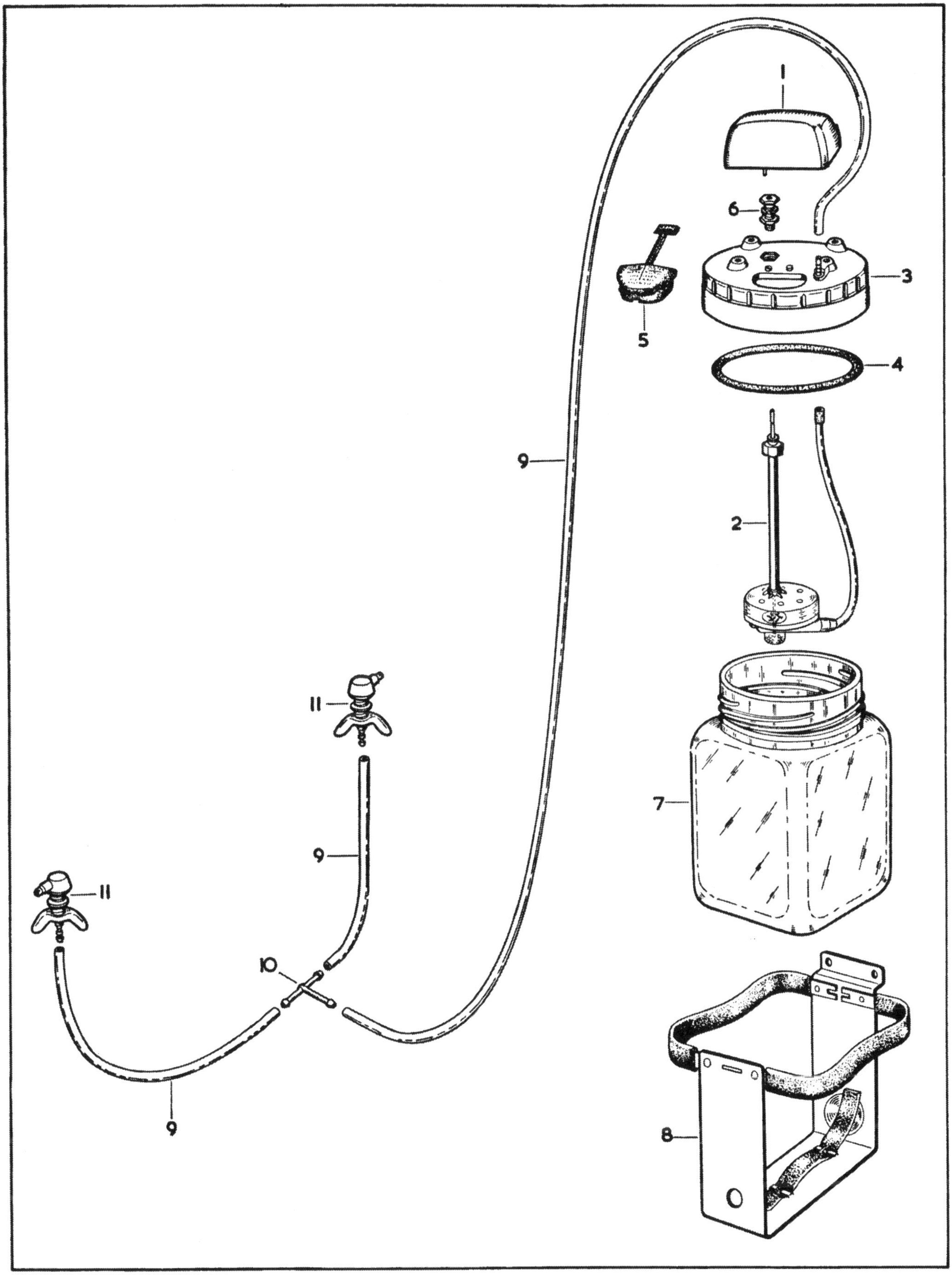

PLATE 56

WINDSCREEN WASHER

FITTED TO CHASSIS No. 1A.8061 and subs. Also to 1K1001 and subs.
1A.20394 and subs. 1K.30001 and subs.

Part No.	Description	Plate No.	No. per Unit	Remarks
C.25438	**WINDSCREEN WASHER UNIT ASSEMBLY (54071313-5.SJ)**		1	
11080	**Motor**, complete with Coupling (54717459)	57–1	1	
11081	**Screw**, securing Motor to Cover (54102956)	57–2	1	
11082	**Coupling**, between Motor and Pump (54718214)	57–3	1	
11083	**Cover and Pump Assembly**, complete with Coupling (54717203)	57–4	1	
11084	**Reservoir** (54713810)	57–5	1	
11085	**Filler Cap**, on Reservoir (54713627)	57–6	1	
11086	**Bracket and Strap Assembly**, holding Washer Unit (54713814)	57–7	1	
11080	**MOTOR, COMPLETE WITH COUPLING (54717459)**	57–1	1	
8043	**Brush Set** (54711077)		1	
11088	**Armature** (54713009)		1	
C.15643	**WASHER TUBE (108" LONG) (54711658)**	57–8	1	
C.15644	**"T" PIECE AT JUNCTION OF FLUID TUBES (295091)**	57–9	1	
C.15645	**JET ASSEMBLY (54711657)**	57–10	2	
C.24782	**BRACKET ASSEMBLY, MOUNTING WINDSCREEN WASHER BRACKET**		1	
UFN.119/L	**Nut**, securing Bracket Assembly to Washer Bracket and to R.H. Front Wing Valance		4	
C.723/A	**Washer**, Shakeproof, under Nuts		4	
C.24751	**Stud**, fixing Washer Bracket at bottom		1	
UFN.119/L	**Nut**, securing Stud to Valance and Washer Bracket on Stud		2	
C.723/A	**Washer**, Shakeproof, under Nuts		2	

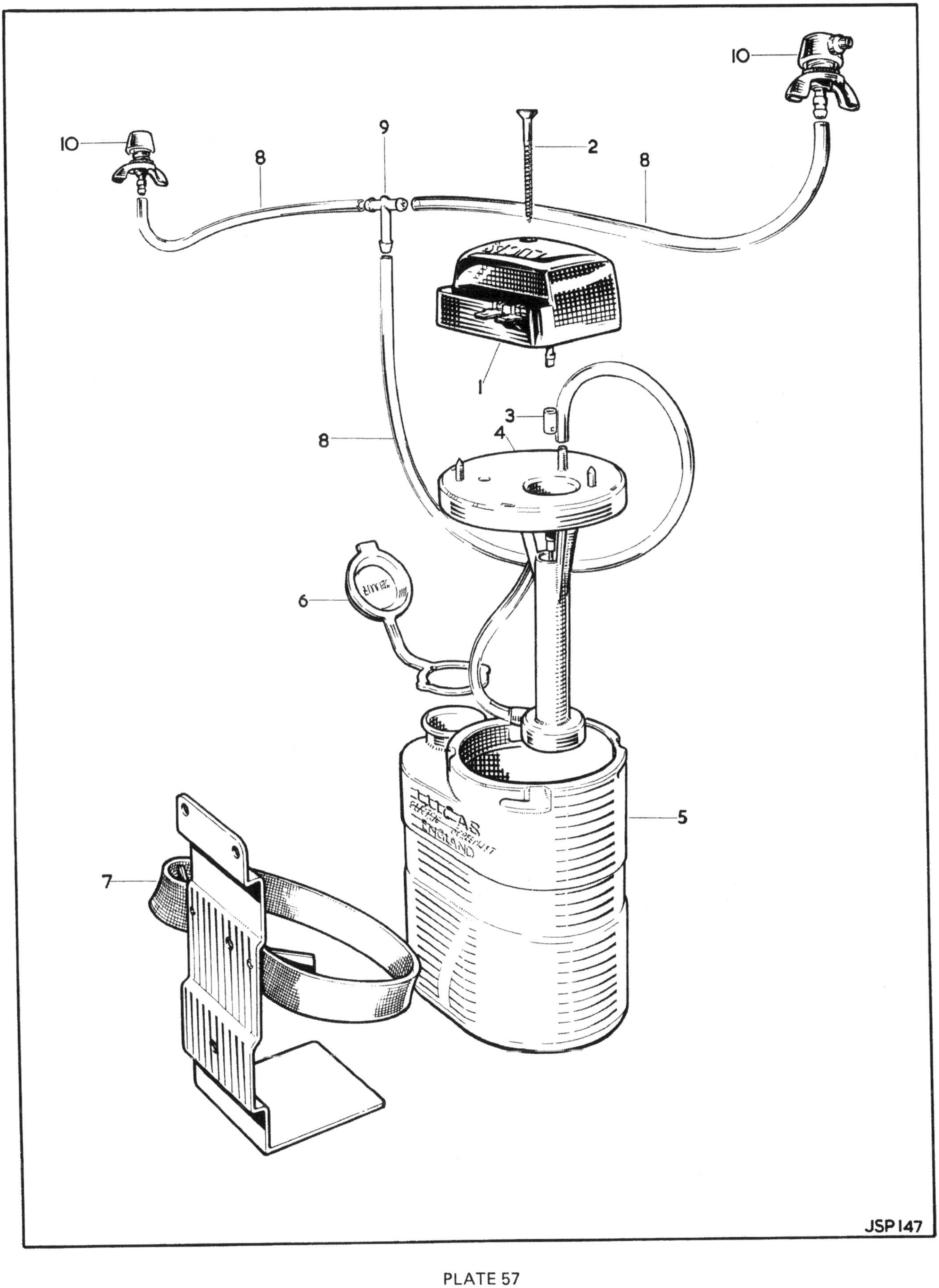

PLATE 57

TOOL KIT

Part No.	Description	Plate No.	No. per Unit	Remarks
C.23297	**Adjustable Spanner**		1	
C.996	**Pliers**		1	
C.11753	**Tyre Pressure Gauge**		1	
C.5444	**Screwdriver**, for adjustment of Distributor Points		1	
C.5587	**Feeler Gauge**		1	
C.993	**Extractor**, for Tyre Valve		1	
C.20713	**Screwdriver**		1	
C.10154	**Screwdriver**, for Battery Terminals		1	
C.20770	**Box Spanner**, for Sparking Plugs		1	
C.20825	**Tommy Bar**, for Box Spanner		1	
C.20715	**Open-ended Spanner**, ¾" x 7/8" A.F.		1	
C.20716	**Open-ended Spanner**, 9/16" x 5/8" A.F.		1	
C.20717	**Open-ended Spanner**, 7/16" x ½" A.F.		1	
C.20718	**Open-ended Spanner**, 11/32" x 3/8" A.F.		1	
C.13269	**Grease Gun**		1	
C.5634	**Sparking Plug** (Champion N.8)		1	FOR 8 : 1 COMPRESSION RATIO ENGINE
C.2954	**Ratchet Lever**, operating Jack		1	
C.2955	**Wheel Brace**		1	FOR CARS FITTED WITH DISC WHEELS
C.10659	**Jack**		1	
C.10792	**Strap**, for Jack		1	
463/A	**Rubber Washer**, for Jack		1	
C.13620	**Bleeder Tube**, complete with Container		1	
C.14927	**Hub Cap Tool**, for fitting and removing Hub Caps		1	FOR CARS FITTED WITH WIRE SPOKE WHEELS
C.27290	**Hammer** (Copper and Rawhide)		1	FOR CARS FITTED WITH WIRE SPOKE WHEELS

Part No.	Description	Plate No.	No. per Unit	Remarks
C.31855	**CAR RADIO ASSEMBLY (RK.DA07/1080)**		1	LONG/MEDIUM WAVEBANDS WITH NEGATIVE OR POSITIVE EARTH APPLICATION
C.31856	**Control Unit** (1080T)		1	
8279	**Aerial** (AW.454/48)		1	
C.29897	**Aerial** (AW.91)		1	Alternatives
C.29899	**Aerial** (Electrically operated) (AW.5)		1	
C.2706	**Grommet**, for Aerial (RMO.3427)		3	
C.22941	**Clip**, for Aerial Control Cable (RMO.1495)		1	
12366	**Loudspeaker Assembly** (10-013)		1	
C.22934	**Lead**, for Loudspeaker (RMR.2440/12)		1	
10044	**Battery Lead** (RMR.2227/12)		1	
12361	**Condenser** (1 mfd) for Dynamo or Alternator (RMO.3861/T11)		1	
12361	**Condenser** (1 mfd) for Petrol Pump (RMO.3861/T11)		1	
12362	**Condenser** (1 mfd) for Coil (RMO.3861/T9)		1	
12363	**Condenser** (.25 mfd) for Oil Gauge Transmitter		1	
12364	**Connector**, for Condenser (RMO.3745)		1	
12367	**Fuse** (1.5 Amp)		1	
C.31862	**CAR RADIO ASSEMBLY (RK/DAO7/1082)**		1	MEDIUM WAVEBAND ONLY
C.31863	**Control Unit** (1082)		1	WITH NEGATIVE OR POSITIVE EARTH APPLICATION
	All other items are as for C.31855			
C.31864	**CAR RADIO ASSEMBLY (RK/DAO7/930)**		1	MEDIUM/SHORT WAVEBANDS WITH NEGATIVE OR POSITIVE EARTH APPLICATION
C.31865	**Control Unit** (930)		1	
C.31866	**Aerial** (Rear Mounted) (AW.93)		1	
12365	**Condenser** (1 mfd) for Time Clock (RMO.3861/T8)		1	
C.12686	**Earth Bond** (RMO.1669)		1	
C.22940	**Extension Lead**, for Condenser (RMR.6016/7")		1	
C.22939	**Connector**, for Extension Lead (RMO.1883)		1	
UFS.119/4R	**Screw**, Set, securing Condenser		1	
UFN.119/L	**Nut**, on Setscrew		1	
C.723/A	**Washer**, Shakeproof, under Nut		1	
AW.102/T	**Washer**, Plain, under Shakeproof Washer		1	
	All other items are as for C.31855			
C.31861	**EXTENSION LOUDSPEAKER KIT (LK/26)**		1	SUPPLIED TO SPECIAL ORDER ONLY
C.12816	**PETROL FILLER CAP (LOCKING TYPE) COMPLETE WITH KEYS**		1	
C.17940	**L.H. WHEEL DISC**		2	FOR USE WITH 5J ROAD WHEELS
C.17940/1	**R.H. WHEEL DISC**		2	FOR USE WITH 5J ROAD WHEELS
C.17923	**RIMBELLISHER**		4	FOR USE WITH 5J ROAD WHEELS
7486	**Clip**, securing Rimbellishers to Road Wheels		12	
C.16684	**TOWING BAR ASSEMBLY, COMPLETE**		1	
8308	**Chromium Cap**, for head of securing Bolts at centre of Rear Bumper		1 pair	
BD.18103	**L.H. REAR DOOR LOCK (3/10826)** (COMPLETE WITH REMOTE CONTROL, INCORPORATING CHILD'S SAFETY CATCH)		1	

OPTIONAL EXTRAS

Part No.	Description	Plate No.	No. per Unit	Remarks
BD.18104	**R.H. REAR DOOR LOCK (3/10825)**		1	
	(COMPLETE WITH REMOTE CONTROL, INCORPORATING CHILD'S SAFETY CATCH)			
BD.13348	**Distance Piece,** between Remote Control Plate and Inner Door Panel		6	
AS.102/5D	**Screw,** Set, securing Remote Control Plate to Inner Door Panel		6	
C.723	**Washer,** Snakeproof, on Setscrews		6	
AW.102/T	**Washer,** Plain, behind Shakeproof Washers		6	
BD.13372	**L.H. OUTER ESCUTCHEON, OPERATING CHILD'S SAFETY CATCH**		1	
BD.13373	**R.H. OUTER ESCUTCHEON, OPERATING CHILD'S SAFETY CATCH**		1	
BD.13350	**SPRING, BEHIND OUTER ESCUTCHEONS**		2	
C.17900	**MAIN PETROL TANK ASSEMBLY**		1	**REPLACES C.17565 WHEN AUXILIARY PETROL TANK IS FITTED**
C.990/1	**Filter Assembly,** at bottom of Tank		1	
C.1617	**Gasket,** for Filter		1	
C.17566	**FILLER CAP**		1	
BD.9898	**Rubber Grommet,** for Filler Spout		1	
C.15476	**PETROL TANK ELEMENT UNIT (TF.1104/042)**		1	
C.937	**Gasket,** under Element Unit		1	
NS.313/3H	**Screw,** Set, securing Element Unit		6	
C.2296/9	**Copper Washer,** on Setscrews		12	
C.17883	**AUXILIARY PETROL TANK ASSEMBLY**		1	
3193	**Drain Plug**		1	
C.26310	**Gasket,** for Drain Plug		1	
C.17893	**Mounting Bracket Assembly,** for Auxiliary Tank		1	
C.17895	**Mounting Bracket Assembly,** for Auxiliary Tank		1	
C.17897	**Mounting Bracket Assembly,** for Auxiliary Tank		1	
C.3053	**Mounting Pad,** for Main and Auxiliary Petrol Tanks		12	
C.6390	**Distance Tube,** through Mounting Pads		6	
C.410	**Special Washer,** under Mounting Pads		6	
C.8737/3	**Nut,** Self-Locking, securing Main Petrol Tank		3	
UFB.137/13R	**Bolt,** securing Auxiliary Petrol Tank		3	
C.17904	**HOSE, CONNECTING MAIN AND AUXILIARY PETROL TANKS**		1	
C.2905/5	**Clip,** securing Hose		2	
C.17905	**HOSE FOR VENT PIPE CONNECTION ON AUXILIARY PETROL TANK**		1	
C.2905/1	**Clip,** securing Hose		2	
C.975	**Grommet,** for Vent Pipe		1	

Part No.	Description	Plate No.	No. per Unit	Remarks
BD.9579	**BRACKET ASSEMBLY FOR MOUNTING SPARE WHEEL**		1	REQUIRED ONLY WHEN AUXILIARY PETROL TANK IS FITTED
UFS.119/4R	**Screw**, Set, securing Bracket		8	
AW.102/T	**Washer**, Plain, on Setscrews		8	
C.8667/17	**Nut**, Self-Locking, on Setscrews		8	
C.17906	**SPARE WHEEL CLAMP ASSEMBLY**		1	REQUIRED ONLY WHEN AUXILIARY PETROL TANK IS FITTED
BD.6241	**Clamp Screw**		1	
BD.10904	**Dome Plate**, on Screw		1	
BD.9581	**Distance Piece**, on Screw		1	
BD.5561	**Hand Knob**		1	
C.704/1	**Mills Pin**, securing Hand Knob to Clamp Screw		1	
C.17907	**Tapped Boss**, fixing Spare Wheel Cover		6	
UFS.119/4R	**Screw**, Set, securing Spare Wheel Cover		6	
C.738	**Washer**, Shakeproof, on Setscrews		6	
BD.21942	**ELECTRICALLY HEATED BACKLIGHT (CLEAR GLASS)**		1	FITTED AS STANDARD EQUIPMENT AT CHASSIS No. 1K.1001 and subs. 1K.30001 and subs.
BD.22742	**ELECTRICALLY HEATED BACKLIGHT ("SUNDYM" GLASS)**		1	Alternative to BD.21942
9653	**Fuse**, 15 amp. (188220)		1	
C.20125	**Fuse Connector**		1	
BD.11524	**Grommet**		2	
BD.22655	**Clip**, for Fuse Holder		1	
BD.711/5	**Screw**, securing Clip		1	
C.5204	**Earth Terminal**		1	
BD.1229/7	**Screw**, securing Earth Terminal		1	
C.2744	**SWITCH, CONTROLLING ELECTRICALLY HEATED BACKLIGHT (35659/A-PS.7)**		1	
2836	**Knob**, for Switch (314437)		1	
C.26467	**WARNING LAMP (AMBER) FOR BACKLIGHT CIRCUIT (38728/A-WL.13)**		1	
11333	**Bezel and Window Assembly** (54361476)		1	
987	**Bulb** (987)			
C.25883	**Escutcheon** (marked "Rear Window") for Switch and Warning Lamp		1	
C.26651	**RELAY, DIMMING WARNING LAMP (33285/A-6.RA)**		1	
DAZ.408/8C	**Screw**, Self-Tapping, securing Relay		2	
C.26473	**Connector Cable**, for Backlight Switch		1	R.H. DRIVE CARS ONLY
C.26474	**Connector Cable**, for Backlight Switch		1	L.H. DRIVE CARS ONLY
	SAFETY HARNESS			
BD.34813	**FRONT SEAT BELT (STATIC TYPE) (LAP/DIAGONAL FITTING) COMPLETE WITH ATTACHMENT BOLTS AND EYE BOLTS**		2	Fitted to Chassis No. 1A.1001 and subs. 1A.20001 and subs. 1K.1001 and subs. 1K.30001 and subs.

OPTIONAL EXTRAS

Part No.	Description	Plate No.	No. per Unit	Remarks
11120	**FRONT SEAT BELT (AUTO-LOK) (LAP/DIAGONAL FITTING) COMPLETE WITH ATTACHMENT BOLTS AND EYE BOLTS**		2	Fitted from Chassis No. 1A.1001 to 1A.12239 1A.20001 to 1A.20547
9766	**REAR SEAT BELT (DIAGONAL FITTING) COMPLETE WITH ATTACHMENT BOLTS AND EYE BOLTS**		2	
BD.24579	**"D" MOTIF FOR SEAT BELT BUCKLE**		4	
UCN.111/K	**Nut,** securing 'D' Motif to Buckle		4	
C.720	**Washer,** Shakeproof, under Nut		4	
C.27046	**LOCK AND IGNITION SWITCH ASSEMBLY ON STEERING COLUMN**		1	
C.26851	**Cover** (Rubber) for Switch Terminals		1	
C.20240	**Connector Cable,** for Steering Column Lock		1	
C.2706	**Grommet,** for Connector Cable (859059)		1	
C.16114	**WING MIRROR**		2	
	NOTE: WING MIRRORS ARE FITTED AS STANDARD EQUIPMENT ON CARS EXPORTED TO BELGIUM, DENMARK, FRANCE, GERMANY, HOLLAND, LUXEMBOURG, SWEDEN, AND SWITZERLAND			
C.5201	**FRONT NUMBER PLATE WITH 'SILVER PEAK' DIGITS**		1	
DZZ.406/8C	**Screw,** Self-Tapping, securing Front Number Plate		4	
C.10060	**REAR NUMBER PLATE WITH 'SILVER PEAK' DIGITS**		1	
DZZ.406/6C	**Screw,** Self-Tapping, securing Rear Number Plate		3	
BD.19937	**SLIDING ROOF KIT**		1	
	Comprising the following parts:			
	FRAME, MAIN PANEL, TRIM PANEL, BRACKETS, LOCKING HANDLE AND SEALING RUBBERS			
BD.2191	**Attachment Bracket Assembly**		2	
BD.19942	**Screw,** Set, securing Attachment Brackets		2	
C.723A	**Washer,** Shakeproof, on Setscrews		2	
BD.19943	**Locking Handle**		1	
2175	**Paper Disc,** under Locking Handle		1	
BD.6257	**Locknut,** securing Locking Handle		1	
BD.2189	**Spring Washer,** for Locknut		1	
BD.19945	**Drain Tube,** front		2	
BD.19946	**Drain Tube,** rear		2	
C.2905/2	**Clip,** securing Drain Tubes		4	
BD.19947	**Sealing Rubber,** front		1	
BD.19948	**Sealing Rubber,** rear		1	
M.104/6R	**Woodscrew,** securing Sliding Roof Frame to Body		30	
M.108/8R	**Woodscrew,** securing Stiffener Brackets to Frame		4	
DA.504/8C	**Screw,** Self-Tapping, securing Trim Panel to Main Panel		4	
BD.1814/7	**Pop Rivet,** securing Stiffener Brackets to Main Panel		4	

INTERPRETATION OF CATALOGUE

Attention is drawn to the lay-out of this catalogue. It will be observed that the heading of each assembly is printed in **BOLD BLOCK LETTERS,** whilst the spares relevant to each assembly are **printed in small type and inset to the right.**

The fact that certain items (printed in small type) are not "inset" indicates that they do not form part of any assembly and must, therefore, be ordered as individual items.

ILLUSTRATIONS

To aid identification of spares, "exploded" illustrations are incorporated in this catalogue.

These illustrations may be used in the following manner:—

(a) If a Crankshaft is required, the index shows that the details for the item are printed on page 6 and that it is illustrated on Plate 1. Turning to page 6, the necessary information for the ordering of the part is given, whilst the column headed "Plate No." indicates that reference to item number 43 on Plate 1 will give a photographic view of the component required.

(b) The details for each illustrated item are printed in the script in numerical Plate number order. An item may therefore be first identified photographically. For example, if Friction Pads are required for the Front Brake, a picture of the part may be found on Plate No. 26, together with the Plate reference number. It is now an easy matter to find Plate number "26-2" in the script and the Part No. and Technical Description are at once apparent.

ORDERING

Customers are advised to make themselves familiar with the method of ordering spare parts in accordance with the explanatory notes which follow.

PUBLICATION D.4

FIRST PUBLISHED OCTOBER 1963
REVISED MAY 1969

INSTRUCTIONS FOR ORDERING SPARE PARTS

To avoid delay through inaccurate interpretation, IT IS OF GREAT IMPORTANCE that the information detailed in the following paragraphs be strictly adhered to when placing orders.

CAR TYPE

Always quote the engine and chassis numbers except when Stock Orders are placed, in which case the Year and Model nust be indicated.

An order for spares in connection with an Automatic Transmission Unit must be accompanied by the **Transmission Serial Number.** If it is desired to replace the Converter Assembly, however, the order for this component must bear the **Serial Number of the Converter,** which will be found stamped on the engine side of the Flywheel, adjacent to the Starter Ring Gear.

DETAILS OF PARTS REQUIRED

When ordering, the most complete information should be given to enable us to deal with your requirements expeditiously and to minimise delay in the despatch of the goods. Therefore we strongly recommend that the following procedure be adopted:—

(1) Identify part with technical description in catalogue.

(2) On Order, quote (a) part number, (b) technical description.

Note: In no case should the illustration plate number be given.

(3) If doubt exists as to the part number or technical description, a dimensioned sketch will assist in the interpretation of the order, or, alternatively, the old part should accompany the order as a pattern. When patterns are sent, it is most important that they be labelled, indicating the name and address of the sender, together with any reference such as order number, etc. Patterns will not be returned unless instructions to that effect accompany the order. In the absence of such instructions, patterns will be scrapped after retention for a period of ONE WEEK.

(4) To order in "sets" is misleading. Please, therefore, **state the specific quantity required.**

SPECIAL ATTENTION

Particular attention is drawn to the fact that when such items as Wings, Doors, Door Glasses, etc., are required, care should be taken to indicate whether they are located on the left-hand or right-hand side of the car. The position is determined by the driver's own left or right-hand when seated in the car.

Orders for Carpets, Casings, etc., should clearly state the precise colour required.

DESPATCH INSTRUCTIONS

The full address of the destination to which the goods are to be sent, together with that of the customer, should be clearly defined.

Whilst every effort will be made to observe customers' wishes as to the mode of transport, it is the normal custom to utilize the quickest possible method, whether it be post, passenger train or road transport, this being dependent upon the size and weight of the package and the destination to which it is to be sent.

PRICES

Current retail prices, shown in the Daimler Parts Master Price List, are subject to alteration without notice and are for delivery at Works. Postage, carriage and packing will be charged extra.

DAMAGE BY FIRE OR ACCIDENT

Whilst every precaution is taken, we do not accept liability for damage which may be caused by fire to any motor car or part thereof entrusted to us for any purpose whatsoever; nor is liability accepted for accidental damage.

INDEX TO PARTS

INDEX TO PARTS

INDEX TO PARTS

INDEX TO PARTS

INDEX TO PARTS

INDEX TO PARTS

INDEX TO PARTS

INDEX TO PARTS

INDEX TO PARTS

OFFICIAL TECHNICAL BOOKS

Brooklands Technical Books has been formed to supply owners, restorers and professional repairers with official factory literature.

Workshop Manuals

Jaguar Service Manual 1946-1948		9781855207844
Jaguar XK 120 140 150 150S & Mk 7, 8 & 9		9781870642279
Jaguar Mk 2 (2.4 3.4 3.8 240 340)	E121/7	9781870642958
Jaguar Mk 10 (3.8 & 4.2) & 420G	E136/2	9781855200814
Jaguar 'S' Type 3.4 & 3.8	E133/3	9781870642095
Jaguar E-Type 3.8 & 4.2 Series 1 & 2	E123/8, E123 B/3 & E156/1	9781855200203
Jaguar E-Type V12 Series 3	E165/3	9781855200012
Jaguar 420	E143/2	9781855201712
Jaguar XJ6 2.8 & 4.2 Series 1		9781855200562
Jaguar XJ6 3.4 & 4.2 Series	E188/4	9781855200302
Jaguar XJ12 Series 1		9781783180417
Jaguar XJ12 Series 2 / DD6 Series 2	E190/4	9781855201408
Jaguar XJ6 & XJ12 Series 3	AKM9006	9781855204010
Jaguar XJ6 OWM (XJ40) 1986-94		9781855207851
Jaguar XJS V12 5.3 & 6.0 Litre	AKM3455	9781855202627
Jaguar XJS 6 Cylinder 3.6 & 4.0 Litre	AKM9063	9781855204638

Owners Workshop Manuals

Jaguar E-Type V12 1971-1974	9781783181162
Jaguar XJ, Sovereign 1968-1982	9781783811179
Jaguar XJ6 Workshop Manual 1986-1994	9781855207851
Jaguar XJ12, XJ5.3 Double Six 1972-1979	9781783181186

Parts Catalogues

Jaguar Mk 2 3.4	J20	9781855201569
Jaguar Mk 2 (3.4, 3.8 & 340)	J34	9781855209084
Jaguar Series 3 12 Cyl. Saloons		9781783180592
Jaguar E-Type 3.8	J30	9781869826314
Jaguar E-Type 4.2 Series 1	J37	9781870642118
Jaguar E-Type Series 2	J37 & J38	9781855201705
Jaguar E-Type V12 Ser. 3 Open 2 Seater	RTC9014	9781869826840
Jaguar XJ6 Series 1		9781855200043
Jaguar XJ6 & Daimler Sovereign Ser. 2	RTC9883CA	9781855200579
Jaguar XJ6 & Daimler Sovereign Ser. 3	RTC9885CF	9781855202771
Jaguar XJ12 Series 2 / DD6 Series 2		9781783180585
Jaguar 2.9 & 3.6 Litre Saloons 1986-89	RTC9893CB	9781855202993
Jaguar XJ-S 3.6 & 5.3 Jan 1987 on	RTC9900CA	9781855204003

Owners Handbooks

Jaguar XK120		9781855200432
Jaguar XK140	E101/2	9781855200401
Jaguar XK150	E111/2	9781855200395
Jaguar Mk 2 (3.4)	E116/10	9781855201682
Jaguar Mk 2 (3.8)	E115/10	9781869826765
Jaguar E-Type (Tuning & prep. for competition)		9781855207905
Jaguar E-Type 3.8 Series 1	E122/7	9781870642927
Jaguar E-Type 4.2 2+2 Series 1	E131/6	9781869826383
Jaguar E-Type 4.2 Series	E154/5	9781869826499
Jaguar E-Type V12 Series 3	E160/2	9781855200029
Jaguar E-Type V12 Series 3 (US)	A181/2	9781855200036
Jaguar XJ (3.4 & 4.2) Series 2	E200/8	9781855201200
Jaguar XJ6C Series 2	E184/1	9781855207875
Jaguar XJ12 Series 3	AKM4181	9781855207868

Carburetters

SU Carburetters Tuning Tips & Techniques	9781855202559
Solex Carburetters Tuning Tips & Techniques	9781855209770
Weber Carburettors Tuning Tips and Techniques	9781855207592

Jaguar - Road Test Books

Jaguar and SS Gold Portfolio 1931-1951	9781855200630
Jaguar XK120 XK140 XK150 Gold Port. 1948-60	9781870642415
Jaguar Mk 7, 8, 9, 10 & 420G	9781855208674
Jaguar Mk 1 & Mk 2 1955-1969	9781855208599
Jaguar E-Type	9781855208360
Jaguar XJ6 1968-79 (Series 1 & 2)	9781855202641
Jaguar XJ12 XJ5.3 V12 Gold Portfolio 1972-1990	9781855200838
Jaguar XJS Gold Portfolio 1975-1988	9781855202719
Jaguar XJ-S V12 1988-1996	9781855204249
Jaguar XK8 & XKR 1996-2005	9781855207578
Road & Track on Jaguar 1950-1960	9780946489695
Road & Track on Jaguar 1968-1974	9780946489374
Road & Track On Jaguar XJ-S-XK8-XK	9781855206298

Available from Jaguar specialists, Amazon and all good motoring bookshops

Brooklands Books Ltd.,
PO Box 904, Amersham,
Bucks., HP6 9JA,
England

www.brooklandsbooks.com

Made in United States
Orlando, FL
04 November 2026

90201136R00186